Marine Biology: A Global Assessment

Marine Biology: A Global Assessment

Editor: Dan Espinoza

Callisto Reference
www.callistoreference.com

Callisto Reference,
118-35 Queens Blvd., Suite 400,
Forest Hills, NY 11375, USA

Visit us on the World Wide Web at:
www.callistoreference.com

ISBN: 978-1-63239-926-7 (Hardback)

The publisher's policy is to use permanent paper from mills that operate a sustainable forestry policy. Furthermore, the publisher ensures that the text paper and cover boards used have met acceptable environmental accreditation standards.

Printed in China.

Cataloging-in-Publication Data

Marine biology : a global assessment / edited by Dan Espinoza.
p. cm.
Includes bibliographical references and index.
ISBN 978-1-63239-926-7
1. Marine biology. 2. Marine sciences. 3. Oceanography.
4. Marine biodiversity. I. Espinoza, Dan.
QH91 .M37 2018
578.77--dc23

Table of Contents

Preface .... **IX**

Chapter 1 **Research Tools to Investigate Movements, Migrations, and Life History of Sturgeons (Acipenseridae) with an Emphasis on Marine-Oriented Populations** **1**
Troy C. Nelson, Phaedra Doukakis, Steven T. Lindley, Andrea D. Schreier, Joseph E. Hightower, Larry R. Hildebrand, Rebecca E. Whitlock, Molly A. H. Webb

Chapter 2 **Growth and Nitrogen uptake Characteristics Reveal Outbreak Mechanism of the Opportunistic Macroalga *Gracilaria tenuistipitata*** **23**
Chao Wang, Anping Lei, Kai Zhou, Zhengyu Hu, Wenlong Hao, Junda Yang

Chapter 3 **Small Changes in Gene Expression of Targeted Osmoregulatory Genes when Exposing Marine and Freshwater Threespine Stickleback (*Gasterosteus aculeatus*) to Abrupt Salinity Transfers** **33**
Annette Taugbøl, Tina Arntsen, Kjartan Østbye, Leif Asbjørn Vøllestad

Chapter 4 **Integrated Assessment of Heavy Metal Pollution in the Surface Sediments of the Laizhou Bay and the Coastal Waters of the Zhangzi Island, China: Comparison among Typical Marine Sediment Quality Indices** **42**
Wen Zhuang, Xuelu Gao

Chapter 5 **Acoustic Telemetry Validates a Citizen Science Approach for Monitoring Sharks on Coral Reefs** **59**
Gabriel M. S. Vianna, Mark G. Meekan, Tova H. Bornovski, Jessica J. Meeuwig

Chapter 6 **Radiocarbon Dating and Wood Density Chronologies of Mangrove Trees in Arid Western Australia** **71**
Nadia S. Santini, Quan Hua, Nele Schmitz, Catherine E. Lovelock

Chapter 7 **Wintering Habitat Model for the North Atlantic Right Whale (*Eubalaena glacialis*) in the Southeastern United States** **79**
Timothy A. Gowan, Joel G. Ortega-Ortiz

Chapter 8 **The Role of the Sedimentary Regime in Shaping the Distribution of Subtidal Sandbank Environments and the Associated Meiofaunal Nematode Communities: An Example from the Southern North Sea** **93**
Michaela Schratzberger, Piers Larcombe

Chapter 9 **Relationships between Degree of Polymerization and Antioxidant Activities: A Study on Proanthocyanidins from the Leaves of a Medicinal Mangrove Plant *Ceriops tagal*** **114**
Hai-Chao Zhou, Nora Fung-yee Tam, Yi-Ming Lin, Zhen-Hua Ding, Wei-Ming Chai, Shu-Dong Wei

Chapter 10 **Long-Term Fluctuations in Circalunar Beach Aggregations of the Box Jellyfish *Alatina moseri* in Hawaii, with Links to Environmental Variability** 124
Luciano M. Chiaverano, Brenden S. Holland, Gerald L. Crow, Landy Blair, Angel A. Yanagihara

Chapter 11 **Cell Turnover and Detritus Production in Marine Sponges from Tropical and Temperate Benthic Ecosystems** 134
Brittany E. Alexander, Kevin Liebrand, Ronald Osinga, Harm G. van der Geest, Wim Admiraal, Jack P. M. Cleutjens, Bert Schutte, Fons Verheyen, Marta Ribes, Emiel van Loon, Jasper M. de Goeij

Chapter 12 **Environmental Quality of Italian Marine Water by Means of Marine Strategy Framework Directive (MSFD) Descriptor 9** 145
Chiara Maggi, Serena Lomiri, Bianca Di Lorenzo, Marco d'Antona, Maria Teresa Berducci

Chapter 13 **Marine Plastic Pollution in Waters around Australia: Characteristics, Concentrations, and Pathways** 156
Julia Reisser, Jeremy Shaw, Chris Wilcox, Britta Denise Hardesty, Maira Proietti, Michele Thums, Charitha Pattiaratchi

Chapter 14 **Marine Litter Distribution and Density in European Seas, from the Shelves to Deep Basins** 167
Christopher K. Pham, Eva Ramirez-Llodra, Claudia H. S. Alt, Teresa Amaro, Melanie Bergmann, Miquel Canals, Joan B. Company, Jaime Davies, Gerard Duineveld, François Galgani, Kerry L. Howell, Veerle A. I. Huvenne, Eduardo Isidro, Daniel O. B. Jones, Galderic Lastras, Telmo Morato, José Nuno Gomes-Pereira, Autun Purser, Heather Stewart, Inês Tojeira, Xavier Tubau, David Van Rooij, Paul A. Tyler

Chapter 15 **Application of Empirical Wave Run-up Formulas to the Polish Baltic Sea Coast** 180
Dominik Paprotny, Paweł Andrzejewski, Paweł Terefenko, Kazimierz Furmańczyk

Chapter 16 **Long-Term Changes in Nutrients and Mussel Stocks are Related to Numbers of Breeding Eiders *Somateria mollissima* at a Large Baltic Colony** 188
Karsten Laursen, Anders Pape Møller

Chapter 17 **Burrows of the Semi-Terrestrial Crab *Ucides cordatus* Enhance CO_2 Release in a North Brazilian Mangrove Forest** 194
Nathalie Pülmanns, Karen Diele, Ulf Mehlig, Inga Nordhaus

Chapter 18 **Egg Size Effects across Multiple Life-History Stages in the Marine Annelid *Hydroides diramphus*** 207
Richard M. Allen, Dustin Marshall

Chapter 19 **Virio- and Bacterioplankton Microscale Distributions at the Sediment-Water Interface** 216
Lisa M. Dann, James G. Mitchell, Peter G. Speck, Kelly Newton, Thomas Jeffries, James Paterson

Chapter 20 **Asexual Propagation of Sea Anemones that Host Anemonefishes: Implications for the Marine Ornamental Aquarium Trade and Restocking Programs** **230**
Anna Scott, Jannah M. Hardefeldt, Karina C. Hall

Permissions

List of Contributors

Index

Preface

Marine biology studies marine organisms, their habitat and physiology. This book studies the fundamental theories and practices related to marine biology. Ecological risk due to pollution and dwindling numbers of aquatic species are causes of concern to marine biologists. Technological advancements in this field have enabled a more concerted effort in understanding the life of aquatic organisms. A number of latest researches have been included in this book on marine biology to keep the readers up-to-date with the global concepts in this area of study. Coherent flow of topics, student-friendly language and extensive use of examples make this book an invaluable source of knowledge.

All of the data presented henceforth, was collaborated in the wake of recent advancements in the field. The aim of this book is to present the diversified developments from across the globe in a comprehensible manner. The opinions expressed in each chapter belong solely to the contributing authors. Their interpretations of the topics are the integral part of this book, which I have carefully compiled for a better understanding of the readers.

At the end, I would like to thank all those who dedicated their time and efforts for the successful completion of this book. I also wish to convey my gratitude towards my friends and family who supported me at every step.

Editor

1

Research Tools to Investigate Movements, Migrations, and Life History of Sturgeons (Acipenseridae), with an Emphasis on Marine-Oriented Populations

Troy C. Nelson[1]*, Phaedra Doukakis[2], Steven T. Lindley[3], Andrea D. Schreier[4], Joseph E. Hightower[5], Larry R. Hildebrand[6], Rebecca E. Whitlock[7], Molly A. H. Webb[8]

1 Fraser River Sturgeon Conservation Society, Vancouver, British Columbia, Canada, **2** Scripps Institution of Oceanography, University of California San Diego, San Diego, California, United States of America, **3** National Marine Fisheries Service, Southwest Fisheries Science Center, Santa Cruz, California, United States of America, **4** Department of Animal Science, University of California Davis, Davis, California, United States of America, **5** U. S. Geological Survey, NC Cooperative Fish and Wildlife Research Unit, Department of Applied Ecology, NC State University, Raleigh, North Carolina, United States of America, **6** Golder Associates Ltd., Castlegar, British Columbia, Canada, **7** Stanford University, Hopkins Marine Station, Pacific Grove, California, United States of America, **8** U. S. Fish and Wildlife Service, Bozeman Fish Technology Center, Bozeman, Montana, United States of America

Abstract

Worldwide, sturgeons (Acipenseridae) are among the most endangered fishes due to habitat degradation, overfishing, and inherent life history characteristics (long life span, late maturation, and infrequent spawning). As most sturgeons are anadromous, a considerable portion of their life history occurs in estuarine and marine environments where they may encounter unique threats (e.g., interception in non-target fisheries). Of the 16 marine-oriented species, 12 are designated as Critically Endangered by the IUCN, and these include species commercially harvested. We review important research tools and techniques (tagging, electronic tagging, genetics, microchemistry, observatory) and discuss the comparative utility of these techniques to investigate movements, migrations, and life-history characteristics of sturgeons. Examples are provided regarding what the applications have revealed regarding movement and migration and how this information can be used for conservation and management. Through studies that include Gulf (*Acipenser oxyrinchus desotoi*) and Green Sturgeon (*A. medirostris*), we illustrate what is known about well-studied species and then explore lesser-studied species. A more complete picture of migration is available for North American sturgeon species, while European and Asian species, which are among the most endangered sturgeons, are less understood. We put forth recommendations that encourage the support of stewardship initiatives to build awareness and provide key information for population assessment and monitoring.

Editor: Athanassios C. Tsikliras, Aristotle University of Thessaloniki, Greece

Funding: Funding for this work was provided by a grant to TCN from the Pacific Ocean Shelf Tracking Project (POST) and their partners, The Census of Marine Life, the Gordon and Betty Moore Foundation, and the Vancouver Aquarium Marine Science Centre. The funders had no role in study design, data collection and analysis, decision to publish, or preparation of the manuscript.

Competing Interests: Larry R. Hildebrand is an employee of Golder Associates Ltd. There are no patents, products in development or marketed products to declare.

* E-mail: troy@frasersturgeon.com

Introduction

There are 27 species of sturgeons and paddlefishes (Order Acipenseriformes) in rivers, lakes, estuaries, near-shore oceanic environments, and inland seas across the northern hemisphere (Table 1). The two species of paddlefishes (Family Polyodontidae) are strictly freshwater in life history while the 25 sturgeon (Family Acipenseridae) species include 16 species that enter into estuaries, oceans, or seas during some part of their life cycle. Sturgeons have five rows of bony scutes and snouts with sensory barbels (Figure 1). All sturgeons spawn in freshwater habitats. The 16 marine-oriented species (species that spend a significant portion of their life history in marine environments) occur on all of the continents to which sturgeons are endemic, including North America, Europe, and Asia.

Sturgeons are generally long-lived fishes that exhibit late onset of maturity, slow growth, and infrequent reproduction [1]. Some species have life spans of well over 100 years, and sexual maturity may be attained as late as 20–25 years of age or more in females [2]. These life-history characteristics are linked to low maximum rates of population growth and underlie intrinsic vulnerability to exploitation-induced decline and a low recovery capacity [3], [4]. Damming of rivers can be particularly detrimental to many sturgeon populations as it can reduce or eliminate spawning and egg/larvae incubation habitats and change important environmental cues relating to flow regimes and hydrographic characteristics [5]. Human activities such as dredging, channelization, dyking, bank stabilization, in-river construction, and shoreline development can affect important juvenile rearing habitats [6]. Other prominent impacts to sturgeon include hybridization, pollution, introduced species, reduced food supply, water diversions, and saltwater intrusion (into important spawning and rearing habitats) [1], [7]. Their value as the source of black caviar (the unfertilized roe of female sturgeon) has led sturgeons (particularly in the Black and Caspian seas) to be the target of intensive legal and illegal fisheries [1], [2]. This has resulted in the collapse of several species and stocks of sturgeon [1]. The combination of inherent life-history challenges, habitat degradation, and exploitation has reduced sturgeon abundance to such low levels that the sturgeons are now considered one of the most endangered groups of animals in the world according to the International Union for Conservation of Nature (IUCN). In 2010, the IUCN reviewed the status of the 25 species (and two subspecies) of sturgeon (Acipenseridae) in their "Red List" (list of threatened species); four species are listed as "threatened," five species as "endangered," and 18 species are "critically endangered" (likely to become extinct in a generation). The IUCN ranking approach is conservative; for example, the IUCN classifies White Sturgeon (all populations) as "least concern," whereas White Sturgeon in Canada are classified "endangered" under the national ranking system [8], and in the US, the Kootenai River White Sturgeon is federally listed as "endangered."

Sturgeons are also some of the least-well-known of the major taxa of concern in terms of spatial distribution and abundance, particularly for oceanic and nearshore phases of their life history. Almost all of the sturgeons that enter into saltwater have been understudied with respect to where they go and why (Table 1) and many endangered species have received relatively little scientific attention and study [9]. Researching these information gaps is extremely important for conservation of habitats and distinct populations. Here, we explore these issues while providing case studies and tools for examining movement and distribution of these ancient fishes.

1.1. Rational for Research: Understanding Movement and Migration – and why it Matters

Sturgeons may move from one location or habitat type to another to feed, reproduce, or overwinter. River movement can be complex and include multi-step migrations [10] and include movement between and among rivers suggesting a meta-population structure [11]. Within a species, populations can differ on the timing of migrations into river systems, time spent within the river (holding), and the distance (upstream) from the marine environment where spawning occurs. Directed migrations from overwintering locations to feeding habitats, in some cases timed to intercept specific, localized prey species, may be a population-based behaviour; other populations of the same species may exhibit an entirely different food-related migration pattern based on the abundance/timing of localized prey. Protecting the genetic heritage and diversity of a species requires an understanding of these complexities at both the species and population levels, and an understanding of specific life-stage habitat requirements. Species and populations subject to harvest typically benefit from well-managed harvest regimes that incorporate respective migration information. There may be sex-specific differences in the timing of movements (often males arrive at the spawning grounds before females) so harvest/interception regimes may need to be sensitive to this in order to maintain the male:female sex-ratio balance within the population.

Figure 1. Gulf Sturgeon (*Acipenser oxyrinchus desotoi*). *Photo: Joe Hightower.*

Preserving the evolutionary potential of a species and its ability to respond to environmental change requires understanding the number of distinct populations within a species. Natal homing fidelity is thought to be strong and thus in many cases rivers, or sections of a watershed between natural barriers, often define populations. On the other hand, metapopulations may also exist [11]. If individual populations differ in terms of abundance and reproductive capacity, researchers and fishery managers may want to minimize mortality of individuals from certain river systems while they are in marine environments. Movement studies can provide the necessary information at both population and species levels and can provide the basis for protective schemes under national and international legislation. Characterizing environmental parameters that correlate with behaviour is further useful in evaluating the potential impact of habitat alteration (e.g. changes to the annual hydrograph of a sturgeon-bearing river). In some areas, introduction of non-native sturgeons and hybridization between species (e.g. Scaphirhynchus) due to habitat alteration and/or unintentional release from aquaculture (e.g. in Europe) makes understanding the movement and life history of introduced sturgeons or hybrid populations an important conservation consideration [12], [13].

Knowledge about movement and the spatial distribution of populations can also improve population assessments and help in specifying suitable management or recovery plans. Population assessments are still lacking for most species and are often hindered by an insufficient understanding of spatial distribution. Sturgeon species and/or populations with an anadromous life-history strategy may require assessment models that account for the seasonal movement patterns of different life-history stages exploited by fisheries [14], [15]. Tagging data can be incorporated into spatially explicit models to allow more accurate estimation of fishing mortality rates experienced by different age classes where fishing intensity varies by area [14], [16].

Insufficient data for parameterisation of models for migratory species with complex life histories can lead to large uncertainties in resource assessment and the likely effects of alternative management actions. The lack of data for key parameters of population dynamics has been identified as a contributor to the depleted and endangered status of many sturgeon populations throughout their range [17], [18]. Below, we present recent studies of the Gulf and Green Sturgeon to show how the information gained from these representative studies has been used to direct subsequent activities toward species conservation and the protection of key habitats.

1.2. Case Studies: The Well-studied Gulf and Green Sturgeon

1.2.1. Gulf Sturgeon. The anadromous Gulf Sturgeon, *Acipenser oxyrinchus desotoi*, a subspecies of the Atlantic Sturgeon, *Acipenser oxyrinchus*, occurs in most Gulf of Mexico river systems from the mouth of the Mississippi River to the west coast of Florida (Figure 2). Both mature and immature Gulf Sturgeon

Table 1. Sturgeons (Family Acipenseridae) of the world, their marine distributions, and their respective IUCN* Category.

Species	Distribution	IUCN* Category
Acipenser baerii, Siberian Sturgeon	FW, E/Bays	Endangered
Acipenser brevirostrum, Shortnose Sturgeon	FW, E, Occasionally Coastal	Vulnerable
Acipenser dabryanus, Yangtze Sturgeon	FW	Critically Endangered (pe)
Acipenser fulvescens, Lake Sturgeon	FW	Least Concern
Acipenser gueldenstaedtii, Russian Sturgeon	E, Sea (Black, Caspian, Azov)	Critically Endangered
Acipenser medirostris, Green Sturgeon	E, Coastal (Baja, California to the Bering Sea)	Near Threatened
Acipenser mikadoi, Sakhalin Sturgeon	E, Coastal (Bering Sea, Sea of Okhotsk, Sea of Japan, south Sakhalin Island; range very restricted at present); $$	Critically Endangered
Acipenser naccarii, Adriatic Sturgeon	E, Coastal; Adriatic Sea; range very restricted at present; $$	Critically Endangered (pe)
Acipenser nudiventris, Ship Sturgeon	E, Sea (Black, Caspian, Azov)	Critically Endangered
Acipenser oxyrinchus Acipenser oxyrinchus desotoi, Gulf Sturgeon *Acipenser oxyrinchus oxyrinchus,* Atlantic Sturgeon	E, Coastal (Gulf of Mexico to Quebec)	Near Threatened
Acipenser persicus, Persian Sturgeon	E, Sea (Black, Caspian)	Critically Endangered
Acipenser ruthenus, Sterlet	FW	Vulnerable
Acipenser schrenckii, Amur Sturgeon	E, Coastal (distribution uncertain); $$	Critically Endangered
Acipenser sinensis, Chinese Sturgeon	E, Coastal (Yellow and East China seas; range very restricted at present)	Critically Endangered
Acipenser stellatus, Stellate Sturgeon	E, Sea (Black, Caspian, Azov)	Critically Endangered
Acipenser sturio, European (Baltic) Sturgeon	E, Coastal (Baltic Sea; once throughout Western Europe; range very restricted at present)	Critically Endangered (pe)
Acipenser transmontanus, White Sturgeon	E, Coastal (Aleutian Islands to Monterey California)	Least Concern
Huso huso, Beluga Sturgeon	E, Sea (Black, Caspian, Azov)	Critically Endangered
Huso dauricus, Kaluga Sturgeon	E. Coastal (Sea of Okhotsk, Tatar Strait, Sea of Japan); $$	Critically Endangered
Scaphirhynchus albus, Pallid Sturgeon	FW	Endangered
Scaphirhynchus platorynchus, Shovelnose Sturgeon	FW	Vulnerable
Scaphirhynchus suttkusi, Alabama Sturgeon	FW	Critically Endangered
Pseudoscaphirhynchus hermanni, Dwarf Sturgeon	FW	Critically Endangered
Pseudoscaphirhynchus fedtschenkoi, Syr Darya Sturgeon	FW	Critically Endangered (pe)
Pseudoscaphirhynchus kaufmanni, Amu Darya Sturgeon	FW	Critically Endangered

*International Union for Conservation of Nature.
NOTES: FW = primarily freshwater over most of life history.
E = estuarine.
Sea = inhabits seas but not oceans.
$$ = additional research is required to determine the marine distribution of these species.
pe = possibly extinct in the wild.
Sources from **REFERENCES**: [10], [61], [211–213].

undergo freshwater migrations, typically entering coastal rivers in March or April and outmigrating to the ocean in September or October [19]. The cool-water period of estuarine or marine residency is critical for growth and reproduction, as Gulf Sturgeon do not feed during their freshwater residency.

Early information about Gulf Sturgeon distribution and migration came primarily from commercial fishing [20]. Fishing operations in individual rivers were mostly short-lived (due to overharvest and subsequent fishery closures, as is typical for many in-river sturgeon fisheries), but the catches did provide some insights into the timing and extent of migration. More detailed information came from surveys in the 1980s conducted in response to declining catches and the species' listing by the state of Florida as "threatened" [21], [22]. Marking with conventional tags mostly showed the range and extent of migration (e.g. recaptures of tagged fish by anglers below a dam or by commercial shrimp trawlers in Gulf of Mexico waters), whereas the first use of electronic tags (and manual/mobile tracking) provided insights about holding, staging, and spawning areas [22]; the use of radio tags in this study allowed the authors to characterize occupied riverine habitats in the Apalachicola River and to relate the timing of upstream and downstream migrations to environmental cues.

Radio transmitters worked well for manual tracking of Gulf Sturgeon over long distances in rivers, but these transmitters cannot be detected in brackish or marine water. In the late 1980s a telemetry study was conducted wherein Gulf Sturgeon were tagged with both radio and sonic tags [23]. This was also the first Gulf Sturgeon study to use stationary receivers to detect and log passage events (in this case, to detect sonic tags as fish moved through barrier island passes [23]).

The listing of Gulf Sturgeon in 1991 as a threatened species under the U. S. Endangered Species Act provided a further boost to research activity. For example, radio tracking studies in the Choctawhatchee River were initiated to identify potential spawn-

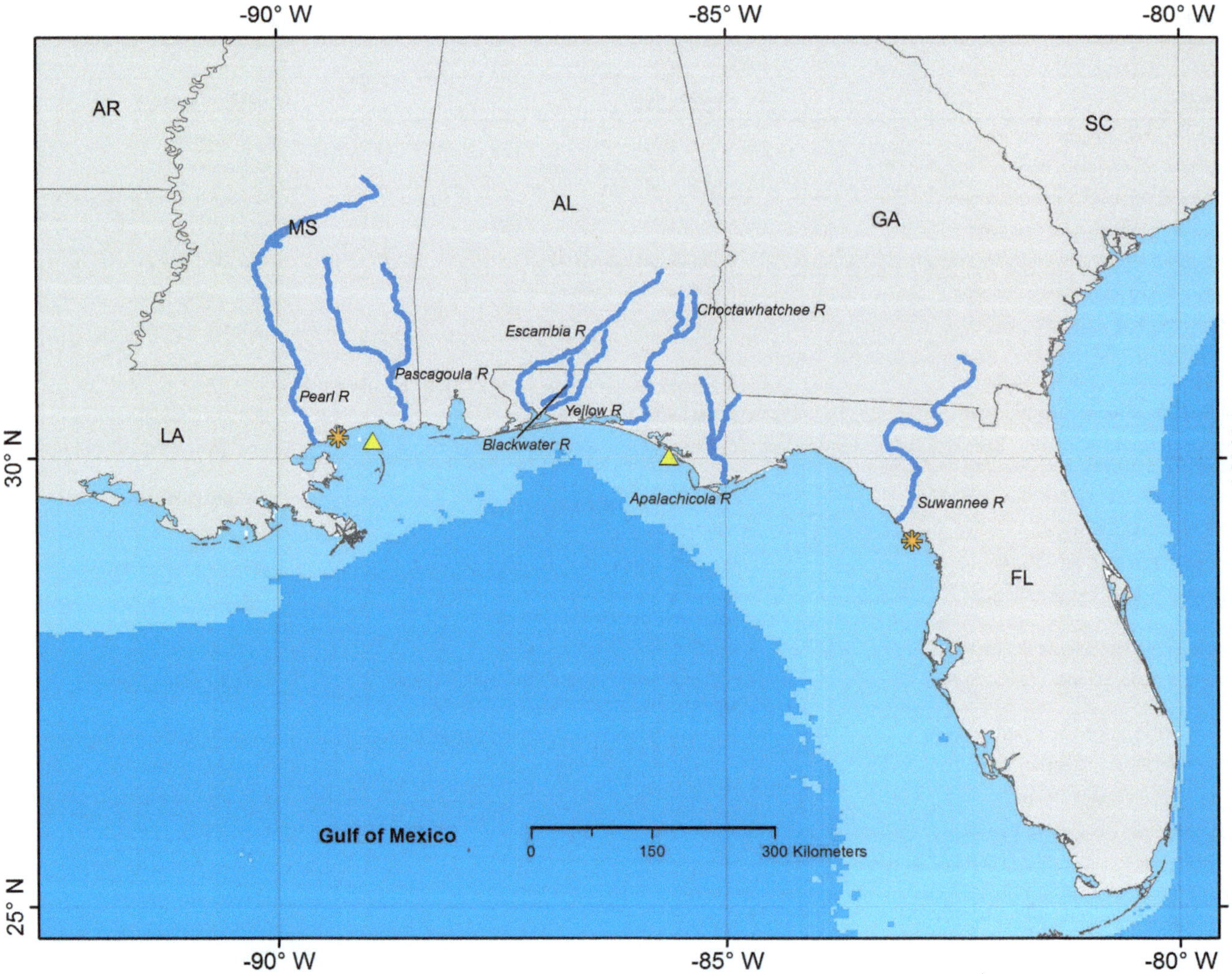

Figure 2. Documented distribution of Gulf Sturgeon in North America, determined from acoustic and archival telemetry projects. The orange asterisks mark the easternmost and westernmost locations of confirmed detections of acoustic-tagged Gulf Sturgeon. Gulf Sturgeon spawn in coastal rivers including the eight shown on this map. Spawning and non-spawning Gulf Sturgeon typically remain in coastal rivers until fall and occupy estuarine and nearshore marine waters during winter. Yellow triangles indicate winter concentration areas for Gulf Sturgeon from two or more river systems. The 100 m isobath is shown as the light blue areas near the coast.

ing sites, with confirmation through the use of artificial substrates to collect the adhesive eggs of Gulf Sturgeon [24]. Deployment of artificial substrates in a grid design provided fine-scale information about spawning habitat in the Suwannee River [25]. Marine habitat studies using sonic tags (and more recently, archival temperature-logging and pop-up archival tags) showed that Gulf Sturgeon sometimes moved long distances along the shoreline and primarily used shallow nearshore areas [26–29]. The fish occupying these marine habitats were often from multiple river systems; for example, [27] reported that Gulf Sturgeon from the Yellow, Choctawhatchee, and Apalachicola rivers were located within a 25-km stretch of coastline (eastern winter concentration area shown on Figure 2). The co-occurrence of Gulf Sturgeon from the Pearl and Pascagoula rivers has been documented [29] in the concentration area off Mississippi (western area shown on Figure 2). Thus, marine and estuarine threats and management efforts may affect more than one population.

Genetic studies have also aided in understanding Gulf Sturgeon migration patterns. For example, it has been shown that within a basin, genetic structure exists at least at the drainage level and possibly at the level of tributary rivers within the basin [30]. The genetic analyses were helpful in interpreting telemetry results since some fish were tagged outside their natal drainage and others were captured or detected in multiple drainages.

These research results formed the basis for the Gulf Sturgeon recovery plan and led to the designation of critical habitats. These important habitats included upper-basin spawning sites with limestone bluffs and outcroppings, estuarine and marine feeding sites with preferred substrates and benthic fauna, and summer resting areas. Genetic results showed strong natal river fidelity, so critical habitat was defined in each of the seven river systems containing currently reproducing populations (Pearl, Pascagoula, Escambia, Yellow/Blackwater, Choctawhatchee, Apalachicola, and Suwannee). This resulted in designation of nearly 2,800 river km as critical habitat for conservation of the species.

1.2.2. Green Sturgeon. In contrast to the relatively well-studied Gulf Sturgeon, the North American Green Sturgeon was little studied until 2002, when the US National Marine Fisheries Service received a petition to list it under the US Endangered Species Act. A severe lack of demographic and basic life-history information hampered the subsequent status review [31]. A particularly troubling unknown was the population origin(s) of Green Sturgeon that form dense aggregations in certain estuaries during summer months. Green Sturgeon were known to use just three rivers for spawning (the Sacramento and Klamath rivers in California, and the Rogue River in Oregon), and to spend much of their lives in marine waters between Alaska and Baja California (Figure 3). The purpose of the summertime estuarine aggregations was unknown, as was the proportion of Green Sturgeon exhibiting this aggregation behaviour. Green Sturgeon in these aggregations are vulnerable to capture in gillnet fisheries that target White Sturgeon and Pacific salmon species, and face environmental threats from activities associated with shellfish aquaculture and industrial activities in the estuaries where they aggregate.

The recent development of new telemetric tagging systems made it feasible to rapidly close some of these information gaps. Initial work focused on Green Sturgeon in the Rogue River, using radio and acoustic tags to learn that Green Sturgeon migrate into rivers in the early spring for spawning in up-river areas, and then hold in deep pools over the summer prior to emigration in the fall when flows rise with the onset of the rainy season [32]. Tagged sturgeon returned to the river to spawn every two to four years [33]. Rogue River fish were also tagged with pop-off archival tags (PAT), which revealed that they remain in fairly shallow water (50–80 m) when in the coastal ocean, and showed that they migrate north to the west coast of Vancouver Island in the fall [34].

A broader study using acoustic tags showed that Green Sturgeon make extensive seasonal migrations among spawning areas, over-summering in various estuaries and bays, and over-wintering areas in the coastal ocean, with many individuals using areas around northern Vancouver Island [35], [36]. Further PAT work, using longer tag deployments, also showed this seasonal migration pattern, and fairly constrained depth and temperature distributions during the winter. Acoustic tags also revealed extensive use of and movement among non-natal estuaries. Green Sturgeon from different populations mixed together in common estuaries, but at different rates. Natal estuaries were used almost exclusively by fish from the associated natal river [36]. Green Sturgeon were also shown to have diverse patterns of migration within and among populations. Within the Sacramento River, acoustic tags revealed that a seasonal water diversion dam was a serious impediment to the spawning migration of Green Sturgeon [37]. Habitat data associated with tag detections was used to gain further insight into freshwater [38] and fine– [39] and coarse-scale [40] marine habitat preferences and seasonal patterns of distribution. Captive rearing of Green Sturgeon is providing important information on salinity tolerance and the timing for successful transition to marine waters, as well as optimal temperature for egg hatching, embryogenesis, and larval and juvenile survival [41–46].

In summary, the application of electronic tag technology to study Green Sturgeon revealed novel, reliable, and useful/relevant information regarding marine migrations and habitat use, spawning locations and periodicity, and both marine and freshwater life history attributes. While tagging revealed clear overall migration patterns, it also revealed a diversity of migratory behaviours that are likely important to consider in the conservation and protection of this species. Complementary genetic studies have determined that two distinct population segments exist within Green Sturgeon that correspond to northern and southern populations, and that estuarine aggregations contain individuals from multiple populations [47]. Data acquired from electronic tagging has already been the basis of many of the conservation regulations put in place after the listing of the southern distinct population segment of Green Sturgeon by the US federal government. Critical habitat for Green Sturgeon was designated in 2009 along the continental shelf of the US from southeast Alaska to central California out to a depth of 110 m, as well as in a number of estuaries and bays in Washington, Oregon, and California. Regulatory biologists are using the acoustic tagging data in their assessments of permit applications for activities such as dredging and dredge spoil disposal, large construction projects within and along the shores of estuaries, and the management of groundfish trawl fisheries in the coastal ocean.

1.3. Case Studies: Less-studied Species in Regions of Europe and Asia

1.3.1. Caspian Sea. The Caspian Sea is one of the most important areas for sturgeons globally and has supplied much of the world's wild caviar. Five of the species that inhabit this region (Stellate, Beluga, Persian, Russian and Ship Sturgeon; Table 1), including those that produce the highly prized Beluga, Sevruga and Osetra caviar, are anadromous. Adult sturgeons move into river systems to reproduce in spring and sometimes winter and fall. Most adults that migrate into rivers in spring are thought to leave the river just after spawning, but those that migrate in the fall and winter may overwinter. The patterns of spawning migrations of Caspian Sea sturgeons differ amongst species, and the damming of rivers has likely altered migration patterns [48].

After hatching, larvae and fingerlings remain within the rivers for several months depending on the species: for some species (e.g. Stellate and Russian), fingerlings may overwinter in river deltas while for others (e.g. Beluga), fingerlings will migrate into the Caspian Sea for rearing. Information regarding the behaviour and movement of juveniles as well as adults within rivers has been gathered through fisheries-dependant surveys and fisheries-independent studies using nets. While a few tagging studies have been attempted (e.g. [49]), none have been conducted on the scale necessary to provide species-specific population and movement/migration information. Such studies would be of further benefit in understanding the reproductive periodicity of the different species.

It is unclear whether adult sturgeon in the Caspian Sea return to their natal rivers for reproduction. Genetic studies using mitochondrial DNA have not shown differentiation amongst river populations [50], but preliminary data from microsatellites suggest that natal homing does occur. Hatchery supplementation and stocking in rivers may be affecting the natural population structure that once existed because hatcheries do not always use broodstock collected in the river system where the hatchery-produced juveniles are released. Detections of genetic haplotypes from the non-native Siberian Sturgeon in tissue samples taken from Caspian Sea Russian Sturgeon have raised questions regarding hatchery practices, non-native releases, and the life history of this potential hybrid (see section 2.4). There is a great need to understand the species and population structure of Caspian Sea sturgeons so that management activities reflect and protect the various populations.

Within the Caspian Sea environment, comprehensive information about the distribution of sturgeons is lacking [48]. Much of what is known has been collected through trawl surveys, which have been limited spatially and geographically. The vertical distribution of sturgeons in the Caspian has been examined through trawl and drift net surveys, but these studies have been similarly limited. Since at-sea trawl survey information is used to generate abundance estimates and harvest quotas, a better understanding of distribution and movement is important. Tagging and tracking studies would greatly complement available

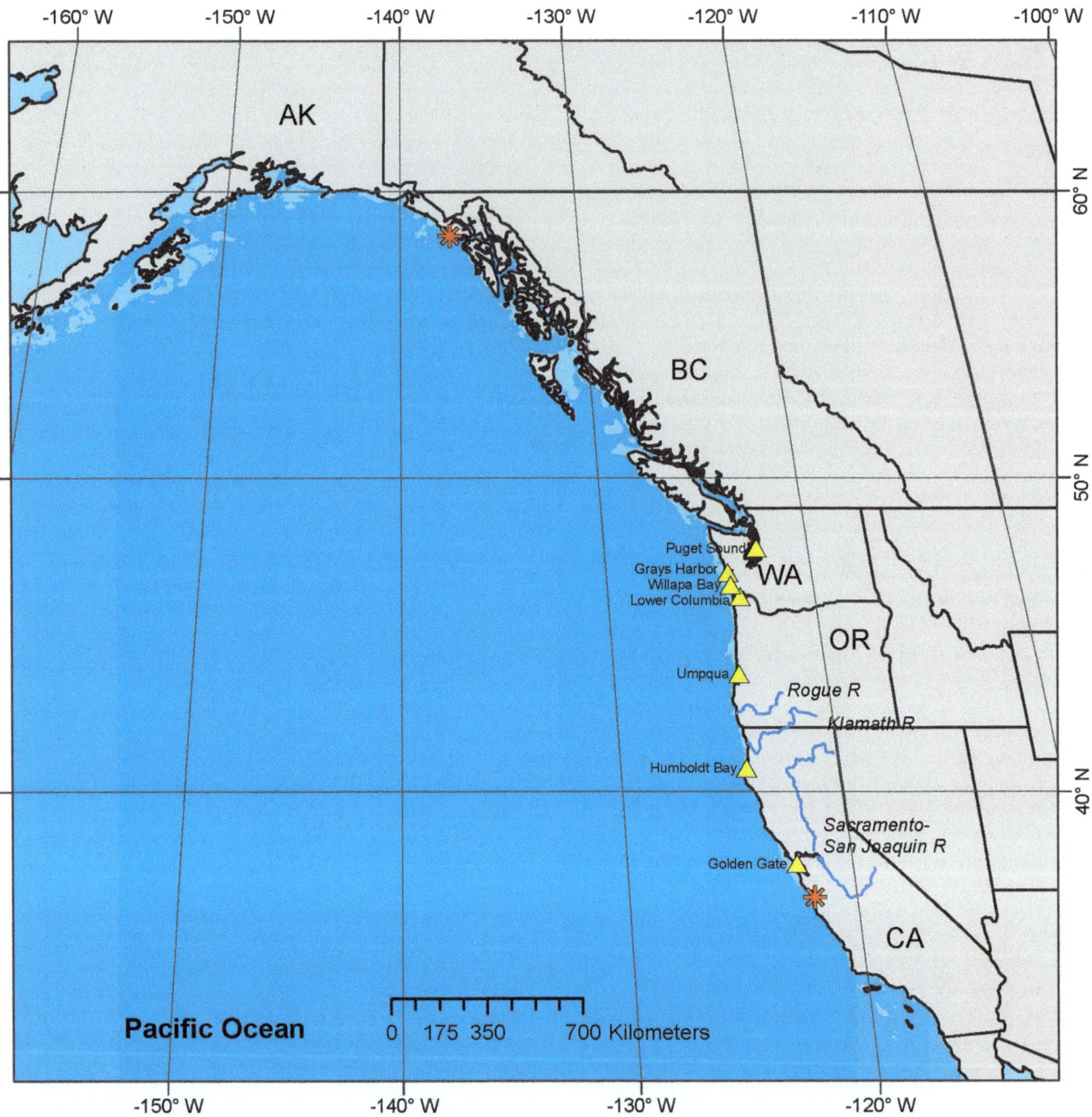

Figure 3. Documented distribution of Green Sturgeon in North America, determined from acoustic telemetry project with fixed receiver array. The orange asterisks mark the northernmost and southernmost locations of confirmed detections of acoustic-tagged Green Sturgeon. Green Sturgeon spawn in California in the Sacramento and Klamath rivers, and in Oregon in the Rogue River (shown in blue). They spend summers in estuaries and bays in California, Oregon, and Washington, and utilize the coastal ocean between southern Alaska and Baja California, Mexico, generally remaining in water less than 100 m deep. Summer aggregation areas are shown as yellow triangles. The 100 m isobath is shown as the light blue areas near the coast.

data from the trawl and net surveys (see [48] for compiled information), allowing better fisheries management and design of protected areas around feeding and spawning areas.

1.3.2. Western Europe Outside of the Black Sea. In Western Europe, sturgeons have been heavily affected by historic fishing and habitat alteration as well as non-native release (see section 2.4); the presence of exotic species and hybrids is particularly problematic in this region. The area is represented by small, sometimes relict populations such as the European (Baltic) Sturgeon *(A. sturio)* and Adriatic Sturgeon (*A. naccarii*; Table 1). For these critically endangered species, the emphasis is on recovery and restoration; distribution and movement data are extremely useful for identifying critical and important habitats.

Restoration projects for European and Atlantic Sturgeon have released marked (mostly tagged with external T-bar tags) juveniles from captive breeding facilities to track subsequent movements and habitat use [51]. Acoustic telemetry has been used to study young European Sturgeon in the Gironde River estuary and the influence of tidal cycles on their movement [52]. [53] used acoustic telemetry to monitor reintroduction trials for American Atlantic Sturgeon (*A. oxyrinchus oxrinchus*) to the Baltic Sea and European Sturgeon to the Elbe River in Germany, providing key information on food selectivity, migration patterns, and potential sources of mortality. This study was important to the success of reintroduction efforts (2,000–5,000 tagged fish released per year) and in guiding planned restoration and habitat protection efforts. Tagging will eventually reveal whether reintroduced animals survive to adulthood and return to their rivers of release to reproduce. This study as well as [54] and [55] not only provide knowledge for species restoration but also insight on the migratory and habitat use patterns of juvenile sturgeon, the least studied life history stage in terms of movement and migration. Rearing of European Sturgeon in captivity also provides information on life history parameters associated with light, temperature, and salinity that are useful for the purpose of inferring population preferences and requirements in the wild [56], [57].

1.3.3. Black Sea. The Black Sea is home to the same commercially important species of sturgeon as found in the Caspian Sea, with the Danube River supporting the largest populations. Sturgeons in this region have also been impacted by fishing and habitat degradation and the focus of scientific research is often on restoration, recovery, and management. In the Black Sea, acoustic tagging studies have been used to examine spawning location and timing for Stellate (A. stellatus) and Russian Sturgeon (A. gueldenstaedtii) in the lower Danube River [58]. A high rate of interception of tagged fish as indicated by reported fishery returns (38% in 1998 and 28% in 1999) limited the success of the effort. Stellate and Russian Sturgeon appeared particularly sensitive to capture and handling, with over 50% of tagged fish subsequently aborting their respective upstream migrations. Tagging is again being attempted in the Black Sea, as is the use of a DIDSON acoustic camera (see section 2.6.2), to investigate movements [59]. The current tagging study may be more successful than the previous effort due to a moratorium on sturgeon fishing currently in place in the Romanian and Bulgarian portions of the Danube River. New research that incorporates microchemistry techniques is focusing on the use of trace elements to understand migratory patterns [60]. Future monitoring and assessment studies that incorporate tagging will be important in understanding whether fishing moratoria are being respected and if sturgeon populations are successfully recruiting and rebounding.

1.3.4. Russian Far East and China. The Russian Far East and China include little-known (and under-studied) anadromous sturgeon species [61], [62]. The Amur (*A. schrenckii*) and Kaluga (*Huso dauricus*) Sturgeon of the Amur River are commercially exploited for meat and caviar. These species are thought to have complex population structures, undertaking considerable migrations within the river (Amur Sturgeon), and marine environment (Kaluga). The Sakhalin Sturgeon inhabits the Russian Far East and is believed to have a primarily marine life history; it is one of the least studied of the sturgeons and is also one of the most critically endangered, with spawning populations possibly reduced to a single river. The Chinese Sturgeon, *A. sinensis*, has benefited from studies aimed at understanding movements in relation to dams and habitat protection [63–65]. As with the Caspian and European species, these species would benefit from studies of migration, movement, and distribution, so that appropriate recovery and management plans can be produced.

Research Techniques

2.1. Tagging and Marking

The application of individually numbered or coded tags to fish prior to release is typically referred to as "tagging." The term "marking," while sometimes used as a substitute for tagging, is typically used to reference batch or group "mark" applications. Subsequent recaptures of tagged fish can yield information on movement, time at large, growth, etc., whereas recaptures of marked fish will provide general information only regarding the entire "mark group." If recapture and reporting rates are sufficiently high, capture-recapture data from tagging studies are very useful for population assessment, providing information on population size, size/age structure, movement rates, fishing/natural mortality, and fish behaviour.

2.1.1. Considerations. Sturgeon are long-lived animals; thus, tags or marks that can be positively identified many years following release are ideal. Although not always feasible, determination of sex and maturational stage of individual sturgeon used in long-term tagging studies can provide greater insights regarding interpretation of subsequent movement and migration patterns. If tagging involves tag implants in the body cavity (typically electronic tags), the process should include visual determination of sex and stage or analysis of plasma concentrations to identify sex [66], [67]; see section 2.1.4 of this paper.

The use of uniquely identifiable tags can provide comparative growth and condition information for long-term studies with adequate sample sizes [68]. Length data (i.e., fork length) collected during initial tag release events and again during subsequent recaptures can be used to calculate daily growth rates for individual fish (based on the number of days at large between the release and subsequent recapture events). Daily growth can be expanded to provide estimates of annual growth, which in turn can be pooled and averaged for size/age groups of fish; these comparative data can provide insights regarding changes in overall population growth rates and condition over time [68].

2.1.2. Marking. Marking sturgeon with traditional methods such as dye marking, tattoos, pigment implantation, or freeze branding is not recommended for studies that intend to report recaptured marks over a long time span (i.e., several years), as these marks cannot be clearly identified on sturgeon within several months or a year of application. Another more-permanent marking approach to be considered, with caution, is scute removal [69], [70]. Scutes are one of the few body parts of sturgeon that can be removed that will not grow back. A combination of scutes removed from pre-determined locations on the sturgeon can translate into information such as location, year, and month. Since scute removal may induce high levels of stress, it is not recommended for use on sturgeon populations that are critically endangered.

Other permanent marks that have been applied to sturgeon include barbel clipping and fin ray removal. Barbel removal is not recommended due to the importance of barbels to sturgeon sensory physiology and the reduction in fitness and increase in mortality it can cause [71]. The total or partial removal of the lead pectoral fin ray has gained popularity and utility for sturgeon aging (fin ray cross sectioning) and can also be used for genetic analyses (fin tissue) and microchemistry analyses (fin ray composition analyses). The removal or clipping of a fin ray can also constitute a short-term (less than 5 years) mark, typically as a secondary mark to a tag application. However, care should be taken to clip the fin ray beyond the articulated base (above the insertion point with the body wall) or mortality rates could increase [72]. As with scute removal, fin ray removal may elevate levels of stress and reduce swimming performance (short term) and is not recommended for use on critically endangered populations.

2.1.3. Conventional Tagging. Tag types for sturgeon fall into three major categories: external tags, internal tags, and electronic (telemetry) tags. For the purposes of this paper, internal tags will be limited to Passive Integrated Transponder (PIT) tags, which are small, individually coded tags that are injected (with a specialized tool) into the body musculature or internal body cavity of sturgeon and are detected with a hand-held electronic tag reader. The use of electronic tags (that can be attached externally or internally) and utility of associated telemetric tag types (radio, acoustic, satellite) is presented in section 2.1.4 of this paper.

External tags – External, non-telemetry tags are typically attached to dorsal or lateral locations on the sturgeon with the intention to minimize the impact or influence of the tag (encourage natural behaviour) and maximize tag retention. Some external tags are applied with an application tool (such as "anchor" or "T" tags, that are applied with a "tagging gun"), while others may be attached by hand, sometimes with the use of an applicator or needle. Tag materials include plastic, PVC, nylon, and metal, and may be available in various colors to attract attention and/or assist with individual or batch identification. External tags are typically labelled with information that includes a unique tag number, and may include contact information for tag reporting, such as an address or phone number. Retention rates for external tags applied to sturgeon vary; some can maintain retentions above 70% for up to 3 years while other external tag application/types may not provide the high levels of tag retention rates required for mark-recapture analyses [69]. T-bar anchor tags can have especially good retention rates [73], [74].

Common external tags used for sturgeon include: the "monel" tag (metal tag that is usually clamped around the front or back of the dorsal fin, operculum, or pectoral fin); the "disc" tag (a flat round plastic tag with a small hole in the center, typically attached to the base of the dorsal fin with stainless wire); and the "anchor" or "T" tag, which is typically a 3–5 cm length of small-diameter PVC tube with a "T" shaped plastic anchor at one end. The "cinch" tag (also known as a "loop" tag) is similar to a long anchor tag except that the ends are attached (forming a loop). Popular attachment points for external tags are typically near the dorsal fin (either anterior or posterior of the fin, alongside the fin, or through the base of the fin) or the edge of the operculum or pectoral fin (popular with monel or similar metal "clamp" tags).

Internal (PIT) tags – A popular and effective tag for sturgeon is the PIT tag, a small, uniquely coded electronic tag that is applied internally via a hand-held applicator (syringe). Long-term retention rates for PIT tags are typically above 95% [69], [73], [75]. Popular PIT tag insertion locations are in the body cavity, base of pectoral fin or dorsal fin (or between base of dorsal fin and lateral line), and behind head plate (left or right of dorsal line). The position behind the head plate (Figure 4) has gained popularity in recent years due to lower tag loss rates and also based on concerns regarding potential human consumption of tagged sturgeon (tags in the "head" area are less likely to be consumed [68]). PIT tags used for sturgeon are typically 2 mm in diameter and 10–14 mm in length. Following application, there is no visible external indication that the fish has been tagged; the fish must be "scanned" with a hand-held electronic PIT tag reader (scanner) to determine if the fish has been tagged. PIT tag readers are typically battery powered, and display the tag number on a small screen. Tags can be detected with most hand-held tag readers from a distance up to about 20 cm, and the signal can be detected through water, flesh, etc.

PIT tags can also be detected with PIT tag antennas that are built into passive underwater apparatus such as tubes, loops, or mats; when a PIT tagged sturgeon passes through or near the antenna, the tag code is logged with time and date in the memory of a receiving unit (connected by cable to the antenna apparatus or array) that is later downloaded. This application may be effective to gather recapture data for constricted locations where tagged sturgeon are forced to pass through a confined area. For some applications, remote PIT tag detection may provide information to calculate overall survival [76]; however, without a secondary detection strategy to document the passage of non-tagged sturgeon (such as a video camera, or test sampling) to determine mark rates, the tag data alone will not allow sufficient data to produce abundance estimates.

Considerations for PIT tags – PIT tags are typically more expensive (in the order of 5–10 times more expensive) per tag than most external tag types but have high retention rates. PIT tag information is transmitted from the tag by way of a specific radio frequency; both the tag and tag reader must be compatible and using the same frequency. Thus, when committing to a long-term research project that uses PIT tags, it is important to determine the long-term availability of both the tags and the compatible tag readers. Also, researchers should determine if other sturgeon researchers that have worked or are working on the target population or species have applied PIT tags to sturgeon. If so, new studies that include PIT tag applications may want to consider using the same tag frequency so that previously tagged sturgeon can be identified.

2.1.4. Telemetry Tagging. The rapid advancement of electronic tag technology (telemetry) has facilitated an explosion of research on sturgeon life history, which in turn has generated novel insights into migratory behaviour, habitat use, and demographic processes. Electronic tagging offers advantages over traditional tagging approaches requiring large sample sizes to overcome low recapture rates. Electronic tags can be categorized by several characteristics. One distinction is whether tags transmit data in real-time to receivers (telemetry tags) or store data to internal memory for transmission at some later date (archival tag). Telemetry tags can be further differentiated by signal transmission mode: acoustic or radio. To date, telemetry tags have been most widely used in sturgeon research, although archival tags have been used with great success in studies of oceanic fish and marine birds.

Acoustic tags transmit identification codes (and potentially other data from pressure or temperature sensors) to hydrophones over a limited range (generally <1 km). Radio tags are similar, with differences related to signal transmission mode: radio waves propagate well through air and freshwater, but poorly in hard or saline water. Because acoustic or radio tagged fish need not be recaptured and can be detected at a distance, strategically placed receivers can collect large amounts of data on tagged fish. Combined acoustic and radio tags (CART) combine both transmission modes, allowing radio receivers to be used in freshwater and acoustic receivers to be used in brackish and saltwater. For both kinds of tags, information about the tagged fish is only collected from fish within the detection area of receivers. This shortcoming can be overcome by archival tags, which record signals from various sensors and store them internally. Typical information recorded by archival tags include pressure (depth), temperature, and light level, which allows estimation of location based on sea surface temperature and the time of sunrise and sunset [77]. Other sensors include electromyograms (EMG) that record muscle contractions, and accelerometers and tilt sensors that provide data on movement and orientation. Data from archival tags can be recovered by recapturing the tagged animal (and downloading data from the tag) or by telemetric transfer of archived data to satellites after pre-programmed release from the tagged animal.

Use and utility of telemetry tags – We searched the primary literature for sturgeon studies reporting the use of electronic tags, and found 55 relevant studies (summarized in Table 2). To date, the most frequent application of electronic tagging has been to study habitat use and movement of wild sturgeon within rivers. A substantial fraction of these studies have

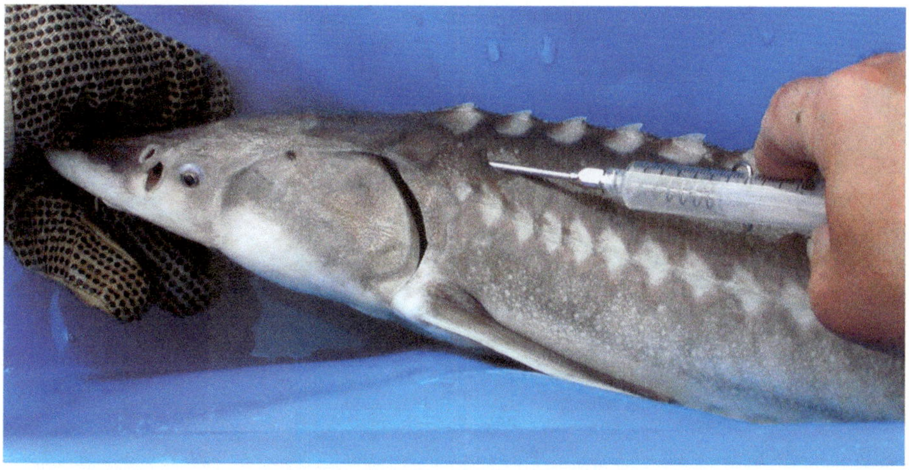

Figure 4. Illustration of the preferred location of PIT tag application on a juvenile White Sturgeon. The PIT tag is injected just beneath the skin, about 1 cm behind the head plate, on the left side of the dorsal scute line. *Photo: Fraser River Sturgeon Conservation Society.*

identified the timing and location of spawning, a topic of obvious interest to those working on conservation of sturgeon. These studies have relied on both acoustic and radio tracking, often in combination, and typically have made use of mobile tracking to detect tagged sturgeon. A related type of study examines the movement and habitat use of cultured sturgeon after release.

The second most frequent use of electronic tags for sturgeon research has been to investigate large-scale migrations (e.g., movement of sturgeon among riverine, estuarine, and marine waters). These studies have used all types of telemetry tags and tracking methods, and most have used acoustic tags, or a combination of acoustic and other tag types. The study of [35] is notable for its use of cross-shelf arrays of receivers to document annual migrations of Green Sturgeon (individuals of which can cover more than 4000 kms per year). This was possible because the continental shelf off the west coast of North America is narrow, and a number of studies were underway using compatible tags and receivers, including the Census of Marine Life's Pacific Ocean Shelf Tracking (POST) program [78].

Movement and habitat-use studies in estuaries ($n = 5$) have relied on acoustic tags and mobile tracking, with the exception of a study of Green Sturgeon in Willapa Bay (Washington) and the Columbia River estuary [79] which used arrays of fixed receivers, and a study of Atlantic Sturgeon in the St. Lawrence River estuary that used mobile tracking and one fixed receiver [80].

The study by [81] used a combination of archival tags, radio tags, and sophisticated data analysis to determine if mature Shovelnose Sturgeon spawned successfully following tagging. The radio tags were used mainly to relocate individual sturgeon (whose state of gonadal development had been determined) bearing the digital storage tags that recorded temperature and depth. The analysis involved fitting a switching model to the depth histories of individual sturgeon; spawning was inferred by observing a period of reduced depth variability. Data provided by electronic tags, especially from archival tags, can be a rich source of information on behaviour that can be analysed with appropriate models that connect behavioural modes to the tag data [82–84].

Electronic tags have also been used to investigate the behaviour of cultured sturgeon following release [85–88], the behaviour of sturgeon near potential fish passage impediments [89], [90], experimental evaluations of habitat suitability [91], and to identify spawning habitat [92], [93]. Both acoustic and electromyogram (EMG) tags have been used to test the response of White Sturgeon to changes in dam operations [94]. EMG tags transmit radio pulses when muscles contract, allowing researchers to measure the physical activity of fish at finer scales than can be determined from tracking techniques that only provide coarse changes in position over time, or in flow fields where swimming speed in water may differ substantially from speed over ground.

The importance of sex determination in telemetry studies – Many biotelemetry studies have been conducted with juvenile and adult sturgeon without identification of sex and stage of maturity. In certain cases, such as identification of general habitat use or spatial distributions of a population, it may not seem imperative to know the sex of individuals. However in many cases, such as when trying to understand why habitat is chosen, or when and why migrations are initiated or ceased, identifying the sex and, in adult populations, the stage of maturity provides greater insight into the behaviour of the animal [24]. It often is stated in the literature that sex and stage of maturity were not determined to reduce stress; it is critical to reduce stress, but trained personnel can determine sex and stage of maturity with very little increased handling time during transmitter application, or blood plasma levels can be examined for retrospective analysis. A review of methods is provided here [95].

If transmitters are attached externally, ultrasonography or endoscopy may be used to determine sex and maturity [96–102]. The primary limitation of these techniques is that it is often difficult to differentiate immature females from males; in addition, the stage of maturity in males cannot be determined as size of testicular lobe does not always confer stage of maturity. Ultrasound and endoscopy are useful tools in the identification of ripe females. When transmitters are implanted surgically, a sterilized otoscope or pen light may be used to visually determine sex and stage of maturity. A larger incision than the incision for insertion of the tag is not required under these circumstances. However, as in the case with ultrasound and endoscopy, the stage of maturity in males cannot be determined visually, though a small biopsy may be collected for histological analysis and stage of maturity determination. Additional limitations of ultrasonography are the potentially high cost of equipment and the moderate level of training required in order to interpret sonograms.

Measurement of circulating sex steroids may be used less invasively to determine sex and stage of maturity [67], [103–107].

Table 2. Summary of recent sturgeon studies that utilized electronic tag technologies to acquire movement information, by species, electronic tag type, and tracking method.

Study Type	Species	Tag Type	Tracking Method	REFERENCE
Habitat use and movements in rivers	*A. brevirostrum*	AC, RA	MT	[214]
	A. fulvescens	AC	MT	[215]
	A. fulvescens	AC, RA	MT	[216]
	A. fulvescens	RA	MT	[217]
	A. gueldenstaedtii	AC	FR, MT	[58]
	A. medirostris	AC	FR	[187]
	A. medirostris	AC, RA	FR, MT	[33], [218]
	A. medirostris	RA	FR, MT	[32]
	A. oxyrinchus desotoi	AC, RA	MT	[24]
	A. oxyrinchus desotoi	RA	MT	[219–221]
	A. sinensis	AC	FR, MT	[63]
	A. sinensis	AC	FR, MT	[64]
	A. sinensis	AC	FR, MT	[65]
	A. stellatus	AC	FR, MT	[58]
	A. transmontanus	AC	MT	[222]
	A. transmontanus	AC	FR	[223]
	A. transmontanus	AC	FR, MT	[78]
	A. transmontanus	AC, RA	FR, MT	[224]
	A. transmontanus	AC	FR	[141]
	S. platorynchus	RA	MT	[225]
Large-scale migration (includes marine)	*A. medirostris*	AC	FR	[35], [39], [40]
	A. medirostris	PAT	AG	[34]
	A. oxyrinchus desotoi	AC	FR	[226]
	A. oxyrinchus desotoi	AC, RA, PAT	MT	[29]
	A. oxyrinchus desotoi	AC, PAT	MT	[27]
	A. oxyrinchus desotoi	AC, RA, CART	MT	[227]
	A. oxyrinchus desotoi	AC, AR, CART	FR, MT	[26]
	A. oxyrinchus desotoi	AC	FR, MT	[228]
	A. transmontanus	AC	FR	[186]
Movements and habitat use within estuaries	*A. medirostris*	AC	FR	[79]
	A. medirostris	AC	MT	[229]
	A. oxyrinchus	AC	FR, MT	[80]
	A. oxyrinchus desotoi	AC	MT	[230]
	A. sturio	AC	MT	[52]
Behaviour of cultured fish after release	*A. naccarii*	AC	MT	[85]
	A. oxyrinchus oxyrinchus	AC	FR, MT	[53]
	A. sturio	AC	FR, MT	[53]
	A. sturio	AC	FR, MT	[54]
	A. transmontanus	AC	FR	[86]
	S. albus	AC	MT	[87]
	S. albus	RA	MT	[88]
Passage through fish ladders	*A. brevirostrum*	RA	FR	[89]
	A. transmontanus	CART	FR	[90]
Habitat use and movements in lakes	*A. brevirostrum*	RA	MT	[231]
	A. fulvescens	RA	MT	[232]
	A. fulvescens	AC	FR	[67]
Identifying spawning habitat	*A. brevirostrum*	AC	MT	[92]
	S. platorynchus	AC, AR	MT	[93]

Table 2. Cont.

Study Type	Species	Tag Type	Tracking Method	REFERENCE
Dam operations (response)	*A. sinensis*	AC	FR, MT	[233]
	A. transmontanus	AC, EMG	MT	[94]
Survival (induced spawning)	*A. oxyrinchus desotoi*	AC	MT	[234]
Cultured fish survival	*A. fulvescens*	AC	FR, MT	[235]
Demographic parameters	*A. oxyrinchus*	AC	FR, MT	[236]
Spawning success	*S. platorynchus*	AC, AR	MT	[81]
Habitat evaluation above barrier	*A. brevirostrum*	RA	MT	[91]
	A. medirostris	AC	FR, MT	[38]
Basic life history	*A. brevirostrum*	CART	FR, MT	[237]
Freshwater habitat, movement	*A. fulvescens*	RA	MT	[238]

Tag Types: AC = Acoustic; RA = Radio; AR = Archival; PAT = Pop-off Archival; CART = Combined Acoustic and Radio; EMG = Electromyogram.
Tracking Method: MT = Mobile Tracking; FR = Fixed Receivers; AG = Archival Geolocation.

Under ideal circumstances, the misclassification rates of assigning fish to classes of sex and maturity (i.e., pre-vitellogenic female, pre-meiotic male, vitellogenic female, etc.) using plasma steroid concentrations would be determined for each species. Interestingly, when classification functions derived for White Sturgeon to predict sex and stage of maturity [104] were applied to a small number of Lake Sturgeon, comparable classification rates were found [106]. Special consideration for the use of this tool should be made in populations that may be exposed to environmental contaminants [107].

Stress associated with capture and handling in telemetry studies – Less is known about the neuroendocrine control of the stress response and roles of allostasis and hormesis in chondrosteans compared to teleosts [108]. However, the cortisol response has been described in several sturgeon species in response to a stressor [108–112], and cortisol has been identified as the primary glucocorticoid in Pallid Sturgeon [113]. Plasma cortisol concentrations (basal and stressed) in sturgeon vary by species [109], [113]; variation in plasma cortisol concentrations may also be influenced by time of day, age, size, season, temperature, and capture and sampling techniques [110], [112], [114], [115]. It is essential to reduce stress (i.e., air exposure, handling time, considering water temperature, etc.) during external attachment or surgical implantation of transmitters. Guidelines for the reduction of stress to fish following capture and during handling/sampling are provided in [95], [116].

2.2. Experimental Design for All Tagging Studies

In order to design a cost-effective and useful tagging program, researchers should compose an experimental design that considers sample size, the costs associated with tagging and recapture activities, and study duration [117]. For example, when making inferences about temporal movement patterns using archival tagging data, the use of representative data sets (covering appropriate spatial and temporal scales) with a large number of reconstructed migrations (ideally 100+) is recommended [118]. With the exception of telemetry studies (where data may be recovered remotely), consideration should be given to the potential number of tags that are expected to be recovered (based on the number of tags deployed and the level of subsequent sampling effort to recover tags) and reported. Low tag reporting rates can lead to low precision in estimates of rates of natural and fishing mortality [119]. Ideally, tagging and marking studies should aim to minimise tag loss and tagging-related mortality and be done in a way that the tag reporting rate is known or can be estimated [119] [122]. Researchers should consider the statistical power of a planned tagging experiment, and whether their sample size will provide adequate power to detect the effect size of interest [123] [124]. Further stratification by length, age, sex, or life-history stage may also be desirable [125]; in this case, the number of releases will need to be scaled up by the number of classes or life-history stages for which parameters will be estimated to achieve the desired level of precision and accuracy [126].

The spatial and temporal distributions of sampling/tag deployment also warrant consideration. Tagging and release in each area combined with availability of fishing/sampling effort by time in each area has been recommended to minimize bias and the standard error of movement rate estimates [126], [127]. The intensity of fishing/sampling effort and the underlying movement rates themselves will also affect the accuracy and precision of movement rate estimates; ideally more tags should be released in areas with low rates of originating and incoming movement [127]. In the context of exploitation rate estimation, tagged fish should be representative of the exploited population, such that each exploitable fish is equally likely to be tagged [128].

2.3. Data Analysis and Modelling

Information from marking and tagging experiments can aid the design of subsequent surveys and the interpretation of the survey data. This is particularly relevant to Caspian Sea sturgeons as total allowable catch estimates are based upon population assessments derived from trawl survey data. Spatially-structured mark-recapture models provide a suitable framework for estimation of rates of residency and movement using tagging data [125–127]. Tagging data also allow inference about population abundance and exploitation rates [128], providing that rates of tag loss and reporting are known or estimable [129], [130]. Electronic tags generate relatively new kinds of data, and methods for the analysis of these data are lagging behind the rapid development of the tags, and the increasing use of large release groups. In the early studies, when tags were more expensive and their utility unproven, typical studies tagged a handful of animals. In this case, data arising from each tag could be presented separately and in full. Newer studies have tagged hundreds or even many thousands of fish (e.g., studies of in-river migration in Pacific salmonids), requiring statistical analyses to estimate demographic parameters or infer behaviours.

Methods for analyzing the rich data streams from archival tags are relatively well-developed. Capture-recapture methods, such as the multi-strata robust design, would seem to be applicable to situations where sturgeon move among many different areas and experience different mortality rates according to their location or state (e.g., maturation status, sex); however, in order to achieve reliable outputs, several challenges must be solved. These include achieving large-enough sample sizes to reliably estimate the many parameters of such models, and dealing with the problem that detections of tagged sturgeon occur over the same time interval as the mortality process. Typical capture-recapture models assume that there is no mortality during the recapture process, which in some cases can be instantaneous.

The data gathered from tagging studies that use simple (internal or external) tags and/or telemetry can potentially provide information on a number of life-history characteristics that are necessary to model population dynamics (such as growth, rate of natural mortality, and spawning periodicity). Using life-history traits to understand how populations may respond to perturbations such as harvest has been advocated as a useful approach to predicting extinction vulnerability and formulating management strategies [131–133]. Population dynamics models can be applied to gain an improved understanding of the status of sturgeon populations and their response to exploitation (historical and/or projected). The majority of population dynamics models for sturgeon species to date can be classified as simulation models, which can be applied to evaluate the sensitivity of population productivity and growth rates to changes in the survival rates of different life stages, or to estimate maximum sustainable yield harvest rates [134–137]. Where a time series of historical catches is available, population dynamics models can be used to reconstruct unfished population abundance over time to estimate depletion (the ratio of current to unfished abundance or biomass) [138–140].

2.4. Genetics

Applying genetic techniques allows a researcher to understand the species that are present in a given area (i.e. native, non-native, hybrid, or aquaculture origin) and the movements of sturgeons on multiple spatial scales and through evolutionary time [see references in this section below]. An important question that can be addressed using genetics is that of where an individual animal reproduces. As a first step in answering this question, reproductively isolated populations of a sturgeon species must be identified through genetic analysis. By knowing the number of isolated populations, appropriate conservation schemes can be created that preserve the evolutionary and ecological potential of a species. The protection of multiple populations can serve to buffer against extinction due to environmental change. Many sturgeons are thought to return to their natal river to reproduce and it is this behaviour that creates genetic structure and separate populations. It is unlikely, on the timescale of interest to managers, that animals from one river system would replace those in another if an individual population were to go extinct; thus, it is important to minimize individual population extinction since translocation of individuals from other populations should only be used as a last resort. A related point is that of identifying the presence of non-native species or hybrids in an area as an important step in conservation planning.

While information on the location of reproduction can be obtained through tagging studies, this approach does not answer the question of whether the animal effectively breeds in an area. Evidence of spawning can be gathered through biochemical means (e.g. measuring hormone levels or gonadal development), the deployment and monitoring of egg-mats or larval traps, or specifically designed electronic tagging studies, but only genetics can pinpoint whether the individual contributes to the next generation. In addition, it may be easier and less costly to use genetics to track breeding. As described below, once the overall genetic structure of a species is characterized, it can be fairly straightforward to identify the population origin of any individual at any location. Combining tagging and genetics can further provide a more comprehensive picture of population structure as defined by present and historic habitat characteristics [141].

Some of the fundamental questions that can be addressed using genetics include:

- Which species or species hybrids exist and reproduce in a given area (especially areas that may have been influenced by captive population release, both intentional and unintentional)?
- How many populations exist within a species? Where do individuals from these populations breed and feed?
- Does a specific area in the ocean or in an estuary include individuals from a single population or multiple populations and, if the latter, which ones?
- Do two or more populations exchange individuals?

Ideally, a genetics study will begin with comprehensive sampling of all potential spawning populations. For some sturgeons (e.g. Shovelnose, Pallid, White, possibly Russian), multiple spawning populations may exist within a single river system. Population differentiation can result from differences in timing or geographic location of spawning and can occur in the absence of any physical barriers separating populations. The samples taken from each population should ideally have corresponding information about the relative age of the animals (larvae, juvenile, adult) and the spawning stage of an adult. Information on life stage of the sample becomes especially important when sampling is conducted in lower reaches of the river and in estuaries because individuals can congregate in estuaries and coastal environments that are not in close proximity to their natal river of origin. When species or hybrid identification is a concern, sampling multiple life-stages can yield insight on whether non-native species or hybrids are effectively breeding in an area.

Sequencing of mitochondrial DNA (mtDNA) is one of the easiest genetic techniques to use for identifying the species present in an area (including non-natives) and studying movement, distribution and population structure and has been extensively applied e.g. [50], [142], [143]. Examples of the use of mtDNA include characterizing the presence of non-native Siberian Sturgeon haplotypes in Russian Sturgeon in the Caspian Sea [144–147] and the presence of non-native sturgeons in rivers in Greece [148]. One or several rapidly evolving gene segments (e.g. control region, cytochrome b) are usually studied, but sequencing mitochondrial genomes is now common. The utility of the latter is debatable given that most of the variation exists within the control region segment. The mtDNA cytochrome b and control region are also useful in differentiating among species [149–151], so researchers working in areas where multiple sturgeon species coexist can conduct species identification with mtDNA. Methods such as sequencing or Restriction Fragment Length Polymorphism (RFLP) analysis can be used to study population structure. The caveat in using mtDNA is that it tracks only the maternal lineage; this is important where species hybridize or where hybrids may be present and/or sex-specific dispersal may occur.

With respect to delineating population structure, one of the best examples of the utility of mtDNA comes from the east coast of

North America. The Atlantic Sturgeon has been particularly well studied and this work has shown fine-scale population structure and spawning site fidelity [142]. The data have been used to develop schemes of population-level protection under the US Endangered Species Act and attributing population origin to individuals captured in the ocean. Given that tagging studies are revealing large-scale oceanic movements at significant distances from natal rivers, this is particularly important because fisheries operating at considerable distances from a natal river could be impacting that population. For example, genetic samples taken from Atlantic Sturgeon captured off of the New York Bight (north eastern US waters) indicated that some of the sturgeon sampled were from breeding populations several hundred kilometres to the south in south-eastern US rivers [152].

Unlike mtDNA, microsatellites are biparentally inherited and genetic studies using these markers would not be biased by sex-specific differences in movements. In addition, microsatellites are highly polymorphic and may reveal population structure on a finer scale than mtDNA. There is less ease in applying this technique across species and species-specific marker development is sometimes necessary. Complications can arise in using microsatellites for polyploid sturgeons (i.e., species with greater than two copies of each chromosome), as some species may possess eight or twelve chromosome copies, however, the use of microsatellites to study populations can be achieved if Mendelian patterns of inheritance of microsatellites are found [153], [154]. General concerns about microsatellites have been raised by many authors [155]. Adriatic, Atlantic, Green, Lake, and White Sturgeon have been the most-studied in terms of the application of this marker; Caspian Sea and Chinese species are also becoming the subject of microsatellite investigations. Microsatellites are being used to understand and monitor hybridization in Scaphirhynchus species (with limited success when successive back-crossing occurs [13]), and hybridization between exotic and endemic species in the Danube River [156].

Microsatellites have been particularly useful in characterizing the marine distribution of Green Sturgeon and these studies, along with those on non-marine oriented sturgeons (White, Shortnose) [153], [154] have demonstrated application in polyploid sturgeons. Within the Green Sturgeon range, two distinct population segments (DPS) have been identified using genetics, corresponding to northern and southern spawning locations [47]. Individuals from the two DPSs often mix in the estuarine habitats of both natal and non-natal river systems. A further application of microsatellites in Green Sturgeon has been the generation river-specific abundance estimates [157]. Microsatellites have been used to characterize the population structure of Atlantic Sturgeon, to design a scheme of DPSs, and to characterize mixed-population fisheries [158]. Regarding the latter, microsatellite analysis revealed that Atlantic Sturgeon captured in US waters off of Virginia and North Carolina originated from populations as far north as Canada and were composed to a large extent of animals originating in the Hudson and Delaware rivers [159].

Many studies use both mtDNA and microsatellite markers. One interesting application, which also used four single-nucleotide polymorphisms (SNPs), is the study of the historical distribution of European and Atlantic Sturgeon. Study of archived and historic specimens revealed that Atlantic Sturgeon colonized an area in the Baltic Sea that was formerly thought to only be inhabited by European Sturgeon [160]. This work not only revised previous assumptions about species distributions, but also allowed restoration efforts to move forward in areas of Europe where sturgeons have been extirpated. Similar applications are ongoing to understand the former distribution of these sturgeons throughout Europe. Another important application has been to detect whether pure or hybrid strains of sturgeons are present in an area, thereby guiding captive breeding initiatives [161].

Emerging approaches include single-copy nuclear genes and SNPs [155], [162]. The application of these approaches to sturgeon has been limited because of the complications associated with polyploidy. Experimentation is ongoing with development of SNPs in Lake Sturgeon, White Sturgeon, some species of Eurasian and US sturgeons. Once developed, SNPs will offer additional power in determining population structure and movement.

There are many different analytical tools in the field of population genetics that can be applied. Measures of genetic differentiation such as genetic distance, F-statistics, analysis of molecular variance (AMOVA), and exact tests can be employed to identify distinct sturgeon populations and measure gene flow between them. These methods are most useful when samples can be obtained from spawning adults or newly hatched larvae in a natal river, where a researcher can be assured that the individuals sampled belong to a particular spawning population. When sampling is conducted in a region of potential mixing between populations (lower river, lake, estuary, ocean, or sea), the origin of each individual examined is unknown and other methods are necessary to examine population structure. Population assignment testing can be used with microsatellite or SNP markers to identify the population of origin of individual sturgeon in an area of potential mixing. There are several types of population assignment tests but they all exploit differences in allele frequencies between populations to assign individuals to their natal population. The software program STRUCTURE [163] may be particularly applicable to sturgeon studies as it can accommodate polyploid microsatellite data. In addition to identifying the populations contributing to a mixed population [47], population assignment tests can be used to evaluate individual dispersal behaviour. The assignment program GENECLASS2 [164] was used to identify two migrant sturgeon that originated from other river systems in the remnant Lake Sturgeon population in the White River, Indiana [165]. Finally, population assignment tests might be used to study population structure at varying hierarchical scales; STRUCTURE was used to examine groups of related Lake Sturgeon populations in the Great Lakes basin [166].

2.5. Microchemistry

Differences in trace elemental profiles between habitats or bodies of water can be exploited to learn more about the migratory behaviour of fishes. Gradients in elements such as Sr, Ba, Ca, Mg, S, and B exist in regions of different salinities, temperatures, and bedrock influences [167]. These elements are incorporated in minute quantities into the calcified structures of fishes, such as otoliths, fin rays or spines, bones, and scales [167]. Differences in the presence or concentrations of trace elements among aquatic habitats create elemental "fingerprints" on calcified structures that can be used to determine where a particular individual originated [168–170]. This information may be used to detect population structure, particularly in species where individuals from different populations mix during non-reproductive times, as has been done with some shell-forming invertebrates [171].

Otoliths are used most often in microchemical analyses of fishes since, unlike scales or skeletal bones, there is no potential for resorption or remodelling of this structure [172]. Although fin rays have the potential for remodelling, several researchers have shown the stability of elemental signatures in fin rays over time, suggesting they are stable structures appropriate for use in microchemical analyses [60], [170], [173].

One advantage of otoliths and fin rays is that changes in elemental composition between regions of incremental growth may be used to reconstruct patterns of movement in these fishes over time [174]. Most work to date has focused on otoliths, although conducting microchemical analyses on fin rays holds much promise for long-lived species such as sturgeons. Microchemical analysis of non-lethally collected fin rays has the potential to reveal age or stage-specific movement behaviour or habitat preferences in threatened or endangered sturgeon species, where otolith collection is undesirable.

2.5.1. Application to Sturgeon Movement. Applications of microchemical techniques to understanding sturgeon movements have been limited to this point, focused primarily on examining changes in marine and freshwater habitat use over the lifetime of an individual. In a study that confirmed the utility of fin rays for examining sturgeon movements [173], no changes in chemical composition due to bone remodelling or fin ray resorption over time were detected in individual White Sturgeon. This study also showed a significant correlation in the deposition of many trace elements between fin rays within individuals over time [173], suggesting that deposition occurs in a predictable way. Since then, several researchers have exploited predictable differences in Sr:Ca ratios between freshwater and marine environments to examine migration of various life stages of sturgeon from freshwater to marine habitats.

Approximately 10% of lower Fraser River White Sturgeon subadults between ages 1–15 showed Sr:Ca ratios that were consistent with migration into the marine environment, although these movements did not appear to be seasonal or periodic [175]. Intermediate concentrations of Sr:Ca ratios in fin rays of 58% of Fraser River adult White Sturgeon examined were suggestive of estuary use between ages 1–15 [175]. Sr:Ca ratios in both otoliths and fin rays were used to detect freshwater-to-saline migrations of subadult and adult Russian Sturgeon [176], [177]. Some individuals were characterized by a single movement into the Caspian Sea (from freshwater), while other adults had patterns consistent with diadromous movements between freshwater and seawater [176], [177]. The ratios of Sr:Ca, Ba:Ca, and Sr:Ba were used to evaluate the age of marine entry of subadult Green Sturgeon [178]; this study also conducted *ex situ* experiments to evaluate whether elemental deposition of Sr, Ba, and Ca in Green Sturgeon fin rays was proportional to environmental concentrations of these elements in freshwater and seawater environments. It was confirmed that ambient ratios of elements in freshwater and saltwater were nearly identical (Sr:Ca) or proportional (Ba:Ca; Sr:Ba) to the ratios found in fin rays of individuals held in freshwater or saltwater, respectively [178]. Wild Green Sturgeon subadults enter brackish estuary habitat between 0.5 and 1.5 years of age, and make their first migration into marine habitat between 2.5–3.5 years of age [178]. The ratios of both Ba:Ca and Sr:Ca ratios aided in interpretation of more complex environmental histories involving transitions between freshwater, brackish estuary, and marine habitats [178]. Although not the goal of these movement studies, some local differences in elemental concentrations [173], [178] suggested the potential for elemental "fingerprinting" in population composition analysis for sturgeon.

2.5.2. Future Research. Microchemistry techniques might be applied to resolve many questions regarding sturgeon movements and habitat use. Although there are some uncertainties associated with deriving age estimates from sturgeon fin rays [179], [180], the ability to examine habitat preferences and movement patterns of approximate age classes will be very useful in enhancing our understanding of sturgeon life history, particularly for those species with access to numerous and complex habitat types. Of particular interest would be the identification of important nursery or rearing habitats for early life stages of sturgeons, as this is unknown for many species. One possible use of elemental "fingerprinting" would be in examining spawning site fidelity in sturgeon. Although homing to natal spawning grounds has been assumed for many species, the long generation times of wild sturgeon have, to date, precluded direct confirmation through tagging studies. Population genetic data for several species [142], [166], [181] show genetic structure between spawning populations, which suggests that sturgeon do exhibit homing behaviour. However, elemental profiles in the core of fin rays of spawning adults might be compared to ambient elemental profiles in spawning habitats, which could confirm the hypothesis that the adults returning to spawn in a particular stream or river originated from there as well. Before these types of analyses are attempted, however, experimental work must establish that larval residency time in natal rivers is long enough for the incorporation of ambient elements in detectable concentrations. Also, it will be important to establish whether trace elements detected in otoliths and fin rays are consistently represented proportionally to ambient trace elements. A study of marine larval fishes [169] found that only one of 12 elements examined showed a similar correlation between ambient concentration and otolith deposition in an open-sea environment. The reported [178] proportional relationship between fin ray trace element concentration in Green Sturgeon and ambient water concentration might have been due to the coarse scale (freshwater vs. saltwater) at which environmental differences were examined.

Microchemical techniques have the potential to provide a great deal of information about sturgeon movement and life history. However, interpretation of elemental profiles in fish bony structures is dependent on an accurate understanding of the factors that influence element deposition. Additional work to ascertain the influence of water temperature, salinity, and the interaction between these parameters on element deposition in sturgeon otoliths and fin rays must be conducted to avoid misinterpretation of elemental profiles [182–184]. In addition, the use of laser ablation-inductively coupled plasma mass spectrometry (LA-ICPMS) techniques on otoliths from other species have shown ontogenetic changes in element deposition [185]; these changes could be quite confounding when reconstructing historical movements and habitat preferences of sturgeon and other long-lived fishes. Diet, too, may affect element deposition in otoliths and fin rays. We also must determine how residency time in a particular habitat affects element deposition [182]. Sturgeon are capable of fairly rapid movements between habitat types [79], [186], [187] and it is uncertain if these movements would be represented accurately in otolith or pectoral fin ray elemental "fingerprints." Also, it will be important to characterize the rate at which elemental profiles within particular locations vary, as rapid rates of naturally or anthropogenically induced environmental change may reduce our ability to accurately reconstruct the movement of individuals over longer timeframes using otoliths and fin rays [188]. These uncertainties should be addressed prior to a broad application of microchemical techniques toward addressing questions regarding sturgeon movement and habitat preferences.

2.6. Observational Technologies

Sturgeon can occupy marine environments or large, deep and typically turbid river systems where direct visual observation of behaviour, movement, or habitat use is difficult and often impossible. Consequently, until recently, what little was known about these attributes of sturgeon ecology were based on anecdotal observations or inferences drawn from capture and tagging studies. In the last two decades, substantial advances in remote sensing technologies have provided researchers with a greater variety of tools that allow direct observation of sturgeon behaviour and habitat use. These technologies can generally be grouped into two categories: those that use underwater cameras to provide light-based images and those that rely on sound waves to produce sonic-based images. Each of these observational categories and their applicability for use in studying sturgeon behaviour and habitat use is discussed below. Although most of the applications encountered in the reviewed literature were used for the study of sturgeon in freshwater systems, in many instances they also have applicability for the study of sturgeon and their habitats in marine and estuarine environments.

2.6.1. Underwater Photography. The use of underwater photography as a tool to examine sturgeon behaviour has only recently begun to be explored. Advantages of using underwater cameras are the ability to directly observe fish in their natural habitats and for some sturgeon species, the absence of any apparent avoidance behaviour of underwater cameras and lights. Disadvantages include limited use in aquatic environments with low water clarity and difficulties in long-term monitoring of fish that are engaged in active feeding or large-scale movements.

Obtaining underwater footage has been facilitated in recent years through the development of small submersible Remote Operated Vehicles (ROVs) and technological advances in low-light digital and video cameras. ROVs have been used in freshwater applications to examine the real-time effects of hydroelectric plant operations on sturgeon feeding behavior [189]. Underwater videography using ROVs has been employed for several years in the upper Columbia River in Canada to document behavior and habitat use by wild White Sturgeon adults and by hatchery-reared and released juveniles [5] as well as to document unusually large (approximately 60,000 fish) aggregations of White Sturgeon in the stilling basin below the spillways at Bonneville Dam (http://videos.oregonlive.com/oregonian/2008/05/sturgeon_ball.html). An ROV also was used to identify critical overwintering habitats for the endangered White Sturgeon in the Nechako River, a tributary to the Fraser River in British Columbia, Canada [190].

Underwater video camera systems have been used to study overwintering habitats of Shortnose Sturgeon in the upper Kennebecasis River, New Brunswick, Canada [191], characterize Lake Sturgeon spawning and substrate preference in the Big Manistee River in Michigan [192], and identify White Sturgeon spawning and early life-stage rearing substrates in the upper Columbia River in Canada [193]. White Sturgeon use of tailrace areas below existing hydroelectric dams in the Columbia River (Canada) was also examined using fixed video cameras and the information generated led to a plan to protect sturgeon during tailrace excavation [194]. White Sturgeon mortality at the Brilliant Expansion power plant was reduced after video monitoring of the powerplant draft tubes and outlets illustrated behavioural responses to reduced flow [195]. Cameras were used to provide real-time data on sturgeon presence during short duration forced outages of the powerplant; video footage recorded after the shut down were reviewed by the dam operators to assess if sturgeon have entered the draft tubes. Depending on the results, different start-up protocols were implemented to reduce risks to sturgeon.

2.6.2. Hydroacoustic Technologies. Over the last three decades, active hydroacoustic techniques have developed and proven to be a relatively easily applied method of unobtrusively evaluating fish populations in freshwater and marine environments [196]. The principles of hydroacoustic assessment of fish are provided in [197] and [198]. Hydroacoustic techniques have a very high sampling power and do not affect fish health, behaviour, or the environment being monitored and have been successfully applied to a variety of fisheries evaluations, including both mobile and stationary assessments of aquatic systems.

Mobile survey hydroacoustic techniques are generally conducted from a boat, traversing predetermined transects in a body of water, and sampling fish and bottom characteristics [196]. Sampled fish produce characteristic acoustic signals that can be processed using specialized software to produce estimates of fish density, abundance, behaviour, and size distribution. Sonars and sounders have been developed that can be used to characterize sea, lake and river bottoms and profiles of the upper layers of the ocean bottom. Advanced substrate classification analysis can be achieved using calibrated (scientific) echosounders and parametric or fuzzy-logic analysis of the acoustic data. Side-scan sonars can be used to derive detailed maps of the topography of an area by moving the sonar across and just above the bottom. Low frequency sonars have been used for continental shelf wide surveys while high frequency sonars are used for more detailed surveys of smaller areas. Various synthetic aperture sonars (SAS), which combine a number of acoustic pings to form an image with much higher resolution than conventional sonars, are under active development (http://www.hydro-international.com/issues/articles/id920- Synthetic_Aperture_Sonar_Challenges.html). This technology has become commercially viable and the technique is well suited for towed or remotely operated underwater vehicles. SAS is expected to replace traditional side-scan sonars for many applications in the future.

All acoustic systems have some sampling limitations with respect to their ability to resolve targets very close to boundaries, such as the bottom. Sturgeon are primarily benthic oriented and often in close proximity to the bottom. In addition, research describing sturgeon target strength, or the amount of acoustic energy reflected from the fish, is limited. To be effectively detected, a fish must return target strength values greater than the surrounding background noise levels. The primary reflecting structure in most fish is the swim bladder, although bones, scutes and other body structures do provide some contribution [199]. The sturgeon swim bladder is the primary acoustic reflecting structure and is located just below the spine, some distance from the ventral surface of the fish. This may aid in detecting these fish on the bottom with a down-looking acoustic system, as there is some inherent separation between the upper surface of the bladder and the bottom itself due to fish morphology [196].

Split Beam – The main advantages of split-beam over other hydroacoustic techniques are improvements in location within the acoustic beam and in minimized susceptibility to noise [200]. Given identical levels of bias in angular resolution, the split-beam system can locate fish within the beam with much greater resolution than single-beam, dual-beam, or sidescan systems [201], [202]. The three-dimensional location of each fish is known for each ping (i.e., each ensonification). This improved spatial resolution results in improved target strength estimates. More accurate target strength estimates allow more accurate spatial expansions, resulting in more accurate estimates of fish abundance and/or biomass. With split-beam target tracking, individually measured echoes may or may not be retained, depending on selection criteria that discriminates fish echoes from other echoes. Selected echoes are tracked to group all echoes from one individual fish. Mean target strength is calculated from the group of echoes from one fish, based on individual-echo target strength measurements made using the split-beam method. As with the other techniques, the signal can be echo integrated, to provide biomass estimates, if desired.

Split-beam hydroacoustics was used to detect Shortnose Sturgeon in the Delaware River [196]. The study was conducted by measuring Shortnose Sturgeon target strength (using net-captured fish) and the range from the bottom at which sturgeon could be acoustically resolved. The authors concluded that Shortnose Sturgeon could be readily detected by a split-beam hydroacoustic system using a combination of attributes (target strength, position relative to the bottom, and echo envelope shape). The demersal distribution of Shortnose Sturgeon is well established and also appeared to be a useful metric for distinguishing these fish from other species.

Fixed-location split-beam sonar technology was used successfully to identify adult Lake Sturgeon as they moved upstream and downstream for spawning in the Sturgeon River, Michigan [203]. Data collected included direction of movement, swimming speed, range from transducer, time and date of passage, and target strength. The Lake Sturgeon spawning population size was estimated and results showed that spilt-beam sonar can be applied to spawning assessments, without the stress of actually handling the large, pre-spawning fish.

Side Scan – Trawling and side scan sonar analysis was used to document an area of consistently high Lake Sturgeon density in Lake St. Clair near the St. Clair River delta [204]. Side scan sonar was used to estimate the abundance of Lake Sturgeon in a 255-ha section of the lake and the data were used to enhance protection and habitat restoration efforts for Lake Sturgeon in this and other Great Lakes connecting waters.

On the lower Missouri River, side-scan sonar data were collected in areas that contained suitable habitats for both Shovelnose Sturgeon and Pallid Sturgeon [205]. Hydroacoustic data sets were collected at the reach scale (mean reach length, 2.4 kilometres) to include the immediate vicinity of a targeted sturgeon location as well as the full range of adjacent habitats. The images obtained were useful for visualizing channel substrate and detecting the presence of adult sturgeon.

Researchers in Virginia used side-scan sonar to characterize the benthic habitats of Atlantic Sturgeon in the James River and identify habitat attributes that may be required to sustain viable sturgeon populations (http://www.thsoa.org/hy09/0512P_02.pdf). Areas of high-frequency sturgeon occurrence (as determined through concurrent telemetry studies) were targeted using a side-scan sonar system. Numerous habitat features were identified and the data will be integrated with tracking data and habitat imagery to identify essential Atlantic Sturgeon habitats.

Broadband – Experiments were conducted in the tidal Delaware River to determine if Shortnose Sturgeon could be detected by broadband sonar and, if so, to develop classifiers that could differentiate Shortnose Sturgeon from co-occurring fish species [206]. The false-positive rate of incorrect identification of a Sturgeon was 16.5%. Notwithstanding this potential problem, the authors concluded that the results of this preliminary study were promising, and further investigations to improve classifier performance were warranted.

To assist in identifying potential Lake Sturgeon habitat in the lower Bad River complex, a digital sonar system combined with a global positioning system was used to provide georeferenced data, and specialized sonar, bottom typing, GIS and statistical software to acoustically map bottom substrate types, locations, and bathymetry [207]. Ground-truth data were developed from both petite Ponar bottom samples and associated acoustic data which were processed with bottom-typing software; these data were used to produce substrate models and maps.

Fish finders – Many commercial and recreational sturgeon anglers use high quality hydroacoustic gear or "fish finders" to locate sturgeon in estuarine and marine environments along the east and west coast of North America. For sturgeon research, a fish finder with the following features is best: multiple zoom settings, bottom lock, and split-screen option (http://www.nwfish.com/Sturgeon/fish_finders_101.htm). Units that allow setting the window size for a specific number of feet while in bottom track will provide higher signal resolution that will show any irregularities on the bottom as well as fish holding right on the bottom. Sturgeon that are suspended some distance above the bottom show up quite well on most high-resolution fish finders (http://www.nwfish.com/images/graphics/ffshotsturg1.jpg). Sturgeon holding tight to a hard bottom are difficult to identify and are usually represented by a "spike" or "bump." (http://www.nwfish.com/images/graphics/ffshotsturg2%20.jpg).

DIDSON™ – In large rivers, a common approach for estimating population size of anadromous fish is to count upstream-migrating fish at a fixed site, using split-beam hydroacoustic equipment. A disadvantage of split-beam sonar in this application is that it generally does not provide sufficient information to allow species identification. A newer technology that can be used to count upstream migrants is the DIDSON™ (dual-frequency identification sonar), a high-definition imaging sonar that provides near-video-quality images. When used at a range of 5–10 m, video files clearly show body shape and swimming behaviour of individual fish. Split-beam gear provides more precise information about fish position, but DIDSON™ data are much easier to interpret, can be used to identify sturgeon to genus, and allow for on-screen measuring of fish lengths. Initial field trials showed potential for utilizing these technologies to determine habitat, identify sturgeon, and estimate densities (http://cars.desu.edu/aqua-sci/Abstracts/LB_et_al_acoustic.pdf). DIDSON™ has also been used to document White Sturgeon presence and activity in the vicinity of power plant outlets [208].

Restoration and management of the lower Missouri River to support recovery of the endangered Pallid Sturgeon required quantifying habitats to isolate specific habitats that may present bottlenecks to reproduction and survival [209]. The approach taken involved intensive reach-scale hydroacoustic mapping using a suite of multi-beam bathymetry, Acoustic Doppler Current Profiler, high-resolution side scan sonar, and DIDSON™ imagery, combined with intensive telemetric tracking. This approach provided measures of habitat availability and selection variables at sub-meter to bedform scales, commensurate with the scale at which fish occupied these habitats. The DIDSON™ imagery indicated that during spawning, sturgeon occupy the lee slopes of dunes facing upstream (presumably to minimize energy expenditure) but episodically move out of dune fields and into deep, fast water over coarse substrate (presumably to release eggs and milt). This multi-scale, multi-instrument remote-sensing approach was essential for improving understanding of the linkages between life stages of a rare fish and its environment.

2.6.3. Future Research. At present, observational techniques for sturgeon research have primarily been used to provide information on sturgeon behaviour and habitat use. These techniques enable researchers to directly observe sturgeon in their natural environment in a manner that does not influence their behaviour and have substantially increased our knowledge of how sturgeons interact with each other and their environment. Most importantly, these techniques have identified critical sturgeon habitats that in turn have resulted in the protection of those habitats through direct management regulations. As these observational techniques are developed further, their potential

Table 3. Schematic of techniques used to study different aspects of the biology of sturgeons.

	RESEARCH TECHNIQUE				
RESEARCH FOCUS	Genetics	External Tags or PIT Tags	Electronic Tags	Microchemistry*	Observational
Marine distribution	3	2	1	4	
Population structure and identification	1	3	2	3*	
Habitat characterization		4	2	2	1
In-river distribution		2	1	4	3
Population abundance/modelling	2	1	3		3
Response to disturbance (e.g. dam or dredging operation)		3	2		1
Life-history characterization (e.g. spawning periodicity, age at maturity, growth)		2	1		

*Microchemistry techniques, while not currently applied on a broad level to study population structure and identification, are building in popularity and applicability and may well rank alongside genetic and electronic.
tagging in the near future.
Research techniques are ranked 1-4 in order of use/utility by sturgeon researchers, with a rank of 1 being highest in terms of applicability. Techniques assigned the same number are of equal utility.

uses will continue to expand. Direct observation of sturgeon behaviour at dams and existing fish passage facilities, either through video or sonic imagery, can be used to provide information that may lead to the development of appropriate passage facilities for anadromous sturgeon. As sonar techniques are refined and new algorithms to process and interpret digital signals are developed, these techniques have the potential to be used to identify individuals and provide assessments of sturgeon distribution and abundance over large areas.

Results and Discussion

The marine life history and distribution of many sturgeons remains a mystery. However, new technologies developed over the past two decades have greatly increased our knowledge base. Refined methods of analyzing DNA provide fine-scale information about genetic structure at the river basin or even sub-basin level. Improvements in telemetry equipment and observational techniques have allowed researchers to identify and characterize key habitats, from spawning sites at the upstream extent of migration to overwintering sites and migratory paths in coastal ocean waters. Telemetry and tagging studies have also shown that estuarine and marine sites used for foraging or overwintering may contain sturgeon from multiple populations. Tagging, genetic, and observational studies have revealed that sturgeon can travel considerable distances from their natal rivers, but that they generally inhabit coastal, shelf areas during their migrations. Also, individuals generally seem to return to natal rivers for reproduction. Taken as a whole, this information shows that sturgeons have complex life-history strategies that are reliant on specific habitats for different life stages. As such, fully protecting a given species requires a comprehensive approach that includes research and conservation efforts in riverine and marine environments.

Yet even with these technological advancements, there are few sturgeon species for which we have a fairly comprehensive understanding of respective life histories, particularly in the marine environment. For the 16 marine-oriented sturgeons, only two (Green and Gulf Sturgeon) have been well studied with respect to their marine distribution (see: Table 1; section 1.2), and even these species would benefit from additional study of the movements and habitat use of young juveniles during their initial years in marine environments. Current studies of Atlantic Sturgeon in North America are now providing an understanding of the marine life history of this species. For the remaining species, many of which are classified as Critically Endangered and some of which are still subject to commercial and recreational fishing, there is a pressing need to apply many of the tools discussed here. As illustrated in Table 2, only two species of sturgeon (Adriatic and European) outside of North America have benefited from electronic tagging studies and our review has indicated few other studies where other techniques have been applied to non-North American taxa. European and Asian species are clearly in need of studies that utilize the techniques presented. Thus, while we have numerous tools in our management toolbox, we are lagging in our application of these tools to study most sturgeon species.

The research tools and techniques reviewed can be used individually or in combination. In Table 3 we present a suite of common research questions (research focus) and rank each of the core research techniques in the context of achieving/producing the desired outcomes. As illustrated for many species, tagging, telemetry, genetics and sometimes observational techniques are often used in combination. Tagging and genetics are complementary tools for characterizing movement on an immediate and fine ecological scale (tagging) as well as an evolutionary scale (genetics). Tagging can also include multiple methods, with simple internal or external tags combined with electronic tags. Population dynamics models may help to direct both core study design and specific elements such as the number of tags deployed, the level of sampling effort, etc. Microchemical techniques may be used to complement movement studies and to add information on contaminants; however, the authors recommend that additional background research may be needed to correctly apply and interpret microchemistry results, and additional work is required to validate some aspects of the approach. Observational techniques are most useful in characterizing habitat uses within marine or freshwater environments, behaviour, and small scale movements, although recent advances have shown promise for use in assessing abundance and distribution as well.

Tagging, telemetry studies, and photographic imagery can be useful ways to engage the public in research and raise awareness about the conservation status of sturgeons. Stewardship-based population monitoring and assessment programs that train individuals to inspect captured fish for the presence of a tag, apply tags to untagged fish, and collect basic information (fish length, girth, location, date) have been successful [75]. Telemetry studies, particularly those using comparatively expensive satellite tags, can be accomplished through individual or corporate sponsorship of a tag (and an individual fish); subsequent movement data generated can be shared with the sponsoring partner and with the public through a dedicated website. Imagery of sturgeon behaviour in the wild is a powerful visual tool that can be used to enhance public outreach programs and galvanize public support for population and habitat conservation, protection, and recovery. Elements of stewardship and public outreach programs can also be incorporated into many areas of school-based educational curriculum, including ecology, biology, natural history, environmental studies, and social studies: http://hsbc.frasersturgeon.com/.

Data collected by the methods described here can provide useful insights into the life history of sturgeon populations, and enable management and conservation. For example, information regarding temporal and spatial movements, migrations, and periodic residency, coupled with a general knowledge of life history events such as spawning and in-river overwintering, could provide a high level of confidence regarding the location and extent of proposed habitat protection for population conservation and recovery. Genetic data on population structure is essential for identifying particularly vulnerable or endangered populations and setting appropriate management actions to conserve them. Gaining an understanding of life-stage-specific movement patterns and spatial distribution is also required to properly assess the impacts of anthropogenic activities such as fishing, dredging, and gravel extraction and to specify management actions accordingly.

All of the techniques discussed require a certain amount of specialized training. Protocols for proper handling, short-term holding, tag applications, and tissue sampling have been produced for some species [210]; many of these established protocols are transferable to other species of sturgeon. Given the widespread use of many techniques, assistance can be readily obtained via contact with experienced researchers. Collaborative research is likely the best approach given that many techniques require up-front purchasing of expensive equipment and a steep learning curve. We hope that this review sparks interest and enthusiasm amongst researchers, especially those outside of North America, and provides some necessary tools and references for forming new studies.

Acknowledgments

The authors thank the following reviewers for their input on draft versions of this document: Ronald Bruch, Gordon Glova, Seth Miller, and Marvin Rosenau. In addition, we thank two anonymous reviewers and the PLOS ONE Academic Editor for their helpful comments on the manuscript. This work is a contribution to the Census of Marine Life.

The use of trade, product, industry or firm names or products or software or models, whether commercially available or not, is for informative purposes only and does not constitute an endorsement by the U.S. Government or the U.S. Geological Survey.

Author Contributions

Conceived and designed the experiments: TCN. Performed the experiments: TCN PD STL ADS JEH LRH REW MAHW. Analyzed the data: TCN PD STL. Wrote the paper: TCN PD STL ADS JEH LRH REW MAHW.

References

1. Pikitch EK, Doukakis P, Laucks L, Chakrabarty P, Erickson DL (2005) Status, trends and management of sturgeon and paddlefish fisheries. Fish Fish 6: 233–265.
2. Billard R, Lecointre G (2001) Biology and conservation of sturgeon and paddlefish. Rev Fish Biol Fisher 10: 355–392.
3. Hutchings JA (2001) Conservation biology of marine fishes: perceptions and caveats regarding assignment of extinction risk. Can J Fish Aquat Sci 58: 108–121.
4. Dulvy NK, Ellis JR, Goodwin NB, Grant A, Reynolds JD, et al. (2004) Methods of assessing extinction risk in marine fishes. Fish Fish 5: 255–276.
5. Hildebrand L, McLeod C, McKenzie S (1999) Status and management of white sturgeon in the Columbia River in British Columbia, Canada: an overview. J Appl Ichthyol 15: 164–172.
6. Glova G, Nelson T, English K, Mochizuki T (2009) A further report on juvenile white sturgeon habitat use in the lower Fraser River, 2008–2009. Report prepared by LGL Limited, Sidney, BC, for the Fraser River Sturgeon Conservation Society, Vancouver, BC. Available: http://www.frasersturgeon.com/media/juvenilesturgeonhabiat2008-09_reduced.pdf. Accessed 22 November 2012.
7. Munro J, Edwards RE, Kahnle AW (2007) Synthesis and Summary. In: Munro J, Hatin D, Hightower JE, McKown K, Sulak KJ, et al. editors. Anadromous sturgeons: habitats, threats and management. Bethesda, MD: American Fisheries Society Symposium 56: 1–15.
8. Committee on the Status of Endangered Wildlife in Canada (2003) COSEWIC assessment summary – white sturgeon. COSEWIC. Ottawa. Available: http://www.registrelep-sararegistry.gc.ca/virtual_sara/files/cosewic/as_white_sturgeon_e.pdf. Accessed 22 November 2012.
9. Jarić I, Gessner J (2012) Analysis of publications on sturgeon research between 1996 and 2010. Scientometrics 90: 715–735.
10. Bemis WE, Kynard B (1997) Sturgeon rivers: an introduction to acipenseriform biogeography and life history. Environ Biol Fish 48: 167–183.
11. Dionne PE, Zydlewski GB, Kinnison MT, Zydlewski J, Wippelhauser GS (2013) Reconsidering residency: characterization and conservation implications of complex migratory patterns of shortnose sturgeon (*Acispenser brevirostrum*). Can J Fish Aquat Sci 70: 119–127.
12. Arndt G-M, Gessner J, Raymakers C (2002) Trends in farming, trade and occurrence of native and exotic sturgeons in natural habitats in central and western Europe. J Appl Ichthyol 18: 444–448.
13. Schrey AW, Boley R, Heist EJ (2011) Hybridization between pallid sturgeon *Scaphirhynchus albus* and shovelnose sturgeon *Scaphirhynchus platorynchus*. J Fish Biol 79: 1828–1850.
14. Michielsens CGJ (2003) Bayesian decision theory for fisheries management of migratory species with multiple life histories. PhD Thesis, Imperial College, University of London.
15. Cunningham CL, Reid DG, McAllister MK, Kirkwood GP, Darby CD (2007) A Bayesian state-space model for mixed-stock migrations, with application to northeast Atlantic mackerel (Scomber scombrus). Afr J Mar Sci 29: 347–357.
16. Sibert JR, Fournier DA (1994) Evaluation of advection-diffusion equations for estimation of movement patterns from tag recapture data. In: Shomura RS, Majkowski J, Lanai S, editors. Proceedings of the first FA0 expert consultation on interactions of Pacific Ocean tuna fisheries. Volume 1 – Summary report and papers on interaction. FA0 Fisheries Tech Paper No 336/1: 108–121.
17. Rochard E, Castelnaud G, Lepage M (1990) Sturgeons (Pisces: Acipenseridae); threats and prospects. J Fish Biol 37: 123–132.
18. Birstein VJ (1993) Sturgeons and paddlefishes: threatened fishes in need of conservation. Conserv Biol 7: 773–787.
19. US Fish and Wildlife Service and the Gulf States Marine Fisheries Commission (1995) Gulf sturgeon recovery plan. US Fish and Wildlife Service, Atlanta, Georgia.
20. Huff JA (1975) Life history of Gulf of Mexico sturgeon, *Acipenser oxyrhynchus desotoi*, in the Suwannee River, Florida. Fla Mar Res Pub 16.
21. Wooley CM, Crateau EJ (1982) Observations of Gulf of Mexico sturgeon (*Acipenser oxyrhynchus desotoi*) in the Appalachicola River, Florida. Fla Sci 45: 244–248.
22. Wooley CM, Crateau EJ (1985) Movement, microhabitat, exploitation, and management of Gulf of Mexico sturgeon, Apalachicola River, Florida. N Am J Fish Manage 5: 590–605.
23. Odenkirk JS (1989) Movement of Gulf of Mexico sturgeon in the Apalachicola River, Florida. Proceedings of the Annual Conference, Southeastern Association of Fish and Wildlife Agencies 43: 230–238.

24. Fox DA, Hightower JE, Parauka FM (2000) Gulf sturgeon spawning migration and habitat in the Choctawhatchee River system, Alabama-Florida. T Am Fish Soc 129: 811–826.
25. Sulak KJ, Clugston JP (1998) Early life history stages of Gulf sturgeon in the Suwannee River, Florida. T Am Fish Soc 127: 758–771.
26. Edwards RE, Sulak KJ, Randall MT, Grimes CB (2003) Movements of Gulf sturgeon (*Acipenser oxyrinchus desotoi*) in nearshore habitat as determined by acoustic telemetry. Gulf Mex Sci 21: 59–70.
27. Edwards RE, Parauka FM, Sulak KJ (2007) New insights into marine migration and winter habitat of Gulf sturgeon. In: Munro J, Hatin D, Hightower JE, McKown K, Sulak KJ, et al. editors. Anadromous sturgeons: habitats, threats and management. Bethesda, MD: American Fisheries Society Symposium 56: 183–196.
28. Parkyn DJ, Murie JE, Harris DE, Colle JD, Holloway J (2007) Seasonal movements of Gulf of Mexico, sturgeon in the Suwannee River and estuary. In: Munro J, Hatin D, Hightower JE, McKown K, Sulak KJ, et al. editors. Anadromous sturgeons: habitats, threats and management. Bethesda, MD: American Fisheries Society Symposium 56: 51–68.
29. Ross ST, Slack WT, Heise RJ, Dugo MA, Rogillio H, et al. (2009) Estuarine and coastal habitat use of Gulf sturgeon (*Acipenser oxyrinchus desotoi*) in the north-central Gulf of Mexico. Estuar Coast 32: 360–374.
30. Dugo MA, Kreiser BR, Ross ST, Slack WT, Heise RJ, et al. (2004) Conservation and management implications of fine-scale genetic structure of Gulf sturgeon in the Pascagoula River, Mississippi. J Appl Ichthyol 20: 243–251.
31. Adams PB, Grimes CB, Lindley ST, Moser ML (2002) Status review for North American green sturgeon, *Acipenser medirostris*. National Marine Fisheries Service, Southwest Fisheries Science Center, Santa Cruz, CA.
32. Erikson DL, North JA, Hightower JE, Weber J, Lauck L (2002) Movement and habitat use of green sturgeon, *Acipenser medirostris*, in the Rogue River, Oregon. J Appl Ichthyol 18: 565–569.
33. Erickson DL, Webb MAH (2007) Spawning periodicity, spawning migration, and size at maturity of green sturgeon, *Acipenser medirostris*, in the Rogue River, Oregon. Environ Biol Fish 79: 255–268.
34. Erickson DL, Hightower JE (2007) Oceanic distribution and behavior of green sturgeon (*Acipenser medirostris*). In: Munro J, Hatin D, Hightower JE, McKown K, Sulak KJ, et al. editors. Anadromous sturgeons: habitats, threats and management. Bethesda, MD: American Fisheries Society Symposium 56: 197–211.
35. Lindley ST, Moser ML, Erickson DL, Belchik M, Welch DW, et al. (2008) Marine migration of North American green sturgeon. T Am Fish Soc 137: 182–194.
36. Lindley ST, Erickson DL, Moser ML, Williams G, Langness OP, et al. (2011) Electronic tagging of green sturgeon reveals population structure and movement among estuaries. T Am Fish Soc 140: 108–122.
37. Heublein JC, Kelly JT, Crocker CE, Klimley AP, Lindley ST (2009) Migration of green sturgeon, *Acipenser medirostris*, in the Sacramento River. Environ Biol Fish 84: 245–258.
38. Mora EA, Lindley ST, Erickson DL, Klimley AP (2009) Do impassable dams and flow regulation constrain the distribution of green sturgeon in the Sacramento River, California? J Appl Ichthyol 25: 39–47.
39. Huff DD, Lindley ST, Rankin PS, Mora EA (2011) Green sturgeon physical habitat use in the coastal Pacific Ocean. PLoS One 6: e25156.
40. Huff DD, Lindley ST, Wells BK, Chai F (2012) Green sturgeon distribution in the pacific ocean estimated from modeled oceanographic features and migration behavior. PLoS One 7: e45852.
41. Allen PJ, Hodge B, Werner I, Cech JJ Jr (2006) Effects of ontogeny, season, and temperature on the swimming performance of juvenile green sturgeon (*Acipenser medirostris*). Can J Fish Aquat Sci 63: 1360–1369.
42. Allen PJ, Nicholl M, Cole S, Vlazny A, Cech JJ Jr (2006) Growth of larval to juvenile Green Sturgeon in elevated temperature regimes. T Am Fish Soc 135: 89–96.
43. Allen PJ, McEnroe M, Forostyan T, Cole S, Nicholl MM, et al. (2011) Ontogeny of salinity tolerance and evidence for seawater-entry preparation in juvenile green sturgeon, *Acipenser medirostris*. J Comp Physiol B 181: 1045–1062.
44. Allen PJ, Cech JJ Jr (2007) Age/size effects on juvenile green sturgeon, *Acipenser medirostris*, oxygen consumption, growth, and osmoregulation in saline environments. Environ Biol Fish 79: 211–229.
45. Van Eenennaam JP, Linares-Casenave J, Deng X, Doroshov S (2005) Effect of incubation temperature on green sturgeon embryos, *Acipenser medirostris*. Environ Biol Fish 72: 145–154.
46. Werner I, Linares-Casenave J, Van Eenennaam JP, Doroshov SI (2007) The effect of temperature stress on development and heat-shock protein expression in larval green sturgeon (*Acipenser medirostris*). Environ Biol Fish 79: 191–200.
47. Israel JA, Bando J, Anderson EC, May B (2009) Polyploid microsatellite data reveal stock complexity among estuarine North American green sturgeon (*Acipenser medirostris*). Can J Fish Aquat Sci 66: 1491–1504.
48. Khodorevskaya RP, Ruban GI, Pavlov DS (2009) Behavior, migrations, distribution and stock of sturgeons in the Volga-Caspian basin. World Sturgeon Conservation Society Special Publication No 3. Books on Demand GmbH. Norderstedt, Germany.
49. Doukakis P, Erickson D, Baimukhanov M, Bokova Y, Erbulekov S, et al. (2008) Field and genetic approaches to enhance knowledge of Ural River sturgeon biology. In: Lagutov V, editor. Rescue of sturgeon species in the Ural River basin. Amsterdam: Springer. 277–292.
50. Doukakis P, Birstein VJ, DeSalle R (2005) Intraspecific structure within three caviar-producing sturgeons (*Acipenser gueldenstaedtii*, *A stellatus*, and *Huso huso*) based on mitochondrial DNA analysis. J Appl Ichthyol 216: 457–460.
51. Williot P, Rochard E, Desse-Berset N, Kirschbaum F, Gessner J (2011) Biology and conservation of the European sturgeon *Acipenser sturio* L 1758. The Reunion of the European and Atlantic Sturgeons. Berlin: Springer.
52. Taverny C, Lepage M, Piefort S, Dumont P, Rochard E (2002) Habitat selection by juvenile European sturgeon (*Acipenser sturio*) in the Gironde estuary (France). J Appl Ichthyol 18: 536–541.
53. Gessner J, Arndt GM, Tiedemann R, Bartel R, Kirschbaum F (2006) Remediation measures for the Baltic sturgeon: status review and perspectives. J Appl Ichthyol 22: 23–31.
54. Acolas ML, Rochard E, Le Pichon C, Rouleau E (2012) Downstream migration patterns of one-year-old hatchery-reared European sturgeon (*Acipenser sturio*). J Exp Mar Biol Ecol 430–431: 68–77.
55. Kolman R, Kapusta A, Duda A, Wiszniewski G (2011) Review of the current status of the Atlantic sturgeon *Acipenser oxyrinchus oxyrinchus* Mitchill 1815, in Poland: principles, previous experience, and results. J Appl Ichthyol 27: 186–191.
56. Staaks G, Kirschbaum F, Williot P (1999) Experimental studies on thermal behavior and diurnal activity rhythms of juvenile European sturgeon (*Acipenser sturio*). J Appl Ichthyol 15: 243–247.
57. Williot P, Rochard E, Rouault T, Kirschbaum F (2009) *Acipenser sturio* recovery research actions in France. Chapter 15 In: Carmona R, Domezain A, García Gallego M, Hernando A, Rodríguez F, et al., editors. Biology, conservation and sustainable development of sturgeons. Fish and Fisheries Series 29: 247–263. Amsterdam: Springer.
58. Kynard B, Suciu R, Horgan M (2002) Migrations and habitats of diadromous Danube River sturgeons in Romania: 1998–2000. J Appl Ichthyol 18: 529–535.
59. Ustaoglu Tiril S, Zengin M, Akbulut B, Memis D, Alagil F, et al. (2011) A participatory approach to tagging and monitoring as an initial step in developing a sturgeon conservation strategy along the Turkish Black Sea Coast. J Appl Ichthyol 27: 411–414.
60. Jarić I, Lenhardt M, Pallon J, Elfman M, Kalauzi A, et al. (2011) Insight into Danube sturgeon life history: trace element assessment in pectoral fin rays. Environ Biol Fish 90: 171–181.
61. Shmigirilov AP, Mednikova AA, Israel JA (2007) Comparison of biology of the Sakhalin sturgeon, Amur sturgeon, and kaluga sturgeon from the Amur River, Sea of Okhotsk, and Sea of Japan biogeographic province. Environ Biol Fish 79: 383–395.
62. Erickson D, Kappenman K, Webb M, Ryabinin N, Shmigirilov A, et al. (2007) Sturgeon conservation in the Russian Far East and China. Endangered Species Bulletin XXXII: 28–32.
63. Yang D, Kynard B, Wei Q, Chen X, Zheng W, et al. (2006) Distribution and movement of Chinese sturgeon, *Acipenser sinensis*, on the spawning ground located below the Gezhouba Dam during spawning seasons. J Appl Ichthyol 22 (Suppl 1): 145–151.
64. Wang CY, Wei QW, Kynard B, Du H, Zhang H (2012) Migrations and movements of adult Chinese sturgeon *Acipenser sinensis* in the Yangtze River, China. J Fish Biol 81: 696–713.
65. Wang C, Kynard B, Wei Q, Du H, Zhang H (2012) Spatial distribution and habitat suitability indices for non-spawning and spawning adult Chinese sturgeons below Gezhouba Dam, Yangtze River: Effects of river alterations. J Appl Ichthyol 29: 31–40.
66. Bruch RM, Dick T, Choudhury A (2001) A field guide for the identification of stages of gonad development in lake sturgeon, *Acipenser fulvescens* Rafinesque, with notes on lake sturgeon reproductive biology and management implications. WI Dept Nat Res Oshkosh and Sturgeon for Tomorrow publication. Graphic Communications Center, Inc., publisher. Appleton, WI.
67. Shaw SL, Chipps SR, Windels SK, Webb MAH, McLeod DT (2013) Influence of sex and reproductive status on seasonal movement of Lake Sturgeon in Namakan Reservoir, Minnesota–Ontario. T Am Fish Soc 142: 10–20.
68. Nelson TC, Gazey WJ, English KK, Rosenau ML (2013) Status of White Sturgeon in the Lower Fraser River, British Columbia. Fisheries 38: 197–209.
69. Rien TA, Beamesderfer RCP, Foster CA (1994) Retention, recognition, and effects on survival of several tags and marks for white sturgeon. Calif Fish Game 80: 161–170.
70. Kimberly DR, Bohl R, Bolden S, Fox D, Hager C, et al. (2010) Atlantic sturgeon research techniques. NOAA Tech Memo NMFS NE 215. 19 p.
71. Bordner CE, Doroshov SI, Hinton DE, Pipkin RE, Fridley RB, et al. (1990) Evaluation of marking techniques for juvenile and adult white sturgeons reared in captivity. In: Parker NC, Giorgi AE, Heidinger RC, Jester DB Jr, Prince ED, et al. editors. Fish-marking techniques. Bethesda, MD: American Fisheries Society. 293–303.
72. Kohlhorst DW (1979) Effect of first pectoral fin ray removal on survival and estimated harvest rate of white sturgeon in the Sacramento-San Joaquin estuary. Calif Fish Game 65: 173–177.
73. Hamel MJ, Hammen JJ, Pegg MA (2012) Tag retention of T-bar anchor tags and passive integrated transponder tags in Shovelnose sturgeon. N Am J Fish Manage 32: 533–538.

74. RL&L Environmental Services (2000) Fraser River white sturgeon monitoring program: comprehensive report (1995–1999). Report by RL&L Environmental Services Ltd, for BC Fisheries, Victoria, BC.
75. Ward DL (2000) Retention of passive integrated transponder tags applied to white sturgeon in Columbia River reservoirs. Apx A In: White sturgeon mitigation and restoration in the Columbia and Snake rivers upstream from Bonneville Dam. Annual Progress Report 1998. Bonneville Power Administration Report DOE/BP-00000140-1.
76. Hewitt DA, Janney EC, Hayes BS, Shively RS (2010) Improving inferences from fisheries capture-recapture studies through remote detection of PIT tags. Fisheries 35: 217–231.
77. Teo SLH, Blackwell S, Boustany A, Walli A, Weng K, et al. (2004) Validation of geolocation estimates based on light level and sea surface temperature from electronic tags. Mar Ecol-Prog Ser 283: 81–98.
78. Payne J, Andrews K, Chittenden C, Crossin G, Goetz F, et al. (2010) Tracking fish movements and survival on the northeast Pacific shelf. Chapter 14. In: McIntyre AD, editor. Life in the world's oceans: diversity, distribution, and abundance. Oxford, UK: Wiley-Blackwell. 269–290.
79. Moser ML, Lindley ST (2007) Use of Washington estuaries by subadult and adult green sturgeon. Environ Biol Fish 79: 243–253.
80. Hatin D, Fortin R, Caron F (2002) Movements and aggregation areas of adult Atlantic sturgeon (*Acipenser oxyrinchus*) in the St Lawrence River estuary, Quebec, Canada. J Appl Ichthyol 18: 586–594.
81. Holan SH, Davis GM, Wildhaber M, DeLonay A, Papoulias D (2009) Hierarchical Bayesian Markov switching models with application to predicting spawning success of shovelnose sturgeon. J Royal Stat Society 58: 47–64.
82. Jonsen ID, Myers RA, Flemming JM (2003) Meta-analysis of animal movement using state-space models. Ecology 84: 3055–3063.
83. Jonsen ID, Fleming JM, Myers RA (2005) Robust state-space modeling of animal movement data. Ecology 86: 2874–2880.
84. Jonsen ID, Myers RA, James MC (2007) Identifying leatherback turtle foraging behaviour from satellite-telemetry using a switching state-space model. Mar Ecol-Prog Ser 337: 255–264.
85. Bronzi P, Garofalo E, Dalla VZ, Battaglia A, Ansferri S, et al. (2006) Telemetry test trials on the behaviour of sub-adults of *Acipenser naccarii* reared in captivity and released into the River Po (Italy). J Appl Ichthyol 22: 152–159.
86. Neufeld MD, Rust PJ (2009) Using passive sonic telemetry methods to evaluate dispersal and subsequent movements of hatchery-reared white sturgeon in the Kootenay River. J Appl Ichthyol 25: 27–33.
87. Jordan GR, Klumb RA, Wanner G, Stancill W (2006) Poststocking movements and habitat use of hatchery-reared juvenile pallid sturgeon in the Missouri River below Fort Randall Dam, South Dakota and Nebraska. T Am Fish Soc 135: 1499–1511.
88. Snook VA, Peters EJ, Young LJ (2002) Movements and habitat use by hatchery-reared pallid sturgeon in the lower Platte River, Nebraska. Am Fish S S 28: 161–174.
89. Cooke DW, Leach SD, Isely JJ (2002) Behavior and lack of upstream passage of shortnose sturgeon at a hydroelectric facility and navigation lock complex. Am Fish S S 28: 101–110.
90. Parsley MJ, Wright CD, van der Leeuw B, Kofoot E, Peery C, et al. (2007) White sturgeon (*Acipenser transmontanus*) passage at the Dallas Dam, Columbia River, USA J Appl Ichthyol 23: 627–635.
91. Finney ST, Isely JJ, Cooke DW (2006) Upstream migration of two pre-spawning shortnose sturgeon passed upstream of Pinopolis Dam, Cooper River, South Carolina. Southeast Nat 5: 369–375.
92. Kieffer M, Kynard B (1996) Spawning of the shortnose sturgeon in the Merrimack River, Massachusetts. T Am Fish Soc 125: 179–186.
93. DeLonay AJ, Papoulias DM, Wildhaber ML, Annis ML, Bryan JL, et al. (2007) Use of behavioral and physiological indicators to evaluate Scaphirhynchus sturgeon spawning success. J Appl Ichthyol 23: 428–435.
94. Geist D, Brown R, Cullinan V, Brink SR, Lepla K, et al. (2005) Movement, swimming speed, and oxygen consumption of juvenile white sturgeon in response to changing flow, water temperature, and light level in the Snake River, Idaho. T Am Fish Soc 134: 803–816.
95. KahnJ, Mohead M (2010) A protocol for use of Shortnose, Atlantic and Gulf and Green Sturgeons. NOAA Tech Memo NMFS-OPR-45. 62p.
96. Vajhi AR, Moghim M, Veshkini A, Masoudifard M (2001) Sex and maturity determination by ultrasonography in Persian sturgeon (*Acipenser gueldenstaedti persicus*). Abstract. 4th International Symposium on Sturgeon, Oshkosh, Wisconsin USA, 8–13. July 2001.
97. Kynard B, Kieffer M (2002) Use of a borescope to determine the sex and egg maturity stage of sturgeons and the effect of borescope use on reproductive structures. J Appl Ichthyol 18: 505–508.
98. Moghim M, Vajhi AR, Veshkini A, Masoudifard M (2002) Determination of sex and maturity in *Acipenser stellatus* by using ultrasonagraphy. J Appl Ichthyol 18: 325–328.
99. Colombo RE, Wills PS, Garvey JE (2004) Use of ultrasound imaging to determine sex of shovelnose sturgeon. N Am J Fish Manage 24: 322–326.
100. Chebanov MS, Chmyr YN (2005) A field guide on early sexing and staging maturity in live sturgeons by using ultrasound techniques. Krasnodar, Russian Federation, Federal Centre of Selection and Genetics for Aquaculture.
101. Wildhaber ML, Papoulias DM, DeLonay AJ, Tillitt DE, Bryan JL, et al. (2005) Gender identification of shovelnose sturgeon using ultrasonic and endoscopic imagery and the application of the method to pallid sturgeon. J Fish Biol 67: 114–132.
102. Divers SJ, Boone SS, Hoover JJ, Boysen KA, Killgore KJ, et al. (2009) Field endoscopy for identifying gender, reproductive stage and gonadal anomalies in free-ranging sturgeon (Scaphirhynchus) from the lower Mississippi River. J Appl Ichthyol 25: 68–74.
103. Ceapa C, Williot P, Le Menn F, Davail-Cuisset B (2002) Plasma sex steroids and vitellogenin levels in stellate sturgeon (*Acipenser stellatus* Pallas) during spawning migration in the Danube River. J Appl Ichthyol 18: 391–396.
104. Webb MAH, Feist GW, Schreck CB, Foster EP, Fitzpatrick MS (2002) Potential classification of sex and stage of gonadal maturity of wild white sturgeon using blood plasma indicators. T Am Fish Soc 131: 132–142.
105. Malekzadeh Viayeh R, Webb MAH, Hallajian A, Kazemi R (2007) Biochemical and morphometric parameters as indicators of sex and gonadal stage of maturity in Persian sturgeon, *Acipenser persicus*. J Appl Ichthyol 22: 364–368.
106. Craig JM, Papoulias DM, Thomas MV, Annis ML, Boase J (2009) Sex assignment of lake sturgeon (*Acipenser fluvescen*) based on plasma sex hormone and vitellogenin levels. J Appl Ichthyol 25: 60–67.
107. Webb MAH, Doroshov SI 2011 Importance of environmental endocrinology in fisheries management and aquaculture of sturgeons. Gen Comp Endrocrinol 170: 313–321.
108. Schreck CB (2010) Stress and fish reproduction: the role of allostatsis and hormesis. Gen Comp Endocrinol 165: 549–556.
109. Barton BA, Bollig H, Hauskins BL, Jansen CR (2000) Juvenile pallid sturgeon (*Scaphirhynchus albus*) and hybrid pallid x shovelnose (*S albus* x *platorynchus*) sturgeons exhibit low physiological responses to acute handling and severe confinement. Comp Biochem Physiol 126A: 125–134.
110. Belanger JM, Son JH, Laugero KD, Moberg GP, Doroshov SI, et al. (2001) Effects of short-term management stress and ACTH injections on plasma cortisol levels in cultured white sturgeon, *Acipenser transmontanus*. Aquaculture 203: 165–176.
111. Bayunova L, Barannikova I, Semenkova T (2002) Sturgeon stress reactions in aquaculture. J Appl Ichthyol 18: 397–404.
112. Lankford SE, Adams TE, Cech JJ Jr (2003) Time of day and water temperature modify the physiological stress response in green sturgeon, *Acipenser medirostris*. Comp Biochem Physiol 135: 291–302.
113. Webb MAH, Allert JA, Kappenman KM, Marcos J, Feist GW, et al. (2007) Identification of plasma glucocorticoids in pallid sturgeon in response to stress. Gen Comp Endocrinol 154: 98–104.
114. Cataldi E, Di Marco P, Mandich A, Cataudella S (1998) Serum parameters of Adriatic sturgeon *Acipenser naccarii* (Pisces: Acipenseriformes): effects of temperature and stress. Comp Biochem Physiol 121: 351–354.
115. Di Marco P, McKenzie DJ, Mandich A, Bronzi P, Cataldi E, et al. (1999) Influence of sampling conditions on blood chemistry values of Adriatic sturgeon *Acipenser naccarii* (Bonaparte, 1836). J Appl Ichthyol 15: 73–77.
116. Kelsch SW, Shields B (1996) Care and handling of sampled organisms. In: Murphy BR, Willis DW, editors. Fisheries Techniques 2nd edition. Bethesda, MD: Am Fish Soc. 121–145.
117. Hunter E, Aldridge JN, Metcalfe JD, Arnold GP (2003) Geolocation of free-ranging fish on the European continental shelf as determined from environmental variables. Mar Biol 142: 601–609.
118. Pollock KH, Hoenig JM, Jones CM (1991) Estimation of fishing and natural mortality when a tagging study is combined with a creel survey or port sampling. Am Fish Soc Monogram 12: 423–434.
119. Martell SJD, Walters CJ (2002) Implementing harvest rate objectives by directly monitoring exploitation rates and estimating changes in catchability. B Mar Sci 70: 695–713.
120. Bacheler NM, Buckel JA, Hightower JE, Paramore LM, Pollock KH (2009) A combined telemetry-tag return approach to estimate fishing and natural mortality rates of an estuarine fish. Can J Fish Aquat Sci 66: 1230–1244.
121. Pedersen MW, Righton D, Thygesen UH, Andersen KH, Madsen H (2008) Geolocation of North Sea cod (*Gadus morhua*) using hidden Markov models and behavioural switching. Can J Fish Aquat Sci 65: 2367–2377.
122. Xiao Y (1996) A framework for evaluating experimental designs for estimating rates of fish movement from tag recoveries. Can J Fish and Aquat Sci 53: 1272–1280.
123. Hilborn R (1990) Determination of fish movement patterns from tag recoveries using maximum likelihood estimator. Can J Fish and Aquat Sci 47: 635–643.
124. Walters CJ, Martell S (2004) Fisheries ecology and management. PrincetonNJ: Princeton University Press. 399 p.
125. Nelson TC, Gazey WJ, English KK, Rosenau ML (2004) Status of white sturgeon in the lower Fraser River: report on the findings of the lower Fraser River white sturgeon monitoring and assessment program 1999–2004. Report prepared for the Fraser River Sturgeon Conservation Society, Vancouver, BC. Available: http://www.frasersturgeon.com/media/2_StatusofWhiteSturgeonintheLowerFraserRiver1999-2004.pdf. Accessed 22 November 2012.
126. Walters CJ, Korman J, McAdam S (2005) An assessment of white sturgeon stock status and trends in the lower Fraser River. Canadian Science Advisory Secretariat Res Doc 2005/066. Available: http://www.dfo-mpo.gc.ca/csas/Csas/DocREC/2005/RES2005_066_e.pdf. Accessed 22 November 2012.

127. Whitlock RE, McAllister M (2009) A Bayesian mark-recapture model for multiple-recapture data in a catch-and-release fishery. Can J Fish Aquat Sci 66: 1554–1568.
128. Thuemler TF (1997) Lake sturgeon management in the Menominee River, a Wisconsin-Michigan boundary water. Environ Biol Fish 48: 311–317.
129. Latour RJ, Hoenig JM, Hepworth DA, Frusher SD (2003) A novel tag-recovery model with two size classes for estimating fishing and natural mortality, with implications for the southern rock lobster (*Jasus edwardsii*) in Tasmania, Australia. ICES J Mar Sci 60: 1075–1085.
130. Pine WE, Pollock KH, Hightower JE, Kwak TJ, Rice JA (2003) A review of tagging methods for estimating fish population size and components of mortality. Fish Res 28: 10–23.
131. Musick JA (1999) Ecology and conservation of long-lived marine animals. Am Fish S S 23: 1–10.
132. Dulvy NK, Sadovy Y, Reynolds JD (2003) Extinction vulnerability in marine populations. Fish Fish 4: 25–64.
133. Boreman J (1997) Sensitivity of North American sturgeons and paddlefish to fishing mortality. Environ Biol Fish 48: 399–405.
134. Beamesderfer RCP, Rien TA, Nigro AA (1995) Differences in the dynamics and potential production of impounded and unimpounded white sturgeon populaitons in the lower Columbia River. T Am Fish Soc. 124: 857–872.
135. DeVore JD, James BW, Tracy CA, Hale DA (1995) Dynamics and potential production of white sturgeon in the unimpounded lower Columbia River. T Am Fish Soc 124: 845–856.
136. DeVore J, James B, Beamesderfer RC (1999) Lower Columbia River white sturgeon current stock status and implications. Washington Department of Fish and Wildlife Rep No 99–08. Olympia, Washington.
137. Pine WE III, Allen MS, Dreitz VJ (2001) Population viability of the Gulf of Mexico sturgeon: inferences from capture-recapture and age-structured models. T Am Fish Soc 130: 1164–1174.
138. Secor DH, Waldman JR (1999) Historical abundance of Delaware Bay Atlantic sturgeon and potential rate of recovery. Am Fish S S 23: 203–216.
139. Walters CJ, Martell SJD, Korman J (2006) A stochastic approach to stock reduction analysis. Can J Fish Aquat Sci 63: 212–223.
140. Whitlock RE, McAllister MK (2012) Incorporating spatial and seasonal dimensions in a stock reduction analysis for lower Fraser River white sturgeon (*Acipenser transmontanus*). Can J Fish Aquat Sci 69: 1674–1697.
141. Nelson RJ, McAdam DSO (2012) Historical population structure of White Sturgeon in the Upper Columbia River detected with combined analysis of capture, telemetry and genetics. J Appl Ichthyol 28: 161–167.
142. Grunwald C, Maceda L, Waldman J, Stabile J, Wirgin I (2008) Conservation of Atlantic sturgeon *Acipenser oxyrinchus oxyrinchus*: delineation of stock structure and distinct population segments. Conserv Genet 9: 1111–1124.
143. Wirgin I, Grunwald C, Carlson E, Stabile J, Peterson DL, et al. (2005) Range-wide population structure of shortnose sturgeon, *Acipenser brevirostrum*, based on sequence analysis of the mitochondrial DNA control region. Estuaries 28: 406–421.
144. Birstein VJ, Doukakis P, DeSalle R (2000) Polyphyly of mtDNA lineages in the Russian sturgeon, *Acipenser gueldenstaedtii*: forensic and evolutionary implications. Conserv Genet 1: 81–88.
145. Jenneckens I, Meyer J-N, Debus L, Pitra C, Ludwig A (2000) Evidence of mitochondrial DNA clones of Siberian sturgeon, *Acipenser baerii*, within Russian sturgeon, *Acipenser gueldenstaedtii*, caught in the River Volga. Ecol Lett 3: 503–508.
146. Rastorguev S, Mugue N, Volkov A, Barmintsev V (2008) Complete mitochondrial DNA sequence analysis of Ponto-Caspian sturgeon species. J Appl Ichthyol 24: 46–49.
147. Mugue NS, Barmintseva AE, Rastorguev SM, Mugue VN, Barmintsev VA (2008) Polymorphism of the mitochondrial DNA control region in eight sturgeon species and development of a system for DNA-based species identification. Russ J Genet 44: 797–802.
148. Koutrakis E, Sapounidis A, Favre-Krey L, Krey G, Economidis PS (2011) Incidental catches of Acipenseridae in the estuary of the River Evros, Greece. J Appl Ichthyol 27: 366–368.
149. Mugue NS, Barmintseva AE, Rastorguev SM, Mugue VN, Barmintsev VA (2008) Polymorphism of the mitochondrial DNA control region in eight sturgeon species and development of a system for DNA-based species identification. Russ J Genet+ 44: 793–798.
150. Birstein VJ, Doukakis P, Sorkin B, DeSalle R (1998) Population aggregation analysis of caviar producing species of sturgeon and implications for diagnosis of black caviar. Conser Biol 124: 766–775.
151. Birstein VJ, Ruban G, Ludwig A, Doukakis P, DeSalle R (2005) The enigmatic Caspian Sea Russian sturgeon: how many cryptic forms does it contain? Syst Biodivers 32: 203–218.
152. Waldman JR, Hart JT, Wirgin II (1996) Stock composition of the New York Bight Atlantic sturgeon fishery based on analysis of mitochondrial DNA. T Am Fish Soc 125: 364–371.
153. Rodzen JA, May B (2002) Inheritance of microsatellite loci in the white sturgeon (*Acipenser transmontanus*). Genome 45: 1064–1076.
154. Rodzen JA, Famula T, May B (2004) Estimation of parentage and relatedness in the polyploid white sturgeon (*Acipenser transmontanus*) using a dominant marker approach for duplicated microsatellite loci. Aquaculture 232: 165–182.
155. Hauser L, Seeb JE (2008) Advances in molecular technology and their impact on fisheries genetics. Fish Fish 9: 473–486.
156. Ludwig A, Lippold S, Debus L, Reinartz R (2009) First evidence of hybridization between endangered sterlets (*Acipenser ruthenus*) and exotic Siberian sturgeons (*Acipenser baerii*) in the Danube River. Biol Invasions 11: 753–760.
157. Israel JA, May B (2010) Indirect genetic estimates of breeding population size in the polyploid green sturgeon (*Acipenser medirostris*). Mol Ecol 19: 1058–1070.
158. Atlantic Sturgeon Status Review Team (2007) Status review of Atlantic sturgeon (*Acipenser oxyrinchus oxyrinchus*). Report to National Marine Fisheries Service, Northeast Regional Office February 23, 2007. 174 p.
159. Laney RW, Hightower JE, Versak BR, Mangold MF, Cole WW, et al. (2007) Distribution, habitat use and size of Atlantic sturgeon captured during cooperative winter tagging cruises, 1988–2006. Am Fish S S 23: 167–182.
160. Gessner J, Arndt G-M, Ludwig A, Kirschbaum F (2007) Remediation of Atlantic sturgeon in the Baltic Sea: background, status, and perspectives. Am Fish S S 23: 301–317.
161. Reinartz R, Lippold S, Lieckfeldt D, Ludwig A (2011) Population genetic analyses of *Acipenser ruthenus* as a prerequisite for the conservation of the uppermost Danube population. J Appl Ichthyol 27: 477–483.
162. Ogden R, Gharbi K, Mugue N, Martinsohn J, Senn H, et al. (2013) Sturgeon conservation genomics: SNP discovery and validation using RAD sequencing. Mol Ecol doi: 10.1111/mec.12234.
163. Pritchard JK, Stephens M, Donnelly P (2000) Inference of population structure using multilocus genotype data. Genetics 155: 945–959.
164. Paetkau D, Slade R, Burden M, Estoup A (2004) Genetic assignment methods for the direct, real-time estimation of migration rate: a simulation-based exploration of accuracy and power. Mol Ecol 13: 55–65.
165. Drauch AM, Fisher BE, Latch EK, Fike JA, Rhodes OE (2008) Evaluation of a remnant lake sturgeon population's utility as a source for reintroductions in the Ohio River system. Conserv Genet 9: 1195–1209.
166. Welsh A, Hill T, Quinlan H, Robinson C, May B (2008) Genetic assessment of lake sturgeon population structure in the Laurentian Great Lakes. N Am J Fish Manage 28: 572–591.
167. Coutant CC (1990) Microchemical analysis of fish hard parts for reconstructing habitat use: practice and promise. Am Fish S S 7: 574–580.
168. Mulligan TJ, Martin FD, Smucker RA, Wright DA (1987) A method of stock identification based on the elemental composition of striped bass *Morone saxatilis* (Walbaum) otoliths. J Exp Mar Biol Ecol 114: 241–248.
169. Warner RR, Swearer SE, Caselle JE, Sheehy M, Paradis G (2005) Natal trace-elemental signatures in the otoliths of an open-coast fish. Limnol Oceanogr 50: 1529–1542.
170. Clarke AD, Telmer KH, Shrimpton JM (2007) Elemental analysis of otoliths, fin rays, and scales: a comparison of bony structures to provide population and life history information for the Arctic grayling (*Thymallus arcticus*). Ecol Freshw Fish 16: 354–361.
171. Becker BJ, Levin LA, Fodrie FJ, McMillan PA (2007) Complex larval connectivity patterns among marine invertebrate populations. P Natl Acad Sci 104: 3267–3272.
172. Campana SE, Thorrold SR (2001) Otoliths, increments, and elements: keys to a comprehensive understanding of fish populations? Can J Fish Aquat Sci 58: 30–38.
173. Veinott GI, Evans RD (1999) An examination of elemental stability in the fin ray of the white sturgeon with laser ablation sampling-inductively coupled plasma-mass spectrometry (LAS-ICP-MS). T Am Fish Soc128: 352–361.
174. Secor DH, Picolli PM (1996) Age- and sex-dependent migrations of striped bass in the Hudson River as determined by chemical microanalysis of otoliths. Estuaries 19: 778–793.
175. Veinott G, Northcote T, Rosenau M, Evans RD (1999) Concentrations of strontium in the pectoral fin rays of the white sturgeon (*Acipenser transmontanus*) by laser ablation sampling-inductively coupled plasma-mass spectrometry as an indicator of marine migrations. Can J Fish Aquat Sci 56: 1981–1990.
176. Arai T, Miyazaki N (2001) Use of otolith microchemistry to estimate the migratory history of the Russian sturgeon, *Acipenser guldenstadti*. J Mar Biol Assoc UK 81: 709–710.
177. Arai T, Levin AV, Boltunov AN, Miyazaki N (2002) Migratory history of the Russian sturgeon (*Acipenser guldenstadti*) in the Caspian Sea, as revealed by pectoral fin spine Sr:Ca ratios. Mar Biol 141: 315–319.
178. Allen PJ, Hobbes JA, Cech JJ Jr, Van Eenennaam JP, Doroshov SI (2009) Using trace elements in pectoral fin rays to assess life history movements in sturgeon: estimating age at initial seawater entry in Klamath River green sturgeon. T Am Fish Soc 138: 240–250.
179. Rien TA, Beamesderfer RCP (1994) Accuracy of precision of white sturgeon age estimates from pectoral fin rays. T Am Fish Soc 123: 255–265.
180. Jackson ND, Garvey JE, Colombo RE (2007) Comparing aging precision of calcified structures in shovelnose sturgeon. J Appl Ichthyol 23: 525–528.
181. Israel JA, Cordes JF, Blumberg MA, May B (2004) Geographic patterns of genetic differentiation among collections of green sturgeon. N Am J Fish Manage 24: 922–931.
182. Elsdon TS, Gillanders BM (2003) Reconstructing migratory patterns of fish based on environmental influences on otolith chemistry. Rev Fish Biol Fish 13: 219–235.
183. Thresher RE (1999) Elemental composition of otoliths as a stock delineator in fishes. Fish Res 43: 165–204.
184. Secor DH, Rooker JR (2000) Is otolith strontium a useful scalar of life cycles of estuarine fishes? Fish Res 46: 359–371.

185. Fowler AJ, Campana SE, Jones CM, Thorrold SR (1995) Experimental assessment of the effect of temperature and salinity on elemental composition of otoliths using laser ablation ICPMS. Can J Fish Aquat Sci 52: 1431–1441.
186. Welch DW, Turo S, Batten SD (2006) Large-scale marine and freshwater movements of white sturgeon. T Am Fish Soc 135: 386–389.
187. Hublein JC, Kelly JT, Crocker CE, Klimley AP, Lindley ST (2009) Migration of green sturgeon, *Acipenser medirostris*, in the Sacramento River. Environ Biol Fish 84: 245–258.
188. Campana SE, Chouinard GA, Hanson JM, Fréchet A, Brattey J (2000) Otolith elemental fingerprints as biological tracers of fish stocks. Fish Res 46: 343–357.
189. Golder (2009) Observations of white sturgeon behaviour in Waneta eddy during a zero flow event at Waneta Dam. Report prepared for BC HYDRO, Castlegar, BC by Golder Associates Ltd, Castlegar, BC Golder Report 08-1480-0088.
190. RL&L (1997) Fraser River White Sturgeon Monitoring Program Region 7 (Omenica-Peace): 1996 Investigations Report prepared for BC Ministry of Environment, Lands and Parks, Fish and Wildlife Section, Prince George, BC. RL&L Report No 520.
191. Xinhai L, Litvak MK, Hughes-Clarke JE (2007) Overwintering habitat use of shortnose sturgeon (*Acipenser brevirostrum*): defining critical habitat using a novel underwater video survey and modeling approach. Can J Fish Aquat Sci 64: 1248–1257.
192. Chiotti JA, Holtgren JM, Auer NA, Ogren SA (2008) Lake sturgeon spawning habitat in the Big Manistee River, Michigan. N Am J Fish Manage 28: 1009–1019.
193. Golder (2009) Lower Columbia River juvenile white sturgeon detection: 2008 investigations data report. Report prepared for BC HYDRO, Castlegar, BC by Golder Associates Ltd, Castlegar, BC. Golder Report 08-1480-0040F.
194. Golder (2003) Brilliant Dam Expansion project: White sturgeon monitoring program, 2002–2003. Report prepared for Brilliant Expansion Power Corporation, Victoria, BC, by Golder Associates Ltd, Castlegar, BC. Golder Report 0228026F.
195. Golder (2009) Post project white sturgeon monitoring in the Brilliant Expansion tailwater area – Summary Report 2008. Report prepared for Columbia Power Corporation, Castlegar, BC, by Golder Associates Ltd, Castlegar, BC. Golder Report 0814800003F.
196. Nealson PA, Tritt GW (2003) Feasibility assessment of split-beam hydro-acoustic techniques for monitoring shortnose sturgeon in the Delaware River. Prepared for Environmental Research and Consulting, Inc, Chadds Ford, PA, by Hydroacoustic Technology Inc, Seattle, WA. HTI Project P1372.
197. Burczynski JJ (1979) Introduction to the use of sonar systems for estimating fish biomass. Food and Agriculture Organization of the United Nations, Rome. FAO Technical Paper No 191.
198. MacLennan DN, Simmonds EJ (1992) Fisheries acoustics. Fish Fish Series 5. London: Chapman & Hall.
199. Jech JM, Horne JK (1998) Sensitivity of acoustic scattering models to fish morphometry. In: Proceedings of the 135th meeting of the Acoustical Society of America, Seattle, 20–26 August 1998. 1819–1820.
200. Ehrenberg JE, Traynor J, Kaczynski VM (1984) The application of multibeam acoustic techniques to marine resource assessment. In: PACON 84: Proceedings of the Pacific congress on marine technology.
201. Ehrenberg JE, Torkelson TC (1996) Application of dual-beam and split-beam target tracking in fisheries acoustics. ICES J Mar Sci 53: 329–344. Available: http://icesjms.oxfordjournals.org/content/53/2/329.full.pdf+html?sid = 7b7d6334-d254-438f-9481-f91794f5104f. Accessed 22 November 2012.
202. Traynor JJ, Ehrenberg JE (1990) Fish and standard-sphere target-strength measurements obtained with a dual-beam and split-beam echo-sounding system. Rapp P-v Reun Cons int Explor Mer 189: 325–335.
203. Auer NA, Baker EA (2007) Assessment of lake sturgeon spawning stocks using fixed-location, split-beam sonar technology. J Appl Ichthyol 23: 113–121.
204. Thomas MV, Haas RC (2002) Abundance, age structure, and spatial distribution of lake sturgeon (*Acipenser fulvescens*) in the St Clair System. Michigan department of natural resources, Lake St Clair fisheries research station, Harrison Township, MI. Fish Res Rep 2076.
205. Reuter JM, Jacobson RB, Elliott CM, Johnson HE III, DeLonay AJ (2008) Side-scan sonar imagery from: Hydraulic and substrate maps of reaches used by sturgeon (genus Scaphirhynchus) in the lower Missouri River, 2005–07. USGS Data Series 386. Reston, Virginia.
206. Brundage HM, Jung J (2009) Experiments with broadband sonar for the detection and identification of endangered shortnose sturgeon. Mar Technol Soc J 43: 78–82.
207. Cholwek G, Yule D, Eitrem M, Quinlan H, Doolittle T (2005) Mapping potential lake sturgeon habitat in the lower Bad River complex. USGS Lake Superior Biological Station Rep. 21. Available: http://www.glsc.usgs.gov/_files/reports/BadRiverFinalReport.pdf. Accessed 22 November 2012.
208. LGL Ltd, Golder Associates (2009) Sturgeon monitoring at Arrow Lakes generating station in 2008. Report submitted to Columbia Power Corporation, Castlegar, BC by LGL Limited environmental research associates, Sidney, BC, and Golder Associates Ltd, Castlegar, BC. 17 p.
209. Jacobson RB, Delonay A, Vishy C, Elliott CM, Reuter JM, et al. (2009) Multi-scale hydroacoustic remote sensing of sturgeon and their habitats in a large, turbid river. American Geophysical Union, fall meeting 2009. Abstract #H54A-06. Available: http://adsabs.harvard.edu/abs/2009AGUFM.H54A.06J. Accessed 22 November 2012.
210. Kahn J, Mohead M (2010) A protocol for use of shortnose, Atlantic, Gulf, and green sturgeons. US Dep Comm NOAA Tech Memo NMFS-OPR-45.
211. Birstein BJ, Bemis WE, Waldman JR (1997) The threatened status of acipenseriform species: a summary. Environ Biol Fish 48: 427–435.
212. Ruban GI (2005) The Siberian sturgeon *Acipenser baerii* Brandt: species structure and ecology. World Sturgeon Conservation Society – Special Pub No 1. Norderstedt (Germany). 203 p.
213. Artyukhin EN, Vecsei P, Peterson DL (2007) Morphology and ecology of Pacific sturgeons. Environ Biol Fish 79: 369–381.
214. Hall J, Smith T, Lamprecht SD (1991) Movements and habitats of shortnose sturgeon, *Acipenser brevirostrum*, in the Savannah River. Copeia 3: 695–702.
215. Caswell N, Peterson D, Manny BA, Kennedy GW (2004) Spawning by lake sturgeon (*Acipenser fulvescens*) in the Detroit River. J Appl Ichthyol 20: 1–6.
216. Lallaman JJ, Damstra RA, Galarowicz T (2008) Population assessment and movement patterns of lake sturgeon (*Acipenser fulvescens*) in the Manistee River, Michigan, USA. J Appl Ichthyol 24: 1–6.
217. McKinley S, Van Der Kraak G, Power G (1998) Seasonal migrations and reproductive patterns in the lake sturgeon, *Acipenser fulvescens*, in the vicinity of hydroelectric stations in northern Ontario. Environ Biol Fish 51: 245–256.
218. Benson RL, Turo S, McCovey BW (2007) Migration and movement patterns of green sturgeon (*Acipenser medirostris*) in the Klamath and Trinity rivers, California, USA. Environ Biol Fish 79: 269–279.
219. Heise R, Slack W, Ross S, Dugo MA (2004) Spawning and associated movement patterns of Gulf sturgeon in the Pascagoula River drainage, Mississippi. T Am Fish Soc 133: 221–230.
220. Hightower JE, Zehfuss KP, Fox DA, Parauka FM (2002) Summer habitat use by Gulf sturgeon in the Choctawhatchee River, Florida. J Appl Ichthyol 18: 595–600.
221. Foster A, Clugston JP (1997) Seasonal migration of Gulf sturgeon in the Suwannee River, Florida. T Am Fish Soc 126: 302–308.
222. Young W, Scarnecchia D (2005) Habitat use of juvenile white sturgeon in the Kootenai River, Idaho, and British Columbia. Hydrobiologia 537: 265–271.
223. Parsley MJ, Popoff ND, van der Leeuw B, Wright C (2008) Seasonal and diel movements of white sturgeon in the lower Columbia River. T Am Fish Soc 137: 1007–1017.
224. Paragamian VL, Kruse G (2001) Kootenai River white sturgeon spawning migration behavior and a predictive model. N Am J Fish Manage 21: 10–21.
225. Curtis G, Ramsey J, Scarnecchia DL (1997) Habitat use and movements of shovelnose sturgeon in Pool 13 of the upper Mississippi River during extreme low flow conditions. Environ Biol Fish 50: 175–182.
226. Sulak KJ, Randall MT, Edwards R, Summers T, Luke K, et al. (2009) Defining winter trophic habitat of juvenile Gulf sturgeon in the Suwannee and Apalachicola rivermouth estuaries, acoustic telemetry investigations. J Appl Ichthyol 25: 505–515.
227. Rogillio HE, Ruth RT, Behrens E, Doolittle C, Granger W, et al. (2007) Gulf sturgeon movements in the Pearl River drainage and the Mississippi sound. N Am J Fish Manage 27: 89–95.
228. Fox DA, Hightower JE, Parauka FM (2002) Estuarine and nearshore marine habitat use by Gulf sturgeon from the Choctawhatchee River system, Florida. Am Fish Soc Symp 28: 111–126.
229. Kelly JT, Klimley AP, Crocker CE (2007) Movements of green sturgeon, *Acipenser medirostris*, in the San Francisco Bay estuary, California. Environ Biol Fish 79: 281–295.
230. Harris JE, Parkyn DC, Murie DJ (2005) Distribution of Gulf of Mexico sturgeon in relation to benthic invertebrate prey resources and environmental parameters in the Suwannee River estuary, Florida. T Am Fish Soc 134: 975–990.
231. Collins M, Cooke D, Post B, Crane J, Bulak J, et al. (2003) Shortnose sturgeon in the Santee-Cooper reservoir system, South Carolina. T Am Fish Soc 132: 1244–1250.
232. Adams WE, Kallemeyn LW, Willis DW (2006) Lake sturgeon, *Acipenser fulvescens*, movements in Rainy Lake, Minnesota and Ontario. Can Field Nat 120: 71–82.
233. Yang D, Kynard B, Wei Q, Chen X, Zheng W, et al. (2006) Distribution and movement of Chinese sturgeon, *Acipenser sinensis*, on the spawning ground located below the Gezhouba Dam during spawning seasons. J Appl Ichthyol 22: 145–151.
234. Parkyn DC, Murie DJ, Colle DE, Holloway JD (2006) Post-release survival and riverine movements of Gulf of Mexico sturgeon (*Acipenser oxyrinchus desotoi*) following induced spawning. J Appl Ichthyol 22: 1–7.
235. Crossman JA, Forsythe PS, Baker EA, Scribner KT (2009) Overwinter survival of stocked age-0 lake sturgeon. J Appl Ichthyol 25: 516–521.
236. Caro F, Hatin D, Fortin R (2002) Biological characteristics of adult Atlantic sturgeon (*Acipenser oxyrinchus*) in the St Lawrence River estuary and the effectiveness of management rules. J Appl Ichthyol 18: 580–585.
237. Kynard B, Breece M, Atcheson M, Kieffer M, Mangold M (2009) Life history and status of shortnose sturgeon (*Acipenser brevirostrum*) in the Potomac River. J Appl Ichthyol 25: 34–38.
238. Auer NA (1999) Population characteristics and movements of Lake sturgeon in the Sturgeon River and Lake Superior. J Great Lakes Res 25: 282–293.

2

Growth and Nitrogen Uptake Characteristics Reveal Outbreak Mechanism of the Opportunistic Macroalga *Gracilaria tenuistipitata*

Chao Wang[1,2], Anping Lei[1]*, Kai Zhou[3]*, Zhengyu Hu[2], Wenlong Hao[1], Junda Yang[1]

1 College of Life Sciences, Shenzhen University, Shenzhen, China, **2** Institute of Hydrobiology, Chinese Academy of Sciences, Wuhan, China, **3** Shenzhen Marine Environment and Resources Monitoring Center, Shenzhen, China

Abstract

Macroalgae has bloomed in the brackish lake of Shenzhen Bay, China continuously from 2010 to 2014. *Gracilaria tenuistipitata* was identified as the causative macroalgal species. The aim of this study was to explore the outbreak mechanism of *G. tenuistipitata*, by studying the effects of salinity and nitrogen sources on growth, and the different nitrogen sources uptake characteristic. Our experimental design was based on environmental conditions observed in the bloom areas, and these main factors were simulated in the laboratory. Results showed that salinity 12 to 20 ‰ was suitable for *G. tenuistipitata* growth. When the nitrogen sources' (NH_4^+, NO_3^-) concentrations reached 40 μM or above, the growth rate of *G. tenuistipitata* was significantly higher. Algal biomass was higher (approximately 1.4 times) when cultured with NH_4^+ than that with NO_3^- addition. Coincidentally, macroalgal bloom formed during times of moderate salinity (~12 ‰) and high nitrogen conditions. The NH_4^+ and NO_3^- uptake characteristic was studied to understand the potential mechanism of *G. tenuistipitata* bloom. NH_4^+ uptake was best described by a linear, rate-unsaturated response, with the slope decreasing with time intervals. In contrast, NO_3^- uptake followed a rate-saturating mechanism best described by the Michaelis-Menten model, with kinetic parameters $V_{max}=37.2$ μM g^{-1} DM h^{-1} and $K_s=61.5$ μM. Further, based on the isotope ^{15}N tracer method, we found that ^{15}N from NH_4^+ accumulated faster and reached an atom% twice than that of ^{15}N from NO_3^-, suggesting when both NH_4^+ and NO_3^- were available, NH_4^+ was assimilated more rapidly. The results of the present study indicate that in the estuarine environment, the combination of moderate salinity with high ammonium may stimulate bloom formation.

Editor: Anna R. Armitage, Texas A&M University at Galveston, United States of America

Funding: This work was supported by National Natural Science Foundation of China (31170491), Shenzhen Municipal Basic research program (JCYJ20130329114940668) and a project from Shenzhen Marine Environment and Resources Monitoring Center(HHZB11025). The funders had no role in study design, data collection and analysis, decision to publish, or preparation of the manuscript.

Competing Interests: The authors have declared that no competing interests exist.

* Email: aplei@szu.edu.cn (AL); kzhou@tom.com (KZ)

Introduction

Macroalgal blooms have a broad range of ecological impacts and are considered to be harmful algal blooms [1]. Many studies have addressed the potential ecological and environmental consequences of macroalgal blooms [2,3], including uncoupled biogeochemical cycles, degraded intertidal environment, reduced biodiversity, hypoxia or anoxia, destruction of coastal marine habitats (e.g., seagrass) and economic losses of marine industries (e.g., fisheries and tourism) [4,5,6,7,8,9].

In January 2013, red algae bloomed in the OCT North Lake of Shenzhen. Initially, a bloom of macroalgae occurred near the shore (Fig. 1A), then the lake surface became full of floating red macroalgae (Fig. 1B). Just 1 month later, from the surface to the bottom of the lake were filled with this kind of fast growing macroalgae and decomposition happened afterwards (Fig. 1C and D), becoming a serious issue for the local administration. The over-grown red algae were identified as Rhodophyta, Gigartinales, Gracilariaceae, Gracilaria, *Gracilaria tenuistipitata* (Fig. 1E and F). The damage caused by the bloom of *G. tenuistipitata* during this period included tourism and cleanup costs (about $ 50,000) (according to the local management office). Besides this, along with macroalgae decomposition, dead fish were also observed, indicating the lake ecological system has been damaged, to some extent.

Several hypotheses were proposed to explain the occurrence of macroalgal blooms in estuarine areas [10,11,12]. The OCT North Lake is connected to Shenzhen Bay which encountered strong anthropogenic influence, there are ~0.2 million tons/day land-based fresh wastewater input [13], resulting in Shenzhen Bay as perennial brackish water (salinity 8–20‰) with high nitrogen loadings (essentially NH_4^+) leading to eutrophication [14]. Considering the estuarine environment, the effect of salinity on growth of *G. tenuistipitata* was investigated in this study. The ability to uptake and assimilate nutrients rapidly is one of the characteristics of opportunistic species. Concerning mechanisms of algal blooms, the bioavailability of nitrogen is the most important among all of the biogenic elements [2,3]. In coastal areas, nitrogen is available to macroalgae majorly in the forms Dissolved

Figure 1. The *G. tenuistipitata* in OCT North Lake. A,B,C) The *G. tenuistipitata* bloom in OCT North Lake, D) The salvaged *G. tenuistipitata*, E) Samples of *G. tenuistipitata* collected, F) Observation and identification of macroalgae.

Inorganic Nitrogen (DIN): ammonium (NH_4^+) and nitrate (NO_3^-). It is well known that fast growing species are positively affected by increased nutrient availability [15], thus, the effects of NH_4^+ and NO_3^- on macroalgal growth were also investigated in this study.

A profound knowledge of the nitrogen uptake kinetics of *G. tenuistipitata* is important essentially for two reasons: It helps to study the algal potential assimilation efficiency and it is fundamental to assess its potential as a biofilter in estuaries and coastal waters [16,17,18]. There is a set of works studying the uptake kinetics in Marine macroalgae Chlorophyta and Rhodophyta species [19,20,21,22]. The Michaelis-Menten equation was generally used to describe nutrient uptake by phytoplankton over the past few years [23], while the results varied depending on algal species and nutrients. Some macroalgae followed Michaelis-Menten kinetics, some did not, and some species did not show a clear way of assimilating NH_4^+ or NO_3^- [22]. Our work intended to provide more information regarding the nitrogen uptake dynamics of brackish water grown Rhodophyta species *G. tenuistipitata* in an attempt to elucidate the mechanism behind bloom formation. Short-term kinetics of NH_4^+ and NO_3^- uptake were studied for the uptake characteristic of *G. tenuistipitata*. Nitrogen uptake has been studied for a variety of macroalgae, but the N species preference is still discussed [24,25]. Elucidating the uptake strategy of *G. tenuistipitata* is essential because this red alga is the major bloom-forming species in nutrient-rich estuaries [26,27]. One mechanism that may enhance bloom ability is the capability to choose and use different DIN species at the same time [19]. This isotope ^{15}N tracer study was used to quantify uptake and assimilation strategy of NH_4^+ and NO_3^- offered simultaneously in order to assess potential mechanisms that may enhance the bloom ability of *G. tenuistipitata* during irregular pulses of nitrogen sources.

Materials and Methods

Ethics Statement

Permission to investigation, public photos and sampling inside the OCT North Lake was approved through permits from the Overseas Chinese Town Holdings Company(OCT Group)and the Shenzhen government. Our field investigation did not involve endangered or protected species.

Algal material

During the macroalgal bloom, thalli of *G. tenuistipitata* were sampled from OCT North Lake (22°31′46.06″N, 113°58′40.32″E), Shenzhen, in January 2013. Healthy algal thalli collected were transported to the lab in a cool box within 4 h. Then they were flushed several times with artificial seawater to remove epiphytes, slime, sediment and small grazers, and then cultured in nutrient-deplete sterile artificial seawater [28] with vitamins and trace elements added according to the f/2 medium at salinity 12±1‰. Each algal sample was cultured for 3–5 days with bubbling air before the experiment. The cultures were kept at the salinity 12±1‰, temperature 25±0.5°C, with irradiance of 56 µE m^{-2} S^{-1} and photoperiod of 12:12 (Light: Dark).

Salinity effects on growth rates

To determine the variable salinity effects on growth of *G. tenuistipitata*, we evaluated the Relative Growth Rates (RGR) of *G. tenuistipitata* at salinities of 4, 8, 12, 16, 20, 24, 28 and 32 ‰ over a 15-day period. Approximately 0.1 g wet weight of *G. tenuistipitata* was added to 600 mL sterile f/2 enriched artificial seawater in a 1-L glass beaker at conditions described above. Each algal sample was assigned at random. The enriched medium was renewed every four days. Four replicates were used for each treatment. Every 4 days, the *G. tenuistipitata* thalli were blotted dry and the fresh weight determined, and RGR were calculated: $RGR = (\ln w_2 - \ln w_1)/\Delta t$, where w_1 is the initial fresh mass and w_2 is the fresh mass after Δt days.

Nitrogen source effects on growth rates

The effects of nitrogen source on growth rates of *G. tenuistipitata* were tested for 15 days. For the nitrogen-effect experiment, two nitrogen sources, ammonium (NH_4Cl) and nitrate ($NaNO_3$) were applied. Nitrogen (NH_4^+ or NO_3^-) concentrations were adjusted to 5, 10, 20, 40, 60, 80, 100 µM, treatments without nitrogen addition were taken as control groups. 30 µM phosphate (KH_2PO_4) was added to all treatments to avoid P limitation [29]. The fresh algal thalli of 0.1 g was randomly put in a 1-L glass beaker filled with 600 mL nutrient-deplete sterile artifical seawater at conditions described above. All glass material used during the experiment had been acid washed with 10% HCL to prevent possible nutrient contamination. Nutrient-enriched media were renewed every four days. Four replicates were used for each treatment.

Nitrogen source uptake

The nitrogen (NO_3^- and NH_4^+) uptake of *G. tenuistipitata* was determined by monitoring the nutrient depletion of the medium. The nutrient-deplete experimental artificial seawater was obtained from *G. tenuistipitata* culture medium mentioned above. After cultured alga for 5 days then NH_4^+ and NO_3^- concentrations were monitored daily until their concentrations were below the detection limit (0.05 µM N L^{-1}), then culture media were filtered (GF/C) used as nutrient-depleted seawater medium. The uptake experiment was performed in 400 mL medium with nutrients added from stock solutions in 500-mL sterilized glass beakers. Uptake experiments with NO_3^- or NH_4^+ were started with 10, 20, 40, 60, 80, 100 µM respectively. To avoid P shortage, 30 µM PO_4^{3-} (KH_2PO_4) was added in the experiments. In order to determine the initial N concentrations, 2 ml seawater samples were taken before adding the macroalga. Further 2 ml seawater samples were collected at each time interval: 10, 20, 30 and 60 min. Seawater samples from both experiments were filtered (GF/C) and kept at −20°C until analysis (within 1 week). Analysis of NH_4^+ and NO_3^- was carried out using a CleverChem 200 automated discrete analyzer (DeChem-Tech. GmbH, Germany) according to the manufacturer's instructions. Algal material was dried to a constant weight (48 h) in an oven kept at 60°C for dry weight. Uptake rates (V) were calculated from changes in substrate concentrations during each sampling interval using the equation: $V = [(C_0 \times V_0) - (C_t \times V_t)]/(t \times DM)$, where C_0 and C_t are the substrate nutrient concentration, and V_0 and V_t are the volumes before and after a sampling period (t), and DM is algal dry biomass.

For estimation of the kinetic parameters of each N source, uptake rates were plotted together in V vs S curves. The V (uptake rate) against V/S ($S = NH_4^+$ or NO_3^- concentration) linear transformation of the Michaelis-Menten equation was used to determine whether the data could be described by Michaelis-Menten kinetics [30]. Data which showed saturation kinetics were fitted to the Michaelis-Menten equation: $V = (V_{max} \times C)/(Ks + C)$, using enzyme kinetics module (non-linear regression model) to analyze each replicate dataset according to the instructions of SigmaPlot 12.5 (SPSS Inc., Chicago, USA). The Michaelis–Menten equation and parameters, V_{max} (maximum uptake rate) and Ks (substrate concentration at which uptake proceeds at half the maximum rate) were obtained from the analysis.

Nitrogen preferences

After nitrogen starvation of 4–5 days, nitrogen uptake was estimated using stable isotopes as tracers. Uptake experiments were initiated by adding tracer amounts (50 µM N with 10 atom % ^{15}N) of highly enriched (96–99%) ^{15}N-labeled NO_3^- or NH_4^+ to nutrient-deplete artificial seawater mentioned above placed in 30 ml acid-washed glass tubes. NO_3^- or NH_4^+ were added in different ratios, with only one DIN species labeled with ^{15}N per trial. After 1 h or 24 h, the algal thalli were briefly rinsed with deionized water, and dried at 60°C to constant weight (48 h), powdered with a Spex mixer (Spex CertiPrep, Inc., Edison, NJ), and analyzed for TN and ^{15}N enrichment with an isotope ratio mass spectrometer (Europa Scientific Integra). Water samples were filtered (GF/C) and immediately analyzed for ammonium, nitrite, and nitrate. ^{15}N enrichment was calculated based on the methods of Naldi and Wheeler [25].

Data analysis

Separate one-way ANOVA were used to analyze the responses of *G. tenuistipitata* to the different salinity treatments or nutrient (NO_3^- and NH_4^+) sources in terms of RGR. Data were initially examined for homogeneity of variance and F-tests were performed to test the significance of difference among treatments. Tukey test (equal variances) was then used to test the significance of difference between two specific treatments, if significant difference was found among treatments. Statistical analyses were performed with SPSS 18.0 (SPSS Inc., Chicago, USA).

Results

Effects of salinity and nitrogen sources on growth rate

After the 15-day culture, there were significant differences in the RGR associated with salinity (One-way ANOVA, $P<0.01$). RGR of *G. tenuistipitata* was relatively high with salinity ranging from 12 to 20 ‰, and the highest RGR was 4.2%/d at salinity 20 ‰. Biomasses significantly decreased when cultured in low (4 ‰) (Tukey's test, $P<0.05$) or high (>20‰) salinity seawater (Tukey's test, $P<0.05$) (Fig. 2A).

There were significant differences between RGR of *G. tenuistipitata* associated with nitrogen addition (NH_4^+ or NO_3^-) after 15 days (One-way ANOVA, $P<0.01$). RGR was enhanced with increasing nitrogen (NH_4^+ or NO_3^-) concentrations, when the nitrogen (NH_4^+, NO_3^-) was up to 40 μM, however, there was no significant difference in RGR with higher concentrations (40–100 μM) ($P>0.05$, Fig. 2B and C), with the highest RGR observed in treatments of 100 μM nitrogen (NH_4^+ or NO_3^-). Interestingly, the algal biomass was higher (approximately 1.5 times) in algae cultured with NH_4^+ addition than that with NO_3^- addition, respectively (Fig. 2B and C).

Nitrogen (NH_4^+or NO_3^-) uptake

During the 1 h experiment, uptake rates ($V_{0-1\ h}$) of *G. tenuistipitata* increased with NH_4^+ concentration levels. Uptake efficiency decreased, as suggested by a gradual decline in the uptake rates of NH_4^+ over sequential time intervals. The *V-S* of NH_4^+ was linearly positive correlated, best described by rate-unsaturated model (Fig. 3A, B, C and D). The bloom-forming red alga showed the highest uptake rates about 32 μM g^{-1}DM h^{-1} for the first 10 min at the concentration of 100 μM.

High NO_3^- uptake rates (V) of *G. tenuistipitata* were obtained from 0 to 10 min. Uptake rates for the NO_3^- decreased over the time intervals. With increasing concentrations of NO_3^-, the uptake rates showed a tendency to saturation (Fig. 4A, B, C and D), and the *V-S* of NO_3^- was fitted to Michaelis-Menten model. Kinetic parameters (V_{max} and K_s) following the Michaelis-Menten equation for NO_3^- uptake by *G. tenuistipitata* at different time intervals during the 1 h experiment were shown in Table 1. The highest V_{max} and K_s values in the experiment with NO_3^- were 37.2 μ M g^{-1} DM h^{-1} and 61.5 μM.

Nitrogen uptake preferences

The accumulation of ^{15}N in *G. tenuistipitata* was rapid in 1 h, but decreased over time (Fig. 5A and B). ^{15}N accumulation from sole source N cultures of NH_4^+ increased faster and during 24 h time interval reached an enrichment twice that for ^{15}N of NO_3^- (51.52 vs. 26.32 N enrichment, Fig. 5B). $^{15}NH_4^+$ accumulation by *G. tenuistipitata* exceeded $^{15}NO_3^-$ enrichment in all the two sources N cultures, regardless of the ratio of NH_4^+ versus NO_3^- supplied and time. When the NH_4^+ supply was lower than that of NO_3^- (20%:80%, 40%:60%), more $^{15}NH_4^+$ was accumulated. These results revealed that when NH_4^+ and NO_3^- were supplied simultaneously, *G. tenuistipitata* preferentially assimilate NH_4^+.

Discussion

Salinity in the estuarine or coastal areas is a reflection of the balance of land source flows and seawater from tidal flux. High tide and water exchange have changed the balance of freshwater and seawater and caused a change of salinity within the OCT North Lake. Salinity is an important influence on the growth, photosynthesis, and respiration of *Gracilaria* spp. [31,32]. Bunsom found that salinity was the most crucial factor affecting *G.*

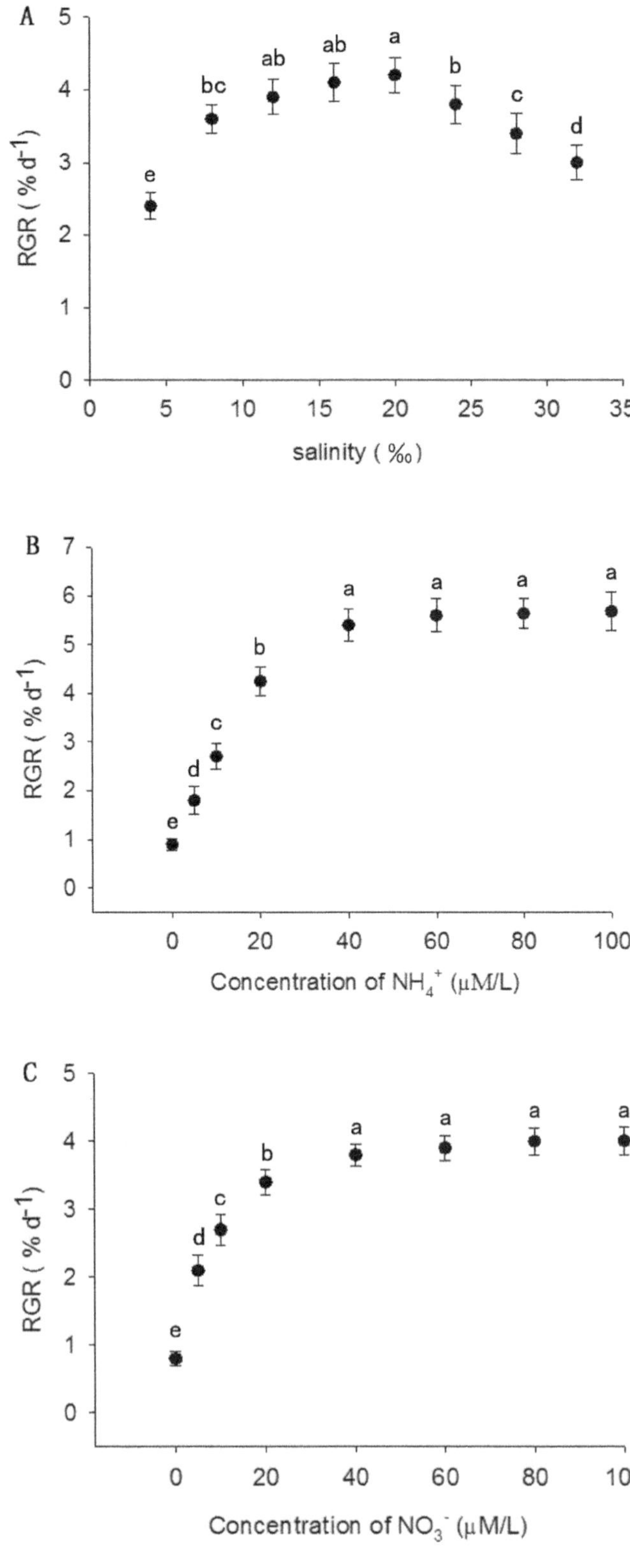

Figure 2. Effects of A) salinity, B) ammonium (NH_4^+) and C) nitrate (NO_3^-) on relative growth rate (RGR) of *G. tenuistipitata* after the 15-day culture. A) f/2 medium was enriched, B and C) Phosphate was enriched in all treatments (30 μM). Means and SD are shown (n = 4).

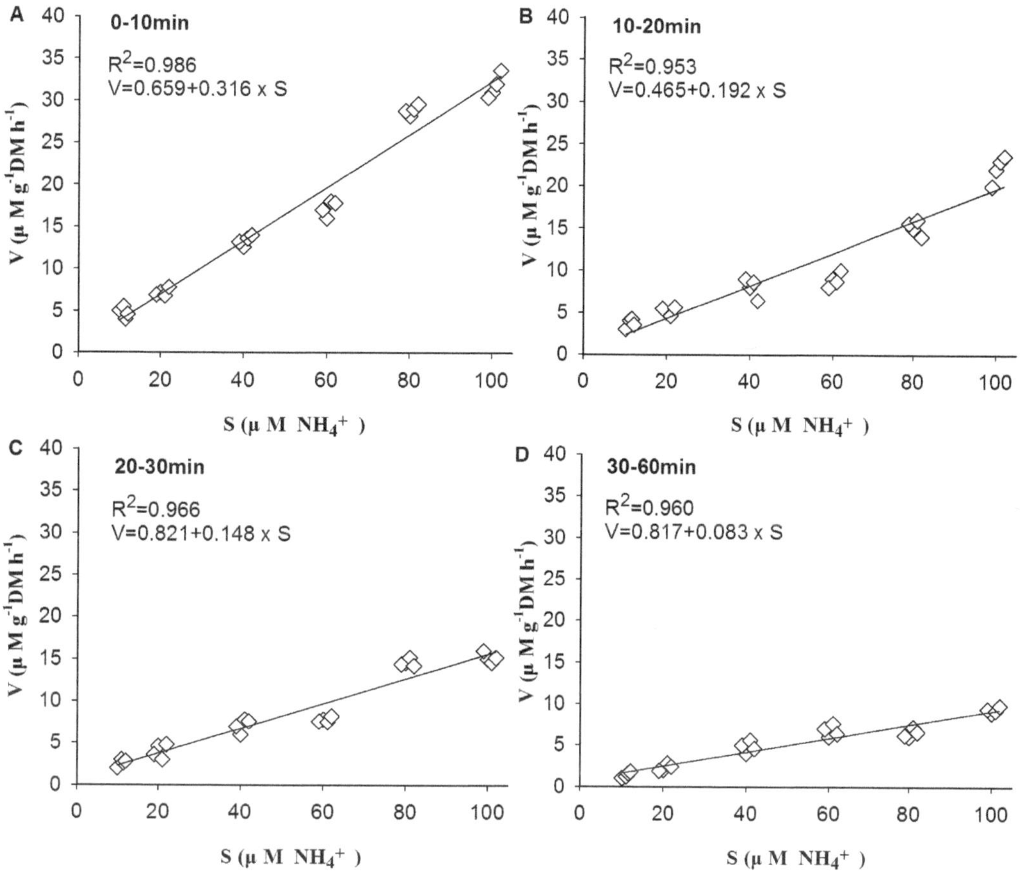

Figure 3. Ammonium (NH_4^+) uptake rates by *G. tenuistipitata* at different time intervals and different substrate concentration. Phosphate (30 μM) was added in all treatments.

tenuistipitata growth and the highest growth rate was at 25‰ [33], and it can tolerate salinities as higher as 39‰ [31]. In this study, the highest growth rate of *G. tenuistipitata* was at 20‰, instead of 25‰, perhaps due to perennial exposure to brackish water. Generally, lower salinities often inhibit growth of seaweeds, affect branching patterns and promote changes in their chemical composition [34,35]. Our results showed the lowest growth rate was at 4‰, which was consistent with the result of low salinity growth inhibition. This study revealed that *G. tenuistipitata* from Lake Shenzhen grew well across a wide range of salinities (4–32‰). At the bloom period in January 2013, the salinity was 12.6‰, which was in the favorable range of this macroalga. Adaptability for low salinity might be one of the mechanisms of macroalgal bloom.

Nitrogen is one of the most important limiting factors for phytoplankton growth in aquatic systems [15,36]. NH_4^+ and NO_3^- are the principal forms of DIN in estuarine environments where nutrient supply is variable and unpredictable [37,38]. NH_4^+ is primarily supplied to estuaries by sewage spills and treated wastewater [39]; NO_3^- is the predominant source in riverine [40] and groundwater inputs [41]. The algae of the genus *Gracilaria* are characterized by a simple thallus with a high surface area to volume ratio, allowing for a rapid and efficient response to nutrient input, resulting in high rates of nutrient uptake and growth [15,42]. Our studies reported the maximum average RGR of *G. tenuistipitata* was 5.68%/d at 100 μM NH_4^+ and 4.01%/d at 100 μM NO_3^-, compared with an RGR in control groups of less than 0.9%/d. The growth of *G. tenuistipitata* became faster with increased DIN (NH_4^+or NO_3^-) concentration. Nutrient enrichment, due to high N inputs via anthropogenic injection and regenerative processes, has contributed to enhanced growth of opportunistic macroalgae [2]. During the bloom period, in the OCT North Lake, nitrogen in shallow waters is often in the form of NH_4^+ with average concentration of 60 μM, and NO_3^- of 35 μM, P (PO_4^{3-}) of 5 μM. DIN concentration was in the range of significantly positive stimulating this alga growth, while P could not meet this algal need according to the Redfield N (16): P (1) ratio, indicating that P was limiting during the bloom.

Opportunistic macroalgae, often have higher nitrogen uptake rates than other more specialized algae [43]. A high uptake rate allows these opportunistic macroalgae to use nutrients efficiently [44]. With respect to uptake rate, some species followed the classic Michaelis-Menten kinetics for N uptake [22,23], while others have exhibited sustained linear, rate-unsaturated uptake mainly due to eutrophic/oligotrophic level [45,46,22,48,49]. Normally, NO_3^- uptake shows saturation kinetics [47]. Smit [48] investigated nitrogen uptake by *Gracilaria gracilis* and concluded that NO_3^- uptake followed a rate-saturating mechanism was best described by the Michaelis-Menten model. A set of studies have used extraordinarily high NH_4^+ concentration (500 μM above levels) as Table 2 shows, which is toxic to many marine organisms [50], but the results were still positively accorded with Michaelis-Menten or linear models. The max DIN concentration (100 μM) in this study was based on the in situ environmental concentrations, and the principal finding about this long-term brackish water lived macroalga for the first time that the NH_4^+ uptake of *G.*

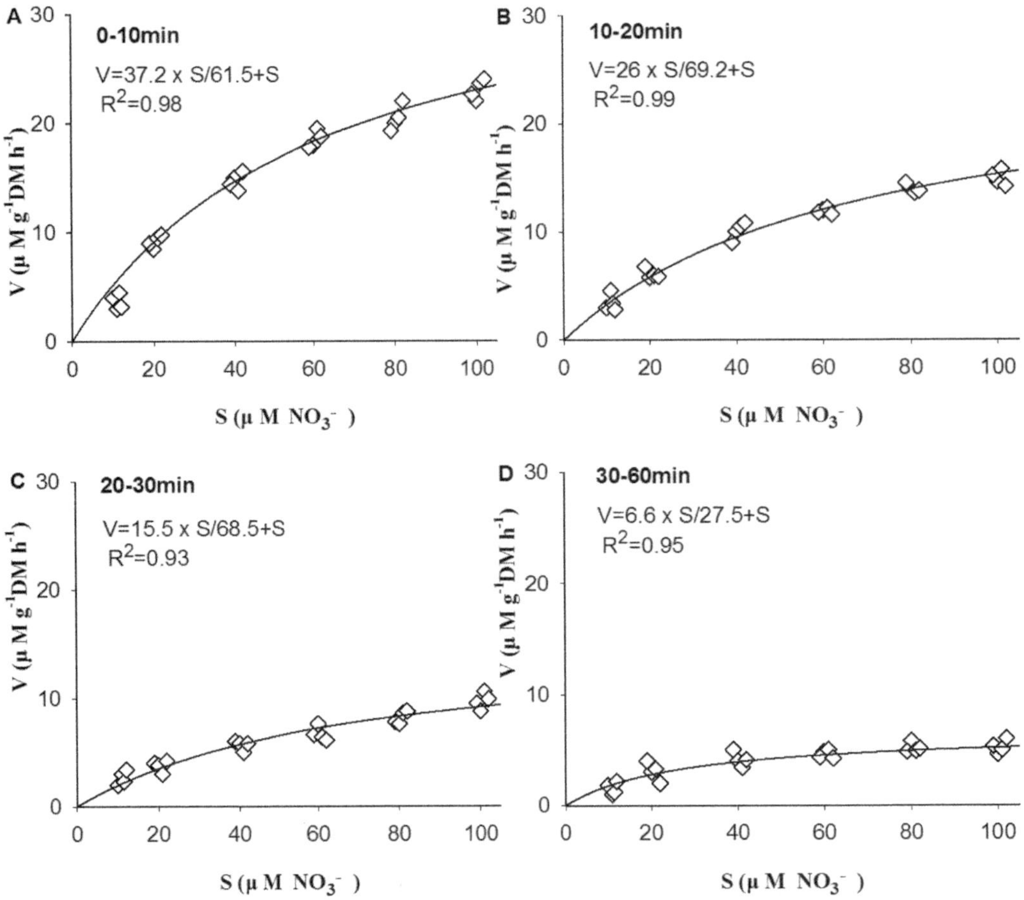

Figure 4. Nitrate (NO_3^-) uptake rates by *G. tenuistipitata* at different time intervals and different substrate concentration. Phosphate (30 μM) was added in all treatments.

tenuistipitata was linear, while the uptake kinetics of NO_3^- were distinctly different from those of NH_4^+. The uptake rate of NO_3^- over time fitted Michaelis-Menten kinetics. This indicated high uptake rate and affinity capability for NH_4^+. During the bloom, NH_4^+ was abundant in OCT North Lake, which might be another reason for macroalgal bloomed.

Generally, nitrogen uptake in macroalgae depends on intracellular levels [15] and environmental nitrogen concentrations [51]. Biological factors such as metabolism, morphology, tissue type, the age of the algae, and nutritional history [42,52,53], all influence nitrogen uptake by macroalgae. The DIN uptake pattern can be explained by different nitrogen assimilation mechanisms. The process of nitrogen uptake and assimilation in macroalgae involves transport from the water across the cell membrane and then assimilation into organic compounds, followed by incorporation into proteins and macromolecules for growth [40]. For NO_3^-, there is the additional step of reduction to NH_4^+ by nitrate reductase after uptake [54]. One explanation for NH_4^+ affinity may be that energy required for nitrate reduction could be saved [53].

Two parameters of Michaelis-Menten kinetics, namely V_{max} and K_s, contain ecologically relevant information relating to the nutrient uptake ability of a species under conditions of varying nutrient availability. In this study, the V_{max} was 37.2 and K_s was 61.5 for uptake of NO_3^- by *G. tenuistipitata*in Table 1. While, the V_{max} declined over time intervals, indicating initially very high uptake rates and lower sustained uptake rates. Previous work addressed the uptake differences between Chlorophyta and

Table 1. Kinetic parameters (V_{max} and K_s) of the Michaels-Menten equation obtained from the rates for uptake of Nitrate (NO_3^-) by *G. tenuistipitata* at different time intervals.

	Time interval (min)	V_{max} (SE)(μM g^{-1} DM h^{-1})	K_s(SE)(μM)	R^2
Nitrate (NO_3^-)	0–10	37.2 (1.85)	61.5 (6.26)	0.98
	10–20	26 (0.97)	69.2 (5.02)	0.99
	20–30	15.5 (0.2)	68.5 (1.85)	0.93
	30–60	6.6 (0.25)	27.5 (3.07)	0.95

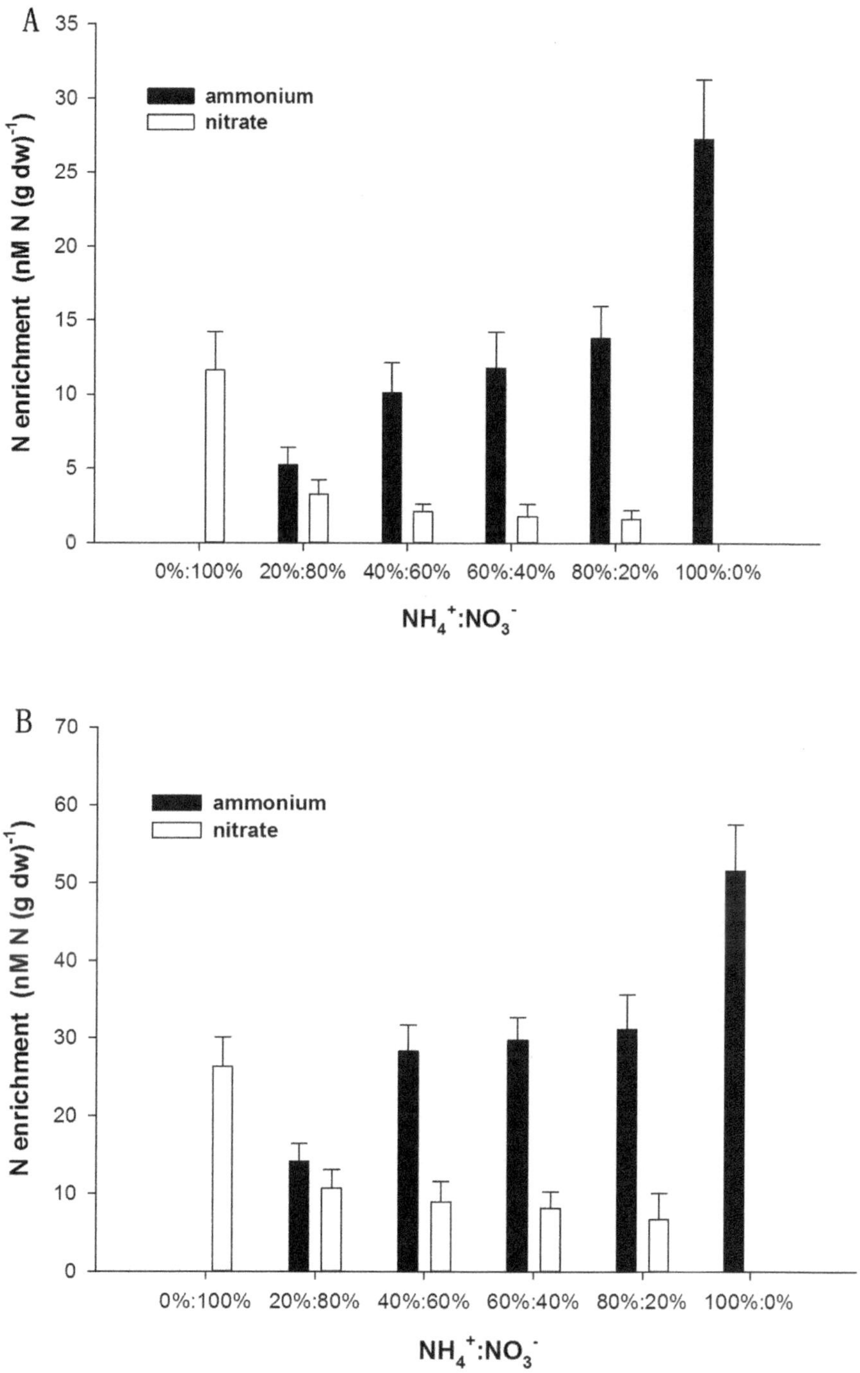

Figure 5. Ammonium (NH_4^+) and nitrate (NO_3^-) enrichment by *G. tenuistipitata* at different time intervals. A) 1 h and B) 24 h. The culture media contained 50 μM total N, but different proportions of NH_4^+: NO_3^-. Phosphate (10 μM) was added in all treatments.

Rhodophyta species in the marine or estuarine environment [19,20,21,22,48,49,55,56]. Commonly, Chlorophyta follow Michaelis-Menten kinetics for NH_4^+ and NO_3^- [20,21,56], Rhodophyta differ depending on the species [19,22,48]. Higher V_{max} were found in Chlorophyta for NO_3^- [20,21]. *G. tenuistipitata* has the relatively higher V_{max} in Rhodophyta species, implying higher assimilating and storing capability for NO_3^-. However, the V_{max}/K_s value was 0.6, less than 1, indicating the competitive advantage for NO_3^- is not so obvious compared with other Rhodophyta species. During the bloom, NO_3^- concentration was relatively low, while NH_4^+ was abundant. What's more, resilience to low salinities might be a principal reason for bloom of *G. tenuistipitata*.

Environmental factors, including temperature, salinity, nutrient levels and other relevant competitors, are principally responsible for regulating macroalgal bloom occurrence [6,7]. Because Shenzhen Bay is in the subtropical zone, the water temperature is relatively constant between day and night, without great changes between different months, so our research did not focus on its influence on algal RGR, with attention to the effect of salinity and DIN. Besides, the competition for resources (especially nutrients) is

Table 2. Comparison of V_{max} (μM g^{-1} DM h^{-1}), K_S (μM) values and substrate range (μM) for several Chlorophyta and Rhodophyta species in the marine or estuarine environment.

	Species	NH^{4+}			NO^{3-}			Reference
		V_{max}	K_s	S. range	V_{max}	K_s	S. range	
Chloro-phyta	*Ulva lactuca*	450	85	~1200	116	34	~50	[20]
	Ulva prolifera	284.6	25.1	~250	124.3	15.1	~250	[21]
	Ulva linza	250.2	37	~250	109.1	23	~250	[21]
	Cladophora montagneana	130	20.7	~357	42.1	1.4	~357	[56]
Rhodo-phyta	*Gracilaria foliifera*	23.8	1.6	~15	9.7	2.5	~15	[55]
	Gracilaria pacifica	21.5	50.9	~15	6.0	26.8	~30	[19]
	Gracilaria gracilis	—	—	~40	35	5.6	~50	[48]
	Gracilaria vermiulophylla	—	—	~160	+	+	~400	[22]
	Catenella nipae	550	692	~1200	8.3	5	~50	[20]
	Gracilaria tenuistipitata	—	—	~500	*	*		[49]
	Gracilaria tenuistipitata	—	—	~100	37.2	61.5	~100	Present study

— means linear, rate-unsaturated response.
+means not clear.
*means no report.

ubiquitous in marine phytoplankton [55], and the chemical defense system is another important potential mechanism for algal competition. Allelopathy, one type of chemical defense competition between plants, can be a potent influential factor. Xu et al. [57] found allelopathic interactions happened between green tide-forming species *Ulva prolifera* and macroalga *Gracilaria lichvoides*, indicating that some allelochemicals released by both of the macroalgae could account for the physiological inhibition of growth. In our study, there is only one species, namely *G. tenuistipitata*, lived and bloomed in the OCT North Lake until 2014, no other macroalga was found. So we neglect the competing role among macroalgae.

From our ^{15}N tracer study, we found the opportunistic macroalga *G. tenuistipitata* actively utilizes multiple DIN species simultaneously, and prefers the most energetically efficient (NH_4^+) when available. Other previous reports have also found NH_4^+ uptake preference. D'Elia and DeBoer found the priority of NH_4^+ uptake, even in very low concentrations (~5 μM), and its presence seriously deceased the uptake of NO_3^- by *Gracilaria foliifera* (Forsskal) *Borgesen* [58], while Hanisak and Harlin determined that only 1 μM NH_4^+ was enough to decrease NO_3^- uptake in *Codium fragile* (van Goor) *Silva* [59]. Haines and Wheeler showed that NO_3^- uptake by *Hypnea musciformis* (Wulfen) *Lamouroux* was reduced by 50% in the presence of about 18 μM NH_4^+ [60]. Our study also found a decrease of NO_3^- uptake when NH_4^+ was present. The result suggested that when NH_4^+ and NO_3^- were supplied in equal or less ratio concentrations simultaneously, more NH_4^+ was taken up and assimilated.

From the present study, three key conclusions were obtained: *i*) the growth of *G. tenuistipitata* was regulated by salinity and N availability in environment. Its tolerance of relatively low salinity in estuarine environment seems to be a principal factor for its bloom rather than the other species. Its high growth rate and extraordinary capabilities for utilizing nutrients, like DIN likely contribute to macroalgal outbreaks; *ii*) The DIN: NH_4^+ and NO_3^- uptake kinetics of *G. tenuistipitata* are different. NH_4^+ uptake demonstrated a rate-unsaturated response, but NO_3^- uptake followed a rate-saturating mechanism which fit the Michaelis-Menten model, with kinetic parameters V_{max} (37.2 μ M g^{-1} DM h^{-1}) and K_s (61.5 μ M), indicating a higher uptake efficiency of NH_4^+ than NO_3^-; *iii*) The strategy of *G. tenuistipitata* assimilating different forms of DIN showed that NH_4^+ incorporated faster than NO_3^-. We propose that when NH_4^+ and NO_3^- were supplied simultaneously, *G. tenuistipitata* preferentially assimilated NH_4^+. One mechanism that may enhance bloom ability is the strategy to choose and use different DIN in estuarine environments.

Author Contributions

Conceived and designed the experiments: CW AL KZ ZH. Performed the experiments: CW WH JY. Analyzed the data: CW AL. Contributed reagents/materials/analysis tools: CW AL WH JY. Wrote the paper: CW.

References

1. ECOHAB (1995) The Ecology and Oceanography of Harmful Algae Blooms. WHOI, Woods Hole, MA: A National Research Agenda.
2. Fletcher RL (1996) The occurrence of "green tides": a review. In: Schramm. W PHNPHE, editor. Marine benthic vegetation Recent changes and the effects of eutrophication. Berlin, Germany: Springer. pp. 7–43.
3. Valiela I, McClelland J, Hauxwell J, Behr PJ, Hersh D, et al. (1997) Macroalgal blooms in shallow estuaries: Controls and ecophysiological and ecosystem consequences. Limnology and Oceanography 42: 1105–1118.
4. Wang C, Yu RC, Zhou MJ (2011) Acute toxicity of live and decomposing green alga *Ulva prolifera* to abalone Haliotis discus hannai. Chinese Journal of Oceanology and Limnology 29: 541–546.
5. Wang C, Yu RC, Zhou MJ (2012) Effects of the decomposing green macroalga *Ulva prolifera* on the growth of four red-tide species. Harmful Algae 16: 12–19.
6. Hernandez I, Peralta G, PerezLlorens JL, Vergara JJ, Niell FX (1997) Biomass and dynamics of growth of Ulva species in Palmones river estuary. Journal of Phycology 33: 764–772.
7. McGlathery KJ (2001) Macroalgal blooms contribute to the decline of seagrass in nutrient-enriched coastal waters. Journal of Phycology 37: 453–456.
8. Nelson TA, Lee A (2001) A manipulative experiment demonstrates that blooms of the macroalga *Ulvaria obscura* can reduce eelgrass shoot density. Aquatic Botany 71: 149–154.
9. Franz DR, Friedman I (2002) Effects of a macroalgal mat (*Ulva lactuca*) on estuarine sand flat copepods: an experimental study. Journal of Experimental Marine Biology and Ecology 271: 209–226.
10. Kirst GO (1995) Influence of salinity on algal ecosystems. In: Eds. W. Wiessner ESaCS, editor. In 'Algae, Environment and Human Affairs. England: Biopress: Bristol. pp. 123–142.
11. Cloern JE (1987) Turbidity as a control on phytoplankton biomass and productivity in estuaries. Continental Shelf Research 7: 1367–1381.
12. Chan TU, Hamilton P (2001) Effect of freshwater flow on the succession and biomass of phytoplankton in a seasonal estuary. Marine and Freshwater Research 52: 869–884.
13. Liu G, Lai LW (2013) the Shenzhen bay is sick. Jing newspaper. Available: http://jb.sznews.com/html/2013-10/20/content_2655847.htm
14. Zhang J, Zhang Y, Zhou K, Zhang J, Li X (2010) Evaluation on temporal and spatial distribution of nutrients and potential eutrophication in Shenzhen Bay. Ecology and Environmental Sciences. 19(2): 253–261.
15. Pedersen MF, Borum J (1996) Nutrient control of algal growth in estuarine waters. Nutrient limitation and the importance of nitrogen requirements and nitrogen storage among phytoplankton and species of macroalgae. Marine Ecology Progress Series 142: 261–272.
16. Carneiro MAA, Freire FAdM, Marinho-Soriano E (2011) Study on biofiltration capacity and kinetics of nutrient uptake by *Gracilaria cervicornis* (Turner) *J. Agardh* (Rhodophyta, Gracilariaceae). Revista Brasileira De Farmacognosia-Brazilian Journal of Pharmacognosy 21: 329–333.
17. Marinho-Soriano E, Panucci RA, Carneiro MAA, Pereira DC (2009) Evaluation of *Gracilaria caudata J. Agardh* for bioremediation of nutrients from shrimp farming wastewater. Bioresource Technology 100: 6192–6198.
18. Yang YF, Fei XG, Song JM, Hu HY, Wang GC, et al. (2006) Growth of *Gracilaria lemaneiformis* under different cultivation conditions and its effects on nutrient removal in Chinese coastal waters. Aquaculture 254: 248–255.
19. Thomas TE, Harrison PJ (1987) Rapid ammonium uptake and nitrogen interactions in 5 intertidal seaweeds grown under field conditions. Journal of Experimental Marine Biology and Ecology 107: 1–8.
20. Runcie JW, Ritchie RJ, Larkum AW (2003) Uptake kinetics and assimilation of inorganic nitrogen by *Catenella nipae* and *Ulva lactuca*. Aquatic Botany 76: 155–174.
21. Luo MB, Liu F, Xu ZL (2012) Growth and nutrient uptake capacity of two co-occurring species, *Ulva prolifera* and *Ulva linza*. Aquatic Botany 100: 18–24.
22. Abreu MH, Pereira R, Buschmann AH, Sousa-Pinto I, Yarish C (2011) Nitrogen uptake responses of *Gracilaria vermiculophylla* (Ohmi) *Papenfuss* under combined and single addition of nitrate and ammonium. Journal of Experimental Marine Biology and Ecology 407: 190–199.
23. Smith SL, Yamanaka Y, Pahlow M, Oschlies A (2009) Optimal uptake kinetics: physiological acclimation explains the pattern of nitrate uptake by phytoplankton in the ocean. Marine Ecology Progress Series 384: 1–12.
24. Lotze HK, Schramm W (2000) Ecophysiological traits explain species dominance patterns in macroalgal blooms. Journal of Phycology 36: 287–295.
25. Naldi M, Wheeler PA (2002) N-15 measurements of ammonium and nitrate uptake by *Ulva fenestrata* (chlorophyta) and *Gracilaria pacifica* (rhodophyta): Comparison of net nutrient disappearance, release of ammonium and nitrate, and N-15 accumulation in algal tissue. Journal of Phycology 38: 135–144.
26. Raffaelli D, Balls P, Way S, Patterson IJ, Hohmann S, et al. (1999) Major long-term changes in the ecology of the Ythan estuary, Aberdeenshire, Scotland; how important are physical factors? Aquatic Conservation-Marine and Freshwater Ecosystems 9: 219–236.
27. Valiela I, Foreman K, Lamontagne M, Hersh D, Costa J, et al. (1992) Couplings of watersheds and coastal waters-sources and consequences of nutrient enrichment in waquoit bay, Massachusetts. Estuaries 15: 443–457.
28. Guillard RRL (1975) Culture of phytoplankton for feeding marine invertebrates. In: Smith WL, Chanley MH, editors. Culture of Marine Invertebrate Animals. New York: Plenum Press. pp. 26–60.
29. Bjornsater BR, Wheeler PA (1990) Effect of nitrogen and phosphorus supply on growth and tissue composition of *Ulva-fenestrata* and *Enteromorpha-intestinalis* (Ulvales,Chlorophyta). Journal of Phycology 26: 603–611.
30. Dowd JE, Riggs DS (1965) A comparison of estimates of michaelis-menten kinetic constants from various linear transformations. Journal of Biological Chemistry 240: 863–&.
31. Israel A, Martinez-Goss M, Friedlander M (1999) Effect of salinity and pH on growth and agar yield of *Gracilaria tenuistipitata* var. *liui* in laboratory and outdoor cultivation. Journal of Applied Phycology 11: 543–549.

32. He LH, Wu M, Qian PY, Zhu MY (2002) Effects of co-culture and salinity on the growth and agar yield of *Gracilaria tenuistipitata* var *liui* Zhang *et* Xia. Chinese Journal of Oceanology and Limnology 20: 365–370.
33. Bunsom C, Prathep A (2012) Effects of salinity, light intensity and sediment on growth, pigments, agar production and reproduction in *Gracilaria tenuistipitata* from Songkhla Lagoon in Thailand. Phycological Research 60: 169–178.
34. Ekman P, Yu SK, Pedersen M (1991) Effects of altered salinity, darkness and algal nutrient status on floridoside and starch content, alpha-galatosidase activity and agar yield of cultivated *gracilaria sordida*. British Phycological Journal 26: 123–131.
35. Choi HG, Kim YS, Kim JH, Lee SJ, Park EJ, et al. (2006) Effects of temperature and salinity on the growth of *Gracilaria verrucosa* and *G-chorda*, with the potential for mariculture in Korea. Journal of Applied Phycology 18: 269–277.
36. Kim JK, Kraemer GP, Neefus CD, Chung IK, Yarish C (2007) Effects of temperature and ammonium on growth, pigment production and nitrogen uptake by four species of *Porphyra* (Bangiales, Rhodophyta) native to the New England coast. Journal of Applied Phycology 19: 431–440.
37. Pinchetti JLG, Fernandez ED, Diez PM, Reina GG (1998) Nitrogen availability influences the biochemical composition and photosynthesis of tank-cultivated *Ulva rigida* (Chlorophyta). Journal of Applied Phycology 10: 383–389.
38. Fong P, Donohoe RM, Zedler JB (1994) Nutrient concentration in tissue of the macroalga enteromorpha as a function of nutrient history-an experimental evaluation using field microcosms. Marine Ecology Progress Series 106: 273–281.
39. Waite TD, Mitchell R (1972) Role of benthic plants in fertilized estuary. Journal of the Sanitary Engineering Division-Asce 98: 763–&.
40. McGlathery KJ, Pedersen MF, Borum J (1996) Changes in intracellular nitrogen pools and feedback controls on nitrogen uptake in *Chaetomorpha linum* (chlorophyta). Journal of Phycology 32: 393–401.
41. McClelland JW, Valiela I (1998) Changes in food web structure under the influence of increased anthropogenic nitrogen inputs to estuaries. Marine Ecology Progress Series 168: 259–271.
42. Pedersen MF (1994) Transient ammonium uptake in the macroalga *Ulva-lactuca* (Chlorophyta) - nature, regulation, and the consequences for choice of measuring technique. Journal of Phycology 30: 980–986.
43. Pedersen MF, Borum J (1997) Nutrient control of estuarine macroalgae: growth strategy and the balance between nitrogen requirements and uptake. Marine Ecology Progress Series 161: 155–163.
44. Phillips JC, Hurd CL (2004) Kinetics of nitrate, ammonium, and urea uptake by four intertidal seaweeds from New Zealand. Journal of Phycology 40: 534–545.
45. Touchette BW, Burkholder JM (2000) Review of nitrogen and phosphorus metabolism in seagrasses. Journal of Experimental Marine Biology and Ecology 250: 133–167.
46. Wen SS, Zhang HY, He WH, Zhang QJ, Xu SN, et al. (2008) Study on NH_4-N removing efficiency and kinetics in *Gracilaria asiatica*. Journal of Fisheries of China 32(5): 754–803.
47. Delia CF, Deboer JA (1978) Nutritional studies of 2 red algae.2. kinetics of ammonium and nitrate uptake. Journal of Phycology 14: 266–272.
48. Smit AJ (2002) Nitrogen uptake by *Gracilaria gracilis* (Rhodophyta): Adaptations to a temporally variable nitrogen environment. Botanica Marina 45: 196–209.
49. Li J, Huang L, Guo F, Cai A, Zheng Y (2007) Absorption of N, P Nutrients by *Gracilaria tenuistipitata* and its resistant effect on the outbreak of red tide. Journal of Xiamen U niversity (Natural Science) 46(2): 221–225.
50. Randall DJ, Tsui TKN (2002) Ammonia toxicity in fish. Marine Pollution Bulletin 45: 17–23.
51. Harrison PJ, Hurd CL (2001) Nutrient physiology of seaweeds: Application of concepts to aquaculture. Cahiers De Biologie Marine 42: 71–82.
52. Neori A, Chopin T, Troell M, Buschmann AH, Kraemer GP, et al. (2004) Integrated aquaculture: rationale, evolution and state of the art emphasizing seaweed biofiltration in modem mariculture. Aquaculture 231: 361–391.
53. Rosenberg G, Ramus J (1984) Uptake of inorganic nitrogen and seaweed surface-area-vloume ratios. Aquatic Botany 19: 65–72.
54. Hurd CL, Berges JA, Osborne J, Harrison PJ (1995) An in-vitro nitrate reductase assay for marine macroalgae - optimization and characterization of the enzyme for *fucus-gardneri* (Phaeophyta). Journal of Phycology 31: 835–843.
55. Zhou MJ, Zhu MY(2006) Progress of the Project"Ecology and Oceanography of Harmful Algal Blooms in China". Advances in Earth Science 7: 673–679.
56. Gordon DM, Birch PB, McComb AJ (1981) Effects of inoganic phosphorus and nitrogen on the growth of estuarine *cladophora* in culture. Botanica Marina 24: 93–106.
57. Xu D, Gao ZQ, Zhang XW, Fan X, Wang YT (2012) Allelopathic Interactions between the Opportunistic Species *Ulva prolifera* and the Native Macroalga *Gracilaria lichvoides*. Plos one 10.1371/journal.pone.0033648.
58. Delia CF, Deboer JA (1978) Nutritional studies of 2 red algae.2. kinetics of ammonium and nitrate uptake. Journal of Phycology 14: 266–272.
59. Hanisak MD, Harlin MM (1978) Uptake of inoganic nitrogen by *codium-fragile* subsp *tomentosoides* (Chlorophyta). Journal of Phycology 14: 450–454.
60. Haines KC, Wheeler PA (1978) Ammonium and nitrate uptake by marine macrophytes *hypnea-musciformis* (Rhodophyta) and *macrocystis-pyrifera* (Phaeophyta). Journal of Phycology 14: 319–324.

3

Small Changes in Gene Expression of Targeted Osmoregulatory Genes When Exposing Marine and Freshwater Threespine Stickleback (*Gasterosteus aculeatus*) to Abrupt Salinity Transfers

Annette Taugbøl[1]*, Tina Arntsen[1], Kjartan Østbye[1,2], Leif Asbjørn Vøllestad[1]

1 Centre for Ecological and Evolutionary Synthesis (CEES), Department of Biosciences, University of Oslo, Blindern, Norway, 2 Hedmark University College, Department of Forestry and Wildlife Management, Campus Evenstad, Elverum, Norway

Abstract

Salinity is one of the key factors that affects metabolism, survival and distribution of fish species, as all fish osmoregulate and euryhaline fish maintain osmotic differences between their extracellular fluid and either freshwater or seawater. The threespine stickleback (*Gasterosteus aculeatus*) is a euryhaline species with populations in both marine and freshwater environments, where the physiological and genomic basis for salinity tolerance adaptation is not fully understood. Therefore, our main objective in this study was to investigate gene expression of three targeted osmoregulatory genes (Na^+/K^+-ATPase (ATPA13), cystic fibrosis transmembrane regulator (CFTR) and a voltage gated potassium channel gene (KCNH4) and one stress related heat shock protein gene (HSP70)) in gill tissue from marine and freshwater populations when exposed to non-native salinity for periods ranging from five minutes to three weeks. Overall, the targeted genes showed highly plastic expression profiles, in addition the expression of ATP1A3 was slightly higher in saltwater adapted fish and KCNH4 and HSP70 had slightly higher expression in freshwater. As no pronounced changes were observed in the expression profiles of the targeted genes, this indicates that the osmoregulatory apparatuses of both the marine and landlocked freshwater stickleback population have not been environmentally canalized, but are able to respond plastically to abrupt salinity challenges.

Editor: Vincent Laudet, Ecole Normale Supérieure de Lyon, France

Funding: The study was supported by the Norwegian Research Council. The funders had no role in study design, data collection and analysis, decision to publish, or preparation of the manuscript.

Competing Interests: The authors have declared that no competing interests exist.

* Email: annette.taugbol@ibv.uio.no

Introduction

The ability to respond rapidly to environmental change is beneficial in variable environments, making phenotypically plastic organisms better adapted in unstable and unpredictable environments [1]. The capacity of an organism to respond to its environment is facilitated by the environmentally induced alteration of gene and protein expression. While the evolution of plasticity depends on the trait(s) in question and the source of environmental variation, there is a general acceptance that the ability to be plastic may be constrained by a variety of costs underlying the plastic responses [2]. As such, evolutionary theory predicts loss of plasticity after periods of environmental stability, when environmental constancy eliminates or weakens the source of selection that was formerly important for its maintenance, given that the cost for the trait is high [3], or through environmentally induced genetic assimilation [4] which reduces the environmental influence on trait expression.

Phenotypic plasticity of a trait is generally assumed to be under selection when a single organism is exposed to several environments during its lifetime which each select for different trait values. Most fish species are stenohaline, living either in fresh or salt water [5,6] where they are exposed to the same type of osmoregulatory challenge during their lifetime. For fish living in marine waters, the concentration of ions is much higher in the water compared to the environment inside the cell, and surrounding ions diffuse into the cell while water is lost. The situation is reversed for a freshwater fish, where the surroundings are ion depleted, making the fish passively loose ions and gain water. In order to maintain a relatively stable internal osmotic environment, fish counteract these effects by a variety of specialized physiological mechanisms, mainly in the gills [7] and the kidney [8], and these genetic adaptations can limit movement between salinities. Only a very limited number of species are truly euryhaline [6], being able to osmoregulate in a wide variety of salinity environments. Even fewer can tolerate extreme changes in osmolality over short time scales, such as the killifishes (*Fundulus spp.*) [9] and the threespine stickleback (*Gasterosteus aculeatus*) [10,11].

The threespine stickleback (order, Gasterosteiformes, family Gasterosteidae; hereafter stickleback) is a small fish that was originally a marine species [12]. However, since the last glaciation, sticklebacks have colonized a large number of brackish and freshwater systems throughout the northern hemisphere and are now occupying an extremely wide haloniche [11,13]. Many of the newly formed freshwater populations have become landlocked due to the isostatic uplifting of the land following deglaciation, and the stickleback in these habitats have consequently been separated from the sea for thousands of years. If the costs of having a plastic osmoregulatory machinery is high, it is expected that these landlocked freshwater stickleback populations should have lost the ability to osmoregulate efficiently in saltwater. However, studies suggest that freshwater populations of stickleback still possess the osmoregulatory machinery enabling them to handle abrupt changes in salinity [10,14], despite having been separated from the marine environment for up to 10–18 000 years [12]. This indicates that during adaptation to freshwater environments, the osmoregulatory physiology of landlocked sticklebacks has not been environmentally assimilated, or alternatively, the functionality of the osmoregulatory apparatus and its genomic architecture may not be open for selective change due to pleiotropic gene-interactions and is thus expected to remain similar in freshwater and marine populations.

One way to test if fish are adapted to a particular haloniche is to expose individuals to salinity challenges by transferring individuals from the original salinity to a test-salinity, tracking the expression of relevant genes through time. Earlier experiments show that stickleback easily tolerate transfers from freshwater to fully marine salinity, as well as the reverse [10,14]. However, it is not clear if the same osmoregulatory machinery is functioning at all times. The aim of this study was to assess the effect of experimental manipulation of salinity on the expression of genes important for osmoregulation. For this purpose, we collected adult fish from one marine and one freshwater site and exposed fish from each population to either 0 or 30 PSU (practical salinity units) for periods varying from 5 minutes to 3 weeks. The expression of four candidate genes was then followed through time (Fig. 1); three of the included genes are related to ion-pumps recognized to be under selection in marine-freshwater gradients (Na^+/K^+-ATPase (ATP1A3), cystic fibrosis transmembrane regulator (CFTR), voltage gated potassium channel gene (KCNH4)), and one is a stress related heat shock protein gene gene (HSP70), also associated with osmotic stress.

The objective of the study was to i) assess how target genes were expressed in the native salinity (assuming adaptation), ii) assess how target genes were affected when freshwater adapted fish were exposed to saltwater and iii) assess how target genes were affected when saltwater adapted fish were exposed to freshwater. As the osmoregulatory challenges are opposite in freshwater and marine environments, with ion secretion needed in saltwater and ion uptake needed in freshwater, we expect that osmoregulatory genes upregulated in freshwater will be downregulated in saltwater, and vice versa. We further expect the stress-related gene to have elevated expression levels in the beginning of the exposure for both groups due to handling and the physiological challenges associated with changing gene expression.

Materials and Methods

Fish and maintenance conditions

Adult stickleback were captured at two locations near Oslo, Norway (Figure 1a, b), during May and June 2010 and 2011. Fish from the marine population are known to breed there, and are not migratory, as many populations are known to be elsewhere [15,16,17]. The marine site (Sandspollen; 59° 39′ 58″N; 10° 35′ 11″ E) has a salinity that fluctuates between 22–29 PSU, while the freshwater pond (Glitredammen; 59° 55′ 53″N; 10° 29′ 55″ E; elevation 82.8 m above sea level) is stable at 0 PSU. The marine fish is comprised only of the completely plated morph (having a full row of lateral plates along its body flank), while the freshwater population only has the low plated morph (with lateral plates in the front region only). The two locations are geographically isolated by approximately 35 km by shortest distance through water, where about 8.5 km is through the river Sandvikselva that contains several steep waterfalls and dams. The age of the lake has been estimated at 7800 years before present using the program Sealevel32 [18]. The program uses information on postglacial land uplift and water level rise to estimate lake age. Downstream movement of fish from Glitredammen is possible, but upstream movement from the sea is impossible.

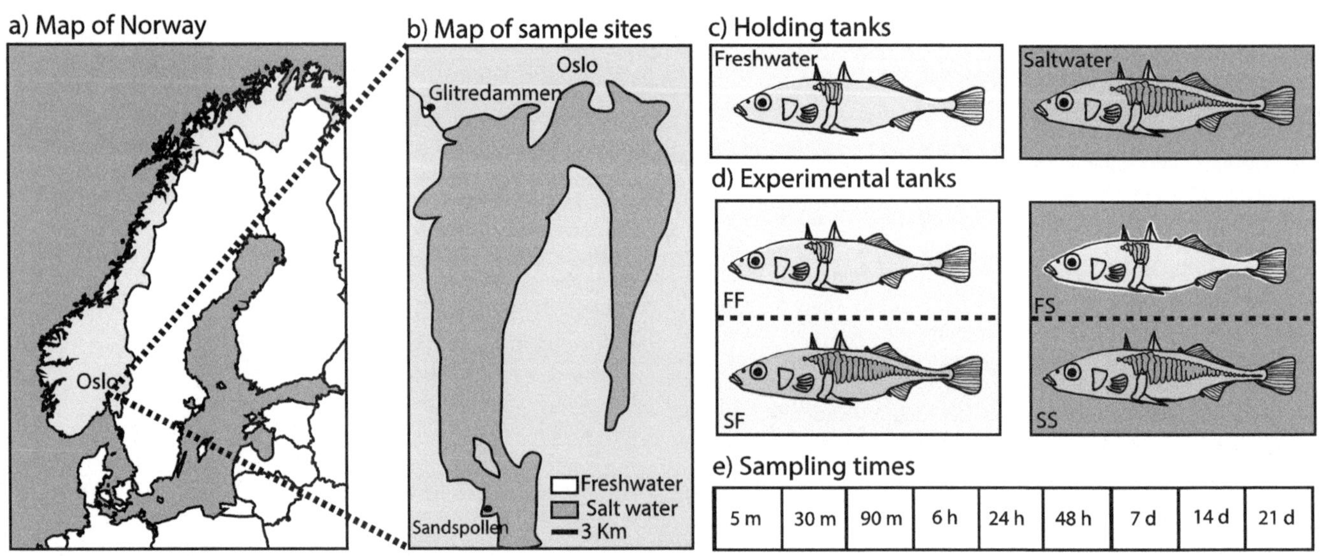

Figure 1. Study area and experimental design. a) Map of Norway showing the position of the sampling sites b) Locations of the two sampling sites, Glitredammen (freshwater) and Sandspollen (marine) c) Wild caught fish from both sampling locations were taken into the lab and placed in holding tanks of their native salinity for a minimum of three weeks. After acclimation, two groups of eight fish from both populations were exposed to either saltwater (30 PSU) or freshwater (0 PSU). The exposure tanks were divided in two by a perforated wall, so that both populations could be exposed to the same water quality at the same time d) Exposure times before the fish was collected and gill tissue was sampled.

After capture, the fish were transported to the aquarium facility at the University of Oslo and acclimated to holding conditions in their native salinity for minimum three weeks prior to the experiment. Two glass holding-tanks (500L) with either salt (30 PSU) or fresh water (0 PSU) were used for acclimation (Figure 1c), using biologically activated canister filters (EHEIM professional 3600), and UV-filtration. The acclimation tanks were covered with black plastic in front and on the sides to reduce visual stress. Further, to reduce potential male nesting behavior, the tanks were not equipped with any environmental enrichment, leaving the tanks free of sand and vegetation. The temperature in the tanks was maintained at room temperature (approx. 20°C) and the light regime was set at a 12:12 light:dark cycle. The fish were fed two times a day with frozen red bloodworms throughout the acclimation and exposure period.

Experimental design and protocol

The experimental setup consisted of 80 L tanks; the tanks were either filled with 0 PSU water or with 30 PSU water (Figure 1c), and covered with black sheets to reduce visual stress. A grey plastic wall divided each experimental tank into two 40 L compartments, where perforation ensured water movement between compartments (Figure 1d).

At the start of an experiment, 8 fish that appeared healthy were collected from each holding tank and placed directly in either the 30 or the 0 PSU experimental tank. Saltwater fish were therefore tested in either saltwater (SS; control) or in freshwater (SF). The freshwater fish were also tested in saltwater (FS) or in freshwater (FF; control) (Figure 1d).

The fish were exposed for different time periods, lasting between 5 minutes and 3 weeks (Figure 1e). The time periods were selected to cover short-term effects as well as long-term changes in gene expression. After each experiment, the fish were quickly netted out of the experimental tanks, immediately killed by a swift blow to the head and was directly processed for tissue collection.

Ethics statement

The experiment was approved by the Norwegian Animal experimentation and care committee (permit no ID 2705) and all efforts were made to minimize suffering.

Candidate gene expression

Candidate genes for osmoregulation were selected based on published studies on divergence in gene expression between marine and freshwater sticklebacks [19], studies identifying outlier regions in DNA sequences between marine and freshwater sticklebacks [20], and preliminary Illumina RNA-sequencing results (Table 1).

The targeted Na^+/K^+-ATPase gene, ATP1A3, has displayed salinity dependent regulation in fish when acclimated to different salinities, including killifish, *Fundulus heteroclitus* [21]. ATP1A3 is a plasma membrane protein that helps the establishment and maintenance of the electrochemical gradients of sodium and potassium ions across the plasma membrane by coupling the exchange of two extracellular K+ ions for three intracellular Na+ ions to the hydrolysis of one molecule of ATP [22], thereby ensuring a relatively constant osmolarity of cells and blood plasma. The protein is powering salt secretion in saltwater fish and salt absorption in freshwater fish [23].

The cystic fibrosis transmembrane regulator, CFTR, is an apical membrane anion channel involved in chloride secretion, and establishes an electrical driving force for trans-epithelial sodium secretion that generate the osmotic driving force for water flow, yielding an isotonic secretory product. As a candidate gene for saltwater adaptation [24], previous studies have also shown an upregulation of chloride cells and CFTR expression in a Hawaiian goby (*Stenogobius hawaiiensis*) [25] and in killifish [26] exposed to salt water.

While able to tolerate a wide range of salinity, whole genome sequencing of marine and freshwater sticklebacks have identified several chromosomal regions that have undergone parallel selection after freshwater invasion, indicating adaptive divergence and evolutionary change across the marine-freshwater boundary [20]. One identified region differing between marine and freshwater sticklebacks was an inversion with alternative functional exons of the voltage gated potassium channel gene, KCNH4, on either side, suggesting marine and freshwater specific isoforms [20]. However, although small, parallel changes in the sequences of genes may result from similar selection pressure across environments [27,28,29], it is the functional gene products and its regulation through expression that gives rise to the phenotype. Therefore, when candidate regions or loci linked to adaptive divergence have been identified, the regions should be tested in function, such as their role in gene regulation in a relevant ecological setting. The primer pairs in this study does not distinguish between marine and freshwater isoforms as the spanning ends of the inversion are identical down to a few base-pairs and the mutations are seemingly located within introns of the gene.

Both the physical handling when fish are transferred between tanks and the changes in water salinity are stressful, thus a stress-related heat-shock protein, HSP70, known to be affected by osmotic stress [30] was also included in the study.

To control for variation in expression levels not due to the experimental treatment, we used two reference genes, Elongation factor 1 alpha (EF1α) and Gluceraldehyde-3-phosphate dehydrogenase (GADPH). EF1α has been used successfully as reference gene in a similar study on sticklebacks [19] as well as in other gene expression studies on fish [26]. GADPH is a commonly used reference gene and has been stably expressed in a wide array of studies spanning predator cues [31] to exposure to offshore produced water [32].

Gene specific primers for target genes CFTR and reference genes GADPH and EF1α were previously designed and optimized (Table 1). Primers for additional target genes (ATP1A3, KCNH4 and HSP70) were designed based on genetic sequences from the Enseml genome browser and NCBI Primer-Blast (Table 1).

Tissue collection, RNA isolation, cDNA synthesis and qPCR

The gill plays an important role in the maintenance of blood ion and acid–base balance in both freshwater- and seawater acclimated fish [7,33,34]. After each fish was sacrificed, gill samples were immediately collected using sterilized tweezers and stored in RNA*later* (Ambion RNA, Life Technologies) according to the manufacturers protocol. The sampled fish were stored individually in 70% EtOH. The mRNA was isolated from the gill samples from each fish separately, using the mRNA direct kit (Invitrogen) as described by the manufacturer. The mRNA concentration and purity was quantified using Bioanalyzer (Agilent 2100 Bioanalyzer) and the Agilent RNA 6000 Pico Kit (Agilent Technologies) according to the protocol, and all samples were diluted down to 0.125 μg/μL before cDNA synthesis. The cDNA was prepared using the Superscript VILO cDNA synthesis kit (Invitrogen by Life Technologies) as described by the manufacturer, and the concentration was checked spectrophotometrically using Nanodrop (NanoDrop Teqnologies INC).

Table 1. Primers used for qPCR expression analysis of threespine stickleback genes.

Target	Gene name	Ensembl gene ID	Primer sequences	°C	E (%)	Reference
ATP1A3b	ATPase, Na+/K+ transporting, alpha 3b polypeptide	ENSGACG00000009524	F: AGCCGAGATCCCCTTCAACTCCA	60	99.07	*This study*
			R: GCTCCTTCCCCTGCACCAGGA			
CFTR	Cystic fibrosis transmembrane conductance regulator	ENSGACG00000009039	F: GCAGGCCTCTTCTTCACCAA	58	98.51	McCairns et al. (2009)
			R: TCCAGATAGAGGCTGATGTTCTTG			
KCNH4	Potassium voltage-gated channel subfamily H member 4	ENSGACG00000008648	F: CACAGTGACCTCTCTGGTGC	60	99.29	*This study*
			R: AGACATGAGCAGGGTCAGGA			
HSP70	Heat shock protein 70	ENSGACG00000013048	F: ATCGGTATTGACCTGGGCAC	60	99.20	*This study*
			R: GGTATCGGTGAACGCCACAT			
Reference						
EF1α	Elongation factor 1	ENSGACG00000002182	F: CATTGTCACTTACCTGAATCACATGA	60	99.26	McCairns et al. (2009)
			R: TGTGGCATTTAACAACATTTCCA			
GADPH	Glyceraldehyde-3-phosphate dehydrogenase	ENSGACG00000005864	F: CAAACCGTTGGTGACAGTATTTG	60	99.9	Sanago et al. (2011)
			R: GCACTGAGCATAAGGACACATCTAA			

The cDNA concentration was diluted down to 15 ng/μL (±1.5 ng/μL) prior to qPCR amplification after testing for optimization (standard curves, two-fold serial dilutions on pooled cDNA) and association curves for each primer pair (all primer pairs tested on dilution curves at 58 and 60°C). Primer efficiencies were calculated using the formula $E = (10^{-1/\text{slope}})-1$. All primer pairs had efficiencies between 95–100% and presented a single product, confirmed with a melting curve.

The qPCR reaction was performed on a Lightcycler 480 (Roche) using SYBR Green PCR Master Mix (Roche). Each 20 μL reaction contained 5.0 μL of the optimized concentrated cDNA, 1.0 μL of each primer, 10.0 μL of SYBR Green and 3.0 μL H2O. The thermocycle program included an enzyme activation step for 5 min, followed by 45 cycles of 95°C for 10 s, 58/60°C for 20 s and 72°C for 20 s. After the amplification phase, a dissociation curve was generated to confirm the presence of a single amplicon. The individual samples were run on duplicated plates, along with three negative reverse-transcriptase controls and an eightfold serial dilution to calibrate plate variation between runs. The obtained cycle threshold (C_q) values for the individuals were adjusted for plate efficiency, and duplicated reactions that differed by more than 0.5 C_q values were checked manually and removed from the analysis.

Statistical analysis

The target gene C_q values were normalized using the mean of the two control genes. Both the EF1α and GADPH were relatively stably expressed across the various time points but did differ somewhat between treatments. The C_q values for GADPH were slightly higher in the SS and SF treatments than in the other treatments ($F_{3,\ 237} = 6.64$, $P<0.001$), but did not differ between treatments for EF1α ($F_{3,\ 235} = 1.43$, $P = 0.236$). The relative expression levels were expressed as the individuals normalized C_q-values of the target transcript, and expressed relative to the mean values of a control group, here set to the 5 min exposure, for each treatment group (SS, SF, FF, FS). This method gives the fold change in expression relative to the control [35].

Variation in fold change of the expression of the different target genes was tested using general linear models. Each treatment group was tested at 9 different time points, where time can be classified both as a continuous variable (in minutes) as well as an ordinal factor (1–9). Preliminary analyses indicated that using continuous time was the better modeling approach and was therefore used in the model, expressed on a log-scale. To account for non-linear effects we included a squared term for time. The general model structure was thus:

$$Y = \text{Treatment} + \text{Time} + \text{Time}^2 + \text{Treatment} * \text{Time} + \text{Treatment} * \text{Time}^2$$

where treatment is the four different treatment types (SS, SF, FF, FS). The best (most parsimonious) model was selected using backward selection, using the Bayesian information criterion (BIC). BIC puts a heavier penalty on parameter number than the more commonly used Akaike information criterion (AIC).

Results

In general all experimental fish handled the transfer to the experimental water qualities well, both when transferred to the control water quality (groups SS and FF) and to the treatment salinity (SF, FS). A total of 9 fish from the freshwater population died across treatments during the experiment (6 in FF and 3 in FS), whereas no marine fish died. A total of 288 sticklebacks were used throughout this study.

We used the gene expression levels at 5 minutes of exposure as the control against which all fold level changes in expression was compared. Overall there were only minor differences in C_q – levels among the various treatment groups for the target genes (Table 2, Figure 2, Appendix S1).

The model that best explained variation in fold change in ATP-expression contained the ln-time and treatment factor ($F_{4,\ 204} = 14.636$; $P<0.0001$). Overall there was a tendency for the relative expression to increase with time, and the time-adjusted

Table 2. Gene expression of four osmoregulatory genes (see Table 1) (C_q-values; mean ± se) for marine and freshwater stickleback after 5 min exposure to either salt or freshwater.

Treatment	CFTR	ATP	KCNH4	HSP70
SS	1.15 3±0.009 (7)	1.111±0.009 (7)	1.415±0.013 (7)	1.272±0.020 (7)
SF	1.133±0.013 (4)	1.121±0.010 (6)	1.429±0.014 (6)	1.257±0.022 (6)
FF	1.158±0.009 (8)	1.126±0.008 (8)	1.413±0.012	1.196±0.019 (8)
FS	1.148±0.011 (6)	1.121±0.010 (6)	1.374±0.014 (6)	1.215±0.022 (6)
	$F_{3,\,21}=0.836$, $P=0.489$	$F_{3,\,23}=1.383$, $P=0.273$	$F_{3,\,23}=2.836$, $P=0.061$	$F_{3,\,23}=3.028$, $P=0.050$

Summary statistics from an analysis of variance for each gene is also given. Treatments are: SS (saltwater fish in saltwater), SF (saltwater fish in freshwater), FF (freshwater fish in freshwater) and FS (freshwater fish in saltwater).

mean fold change was slightly larger for the SS and FS (SS: 1.547±0.093; FS: 1.555±0.105) than for the FF and SF treatments (SF: 1.283±0.094; FF: 1.193±0.105). To further examine how the fold-change varied with treatment we re-ran the analyses grouping the fish into those tested in freshwater (FF, SF) and those tested in saltwater (SS, FS). Overall, the best model using this model-structure fit better to the data than the best model using treatment group (BIC treatment group =478.0; BIC treatment water type =466.0). The best model contained the interaction with time and water treatment type.

The model that best explained variation in fold change in CFTR-expression contained the ln-time and treatment factor ($F_{4,\,206}=13.041$; $P<0.0001$). Overall there was a tendency for the relative expression to increase with time, and the time-adjusted mean fold change was larger for the two control treatments (SS: 1.592±0.087; FF: 1.427±0.099) than for the transfer treatments (SF: 0.903±0.088; FS: 1.155±0.094). The transfer treatments and the control treatments differed significantly (Tukey HSD post hoc test, $P<0.05$).

The model that best explained variation in fold change in KCNH-expression contained the ln-time ($F_{1,207}=14.937$; $P=0.0001$) and treatment factor ($F_{3,207}=14.593$; $P<0.0001$). Overall there was a tendency for the relative expression to increase with time (0.087±0.023). To further examine how the fold-change varied with treatment we reran the analyses grouping the fish into those tested in freshwater (FF, SF) and those tested in saltwater (SS, FS). Overall the best model using this model structure fit better to the data than the best model using treatment group (BIC treatment group =527.1; BIC treatment water type =537.6). The best model contained time ($F_{1,209}=14.944$; $P<0.0001$ and water treatment type ($F_{1,209}=43.995$; $P<0.0001$). The KCNH-expression was significantly elevated in freshwater (1.488±0.079), whereas it was significantly decreased in saltwater (0.756±0.077).

The model that best explained variation in fold change in HSP70-expression only contained the treatment factor ($F_{3,\,203}=2.688$; $P=0.048$). Overall, mean fold change was larger than one for fish originating from saltwater (SS: 1.286±0.181; SF: 1.526±0.182) whereas if was smaller than one for the fish originating from freshwater (FF: 0.865±0.209; FS: 0.902±0.205). The fish from freshwater and saltwater differed significantly in expression level (Tukey HSD post hoc test, $P<0.05$). However, despite the expression levels being quite different between the ecotypes, the explanatory power of the model was low ($R^2=0.038$).

Discussion

Teleost fishes maintain nearly constant internal osmotic concentration and have osmoregulatory machinery fine-tuned to the external salinity in either salt or fresh water. However, some species can tolerate a wide range of environmental salinity, also with only short or no acclimation. Our main objective in this study was to investigate the expression of three relevant osmoregulatory genes and one stress related gene in marine and freshwater threespine stickleback, when exposed to non-native salinity for various periods. The study showed that both populations were capable of handling a direct transfer to a new and very different salinity. Survival was high in all treatments and the variation in gene expression was relatively small. This suggests that the capacity for osmoregulation in a wide range of salinity regimes has not been lost in either population. And yet, interestingly, no pronounced changes were observed in the expression profiles of the genes targeted in this study. This suggests that the ability of these fish to reverse the direction of their osmoregulation has not been canalized or lost.

Expression of ATP1A3 was elevated in saltwater

In this study, the ATP1A3 expression was lower in the FF treatment group at all time points, with a mean difference of 0.03 on an exponential scale compared to SS. Further, the SF group did show a weak down-regulation of ATP1A3 compared to SS and the expression levels stabilized around the FF values after approximately 24 hours. Comparing the FF and the FS, the FF had a consistently lower expression of ATP1A3, equally as the SS group.

As the Na^+/K^+-ATPase transporters both secrete and absorb salt in order to obtain a nearly constant internal osmotic concentration when in marine- and freshwater, respectively [7], it might be expected that the long-term expression of the protein stabilize on equal levels. Nevertheless, when a fish experiences a change in external osmolarity, the expression level is expected to shift in order to handle the novel osmotic and ionic stress: it is therefore surprising that we see so little change in the overall expression on the shorter time-scale in this study. Overall, the main finding of ATP1A3 being less expressed in freshwater is in accordance with most research on salmonid fish, where gill Na^+/K^+-ATPase activity is higher in seawater acclimated fish and decreases following migration into freshwater [36,37]; other fish species also show this pattern, including sea bass, *Dicentrarchus labrax* [38] and flounder, *Platichthys flesus* [39]. However, yet other studies have shown that the expression levels for the equivalent ATP isoform in Atlantic salmon (*Salmo salar*) did not change as a result of freshwater exposure [40], or as in killifish,

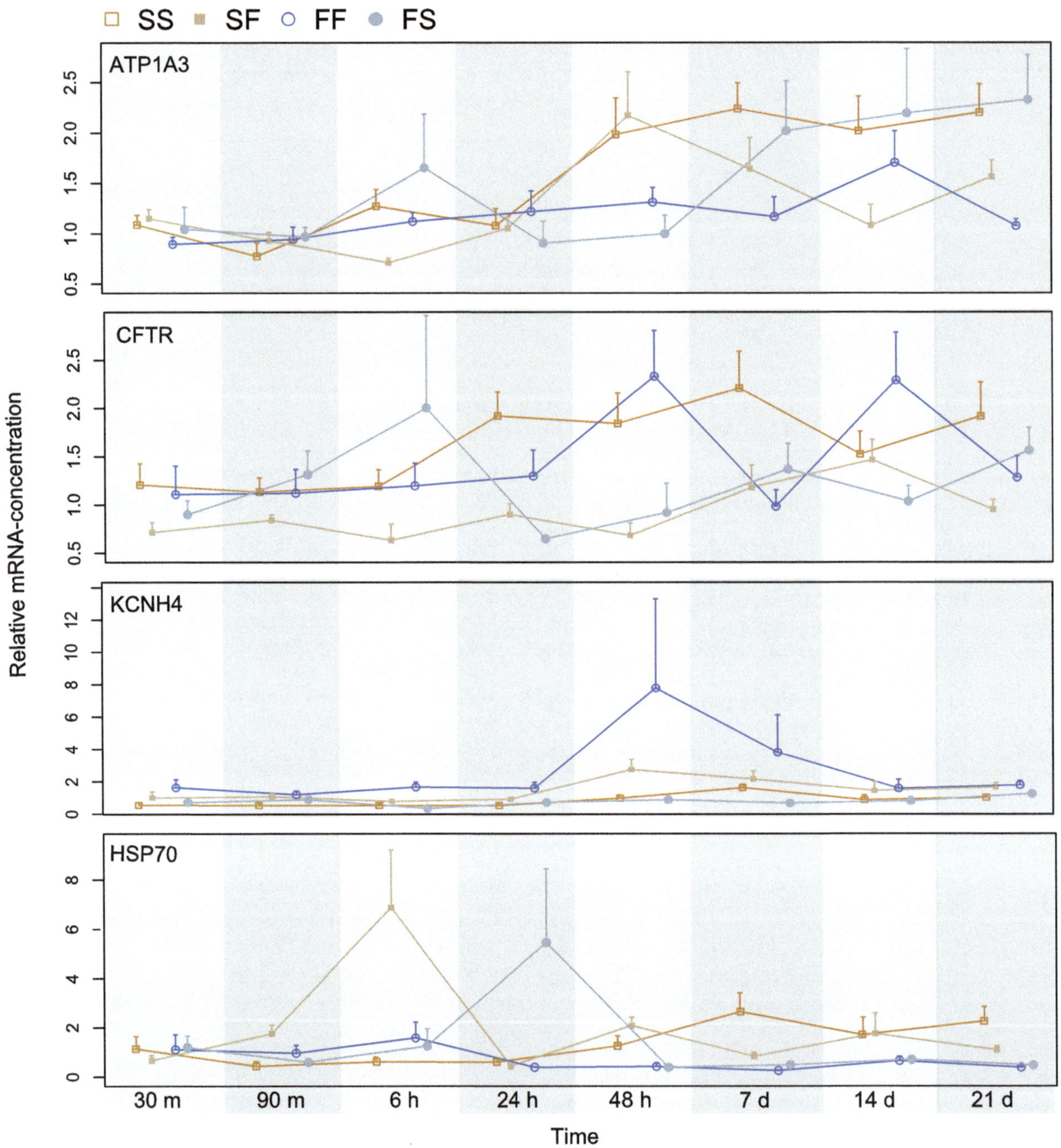

Figure 2. Relative mRNA-expression for the four targeted genes, ATP1A3, CFTR, KCNH4 and HSP70 in saltwater control (SS, dark orange), saltwater fish exposed to freshwater (SF, light orange), freshwater control (FF, dark blue) and freshwater fish exposed to saltwater (FS, light blue) for exposure periods relative to the 5 minute exposure. Values between 0–1 indicate lower expression, and values over 1 indicate higher expression relative to the expression at 5 minutes.

where ATP1A3 was up-regulated in freshwater [21]. Overall, previous reviews on fish osmoregulation state that the role of gill Na^+/K^+-ATPase is uncertain [41], unclear [7], or that the energy required for sodium uptake can be generated only by Na^+/K^+-ATPase [42]. It should however be noted that the cellular localization of the Na^+/K^+-ATPase transporter in this study is not known, and a histological analysis of the gene localization during a salinity challenge could provide additional information on the osmoregulatory function in stickleback.

No major changes in CFTR expression between treatments

Overall, only small changes in CFTR expression were observed across the treatments in the present study. CFTR had a slightly higher expression level in SS compared to FF, but only in the early time-periods (5 min to 24 hours), where the overall expression was reduced in both treatments compared to T = 5 min. After the first day, the expression stabilizes for both groups. Transferring marine fish into freshwater (SF) lead to an overall reduced expression of CFTR compared to the SS treatment, but this trend was not consistent across all time points. The freshwater fish transferred to saltwater (FS) had a higher expression of CFTR compared to FF for the first 6 hours, but was generally stable across all time points, indicating no major change in expression.

CFTR is thought to be central in the ion excretion at the gills of marine fish, as it is involved in the passive transport of chloride ions [24], and as such the expression is expected to decrease following freshwater acclimation. Anticipated increased expression of CFTR in apical membranes in response to transfer to saltwater have been illustrated in Atlantic salmon [43] and eel, *Anguilla anguilla*, [44]. However, contrary to expectations, another study on lab-reared stickleback observed a higher expression of CFTR in freshwater compared to saltwater [19]. Further, when comparing long-term expression levels of CFTR in saltwater exposed Atlantic salmon, the expression levels tended to decline towards the control after 30 days [43]. Although we would have expected a higher difference in the expression of CFTR, especially between the two control groups (SS and FF), there is much confounding evidence of CFTR-expression across taxa [45]; reported expression of CFTR in freshwater spans from not expressed at all in Mosambique tilapia, *Oreochromis mossambicus* [46], diffuse in Killifish [47] to no change in expression in striped bass, *Morone saxatilis* [48], indicating an overall complicated involvement of CFTR in freshwater osmoregulation [47]. An alternative to differential expression of the ion-transporter CFTR could be a redistribution and reuse of CFTR proteins, as Marshall et al. [47] illustrated movement of the protein from an apical location in SW to a more diffuse and basolateral location in FW. This could also be the case for the stickleback in this study, as a rearrangement of CFTR-proteins and hence its activity state would not be picked up by the qPCR analysis.

Much variation, but higher overall expression of KCNH4 in freshwater fish

Comparing the KCNH4-expression between the two control treatments (SS and FF), there was more temporal variation than would have been expected: KCNH4-expression decreased during the first 24 h in the SS treatment, whereas in the FF treatment the expression increased the first 48 h, before both seemed to be stabilized. The transfer of marine fish to freshwater (SF) followed approximately the same curve as for the SS treatment, and the transfer of freshwater fish to saltwater (FS) also demonstrate a down-regulated expression during the first 24 h, before expression increased reaching control levels (5 min exposure) after three weeks. Overall, the expression was higher in FF and FS compared to SS and SF. However, as there was no overall trend in the expression of KCNH4 for any of the groups in this study, this indicates that the gene is involved in other processes than osmoregulation, or that the osmoregulatory function of this transcript is located in other organs than the gill.

KCNH4 is a voltage-gated ion channel protein that is sensitive to voltage changes in the cell membrane and is known to have several functions, including regulation of cell volume, maintaining resting currents and affecting cardiac contractility [49]. Recent genome re-sequencing of 21 individual stickleback from marine and freshwater habitats across their global distribution revealed 81 loci underlying repeated parallel divergence in marine and freshwater ecotypes, including three chromosomal inversions [20]. One of the inversion sites has marine and freshwater specific 3′ exons of the KCNH4-gene, indicating parallel ecological selection on the breaking sites of the inversion, and possibly also directly on KCNH4 [20]. Although it is clear that chromosomal rearrangements can contribute to speciation [50,51,52], it is less evident how they do so and which mechanisms are involved. It is therefore important to identify whether inversion events have led to profound changes in the expression pattern of genes involved in the inversion, in relevant experimental setups.

While we do not have any information on the chromosomal arrangements in these two populations, it is still interesting to quantify the expression of the gene across different environments, as one would expect different expression profiles if the gene is under osmoregulatory selection. Chromosomal inversions are known to alter gene activity, either by causing non-functionalization of the gene, generating alternative splice sites or by altering gene regulatory networks [53,54]. In a study on development under different thermal selection regimes, one population of *Drosophila subobscura* had different expression patterns for loci located within and between inversions, where significant differences in expression tended to be more commonly found inside rather than outside the inversion [55]. The increased expression of KCNH4 in freshwater found in this study indicates a potential regulatory effect of the inversion on chromosome 1, however more studies are needed in order to disentangle the complete effect of the inversion.

Expression of HSP70 was elevated in freshwater adapted fish

HSP70 was identified as a candidate gene for detecting short-term osmotic stressful conditions in stickleback, with higher expression in freshwater compared to saltwater. When comparing SS and FF expression of HSP70, the marine fish had an overall lower expression before stabilizing after the first 24 h. Marine fish exposed to freshwater (SF) had an increased expression compared to SS the first 6 h before normalizing and freshwater fish exposed to saltwater (FS) had an overall lower expression compared to FF.

Capture, handling and crowding are all factors that can initiate a stress response in fish, as can short-term fluctuations in the physical environment. Physiological responses to stressors are complex, but include increased activity of cellular defense mechanisms, such as the up-regulation of HSP-genes [56]. The involvement of HSP70 in the acclimation of fish to salinity changes has been well documented experimentally [57,58,59]. Larsen et al. [59] observed a significant induction of HSP70 in kidney tissue of two populations of flounder, *Platichtys flesus*, when introduced to non-native salinities for both short- and long term exposures. However, the same study also illustrated tissue-specific up-regulation of HSP70, as expression in gill and liver was differentially affected by differences in salinity [59]. Other similar studies on Atlantic cod, *Gadus morhua*, illustrated expression differentiations in both gills and kidney after salinity transplantations during the first 24 hours [60], similar to the result in this study.

Conclusions and perspectives

In our study system, sticklebacks from the marine population are genetically (Østbye et al., unpublished data) and morphologically [61] differentiated from the freshwater population living in the river below the waterfall. However, as the populations are genetically differentiated, a surprising result of this study is how little variation in gene expression was observed when the fish were directly transferred to the contrasting salinity treatments, and additionally, how little it differed between the two populations in their native salinities.

Movement between environments of different salinity is physiologically costly [62], and resident populations experiencing different salinity levels are predicted to be locally adapted to their native habitat as traits promoting euryhalinity are expected to be rapidly lost if they are not under selection [63]. It is likely that adaption to the local environment takes time, but the freshwater fish in this study have been separated from the marine populations for more than 7000 years (between 3500 to 7000 generations assuming a two year or one year life cycle), likely sufficient time for local adaptation given reasonable selection [64,65]. Further, population genetic studies on stickleback have recognized salinity to be a major factor in the distribution of genotypes in systems that exchange migrants [66,67], also across high gene-flow environments such as in the Baltic ocean [68], indicating that adaption to salinity is under selection.

Based on survival alone, the results from this study suggest that over the>3500 generations of adaptions to freshwater environments, the osmoregulatory physiology of landlocked stickleback has not been significantly canalized or experienced strong selection as they have retained their capacity for osmoregulation in saltwater. Additionally the locally adapted marine stickleback can osmoregulate in freshwater, despite originating as a marine species. This indicates that for stickleback, the cost of retaining osmoregulatory plasticity is small, or that the traits promoting euryhalinity in stickleback are under strong selection or pleitropically linked to other traits under strong selection. However, for fish living in marine and freshwater environments, the selection pressure for osmoregulation is still opposite, indicating that the stickleback must have alternative cell-regulating mechanisms for survival in unfamiliar salinities. In this study we only targeted gene expression values by quantifying the amount of mRNA extracted from gill tissue, however, the amount of mRNA does not necessarily imply equal concentrations of the functional proteins as a number of mechanisms can limit or increase the production. Protein expression profiles, also from kidney tissue, could have revealed different results. Other alternative strategies the stickleback may be utilizing could include changes in activity state of the ion-transport proteins by movement within the cell (activation/down-regulation), or by re-using the proteins by reversing the orientation in the cell-membrane, or by modulations following the interaction of other proteins [69,70,71].

Whatever method it is that the stickleback seems to be employing to osmoregulate so effectively, it has created a species that is incredibly well able to colonize new habitats regardless of the salinity they find themselves in, which has been a huge asset to this species in its spread throughout the northern hemisphere. Additional studies targeting the exact genetic and physiological mechanisms for the wide salinity tolerance in marine and freshwater stickleback are needed to understand the stickleback's incredible capacity for ion secretion and absorption. Their ability to adapt immediately to the environmental demands, with no apparent increase in physiological stress, is as unusual as it is intriguing, especially as this has been so evolutionarily important for this widespread species.

Acknowledgments

We thank Anders Herland and Haaken Hveding Christensen for assistance with fish maintenance, Monica Solbakken, Martin Malmstrøm, Nanna Winger Steen, Ave Tooming-Klunderud and Mari Espelund for help in the lab, Kjetill S. Jakobsen, Alexander J. Nederbragt and Bastian Staar for interesting discussions and ideas and Anna Mazzarella for constructive comments to the manuscript.

Author Contributions

Conceived and designed the experiments: AT LAV KØ. Performed the experiments: AT TA KØ. Analyzed the data: AT TA LAV. Wrote the paper: AT LAV.

References

1. West-Eberhard MJ (2005) Phenotypic accommodation: adaptive innovation due to developmental plasticity. Journal of Experimental Zoology Part B-Molecular and Developmental Evolution 304B: 610–618.
2. DeWitt TJ, Sih A, Wilson DS (1998) Costs and limits of phenotypic plasticity. Trends in Ecology & Evolution 13: 77–81.
3. Masel J, King OD, Maughan H (2007) The loss of adaptive plasticity during long periods of environmental stasis. American Naturalist 169: 38–46.
4. Lande R (2009) Adaptation to an extraordinary environment by evolution of phenotypic plasticity and genetic assimilation. Journal of Evolutionary Biology 22: 1435–1446.
5. Schultz ET, McCormick SD (2013) Euryhalinity in an evolutionary context. In: McCormick SD, Farrell AP, Brauner CJ, editors. Fish physiology: Euryhaline fishes. Oxford: Elsevier Science. pp. 477–529.
6. Edwards SL, Marshall WS (2012) Principles and patterns of osmoregulation and euryhalinity in fishes. In: Stephen D. McCormick APF, Colin JB, editors. Fish Physiology: Academic Press. pp. 1–44.
7. Evans DH, Piermarini PM, Choe KP (2005) The multifunctional fish gill: dominant site of gas exchange, osmoregulation, acid-base regulation, and excretion of nitrogenous waste. Physiological Reviews 85: 97–177.
8. Varsamos S, Nebel C, Charmantier G (2005) Ontogeny of osmoregulation in postembryonic fish: A review. Comparative Biochemistry and Physiology a-Molecular & Integrative Physiology 141: 401–429.
9. Griffith RW (1974) Environment and salinity tolerance in the genus *Fundulus*. Copeia 1974: 319–331.
10. Heuts MJ (1947) Experimental studies on adaptive evolution in *Gaserosteus-aculeatus L.* Evolution 1: 89–102.
11. Wootton RJ (1976) The biology of the sticklebacks. New York: Academic Press.
12. Bell MA (1977) Late Miocene marine Threespine stickleback, *Gasterosteus aculeatus*, and its zoogeographic and evolutionary significance. Copeia: 277–282.
13. Bell MA, Foster SA (1994) The evolutionary biology of the threespine stickleback. New York: Oxford University Press.
14. Grøtan K, Østbye K, Taugbøl A, Vøllestad LA (2012) No short-term effect of salinity on oxygen consumption in threespine stickleback (*Gasterosteus aculeatus*) from fresh, brackish, and salt water. Canadian Journal of Zoology 90: 1386–1393.
15. Raeymaekers JAM, Maes GE, Audenaert E, Volckaert FAM (2005) Detecting Holocene divergence in the anadromous-freshwater three-spined stickleback (*Gasterosteus aculeatus*) system. Molecular Ecology 14: 1001–1014.
16. Von Hippel FA, Weigner H (2004) Sympatric anadromous-resident pairs of threespine stickleback species in young lakes and streams at Bering Glacier, Alaska. Behaviour 141: 1441–1464.
17. McPhail JD (1994) Speciation and the evolution of reproductive isolation in the sticklebacks (*Gasterosteus*) of south-western British Columbia. In: Bell AM, Foster JR, editors. The evolutionary biology of the threespine stickleback. Oxford: Oxford University Press. pp. 399–437.
18. Møller JJ (2003) Relative sea level change in Fennoscandia. Net version 3.00. University of Tromsø: Department of Geology, TMU.
19. McCairns RJS, Bernatchez L (2010) Adaptive divergence between freshwater and marine sticklebacks: Insights into the role of phenotypic plasticity from an intergrated analysis of candidate gene expression. Evolution 64: 1029–1047.
20. Jones FC, Grabherr MG, Chan YF, Russell P, Mauceli E, et al. (2012) The genomic basis of adaptive evolution in threespine sticklebacks. Nature 484: 55–61.
21. Whitehead A, Roach JL, Zhang SJ, Galvez F (2012) Salinity- and population-dependent genome regulatory response during osmotic acclimation in the killifish (*Fundulus heteroclitus*) gill. Journal of Experimental Biology 215: 1293–1305.
22. Mobasheri A, Avila J, Cozar-Castellano I, Brownleader MD, Trevan M, et al. (2000) Na+, K+-ATPase isozyme diversity; Comparative biochemistry and physiological implications of novel functional interactions. Bioscience Reports 20: 51–91.

23. Bonting SL (1970) Sodium-pottassium activated adenosine triphosphatase and cation transport. In: Bittar EE, editor. Membranes and ion transport. New York: John Wiley & Sons. pp. 257–363.
24. Silva P, Solomon R, Spokes K, Epstein FH (1977) Ouabain inhibition of gill Na-K ATPase: relationship to active chloride transport. Journal of Experimental Zoology 199: 419–426.
25. McCormick SD, Sundell K, Bjornsson BT, Brown CL, Hiroi J (2003) Influence of salinity on the localization of Na+/K+-ATPase, Na+/K+/2Cl(−)cotransporter (NKCC) and CFTR anion channel in chloride cells of the Hawaiian goby (*Stenogobius hawaiiensis*). Journal of Experimental Biology 206: 4575–4583.
26. Scott GR, Richards JG, Forbush B, Isenring P, Schulte PM (2004) Changes in gene expression in gills of the euryhaline killifish *Fundulus heteroclitus* after abrupt salinity transfer. American Journal of Physiology-Cell Physiology 287: C300–C309.
27. Hoekstra HE (2006) Genetics, development and evolution of adaptive pigmentation in vertebrates. Heredity 97: 222–234.
28. Chan YF, Marks ME, Jones FC, Villarreal G, Shapiro MD, et al. (2010) Adaptive evolution of pelvic reduction in Sticklebacks by recurrent deletion of a Pitx1 enhancer. Science 327: 302–305.
29. Rosenblum EB, Roempler H, Schoeneberg T, Hoekstra HE (2010) Molecular and functional basis of phenotypic convergence in white lizards at White Sands. Proceedings of the National Academy of Sciences of the United States of America 107: 2113–2117.
30. Sørensen J, Kristensen T, Loeschcke V (2003) The evolutionary and ecological role of heat shock proteins. Ecology Letters 6: 1025–1037.
31. Sanogo YO, Hankison S, Band M, Obregon A, Bell AM (2011) Brain transcriptomic response of threespine sticklebacks to cues of a predator. Brain Behavior and Evolution 77: 270–285.
32. Knag AC, Taugbøl A (2013) Acute exposure to offshore produced water has an effect on stress- and secondary stress responses in three-spined stickleback *Gasterosteus aculeatus*. Comparative Biochemistry and Physiology C-Toxicology & Pharmacology 158: 173–180.
33. Krogh A (1937) Osmotic regulation in fresh water fishes by active absorption of chloride ions. Zeitschrift vergleichende Physiologie 24: 656–666.
34. Evans DH (2008) Teleost fish osmoregulation: what have we learned since August Krogh, Homer Smith, and Ancel Keys. American Journal of Physiology-Regulatory Integrative and Comparative Physiology 295: R704–R713.
35. Livak KJ, Schmittgen TD (2001) Analysis of relative gene expression data using real-time quantitative PCR and the 2(T)(-Delta Delta C) method. Methods 25: 402–408.
36. Bystriansky JS, Schulte PM (2011) Changes in gill H+-ATPase and Na+/K+-ATPase expression and activity during freshwater acclimation of Atlantic salmon (*Salmo salar*). Journal of Experimental Biology 214: 2435–2442.
37. Shrimpton JM, Patterson DA, Richards JG, Cooke SJ, Schulte PM, et al. (2005) Ionoregulatory changes in different populations of maturing sockeye salmon *Oncorhynchus nerka* during ocean and river migration. Journal of Experimental Biology 208: 4069–4078.
38. Jensen MK, Madsen SS, Kristiansen K (1998) Osmoregulation and salinity effects on the expression and activity of Na+, K+-ATPase in the gills of European sea bass, *Dicentrarchus labrax* (L.). Journal of Experimental Zoology 282: 290–300.
39. Stagg RM, Shuttleworth TJ (1982) Na+, K+ ATPase, quabain binding and quabain-sensitive oxygen consumption in gills from*Platichthys flesus* adapted to seawater and freshwater. Journal of comparative physiology 147: 93–99.
40. Folmar LC, Dickhoff WW (1980) The parr-smolt transformation (smoltification) and seawater adaptation in salmonids: A review of selected literature. Aquaculture 21: 1–37.
41. Perry SF (1997) The chloride cell: Structure and function in the gills of freshwater fishes. Annual Review of Physiology 59: 325–347.
42. Kirschner LB (2004) The mechanism of sodium chloride uptake in hyperregulating aquatic animals. Journal of Experimental Biology 207: 1439–1452.
43. Singer TD, Clements KM, Semple JW, Schulte PM, Bystriansky JS, et al. (2002) Seawater tolerance and gene expression in two strains of Atlantic salmon smolts. Canadian Journal of Fisheries and Aquatic Sciences 59: 125–135.
44. Wilson JM, Antunes JC, Bouca PD, Coimbra J (2004) Osmoregulatory plasticity of the glass eel of *Anguilla anguilla*: freshwater entry and changes in branchial ion-transport protein expression. Canadian Journal of Fisheries and Aquatic Sciences 61: 432–442.
45. Havird JC, Henry RP, Wilson AE (2013) Altered expression of Na+/K+-ATPase and other osmoregulatory genes in the gills of euryhaline animals in response to salinity transfer: A meta-analysis of 59 quantitative PCR studies over 10 years. Comparative Biochemistry and Physiology D-Genomics & Proteomics 8: 131–140.
46. Hiroi J, McCormick SD, Ohtani-Kaneko R, Kaneko T (2005) Functional classification of mitochondrion-rich cells in euryhaline Mozambique tilapia (*Oreochromis mossambicus*) embryos, by means of triple immunofluorescence staining for Na/K+-ATPase, Na+/K+/2Cl(−) cotransporter and CFTR anion channel. Journal of Experimental Biology 208: 2023–2036.
47. Marshall WS, Lynch EA, Cozzi RRF (2002) Redistribution of immunofluorescence of CFTR anion channel and NKCC cotransporter in chloride cells during adaptation of the killifish *Fundulus heteroclitus* to sea water. Journal of Experimental Biology 205: 1265–1273.
48. Madsen SS, Jensen LN, Tipsmark CK, Kiilerich P, Borski RJ (2007) Differential regulation of cystic fibrosis transmembrane conductance regulator and Na+, K+-ATPase in gills of striped bass, *Morone saxatilis*: effect of salinity and hormones. Journal of Endocrinology 192: 249–260.
49. Gutman GA, Chandy KG, Grissmer S, Lazdunski M, McKinnon D, et al. (2005) International union of pharmacology. LIII. Nomenclature and molecular relationships of voltage-gated potassium channels. Pharmacological Reviews 57: 473–508.
50. Ellegren H, Smeds L, Burri R, Olason PI, Backstrom N, et al. (2012) The genomic landscape of species divergence in *Ficedula* flycatchers. Nature 491: 756–760.
51. Rieseberg LH (2001) Chromosomal rearrangements and speciation. Trends in Ecology & Evolution 16: 351–358.
52. Noor MAF, Grams KL, Bertucci LA, Reiland J (2001) Chromosomal inversions and the reproductive isolation of species. Proceedings of the National Academy of Sciences of the United States of America 98: 12084–12088.
53. Kirkpatrick M, Barton N (2006) Chromosome inversions, local adaptation and speciation. Genetics 173: 419–434.
54. Matzkin LM, Merritt TJS, Zhu CT, Eanes WF (2005) The structure and population genetics of the breakpoints associated with the cosmopolitan chromosomal inversion In(3R)Payne in *Drosophila melanogaster*. Genetics 170: 1143–1152.
55. Laayouni H, Garcia-Franco F, Chavez-Sandoval BE, Trotta V, Beltran S, et al. (2007) Thermal evolution of gene expression profiles in Drosophila subobscura. Bmc Evolutionary Biology 7.
56. Moseley P (2000) Stress proteins and the immune response. Immunopharmacology 48: 299–302.
57. Deane E, Kelly S, Luk J, Woo N (2002) Chronic salinity adaptation modulates hepatic heat shock protein and insulin-like growth factor I expression in black sea bream. Marine Biotechnology 4: 193–205.
58. Fangue NA, Hofmeister M, Schulte PM (2006) Intraspecific variation in thermal tolerance and heat shock protein gene expression in common killifish, Fundulus heteroclitus. Journal of Experimental Biology 209: 2859–2872.
59. Larsen PF, Nielsen EE, Williams TD, Loeschcke V (2008) Intraspecific variation in expression of candidate genes for osmoregulation, heme biosynthesis and stress resistance suggests local adaptation in European flounder (*Platichthys flesus*). Heredity 101: 247–259.
60. Larsen PF, Nielsen EE, Meier K, Olsvik PA, Hansen MM, et al. (2012) Differences in salinity tolerance and gene expression between two populations of Atlantic Cod (*Gadus morhua*) in response to salinity stress. Biochemical Genetics 50: 454–466.
61. Bjærke O, Østbye K, Lampe HM, Vøllestad LA (2010) Covariation in shape and foraging behaviour in lateral plate morphs in the three-spined stickleback. Ecology of Freshwater Fish 19: 249–256.
62. Moyle PB, Cech JJ (1996) Fishes. An introduction to ichthyology. 3rd edition. Upper Saddle River, N.J.: Prentice Hall.
63. Schultz ET, McCormick SD (2012) Euryhalinity in an evolutionary context. In: Stephen D. McCormick APF, Colin JB, editors. Fish physiology: Academic Press. pp. 477–533.
64. Kinnison MT, Hendry AP (2001) The pace of modern life II: from rates of contemporary microevolution to pattern and process. Genetica 112: 145–164.
65. Hendry AP, Kinnison MT (1999) Perspective: The pace of modern life: Measuring rates of contemporary microevolution. Evolution 53: 1637–1653.
66. McCairns RJS, Bernatchez L (2008) Landscape genetic analyses reveal cryptic population structure and putative selection gradients in a large-scale estuarine environment. Molecular Ecology 17: 3901–3916.
67. Taugbøl A, Junge C, Quinn TP, Herland A, Vøllestad LA (2014) Genetic and morphometric divergence in threespine stickleback in the Chignik catchment, Alaska. Ecology and Evolution 4: 144–156.
68. DeFaveri J, Jonsson PR, Merilä J (2013) Heterogeneous genomic differentiation in marine threespine stickleback: Adaptation along an environmental gradient. Evolution 67: 2530–2546.
69. Pertl H, Pockl M, Blaschke C, Obermeyer G (2010) Osmoregulation in Lilium pollen grains occurs via modulation of the plasma membrane H+ ATPase activity by 14-3-3 proteins. Plant Physiology 154: 1921–1928.
70. Szczesnaskorupa E, Browne N, Mead D, Kemper B (1988) Positive charges at the NH2 terminus convert the membrane-anchor signal peptide of cytochrome P-450 to a secretory signal peptide. Proceedings of the National Academy of Sciences of the United States of America 85: 738–742.
71. Hartmann E, Rapoport TA, Lodish HF (1989) Predicting the orientation of eukaryotic membrane-spanning proteins. Proceedings of the National Academy of Sciences of the United States of America 86: 5786–5790.

4

Integrated Assessment of Heavy Metal Pollution in the Surface Sediments of the Laizhou Bay and the Coastal Waters of the Zhangzi Island, China: Comparison among Typical Marine Sediment Quality Indices

Wen Zhuang[1,2,3], **Xuelu Gao**[1]*

1 Key Laboratory of Coastal Environmental Processes and Ecological Remediation, Yantai Institute of Coastal Zone Research, Chinese Academy of Sciences, Yantai, Shandong, China, **2** College of City and Architecture Engineering, Zaozhuang University, Zaozhuang, Shandong, China, **3** University of Chinese Academy of Sciences, Beijing, China

Abstract

The total concentrations and chemical forms of heavy metals (Cd, Cr, Cu, Ni, Pb and Zn) in the surface sediments of the Laizhou Bay and the surrounding marine area of the Zhangzi Island (hereafter referred to as Zhangzi Island for short) were obtained and multiple indices and guidelines were applied to assess their contamination and ecological risks. The sedimentary conditions were fine in both of the two studied areas according to the marine sediment quality of China. Whereas the probable effects level guideline suggested that Ni might cause adverse biological effects to occur frequently in some sites. All indices used suggested that Cd posed the highest environmental risk in both the Laizhou Bay and the Zhangzi Island, though Cd may unlikely be harmful to human and ecological health due to the very low total concentrations. The enrichment factor (EF) showed that a substantial portion of Cr was delivered from anthropogenic sources, whereas the risk assessment code (RAC) indicated that most Cr was in an inactive state that it may not have any adverse effect either. Moreover, the results of EF and geoaccumulation index were consistent with the trend of the total metal concentrations except for Cd, while the results of RAC and potential ecological risk factor did not follow the same trend of their corresponding total metal concentrations. We also evaluated the effects of using different indices to assess the environmental impact of these heavy metals.

Editor: Wei-Chun Chin, University of California, Merced, United States of America

Funding: This study was co-supported by the National Natural Science Foundation of China (41376083), the Department of Science and Technology of Shandong Province (2012GHY11535) and the CAS/SAFEA International Partnership Program for Creative Research Teams (Representative Environmental Processes and Resources Effects in Coastal Zones). The funders had no role in study design, data collection and analysis, decision to publish, or preparation of the manuscript.

Competing Interests: The authors have declared that no competing interests exist.

* E-mail: xlgao@yic.ac.cn

Introduction

Heavy metals in ecosystems have received extensive attention because they are toxic, non-biodegradable in the environment and easy to accumulate and magnify in organisms. Concentrations of heavy metals in aquatic ecosystems have increased considerably due to the inputs of industrial waste, sewage runoff and agriculture discharges [1,2]. In other words, heavy metal pollution may likely go with the rapid economic development [3,4]. The measurements of heavy metals only in the water and in the suspended material are not conclusive due to water discharge fluctuations and low resident time [5]. With a combined action of adsorption, hydrolysis and co-precipitation, only a small part of free metal ions stay dissolved in water, and a large quantity of them get deposited in the sediments [6]. However, when environmental conditions change, sediments may transform from the main sink of heavy metals to sources of them for the overlying waters [1,7]. Therefore the contents of heavy metals in sediments are often monitored to provide basic information for environmental risk assessment [8,9].

In recent decades, various risk assessment indices have been applied to evaluate the environmental risks of metals in marine sediments. Caeiro et al. [10] classified them into three types: contamination indices, background enrichment indices and ecological risk indices. In the present study contamination indices and background enrichment indices were collectively called contamination indices. To assess the metal contamination, the geoaccumulation index (I_{geo}) [5,11] and the enrichment factor (EF) [3,12] are often used. Meanwhile, the risk assessment code (RAC) [13] and the potential ecological risk index (ER) [14] are very popular indices in evaluating the ecological risk posed by heavy metals in sediments. Therefore, these four indices were employed to assess the contamination and ecological risks of the six selected metals (i.e., Cd, Cr, Cu, Ni, Pb and Zn) in the surface sediments of the Laizhou Bay and the coastal waters of the Zhangzi Island (hereafter also referred to as Zhangzi Island for short), China. Numerous sediment quality guidelines (SQGs) also have been developed to deal with environmental concerns. Marine Sediment Quality of China (GB 18668-2002) [9] is one of the SQGs usually used as a general measure of marine sediment contamination in

China, so this guideline was chosen in this study. TEL (threshold effects level) and PEL (probable effects level), which were proved to be effective sediment quality guidelines [15,16], were also used in this study. Many studies have shown that different conclusions or even contradictory conclusions may be drawn by using different risk assessment methods for the same sample or for different elements within the same sample [10,13,17]. Therefore, the relationships among the four index methods and the total concentrations of metals were explored to find out their differences in the environmental risk assessment of heavy metals. Since all the four indices are very popular in evaluating the environmental risks posed by heavy metals in sediments all over the world, we hope to provide useful information about these indices for other researchers to refer to when they carry out similar studies.

To sum up, the purposes of this study are i) to quantify and explain the spatial distribution and fractionations of six heavy metals (Cd, Cr, Cu, Ni, Pb and Zn) in the surface sediments of the Laizhou Bay and the Zhangzi Island; ii) to explore the degree of contamination and the potential ecological risks of these heavy metals to the environment; and iii) to investigate the differences among the used risk assessment indices and the SQGs in the environmental risk assessment of heavy metals.

Materials and Methods

Ethics statement

This study did not involve endangered or protected species and no specific permissions were required for these locations/activities in this study. The specific locations of the present study were shown in Fig. 1.

Study area

The Laizhou Bay (area–7000 km^2, coastline length–320 km, mean depth <10 m, max. depth–18 m) lies in the southern part of the Bohai Sea, accounting for up to 10% of the total area (Fig. 1). It is a semi-closed shallow area with relatively flat seafloor which is formed by the accumulation of riverine suspended matters. There are more than a dozen of rivers running into the Laizhou Bay, among which the Yellow River and the Xiaoqinghe River influence the Laizhou Bay most. Riverine sediment load of the Yellow River and the Xiaoqinghe River began to decrease from the second half of the 20th century, whereas the amount of the contaminants brought by them increased with years [18,19,20]. From the western coast to the eastern coast of the Laizhou Bay, there in turn have Dongying Port, Yangjiaogou Port, Weifang Port and Longkou Port which are important ports in Shandong Province. Due to the abundant seawater resources and underground brine resources in the coastal Laizhou Bay, one of the biggest chemical industrial bases in the world called Weifang Binhai Economic Development Zone is located along its southwestern coast. More than 400 chemical enterprises are located nearby and large amounts of non-purified or insufficiently purified wastewaters are discharged into the Laizhou Bay. The Laizhou Bay can be characterized as a region surrounded by areas with high population growth and rapid economic development in China. The Laizhou Bay used to be one of the most important spawning and breeding grounds for many marine organisms in China. The rapid economic development has brought serious ecological damages to the Laizhou Bay, and fishery resources in the Laizhou Bay are gradually disappearing. The overall sharp decrease of fishery resources and unpredictable nature of the sediment highlight the necessity of the environmental risk assessments of pollutants, especially the heavy metals possessing high affinities for sedimentary materials.

The Zhangzi Island (coastline length–60 km) is located in a national first-class clean sea area in the northern North Yellow Sea which is ~100 km away from Dalian City, Liaoning Province (Fig. 1). The offshore area of the Zhangzi Island is the largest aquaculture base for choice rare seafood in China, which produces conchs, sea cucumbers, scallops, abalones, sea urchins and so on. The only national original seed field for *Patinopecten yessoensis* is located in this area. Fodder-feeding has been banned from seafood farming for many years in this area, and the cultivation of crops on the island is also prohibited. It is thought that natural environment in the Zhangzi Island is generally better than other typical Chinese coastal seas such as the Laizhou Bay. However, sewage (residual feeds, excrement and suspended particles, etc.) discharged during the process of raising seedlings and the frequent operation of ships may result in the deterioration of the local waters and sediment qualities.

Sampling

In this study, a total of 18 surface sediment samples were collected in the Laizhou Bay in October 2011 and 7 surface sediment samples were collected in the Zhangzi Island in November 2011 (Fig. 1). In the Laizhou Bay, sampling sites L1 and L6 were near to the new and old mouths of the Yellow River, respectively; site L18 was near to the estuary of the Xiaoqinghe River. In the Zhangzi Island, the sampling sites were chosen stochastically. Three sites were located in the intertidal zone (Z1–Z3); three were in the coastal waters (Z4–Z6), among which Z4 and Z5 were located in the mariculture areas where sea cucumbers and scallops were farmed, respectively. Site Z7 was in the intertidal zone of an outer island called the Dalian Island. Surface sediment samples (~0–5 cm) were collected by a stainless steel grab sampler and/or a plastic spatula, and were placed in acid-rinsed polyethylene bags. They were transported to the laboratory in a cooler box with ice packs and stored at 4 °C until further treatment.

Analytical methods

The information about the fractionations of metals in the surface sediments was obtained by a sequential extraction procedure reported by Rauret et al. [21]. The four operationally defined geochemical fractions which were separated under this scheme are acid soluble, reducible, oxidizable and residual. The detailed sequential extraction protocol used in this study has been described elsewhere [22].

Previous experiments have shown that sample drying could alter the solid phase distribution of trace elements [23,24]. Furthermore, the elemental concentrations in sediments are highly dependent on the grain size [25,26]. So a drying and grinding treatment could potentially alter the extractability of elements [27]. For the reasons above, wet and unground sediments were used for the sequential extraction procedure in this study to reduce errors.

The mixture of concentrated HF, HNO_3 and $HClO_4$ (5:2:1) [22] was used to digest five randomly selected residues instead of the so-called pseudototal digestion with aqua regia used by Rauret et al. [21]. The total metal concentrations in all samples were obtained by the same method used to get the metal concentrations in the residual fraction. The concentrations of metals in residual fractions were estimated by subtracting the metal concentrations obtained in the first three steps of sequential extraction from the total metal concentrations. The sum of the measured values of the four geochemical fractions accounted for 85–110% of the values from the total digestion experiment.

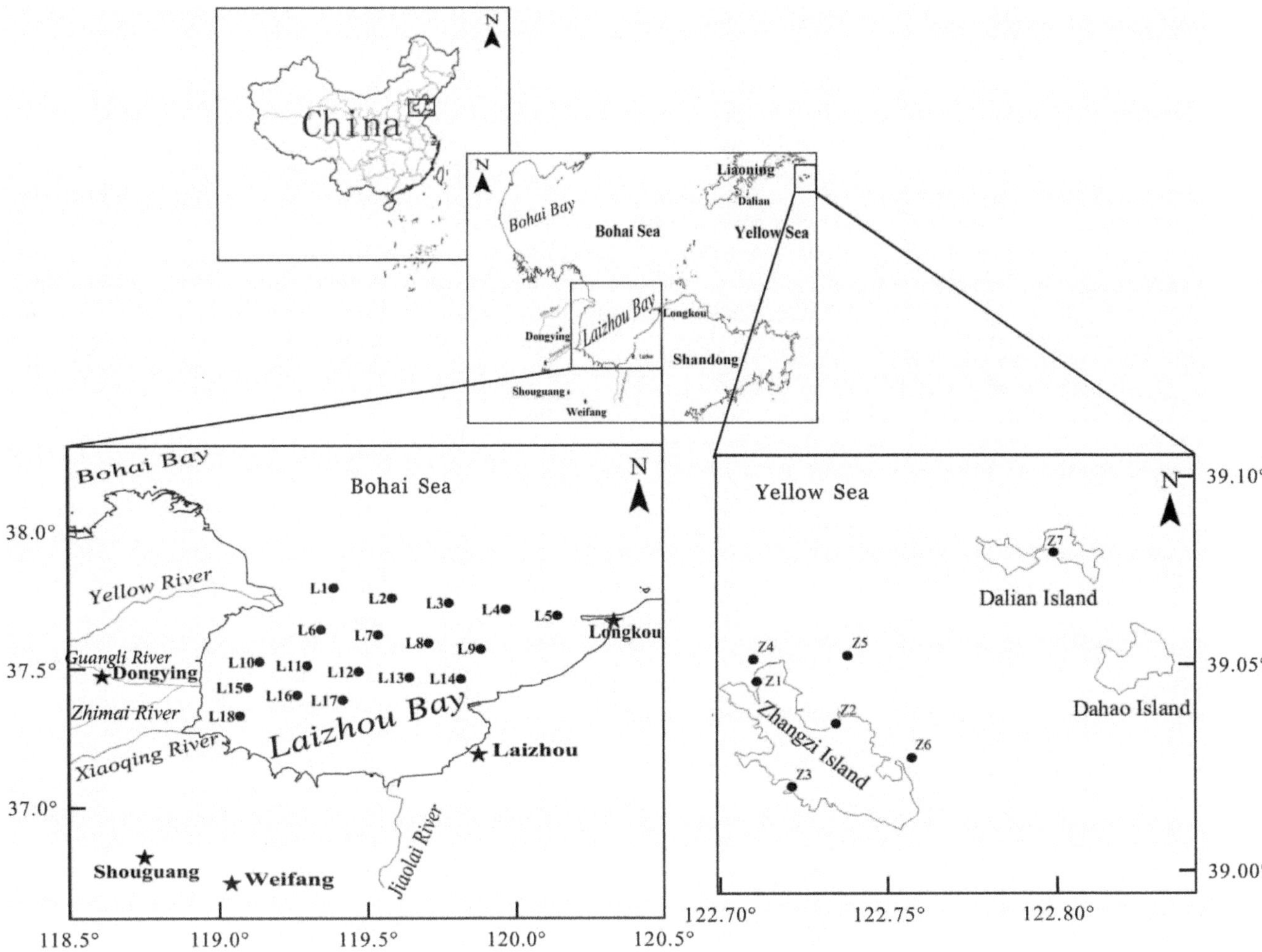

Figure 1. Location of sampling sites in the Laizhou Bay and the coastal Zhangzi Island.

Inductively coupled plasma mass spectrometry (PerkinElmer Elan DRC II) was applied in this work for the determination of Cd, Cr, Cu, Pb, Ni and Zn. In addition, the concentration of Al was analyzed by inductively coupled plasma optical emission spectrometer (PerkinElmer Optima 7000 DV) and the enrichment factor for each element was calculated.

The total organic carbon (TOC) in sediments was obtained by determining the total carbon using an Elementar vario MACRO cube CHNS analyzer after removing the inorganic carbon with 1 M HCl. The substance concentrations of sediments were expressed on the dry weight basis based on the results of moisture contents, which were determined gravimetrically by comparing the weight differences before and after heating an aliquot of sediment at 105 °C until constant weight was obtained. The grain size of samples was analyzed by a Malvern Mastersizer 2000 laser diffractometer capable of analyzing particle sizes between 0.02 and 2000 μm. The percentages of the following three groups of grain sizes were determined: <4 μm (clay), 4–63 μm (silt), and >63 μm (sand) [3,21].

Quality control

The analytical data quality was guaranteed through the implementation of laboratory quality assurance and quality control methods, including the use of standard operating procedures, calibration with standards, analysis of reagent blanks, and analysis of replicates. The precision of the analytical procedures was tested by recovery measurements on the Chinese national geostandard samples (GBW-07333 and GBW-07314). The results were consistent with the reference values, and the differences were all within 10%. The precision of the analytical procedures, expressed as the relative standard deviation (RSD), ranged from 5% to 10%. The precision of the analysis of standard solution was better than 5%. All analyses were carried out in duplicate, and the results were expressed as the mean.

Assessment of sediment contamination and ecological risks

In this study, four different indices were used to assess the degree of heavy metal contamination and ecological risks in the surface sediments of the Laizhou Bay and the Zhangzi Island. For the comparison purpose, the average upper continental crust (UCC) values [29] were chosen as the reference background values in all of the following related indices (Table 1).

1. Enrichment factor (EF). EF is a useful contamination index in determining the degree of anthropogenic heavy metal pollution. The EF for each element was calculated to evaluate anthropogenic influences on heavy metals in sediments using the following formula [3]:

Table 1. The metal guideline values of two different criteria used to distinguish marine sediment quality and the average upper continental crust (UCC) values.

Sediment quality guidelines	Cd	Cr	Cu	Ni	Pb	Zn	Reference
Class I upper limit	0.5	80	35		60	150	[9]
Class II upper limit	1.5	150	100		130	350	[9]
Class III upper limit	5	270	200		250	600	[9]
TEL guideline	0.68	52.3	18.7	15.9	30.2	124	[28]
PEL guideline	4.2	160	108	42.8	112	271	[28]
UCC	0.098	35	25	20	20	71	[29]

Content unit is μg g^{-1} dry weight for all elements.

$$EF = \frac{(C_x/C_{Al})_{Sample}}{(C_x/C_{Al})_{Background}}$$

Where C_x and C_{Al} denote the concentrations of element x and Al in the samples and in UCC, respectively. In this study, Al was used as the reference element for geochemical normalization, because it represents the quantity of aluminosilicates which is generally the predominant carrier phase for metals in coastal sediments and its natural concentration tends to be uniform [30]. According to EF values, each sample falls into one of the seven tiers: i) EF<1 indicates no enrichment; ii) 1<EF<3 is minor enrichment; iii) 3<EF<5 is moderate enrichment; iv) 5<EF<10 is moderately severe enrichment; v) 10<EF<25 is severe enrichment; vi) 25<EF<50 is very severe enrichment; and vii) EF>50 is extremely severe enrichment [12].

2. Geoaccumulation index (I_{geo}). I_{geo} is also a contamination index which is defined by the following equation:

$$I_{geo} = \log_2(C_n/1.5B_n)$$

Where C_n is the measured concentration of metal n; B_n is the geochemical background concentration of metal n. Correction index 1.5 is usually used to characterize the sedimentary and geological characteristics of rocks and other effects [5]. The geoaccumulation index consists of seven classes: $I_{geo}\leq 0$ (Class 0, practically uncontaminated); $0<I_{geo}\leq 1$ (Class 1, uncontaminated to moderately contaminated); $1<I_{geo}\leq 2$ (Class 2, moderately contaminated); $2<I_{geo}\leq 3$ (Class 3, moderately to heavily contaminated); $3<I_{geo}\leq 4$ (Class 4, heavily contaminated); $4<I_{geo}\leq 5$ (Class 5, heavily to extremely contaminated); $5<I_{geo}$ (Class 6, extremely contaminated) [31].

3. Risk assessment code (RAC). RAC which was originally developed by Perin et al [32] is widely used in ecological risk assessments of heavy metals in sediments. RAC is defined as:

$$RAC = Exc\% + Carb\%$$

Exc% and Carb% are percentages of metals in exchangeable and carbonate fractions (i.e., acid soluble fractions in the present study). According to RAC values, each sample falls into one of the five tiers: i) RAC≤1% (no risk); ii) 1%<RAC≤10% (low risk); iii) 10%<RAC≤30% (medium risk); iv) 30%<RAC≤50% (high risk); v) 50%<RAC (very high risk).

4. Potential ecological risk factor (ER). ER was originally developed by Hakanson [33] and is also an index widely used in ecological risk assessments of heavy metals in sediments. According to this methodology, the potential ecological risk index is defined as:

$$ER^i = Tr^i \cdot C_f^i$$

$$C_f^i = C_o^i / C_n^i$$

ER^i is the potential ecological risk factor for a given element i; Tr^i is the toxic-response factor for element i (e.g., Cd = 30, Cu = Pb = Ni = 5, Cr = 2, Zn = 1); C_f^i, C_o^i and C_n^i are the contamination factor, the concentration in the sediment and the background reference level for element i, respectively. According to Hakanson [33] the following tiers are used for the ER^i value: i) ER≤40 (low risk); ii) 40<ER≤80 (moderate risk); iii) 80<ER≤160 (considerable risk); iv) 160<ER≤320 (high risk); v) 320<ER (very high risk).

Sediment quality guidelines

Numerous sediment quality guidelines (SQGs) have been developed to deal with environmental concerns, and two of them were chosen to assess the contamination extent of individual metals in the surface sediments of the Laizhou Bay and the Zhangzi Island (Table 1).

The marine sediment quality of China (GB18668-2002) [9] has defined three grades of marine sediments, in which the contents of five metals (i.e., Cd, Cr, Cu, Pb and Zn) are regarded as parameters used to classify marine sediment quality. According to this criterion, three classes are identified: i) mariculture, nature reserve, endangered species reserve, and leisure activities are suitable; ii) industry and tourism site can be established; iii) only used for harbor.

Threshold effects level (TEL) and probable effects level (PEL) are also sediment quality guidelines which are widely used [28]. TEL is the concentration below which adverse biological effects rarely occur; PEL is the concentration above which adverse biological effects frequently occur.

Based on the fact that heavy metals occur in sediments as complex mixtures, the mean PEL quotient method has been applied to determine the possible biological effect of combined toxicant groups by calculating the mean quotients for a large range of contaminants using the following formula [34]:

$$\text{meanPELquotient} = \sum (C_x/PEL_x)/n$$

Where C_x is the sediment concentration of component x, PEL_x is the PEL for compound x and n is the sum of components. Based on the analyses of matching chemical and toxicity data from over 1000 sediment samples from the USA estuaries, the mean PEL quotients of <0.1 have an 8% probability of being toxic, the mean PEL quotients of 0.11–1.5 have a 21% probability of being toxic, the mean PEL quotients of 1.51–2.3 have a 49% probability of being toxic, and the mean PEL quotients of >2.3 have a 73% probability of being toxic [15].

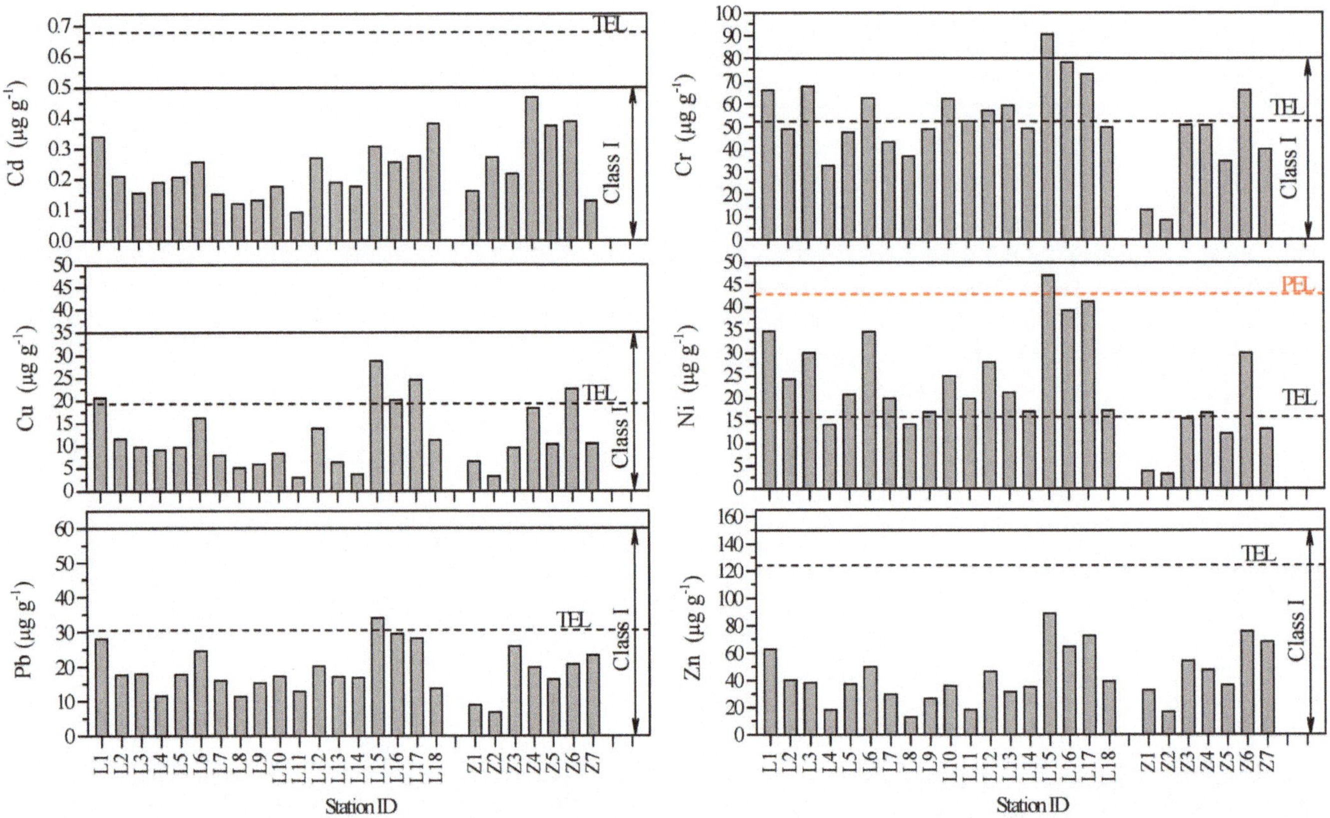

Figure 2. Information of the total concentrations of the studied metals. The spatial variations of studied metals in total concentrations of the surface sediments from the Laizhou Bay and the coastal Zhangzi Island. The horizontal dash lines represent their corresponding TEL or PEL concentrations; the horizontal solid lines represent their corresponding higher boundary values of Class I sediment category of China.

Statistical Analysis

Statistical methods were applied to process the analytical data in terms of the distribution and correlation among the studied parameters. Pearson's correlation coefficient analysis was performed to identify the relationship among heavy metals in sediments and their possible sources. The principal component analysis (PCA) of the normalized variables (Z-scores) was performed to extract significant principal components (PCs) and further reduce the contribution of variables with minor significance. After that factor analysis (FA) was conducted. These PCs were then subjected to varimax rotation to generate varifactors (VFs). The commercial statistics software package SPSS (version 19.0) for Windows was used for statistical analyses mentioned above in the present study.

The agglomerative hierarchical clustering (AHC) analysis was conducted on the normalized data set using Ward's method with Euclidean distances as a measure of similarity to assess the interrelationships among the sampling sites. The XLSTAT software (version 2013) was used in the AHC analysis.

Results and Discussion

Metals in total concentrations

The spatial distribution of heavy metals is shown in Fig. 2 and the related information is summarized in Table 2. Based on the mean concentrations, the target elements in the surface sediments of the Laizhou Bay exhibited the following descending order: Cr (56.7 μg g^{-1}) >Zn (41.5 μg g^{-1}) > Ni (25.9 μg g^{-1}) > Pb (19.4 μg g^{-1}) > Cu (12.0 μg g^{-1}) > Cd (0.22 μg g^{-1}); in the Zhangzi Island the corresponding result was Zn (47.1 μg g^{-1}) > Cr (37.4 μg g^{-1}) > Pb (17.3 μg g^{-1}) > Ni (13.5 μg g^{-1}) > Cu (11.5 μg g^{-1}) > Cd (0.29 μg g^{-1}).

In the Laizhou Bay, the highest concentrations of Cr (90.4 μg g^{-1}), Cu (28.7 μg g^{-1}), Ni (47.1 μg g^{-1}), Pb (30.4 μg g^{-1}) and Zn (88.6 μg g^{-1}) were all found in the surface sediments of site L15, which was about 10 km from the estuary of the Guanglihe River, Dongying City; the highest concentration of Cd (0.38 μg g^{-1}) was found at site L18 which was about 8 km from the estuary of the Xiaoqinghe River, Weifang City. Relatively higher concentrations of all the six metals studied were also found at sites L1, L6, L16 and L17. L1 and L6 were about 10 km from the new and old mouths of the Yellow River, respectively, indicating the contribution to heavy metal content of terrigenous input. The samples from L17 and L16 had the first and the second highest percentages of fine fractions (clay and silt) which demonstrated that the deposition of fine grained materials physically controls the abundance and distribution of metals in sediments [16]. The information about the grain size and TOC in the sediments of this study has been described in detail in Gao et al. [35].

In the Zhangzi Island, the highest concentrations of Cr (62.2 μg g^{-1}), Cu (22.5 μg g^{-1}), Ni (30.0 μg g^{-1}), Pb (25.8 μg g^{-1}) and Zn (75.4 μg g^{-1}) were all found in the surface sediments of site Z6, which was about 0.25 km from the coast and had the highest percentage of fine fractions and TOC [35]. The highest concentration of Cd (0.47 μg g^{-1}) was found in the surface sediments of site Z4 located in the mariculture area about 0.25 km from the coast where sea cucumbers were farmed. The concentrations of the rest of the metals studied at site Z4 were

Table 2. Heavy metal concentrations in the surface sediments of the Laizhou Bay and the coastal Zhangzi Island; and the related values reported for the surface sediments of other marine areas of China are shown for comparison.

Location	Sampling date		Cd	Cr	Cu	Ni	Pb	Zn	References
Laizhou Bay, China	Oct., 2011	Range	0.09–0.38	32.4–90.4	2.9–28.7	14.1–47.1	11.4–34.0	12.8–88.6	Present study
		Mean	0.22	56.7	12.0	25.9	19.4	41.5	
Coastal Zhangzi Island, China	Nov., 2011	Range	0.13–0.47	8.4–65.6	3.3–22.5	3.2–30.0	6.7–25.8	16.2–75.4	Present study
		Mean	0.29	37.4	11.5	13.5	17.3	47.1	
Coastal Shandong Peninsula (Yellow Sea), China	2007		na[a]	57.8	20.0	31.2	28.4	74.7	[36]
Liaodong Bay, China	2009		na	46.4	19.4	22.5	31.8	71.7	[37]
Coastal East China Sea, China	May, 2009		0.30	84.2	33.1	36.1	28.0	102.4	[38]
Coastal Bohai Bay, China	May, 2008		0.22	101.4	38.5	40.7	34.7	131.1	[3]
Intertidal Bohai Bay, China	May, 2008		0.12	68.6	24.0	28.0	25.6	73.0	[39]
Jinzhou Bay, China	Oct., 2009		26.8	na	74.1	43.5	124.0	689.4	[14]
Laizhou Bay, China	May, 2007		0.081	57.1	13.3	19.4	20.2	59.4	[40]
Laizhou Bay, China	May, 2008		0.11	na	15.0	na	11.7	50.8	[41]
Daya Bay, China	Jan., 2006		0.052	na	20.8	31.2	45.7	113	[22]
North Bohai and Yellow Sea, China	Oct., 2008		0.15	47	13	na	25	60	[42]
Changjiang Estuary, China	Apr. and Aug., 2009		0.26	78.9	30.7	31.8	27.3	94.3	[43]
The Atlantic and Cantabric coasts, Spain	2001–2007		na	na	115.0	na	91.0	230.0	[44]
İzmit Bay, Turkey	Apr. 2002		5.10	75.0	66.4	41.2	104.7	961	[45]
Masan Bay, Korea	2004–2005		1.24	67.1	43.4	28.8	44.0	206.3	[46]
Gironde Estuary, France			0.48	78.4	24.5	31.7	46.8	168.0	[47]
San Francisco Bay, USA	Mar. 2000– Mar. 2001		0.14	19	33	33	19	60	[48]
Jade Bay, Germany	2009–2010		0.25	49	7	10	16	43	[49]

[a]na: not available.
Content unit is $\mu g\ g^{-1}$ dry weight for all elements.

also relatively higher than the other sites in the Zhangzi Island except Z6, which reflected the anthropogenic influence (e.g. fishing operations) on heavy metals.

For the comparison purpose, the average UCC values (Table 1) and related values reported about the surface sediments of some of the marine areas in China and other countries were also shown (Table 2). In the surface sediments from the Laizhou Bay, the mean total contents of Cd, Cr and Ni were clearly higher with respect to their corresponding average values in the UCC; the mean total content of Cu was close to its corresponding average value in the UCC. In the Zhangzi Island, only the values of Cd and Cr were higher than their corresponding average values in the UCC. The average concentrations of Cd, Cr, Ni, Pb and Zn in the Laizhou Bay, Cd, Cu, Ni, Pb and Zn in the Zhangzi Island were within the range identified in the other marine areas listed in Table 2. The average concentrations of Cu in the Laizhou Bay and Cr in the Zhangzi Island were lower than all the values in the other studies listed. All the average concentrations of the studied metals in the two areas were far below the values of the Jinzhou Bay in China, the Atlantic and Cantabric coasts in Spain, the İzmit Bay in Turkey and the Masan Bay in Korea which were much heavily polluted coastal zones in the world [14,44,45,46]. All the average concentrations of the studied metals in the Zhangzi Island were close to the values of the Jade Bay in Germany where sediment quality was in good condition [49].

Correlation analyses have been widely used in environmental studies. They provide an effective way of revealing the relationships between multiple variables and parameters by which the factors as well as sources of chemical components could be better understood [5,13,38,50]. The correlation matrix for the parameters studied was shown in Table 3. All the metals were significantly correlated with each other in the surface sediments of the Laizhou Bay, suggesting a major common origin in sediments in this area. The wastewater discharged from industrial sources into the surrounding rivers which runs into the Laizhou Bay could be responsible for this [18,19,20]. It has been reported

Table 3. Pearson correlation matrix for the sediment components.

		Cd	Cr	Cu	Ni	Pb	Zn	%Clay	%Silt	%Sand	%TOC
Laizhou Bay	Cd	1	0.472[c]	0.736[a]	0.539[c]	0.602[b]	0.722[a]	0.458	0.216	−0.331	0.381
	p		0.048	0.000	0.021	0.008	0.001	0.056	0.390	0.180	0.118
	Cr		1	0.786[a]	0.932[a]	0.894[a]	0.883[a]	0.535[c]	0.394	−0.473[c]	0.252
	p			0.000	0.000	0.000	0.000	0.022	0.105	0.047	0.312
	Cu			1	0.917[a]	0.924[a]	0.949[a]	0.759[a]	0.608[b]	−0.700[a]	0.530[c]
	p				0.000	0.000	0.000	0.000	0.007	0.001	0.024
	Ni				1	0.962[a]	0.938[a]	0.746[a]	0.593[b]	−0.685[b]	0.488[c]
	p					0.000	0.000	0.000	0.010	0.002	0.040
	Pb					1	0.965[a]	0.735[a]	0.572[c]	−0.668[b]	0.489[c]
	p						0.000	0.001	0.013	0.002	0.040
	Zn						1	0.708[a]	0.526[c]	−0.628[b]	0.484[c]
	p							0.001	0.025	0.005	0.042
Coastal Zhangzi Island	Cd	1	0.456	0.653	0.501	0.072	0.036	0.716	0.783[c]	−0.774[c]	0.525
	p		0.303	0.112	0.252	0.878	0.939	0.070	0.037	0.041	0.226
	Cr		1	0.884[b]	0.953[a]	0.858[c]	0.865[c]	0.745	0.842[c]	−0.824[c]	0.849[c]
	p			0.008	0.001	0.014	0.012	0.055	0.018	0.023	0.016
	Cu			1	0.928[b]	0.558	0.747	0.880[b]	0.929[b]	−0.926[b]	0.829[c]
	p				0.003	0.193	0.054	0.009	0.003	0.003	0.021
	Ni				1	0.694	0.848[c]	0.884[b]	0.901[b]	−0.906[b]	0.932[b]
	p					0.083	0.016	0.008	0.006	0.005	0.002
	Pb					1	0.830[c]	0.318	0.458	−0.424	0.522
	p						0.021	0.487	0.302	0.343	0.229
	Zn						1	0.550	0.565	−0.567	0.667
	p							0.201	0.187	0.184	0.101

[a]$p<0.001$.
[b]$0.001<p<0.01$.
[c]$0.01<p<0.05$.
n = 18 for the Laizhou Bay and n = 7 for the coastal Zhangzi Island.

that the deposition of fine grained materials and organic matter physically controls the abundance and distribution of metals in sediments [51]. The concentrations of Cu, Ni, Pb, and Zn appeared to be influenced by both the sediment grain size composition and the amount of organic matter; the concentration of Cr appeared to be more influenced by the sediment grain size composition than by the amount of organic matter; the concentration of Cd appeared to be influenced by neither the sediment grain size composition nor the amount of organic matter, perhaps because it is a typical anthropogenic element.

In the surface sediments of the Zhangzi Island, Cr was significantly correlated with the other studied metals except Cd, suggesting a wide origin of Cr in sediments of this area. The concentrations of Cr, Cu and Ni appeared to be influenced by both the sediment grain size composition and the amount of organic matter, and the concentration of Cd appeared to be influenced only by grain size composition. The concentrations of Pb and Zn were significantly correlated, whereas both of them had no significant correlation with grain size composition and the amount of organic matter, which indicated that they had a major common origin in sediments but not the same as the other metals. Zn might be released from the anti-corrosion paints used on ship hulls or from other anthropogenic sources [52,53], and Pb might be released with engine exhaust of ships [54,55,56].

Metal fractionation

The sequential extraction technique is proposed to provide information about the strength and ways of metals associating with sediments and thus predict the possible metal impact on biota in aquatic ecosystems [16,57]. The metals in acid soluble fraction (i.e., the exchangeable and bound to carbonate fractions) are mainly introduced by human activities and are considered to be weakly bound. This fraction may equilibrate with aqueous phase and thus become more rapidly bioavailable and cause environmental toxicity [58]. The reducible fraction (bound to Fe/Mn oxyhydroxides) and the oxidizable fraction (bound to organic matter) can be mobilized when environmental conditions become increasingly reducing or oxidizing, respectively [58]. The detrital fraction which is composed of metals present in the inert fraction, being of lattice origin or primary mineral phases, can be regarded as a measure of contribution by natural sources [59]. The percentages of heavy metal concentrations that were extracted in each step of the sequential extraction procedure used in this study were presented in Fig. 3, and the pearson correlation matrix for metal fractionations with grain size and TOC was shown in Table 4.

Each studied metal displayed the similar compositional characteristics between the surface sediments in the Laizhou Bay and the Zhangzi Island. On average, the residual fraction was the most dominant one for all the studied metals except Cd, indicating the

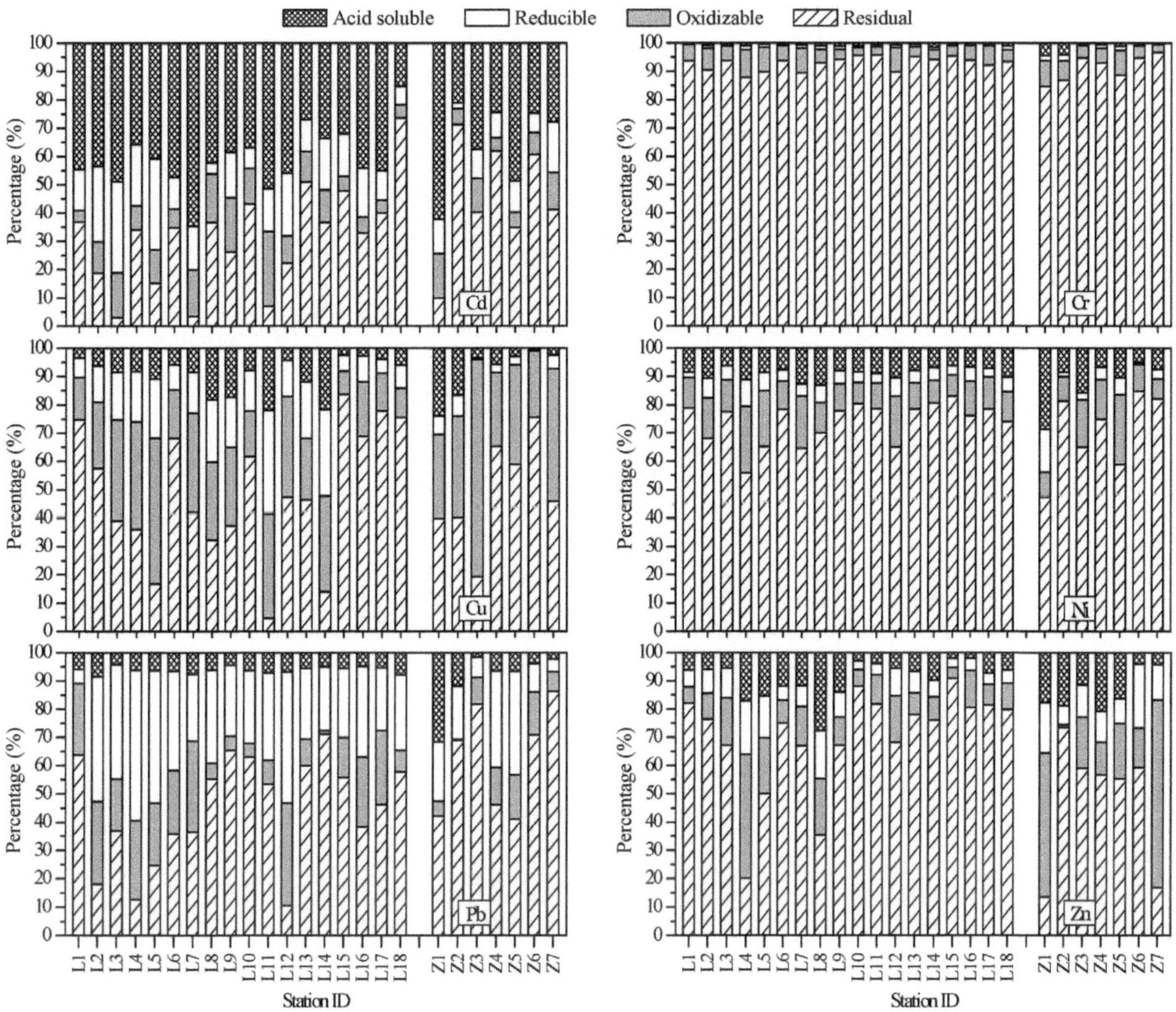

Figure 3. Geochemical phases of the studied metals. The distributions of studied metals in different geochemical phases of the surface sediments from the Laizhou Bay and the coastal Zhangzi Island.

paramount mineralogical origin of these metals; while differences among sampling sites were obvious, which might result from the combined effects of the physicochemical conditions of the sedimentary environment, the intensity of human activities and so on (Fig. 3).

Generally, except for Cd, the relative proportions of metals in the acid soluble fraction were very low, especially for Cr ($\approx$1% of total concentration). In the Laizhou Bay, the very low concentrations of Cr and Cu in the acid soluble fraction could still be partly from mineral sources, because significant positive correlation between this fraction and sand was observed (Table 4). Only Cd had observable contents of the acid soluble fraction with the mean values of 41.0% and 35.2% in the Laizhou Bay and the Zhangzi Island, respectively. The result was similar to the previous study carried out in the Bohai Bay by Gao and Chen [3]. The excessive input of Cd into water induced by phosphorus fertilizer has been widely reported [60,61]. In addition, the presence of Cd could also be a result of road traffic, which has been described as an important source of Cd emission [62].

Pb exhibited the highest proportion within the reducible fraction (5.0–53.2% in the Laizhou Bay; 4.4–36.7% in the Zhangzi Island) among the six studied metals, which might be the result of the higher stability of Pb-oxides, and also could be attributed to the adsorption, flocculation and co-precipitation of heavy metals with the colloids of Fe and Mn oxyhydroxide [63]. The same result was also reported by other researchers [3,16,22,39]. In both of the studied areas, Pb in oxidizable fraction was significantly correlated with clay and silt, indicating this fraction might be mainly from terrestrial source (Table 4).

On average, the proportions of non-residual Cu, Pb and Zn were identified being the highest in the oxidizble fraction in the Laizhou Bay; in the Zhangzi Island, the proportions of non-residual Cu, Zn and Ni were the highest in the oxidizble fraction. In both of the two studied areas, Cu had the highest proportion (8.2–51.5% in the Laizhou Bay; 23.4–76.8% in the Zhangzi Island) among these metals. This result could be explained by the affinity of metals with organic matter, especially humic substances, which are both the components of natural organic matter and chemical actives in complexing metals [64,65]. The partitioning patterns of Zn and Cu were somewhat very similar with each other. The observed non-residual fractions of Zn and Cu might be the result of ZnO and Cu_2O released from the anti-corrosion

Table 4. Pearson correlation matrix for metal fractionations with grain size and TOC.

		Laizhou Bay				Coastal Zhangzi Island			
		F1	F2	F3	F4	F1	F2	F3	F4
Cd	%Clay	0.435	0.344	−0.516[c]	−0.235	−0.266	−0.288	−0.441	0.356
	p	0.071	0.163	0.028	0.347	0.564	0.531	0.322	0.433
	%Silt	0.381	0.452	−0.450	−0.275	−0.323	−0.277	−0.463	0.400
	p	0.119	0.060	0.061	0.270	0.480	0.548	0.295	0.374
	%Sand	−0.421	−0.424	0.498[c]	0.270	0.311	0.283	0.462	−0.392
	p	0.082	0.080	0.035	0.279	0.498	0.539	0.296	0.384
	%TOC	0.398	0.501[c]	−0.403	−0.323	−0.328	−0.272	−0.241	0.355
	p	0.102	0.034	0.097	0.191	0.473	0.556	0.603	0.434
Cr	%Clay	−0.673[b]	−0.511[c]	0.525[c]	−0.324	−0.484	−0.307	−0.184	0.338
	p	0.002	0.030	0.025	0.190	0.271	0.504	0.693	0.458
	%Silt	−0.560[c]	−0.343	0.600[b]	−0.436	−0.586	−0.385	−0.276	0.439
	p	0.016	0.164	0.009	0.070	0.167	0.394	0.550	0.324
	%Sand	0.634[b]	0.432	−0.593[b]	0.406	0.565	0.367	0.253	−0.416
	p	0.005	0.074	0.010	0.095	0.187	0.418	0.584	0.353
	%TOC	−0.566[c]	−0.297	0.671[b]	−0.511[c]	−0.544	−0.561	−0.399	0.520
	p	0.014	0.231	0.002	0.030	0.207	0.190	0.375	0.232
Cu	%Clay	−0.785[a]	−0.651[b]	−0.192	0.542[c]	−0.559	−0.701	−0.429	0.781[c]
	p	0.000	0.003	0.445	0.020	0.192	0.079	0.336	0.038
	%Silt	−0.695[a]	−0.535[c]	0.023	0.369	−0.598	−0.800[c]	−0.296	0.684
	p	0.001	0.022	0.927	0.132	0.156	0.031	0.519	0.090
	%Sand	0.764[a]	0.609[b]	0.070	−0.461	0.594	0.781[c]	0.337	−0.719
	p	0.000	0.007	0.782	0.054	0.160	0.038	0.460	0.069
	%TOC	−0.636[b]	−0.492[a]	0.060	0.319	−0.597	−0.853[c]	−0.070	0.472
	p	0.005	0.038	0.813	0.197	0.157	0.015	0.881	0.285
Ni	%Clay	−0.149	−0.131	0.290	−0.108	−0.485	−0.385	0.186	0.335
	p	0.556	0.604	0.244	0.671	0.270	0.394	0.689	0.463
	%Silt	−0.169	0.144	0.448	−0.266	−0.475	−0.414	0.237	0.316
	p	0.501	0.569	0.062	0.285	0.281	0.356	0.608	0.489
	%Sand	0.168	−0.029	−0.398	0.208	0.483	0.410	−0.226	−0.325
	p	0.506	0.908	0.102	0.407	0.272	0.361	0.626	0.477
	%TOC	0.025	0.106	0.445	−0.310	−0.400	−0.500	0.095	0.366
	p	0.923	0.676	0.064	0.210	0.374	0.254	0.840	0.420
Pb	%Clay	0.083	0.011	0.839[a]	−0.496[c]	−0.366	0.098	0.792[c]	−0.101
	p	0.744	0.965	0.000	0.036	0.420	0.834	0.034	0.830
	%Silt	−0.111	0.264	0.736[a]	−0.581[c]	−0.436	0.134	0.820[c]	−0.095
	p	0.661	0.289	0.000	0.012	0.329	0.774	0.024	0.839
	%Sand	0.031	−0.165	−0.813[a]	0.569[c]	0.421	−0.126	−0.822[c]	0.098
	p	0.903	0.514	0.000	0.014	0.347	0.788	0.023	0.835
	%TOC	0.271	0.258	0.841[a]	−0.660[b]	−0.470	−0.216	0.679	0.201
	p	0.277	0.301	0.000	0.003	0.287	0.642	0.093	0.666
Zn	%Clay	−0.410	−0.183	−0.112	0.249	−0.355	0.584	−0.415	0.392
	p	0.091	0.467	0.658	0.319	0.435	0.169	0.354	0.384
	%Silt	−0.189	0.119	0.216	−0.071	−0.236	0.493	−0.482	0.447
	p	0.453	0.639	0.390	0.780	0.611	0.261	0.273	0.315
	%Sand	0.294	0.009	−0.081	−0.066	0.271	−0.524	0.469	−0.437
	p	0.237	0.973	0.748	0.794	0.556	0.228	0.289	0.327

Table 4. Cont.

	Laizhou Bay				Coastal Zhangzi Island			
	F1	F2	F3	F4	F1	F2	F3	F4
%TOC	−0.204	0.035	−0.034	0.081	−0.482	0.613	−0.413	0.422
p	0.418	0.891	0.894	0.751	0.273	0.144	0.357	0.346

[a]$p<0.001$.
[b]$0.001<p<0.01$.
[c]$0.01<p<0.05$.
n = 18 for the Laizhou Bay and n = 7 for the coastal Zhangzi Island; F1, F2, F3 and F4 represent acid soluble, reducible, oxidizable and residual fraction, respectively.

paints used on ship hulls during the maintenance of ships or from other anthropogenic sources [52,53,66]. The residual fractions of Cu in both the Laizhou Bay and the Zhangzi Island could be from rock weathering source because significant positive correlation between this fraction and clay was observed (Table 4).

The concentration of Cr showed a completely different pattern from the others. Among the non-residual fraction, Cr had the highest content in oxidizable fraction (2.8–9.7% in the Laizhou Bay; 2.6–9.1% in the Zhangzi Island). In the Laizhou Bay, Cr in oxidizable fraction was significantly correlated with clay, silt and TOC, indicating this fraction was mainly from terrestrial source and was influenced by the amount of organic matter (Table 4). Substantial amounts of Cr were found in the residual phases (87.8–95.7% in the Laizhou Bay; 84.8–96.8% in the Zhangzi Island). This suggested that Cr had the strongest associations with the crystalline sedimentary components.

Pearson correlation analysis can partly give information on the sources of the metals in the environment. To explore this topic further, risk indices, principal component analysis and factor analysis were used later in this article.

Risk assessment of heavy metals

1. Risk assessment according to SQGs. The higher boundary values of Class I sediment category of China [9] and the corresponding TEL and PEL concentrations were listed in Table 1 and marked in Fig. 2. All the concentrations of metals at all the sites in both the Laizhou Bay and the Zhangzi Island were below the values of the upper limit for Class I sediment except for Cr in the surface sediments of site L15. This indicated that the sedimentary environments of both the Laizhou Bay and the Zhangzi Island were in good condition according to the marine sediment quality of China.

The data of this study also suggested that no site exceeded the TEL guideline for Cd or Zn in both the Laizhou Bay and the Zhangzi Island. In the case of other metals in the Laizhou Bay, 50%, 78%, 11% and 94% of sites were below the TEL guideline for Cr, Cu, Ni and Pb, respectively; site L15 was even above the PEL guideline for Ni. In the case of other metals in the Zhangzi Island, site Z6 was above the TEL guideline for Cr and Cu; sites Z4 and Z6 were above the TEL guideline for Ni; no site exceeded the TEL guideline for Pb. As shown in Fig. 4, in the surface sediments of the Laizhou Bay and the Zhangzi Island, the combination of the six studied metals of sites Z1 and Z2 might have an 8% probability of being toxic, and these metals might have a 21% probability of being toxic for all the other sites.

The result of the hierarchical cluster analysis of the sampling sites based on the data of metals in total concentrations and SQGs was shown in Fig. 5. Three main different clusters could be observed for the Laizhou Bay. Cluster 1 involved several sites (L1, L15, L16 and L17) near to the Yellow River mouths and in the southwestern Laizhou Bay, respectively, indicating the sites were moderately to heavily contaminated according to the SQGs; Cluster 2 was made up of the sites L4, L7, L8, L9, L11 and L14 which were uncontaminated or less contaminated; Cluster 3 was made up of the rest sites which were moderately contaminated. Two main different clusters could be observed for the Zhangzi Island. Cluster 1 included two sites (Z1 and Z2) in the intertidal zone which were uncontaminated according to the SQGs; Cluster 2 was made up of the rest sites which were moderately contaminated.

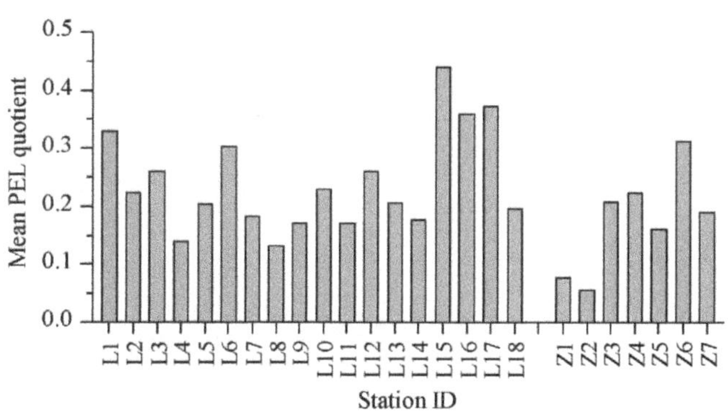

Figure 4. The spatial distribution of mean PEL quotient values. The histogram shows spatial distribution of mean PEL quotient values in the surface sediments of the Laizhou Bay and the coastal Zhangzi Island.

2. Risk assessment according to contamination and ecological risk indices. According to Zhang and Liu [67], EF values between 0.5 and 1.5 indicate that the given metal is entirely derived from crustal materials or natural weathering processes, whereas EF values higher than 1.5 suggest that a significant portion of metal is delivered from non-crustal materials and the sources are more likely to be anthropogenic. The spatial distributions of calculated EFs for each of the studied metals were shown in Fig. 6. The mean EF values of Cd and Cr suggested their enrichments in most surface sediments of the Laizhou Bay and the Zhangzi Island. In the Laizhou Bay, the highest EF value of Cd was recorded at site L18 (5.7) near the estuary of the Xiaoqinghe River; in the Zhangzi Island, the highest EF value of Cd was found at site Z5 (9.6) 1.4 km away from the coast where scallops were farmed; sites Z2, Z4 and Z6 in the Zhangzi Island also had high EF values of Cd within the range of 5 to 10. This indicated that Cd in the surface sediments of these five sites was in moderately severe enrichment [12]. The highest EF values of Cr were recorded at site L15 (2.8) near the Guanglihe River estuary in the Laizhou Bay and at site Z4 (2.7) where sea cucumbers were farmed in the

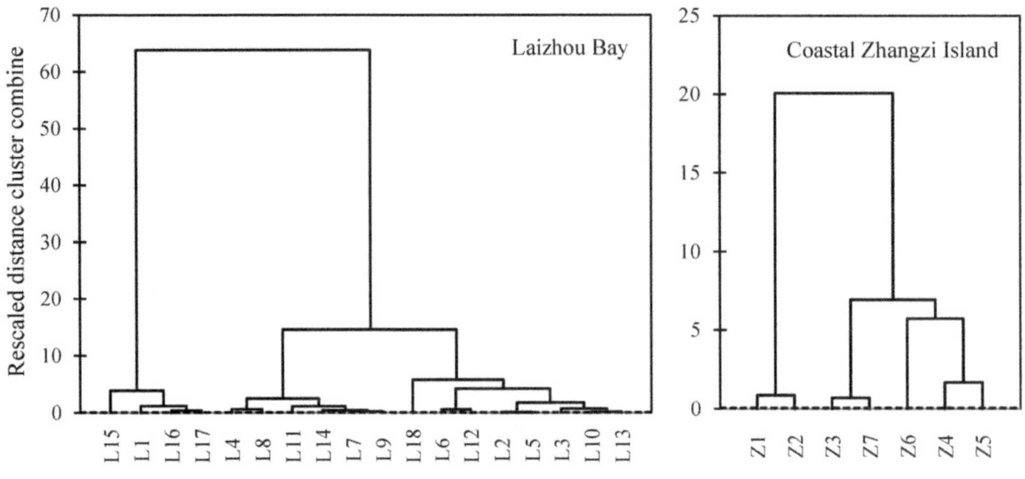

Figure 5. Dendrogram showing clustering of sampling sites in the two studied areas.

Zhangzi Island. According to the EF values, the contribution from anthropogenic sources was negligible for Zn at all sites in the Laizhou Bay and for Cu in all sites of the Laizhou Bay and the Zhangzi Island. Generally, Ni and Pb were slightly enriched in the surface sediments near the Yellow River Estuary and the southwestern Laizhou Bay and sites Z4–Z7 of the Zhangzi Island; Zn was also slightly enriched at Z7 which was near a small wharf.

The spatial distributions of calculated I_{geo} values for each of the studied metals were shown in Fig. 7. Most of the I_{geo} values of Cd were between 0 and 1 which showed that these sites were uncontaminated to moderately contaminated. I_{geo} values of Cd in the surface sediments of the sites near to the mouths of the Yellow River, the Guanglihe River and the Xiaoqinghe River in the Laizhou Bay and all the three sites in the coastal waters of the Zhangzi Island were between 1 and 2, further indicating these sites were more affected by human activities than other sites. The I_{geo} values suggested that Cu and Zn at all sites in both the Laizhou Bay and the Zhangzi Island were in the uncontaminated level, and this was true for Pb except at site L15 where the I_{geo} value of Pb was a little higher than 0. The values of I_{geo} for Cr and Ni at most sites were <0 except at several ones that were near to the Yellow River mouths and in the southwestern Laizhou Bay; for all the sampling sites in the Zhangzi Island, the I_{geo} values indicated that their surface sediments were practically uncontaminated by Ni, and Cr presented the same situation like Ni except at site Z6 which had a condition of slight Cr pollution.

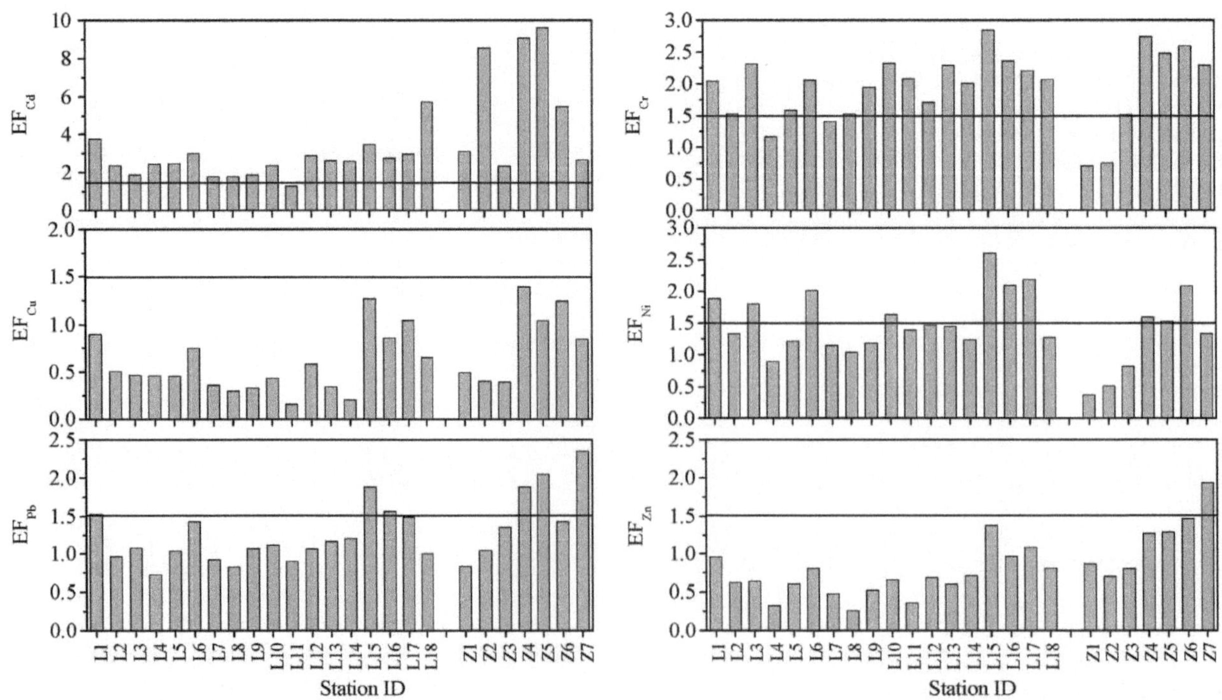

Figure 6. The spatial distributions of EF values. The histogram shows spatial distributions of EF values for heavy metals in the surface sediments of the Laizhou Bay and the coastal Zhangzi Island. The horizontal lines represent EF value of 1.5.

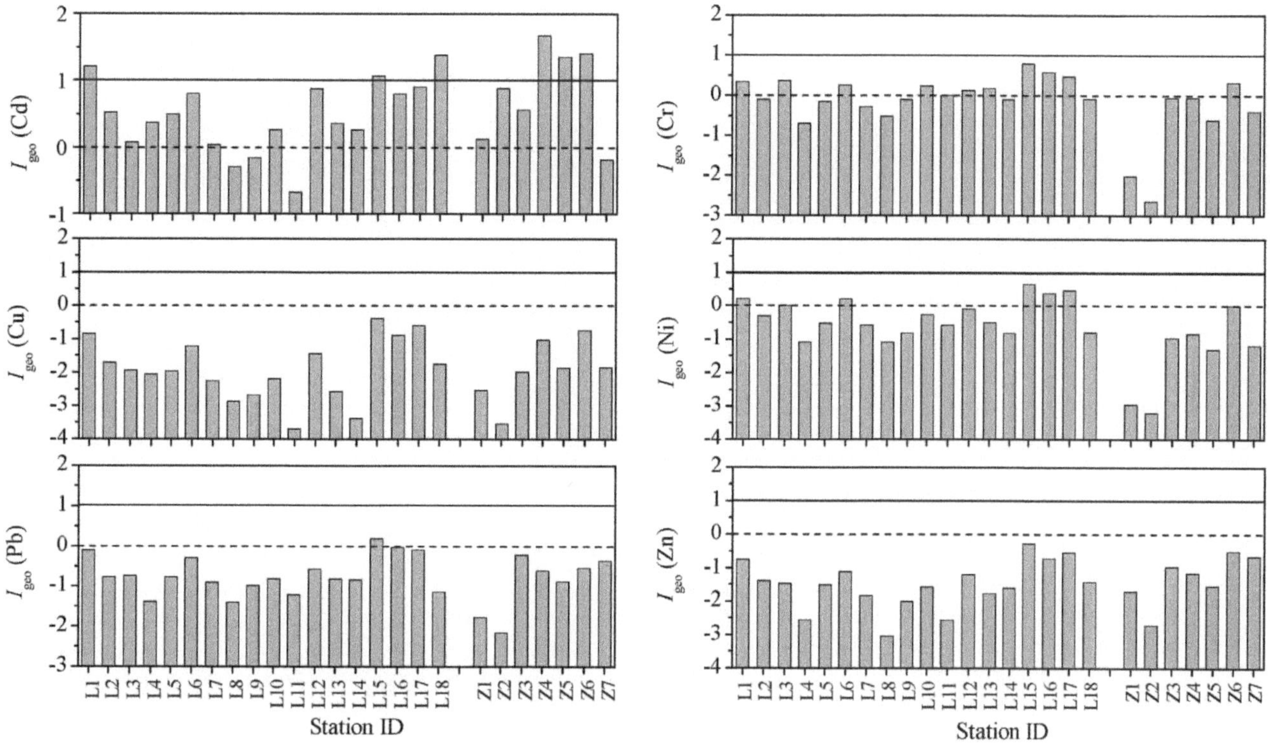

Figure 7. The spatial distributions of I_{geo} values. The histogram shows the spatial distributions of I_{geo} values for heavy metals in the surface sediments of the Laizhou Bay and the coastal Zhangzi Island. The horizontal dash and solid lines represent I_{geo} values of 0 and 1, respectively.

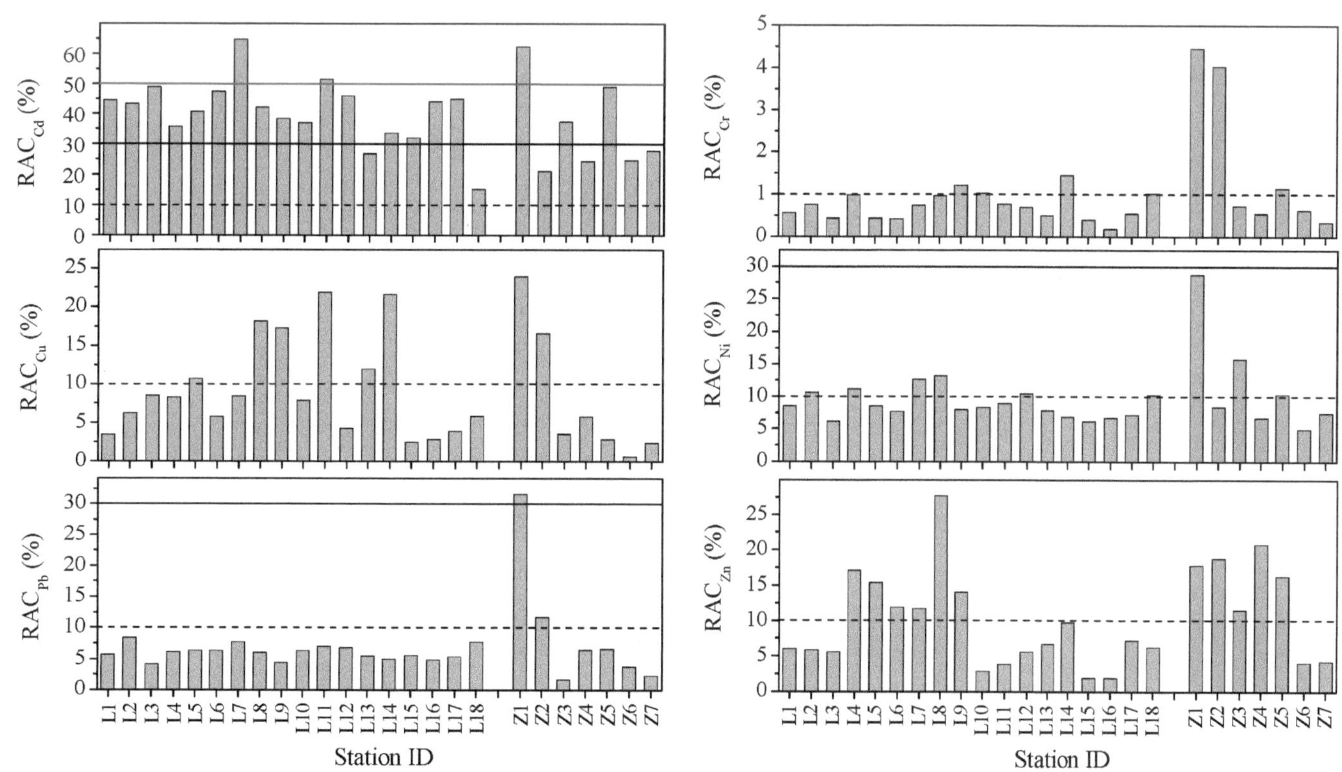

Figure 8. The spatial distributions of RAC values. The histogram shows the spatial distributions of RAC values (%) for heavy metals in the surface sediments of the Laizhou Bay and the coastal Zhangzi Island. The horizontal dash lines represent RAC values (%) of 1 or 10. The horizontal solid lines represent RAC values (%) of 30 or 50.

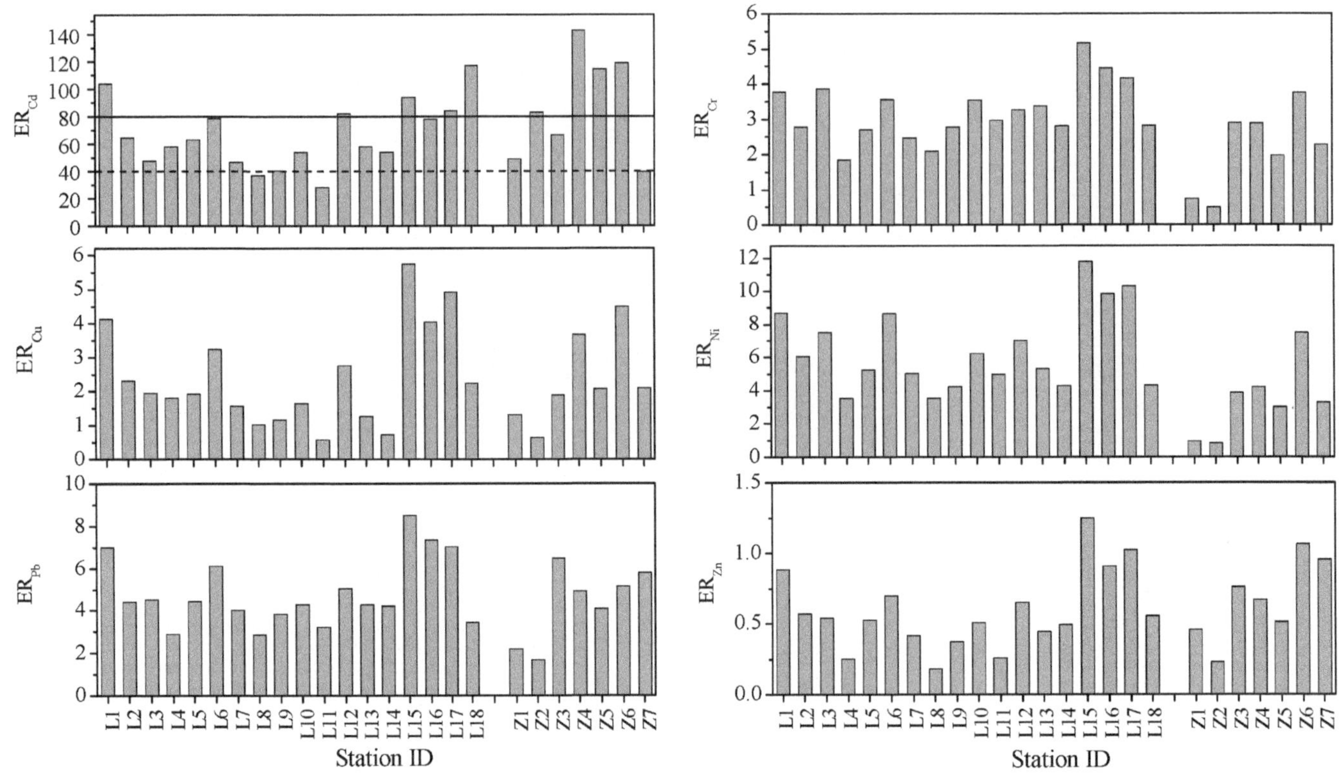

Figure 9. The spatial distributions of ER values. The histogram shows the spatial distributions of ER values for heavy metals in the surface sediments of the Laizhou Bay and the coastal Zhangzi Island. The horizontal dash and solid lines represent ER values of 40 and 80, respectively.

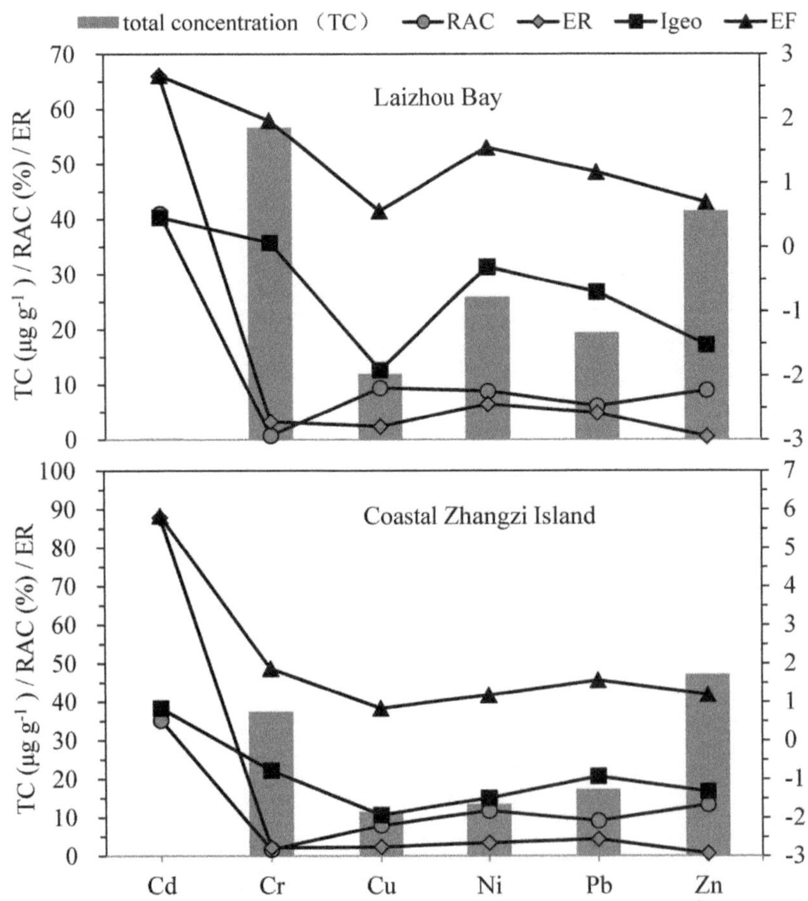

Figure 10. The mean values of EF, I_{geo}, RAC, ER and total concentration for metals.

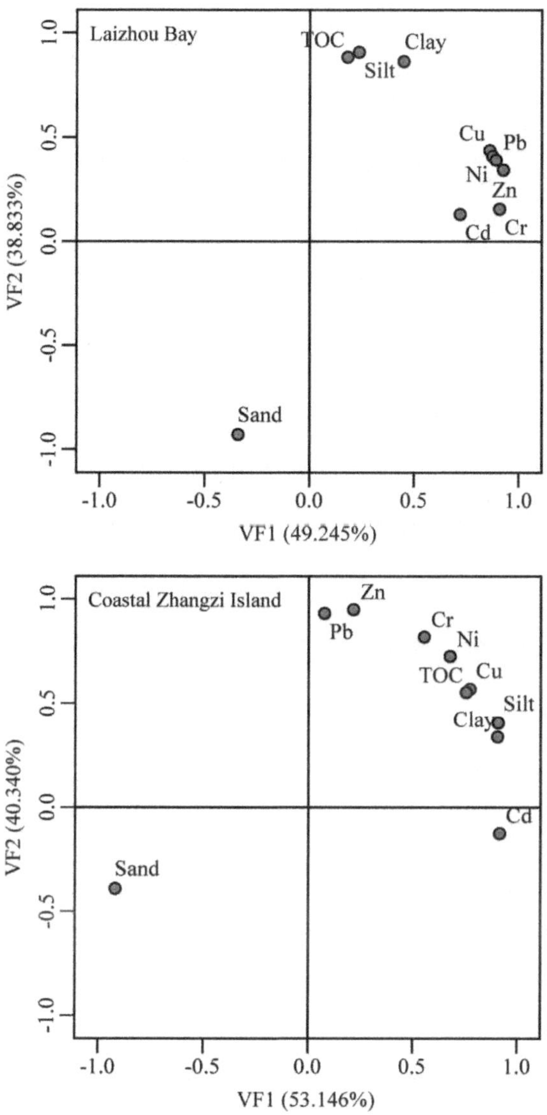

Figure 11. Loading plots of the principal components obtained for the data set.

The spatial distributions of calculated RACs for each of the studied metals were shown in Fig. 8. It showed that most of the sites suffered high risks from Cd (30%<RAC≤50%). Risk from Cd in the surface sediments of sites L7 and L11 in the Laizhou Bay and site Z1 in the Zhangzi Island was very high (RAC>50%). Although the total concentration of Cd in the surface sediments of Z1 was very low, the proportion of acid soluble fraction of it was high; this might be because Z1 was near a seedling factory and was severely impacted by the discharge from that factory. The values of RAC at site Z1 were also higher than the other sites in the Zhangzi Island for Cu (24.0%), Ni (28.8%) and Pb (31.6%). Cr at all the sites had no to low risk; Cu, Ni and Zn at most sites had low to medium risk; Pb at all the sites had low risk except at sites Z1 and Z2 which had high risk and medium risk, respectively.

The spatial distributions of calculated ERs for each of the studied metals were shown in Fig. 9. According to ER index, potential ecological risk posed by Cr, Cu, Ni, Pb and Zn was very low at all the sites. However, the potential ecological risk posed by Cd was obvious at most of the sites. Except for sites L8 and L11 in

Table 5. Loadings of experimental variables on significant principal components for the data sets of the Laizhou Bay and the coastal Zhangzi Island.

	Laizhou Bay		Coastal Zhangzi Island	
	PC1	PC2	PC1	PC2
Cd	**0.639**	0.359	**0.916**	−0.126
Cr	**0.800**	0.460	**0.553**	**0.818**
Cu	**0.944**	0.213	**0.774**	**0.567**
Ni	**0.937**	0.243	**0.678**	**0.725**
Pb	**0.937**	0.268	0.075	**0.930**
Zn	**0.933**	0.327	0.214	**0.949**
Clay	**0.897**	−0.379	**0.906**	0.339
Silt	**0.760**	**−0.549**	**0.909**	0.405
Sand	**−0.853**	**0.498**	**−0.918**	−0.391
TOC	**0.703**	**−0.566**	**0.755**	**0.552**
Eigenvalue	4.925	3.883	5.315	4.034
% Total variance	49.245	38.833	53.146	40.340
Cumulative % variance	49.245	88.078	53.146	93.486

Bold values indicate strong loadings.

the Laizhou Bay and Z7 in the Zhangzi Island, all the other sites at least suffered moderate risk from Cd (40<ER≤80). Sites L1, L15, L17 and L18 in the Laizhou Bay and sites Z4–Z6 suffered considerable risk from Cd (80<ER≤160). The sites which suffered considerable risk from Cd according to ER index were also moderately contaminated according to I_{geo} index.

The mean values of EF, I_{geo}, RAC and ER of the six studied metals and their total concentrations were summarized in Fig. 10. It showed clearly that Cd had the highest potential risk according to all the four indices. However, inconsistent conclusions for the other five metals according to these four indices could be drawn. In addition, the results of EF and I_{geo} were consistent with the trend of the total metal concentrations except for Cd; meanwhile the results of RAC and ER had no consistency with the trend of their corresponding total metal concentrations. The explanation for this might be as the following: i) the EF and I_{geo} indices are calculated mainly based on the concentrations of the total metals and enrichment levels. ii) The ER index is based on the toxic-response factor besides the total concentration. For example, though total concentration of Cd is usually pretty lower than Zn in sediments, but the higher value of toxic-response factor of Cd (Tr = 30) might make it much more toxic than Zn whose toxic-response factor is only 1. iii) The RAC index, which reflects the potential mobility of sedimentary metals, is based on the chemical form of a given metal which has no direct relationship with its total concentration. It is generally accepted that the fractionation can give more information on the bio-availability and bio-toxicity of a certain metal than the total concentration [3,16,38,68]. However, the results of the present study indicated the necessity of further verifying the prediction accuracy of the fractionation based methods.

It is generally known that the mechanisms of these SQS and indices are divergent and they are used to assess the environmental risk from different angles. Neither the total concentration based index nor the chemical fractionation based index alone could be sufficient in revealing the biogeochemical information of heavy metals in sediments. However, the trend of risk level according to the assessment results of them should be consistent or at least should not be opposite to each other; if not, misleading results of risk assessment may even cause misjudgment during the formulation and enforcement of public policies by government. Therefore, the merits and defects should be evaluated before the application of these assessment methods, and it is necessary to use multiple evaluation methods. Different metals may have different risk gradation ranges. For instance, the differences between metal fractionation characteristics of Cd and Cr are often substantial, but the risks of these two metals ranked by RAC indexes are within the same gradation range that this may not be true. We believe that a more comprehensive index to reveal the information of the concentration, the chemical fractionation and the toxic-response factor of heavy metals should be developed. Therefore, we suggest that more toxicity tests, benthic community analyses, bioaccumulation tests, and a combination of these should be carried out to make the risk assessment of sedimentary heavy metals more accurate and reliable.

Principal component analysis/factor analysis

PCA/FA was performed to identify interrelationships of the six studied heavy metals and the major constituents of the sediments (TOC, clay, silt and sand). Table 5 showed that there were two PCs for the surface sediments in both the Laizhou Bay and the Zhangzi Island. The loading plots of the VFs were presented in Fig. 11. These PCs were the ones with eigenvalues larger than 1, and altogether they accounted for 88.1% and 93.5% of the variance in the data of the Laizhou Bay and the Zhangzi Island, respectively.

In the Laizhou Bay, PC1 which explained 49.3% of the total variance was positively related to all the heavy metals and major constituents except sand. Sand was significantly negatively related to PC1. The high loading of clay, silt and TOC with PC1 highlighted the influence of fine grained minerals and organic matter on the distributions of heavy metals in the sediments of the Laizhou Bay, and revealed that these metals were mainly from terrestrial source especially via rivers [68]. PC2 which explained 38.8% of the total variance was only positively related to sand, indicating that sand could hardly capture metal ions.

In the Zhangzi Island, PC1 was positively related to Cd, Cr, Cu, Ni, Clay, Silt and TOC. This PC represented terrestrial sources. PC2 was also positively related to Cr, Cu, Ni and TOC, indicating they were from both terrestrial inputs and biogenic sources [68]. Zn and Pb were only positively related to PC2, indicating their different sources with the other studied metals. Zn and Pb might be released from ships and from biogenic sources. In addition, previous studies have shown that a large amount of Pb is supplied by the precipitation of aerosols in coastal environments [69]. So the precipitation of aerosols might be another important source of Pb in the surface sediments of the Zhangzi Island.

Conclusions

This study investigated the total concentrations and fractionation of heavy metals in the surface sediments from the Laizhou Bay and the Zhangzi Island. The relatively higher concentrations of metals in the Laizhou Bay were mainly distributed near to the new and old mouths of the Yellow River, the mouths of Guanglihe and Xiaoqinghe Rivers, and in the middle of the Bay. The relatively higher concentrations of metals in the Zhangzi Island were mainly distributed in and near the mariculture areas. In the Laizhou Bay, all the metals studied were mainly from terrestrial sources, and especially Cd, Cr and Ni had obvious anthropogenic sources; in the Zhangzi Island, both natural and anthropogenic sources contributed significantly to the metal contents.

The marine sediment quality of China showed that the sedimentary environment in both the Laizhou Bay and the Zhangzi Island were in good condition. TEL/PEL guidelines revealed that adverse biological effects might occur frequently in some areas in the Laizhou Bay and the Zhangzi Island especially from Cr and Ni. Based on the mean PEL quotient, surface sediments of sites Z1 and Z2 had an 8% probability of toxicity, and surface sediments of the rest sites of the two studied areas had a 21% probability of toxicity.

All the four risk assessment indices used in this study revealed an obvious pollution risk by Cd, especially in sites near the river mouths and in the southwestern Laizhou Bay and in the coastal waters of the Zhangzi Island. Nevertheless, contradictory conclusions could be obtained when different indices and SQGs are used. Significant negative correlations between RAC and the other indices and between RAC and the total metal concentration for Cd, Cr, Cu and Ni were found. We suggest that toxicity tests, bioaccumulation tests, and other related experiments should be further carried out in order to make the risk assessment methods more accurate and reliable in the analysis of sedimentary heavy metals.

Acknowledgments

We are thankful to the anonymous reviews and the Academic Editor whose pertinent comments have greatly improved the quality of this paper. We thank Yong Zhang, Peimiao Li, Fengxia Zhou and Jinfeng Zhang for participating in the sample collection.

Author Contributions

Conceived and designed the experiments: WZ XG. Performed the experiments: WZ. Analyzed the data: WZ XG. Contributed reagents/materials/analysis tools: XG. Wrote the paper: WZ XG.

References

1. Prica M, Dalmacija B, Rončević S, Krčmar D, Bečelić M (2008) A comparison of sediment quality results with acid volatile sulfide (AVS) and simultaneously extracted metals (SEM) ratio in Vojvodina (Serbia) sediments. Sci Total Environ 389: 235–244.
2. Yang YQ, Chen FR, Zhang L, Liu JS, Wu SJ, et al. (2012) Comprehensive assessment of heavy metal contamination in sediment of the Pearl River Estuary and adjacent shelf. Mar Pollut Bull 64: 1947–1955.
3. Gao XL, Chen CTA (2012) Heavy metal pollution status in surface sediments of the coastal Bohai Bay. Water Res 46: 1901–1911.
4. Gao XL, Zhou FX, Chen CTA (2014) Pollution status of the Bohai Sea, China: An overview of the environmental quality assessment related trace metals. Environ Int 62: 12–30.
5. Varol M (2011) Assessment of heavy metal contamination in sediments of the Tigris River (Turkey) using pollution indices and multivariate statistical techniques. J Hazard Mater 195: 355–364.
6. Gaur VK, Gupta SK, Pandey SD, Gopal K, Misra V (2005) Distribution of heavy metals in sediment and water of River Gomti. Environ Monit Assess 102: 419–433.
7. van Den Berg GA, Loch JPG, van Der Heijdt LM, Zwolsman JJG (1999) Mobilisation of heavy metals in contaminated sediments in the river Meuse, The Netherlands. Water Air Soil Poll 116: 567–586.
8. Long ER, MacDonald DD, Smith SC, Calder FD (1995) Incidence of adverse biological effects within ranges of chemical concentrations in marine and estuarine sediments. Environ Manage 19: 81–97.
9. SEPA (State Environmental Protection Administration of China) (2002) Marine Sediment Quality (GB 18668–2002). Standards Press of China, Beijing.
10. Caeiro S, Costa MH, Fernandes F, Silveira N, Coimbra A, et al. (2005) Assessing heavy metal contamination in Sado Estuary sediment: An index analysis approach. Ecol Indic 5: 151–169.

11. Porstner U (1989) Lecture Notes in Earth Sciences (Contaminated Sediments). Springer Verlag, Berlin. pp. 107–109.
12. Sakan SM, Dordević DS, Manojlović DD, Predrag PS (2009) Assessment of heavy metal pollutants accumulation in the Tisza river sediments. J Environ Manage 90: 3382–3390.
13. Zhao S, Feng CH, Yang YR, Niu JF, Shen ZY (2012) Risk assessment of sedimentary metals in the Yangtze Estuary: New evidence of the relationships between two typical index methods. J Hazard Mater 241–242: 164–172.
14. Li XY, Liu LJ, Wang YG, Luo GP, Chen X, et al. (2012) Integrated assessment of heavy metal contamination in sediments from a coastal industrial basin, NE China. PLoS ONE 7: e39690.
15. Long ER, MacDonald DD, Severn CG, Hong CB (2000) Classifying probabilities of acute toxicity in marine sediments with empirically derived sediment quality guideline. Environ Toxicol Chem 19: 2598–2601.
16. Sundaray SK, Nayak BB, Lin S, Bhatta D (2011) Geochemical speciation and risk assessment of heavy metals in the river estuarine sediments–a case study: Mahanadi basin, India. J Hazard Mater 186: 1837–1846.
17. Yu GB, Liu Y, Yu S, Wu SC, Leung AOW, et al. (2011) Inconsistency and comprehensiveness of risk assessments for heavy metals in urban surface sediments. Chemosphere 85: 1080–1087.
18. Ma SS, Xin FY, Cui Y, Qiao XY (2004) Assessment of main pollution matter volume into the sea from Yellow River and Xiaoqing River. Mar Fish Res 25: 47–51. (In Chinese with English abstract)
19. Wang HJ, Yang ZS, Saito Y, Liu JP, Sun XX (2006) Interannual and seasonal variation of the Huanghe (Yellow River) water discharge over the past 50 years: Connections to impacts from ENSO events and dams. Global Planet Change 50: 212–225.
20. Wang HJ, Yang ZS, Saito Y, Liu JP, Sun XX, et al. (2007) Stepwise decreases of the Huanghe (Yellow River) sediment load (1950–2005): Impacts of climate change and human activities. Global Planet Change 57: 331–354.
21. Rauret G, López-Sánche JF, Sahuquillo A, Rubio R, Davidson C, et al. (1999) Improvement of the BCR three step sequential extraction procedure prior to the certification of new sediment and soil reference materials. J Environ Monitor 1: 57–61.
22. Gao XL, Chen CTA, Wang G, Xue QZ, Tang C, et al. (2010) Environmental status of Daya Bay surface sediments inferred from a sequential extraction technique. Estuar Coast Shelf S 86: 369–378.
23. Rapin F, Tessier A, Campbell PGC, Carignan R (1986) Potential artifacts in the determination of metal partitioning in sediments by a sequential extraction procedure. Environ Sci Technol 20: 836–840.
24. Hjorth T (2004) Effects of freeze-drying on partitioning patterns of major elements and trace metals in lake sediments. Anal Chim Acta 526: 95–102.
25. Horowitz AJ, Elrick KA (1988) Interpretation of bed sediment traces metal data: methods for dealing with the grain size effect. In: Lichrenberg JJ, Winter JA, Weber CI, Fradkin L, editors. Chemical and Biological Characterization of Sludges, Sediments Dredge Spoils, and Drilling Muds. ASTM Special Technical Publication, Philadelphia 976. pp. 114–128.
26. Howari FM, Banat KM (2001) Assessment of Fe, Zn, Cd, Hg, and Pb in the Jordan and Yarmouk river sediments in relation to their physicochemical properties and sequential extraction characterization. Water Air Soil Poll 132: 43–59.
27. Gilliam FS, Richter DD (1988) Correlations between extractable Na, K, Mg, Ca, P & N from fresh and dried samples of two Aquults. Eur J Soil Sci 39: 209–214.
28. MacDonald DD, Scottcarr R, Calder FD, Long ER, Ingersoll CG (1996) Development and evaluation of sediment quality guidelines for Florida coastal waters. Ecotoxicology 5: 253–278.
29. Taylor SR, McLennan SM (1995) The geochemical evolution of the continental crust. Rev Geophys 33: 241–265.
30. Alexander CR, Smith RG, Calder FD, Schropp SJ, Windom HL (1993) The historical record of metal enrichments in two Florida estuaries. Estuaries 16: 627–637.
31. Müller G (1969) Index of geoaccumulation in sediments of the Rhine River. Geojournal 2: 108–118.
32. Perin G, Craboledda L, Cirillo M, Dotta L, Zanette ML, et al. (1985) Heavy metal speciation in the sediments of Northern Adriatic Sea: a new approach for environmental toxicity determination. In: Lekkas TD, editor. Heavy Metal in the Environment. CEP Consultant, Edinburgh 2. pp. 454–456.
33. Hakanson L (1980) An ecological risk index for aquatic pollution control. A sedimentological approach. Water Res 14: 975–1001.
34. Carr RS, Long ER, Windom HL, Chapman DC, Thursby G, et al. (1996) Sediment quality assessment studies of Tampa Bay, Florida. Environ Toxicol Chem 15: 1218–1231.
35. Gao XL, Li PM, Chen CTA (2013) Assessment of sediment quality in two important areas of mariculture in the Bohai Sea and the northern Yellow Sea based on acid-volatile sulfide and simultaneously extracted metal results. Mar Pollut Bull 72: 281–288.
36. Li GG, Hu BQ, Bi JQ, Leng QN, Xiao CQ, et al. (2013) Heavy metals distribution and contamination in surface sediments of the coastal Shandong Peninsula (Yellow Sea). Mar Pollut Bull 76:420–426.
37. Hu BQ, Li J, Zhao JT, Yang J, Bai FL, et al. (2013) Heavy metal in surface sediments of the Liaodong Bay, Bohai Sea: distribution, contamination and sources. Environ Monit Assess 185: 5071–5083.
38. Yu Y, Song JM, Li XG, Yuan HM, Li N, et al. (2013) Fractionation, sources and budgets of potential harmful elements in surface sediments of the East China Sea. Mar Pollut Bull 68: 157–167.
39. Gao XL, Li PM (2012) Concentration and fractionation of trace metals in surface sediments of intertidal Bohai Bay, China. Mar Pollut Bull 64: 1529–1536.
40. Hu NJ, Shi XF, Liu JH, Huang P, Yang G, et al. (2011) Distributions and impacts of heavy metals in the surface sediments of the Laizhou Bay. Adv Mar Sci 29: 63–72. (In Chinese with English abstract)
41. Luo XX, Zhang R, Yang JQ, Liu RH, Tang W, et al. (2010) Distribution and pollution assessment of heavy metals in surface sediment in Laizhou Bay. Ecol Environ Sci 19: 262–269. (In Chinese with English abstract)
42. Luo W, Lu YL, Wang TY, Hu WY, Jiao WT, et al. (2010) Ecological risk assessment of arsenic and metals in sediments of coastal areas of northern Bohai and Yellow Seas, China. Ambio 39: 367–375.
43. Zhang WG, Feng H, Chang JN, Qu JG, Xie HX, et al. (2009) Heavy metal contamination in surface sediments of Yangtze River intertidal zone: An assessment from different indexes. Environ Pollut 157: 1533–1543.
44. Durán I, Sánchez-Marín P, Beiras R (2012) Dependence of Cu, Pb and Zn remobilization on physicochemical properties of marine sediments. Mar Environ Res 77: 43–49.
45. Pekey H, Doğan G, 2013. Application of positive matrix factorisation for the source apportionment of heavy metals in sediments: A comparison with a previous factor analysis study. Microchem J 106: 233–237.
46. Hyun S, Lee C-H, Lee T, Choi J-W (2007) Anthropogenic contributions to heavy metal distributions in the surface sediments of Masan Bay, Korea. Mar Pollut Bull 54: 1059–1068.
47. Larrose A, Coynel A, Schäfer J, Blanc G, Massé L, et al. (2010) Assessing the current state of the Gironde Estuary by mapping priority contaminant distribution and risk potential in surface sediment. Appl Geochem 25: 1912–1923.
48. Lu XQ, Werner I, Young TM (2005) Geochemistry and bioavailability of metals in sediments from northern San Francisco Bay. Environ Int 31: 593–602.
49. Beck M, Böning P, Schückel U, Stiehl T, Schnetger B et al. (2013) Consistent assessment of trace metal contamination in surface sediments and suspended particulate matter: A case study from the Jade Bay in NW Germany. Mar Pollut Bull 70: 100–111.
50. Zhuang W, Gao XL (2013) Acid-volatile sulfide and simultaneously extracted metals in surface sediments of the southwestern coastal Laizhou Bay, Bohai Sea: Concentrations, spatial distributions and the indication of heavy metal pollution status. Mar Pollut Bull 76: 128–138.
51. Gomes F, Godoy JM, Godoy ML, Carvalho Z, Lopes R, et al. (2009) Metal concentrations, fluxes, inventories and chronologies in sediments from Sepetiba and Ribeira Bays: A comparative study. Mar Pollut Bull 59: 123–133.
52. Goh BPL, Chou LM (1997) Heavy metal levels in marine sediments of Singapore. Environ Monit Assess 44: 67–80.
53. Turner A (2010) Marine pollution from antifouling paint particles. Mar Pollut Bull 60: 159–171.
54. Guillen G, Ruckman M, Smith S, Broach L (1993) Marina impacts in Clean Lake and Galveston Bay. Special Report D7-001A prepared for the Texas Water Commission, Houston, TX.
55. Hinkey LM, Zaidi BR, Volson B, Rodriguez N (2005) Identifying sources and distributions of sediment contaminants at two US Virgin Islands marinas. Mar Pollut Bull 50: 1244–1250.
56. Hinkey LM, Zaidi BR (2007) Differences in SEM–AVS and ERM–ERL predictions of sediment impacts from metals in two US Virgin Islands marinas. Mar Pollut Bull 54: 180–185.
57. Tessier A, Campbell PGC, Bisson M (1979) Sequential extraction procedure for the speciation of particulate trace metals. Anal Chem 51: 844–851.
58. Karbassi AR, Shankar R (2005) Geochemistry of two sediment cores from the west coast of India. Int J Environ Sci Te 1: 307–316.
59. Salmonas W, Förstner U (1980) Trace metal analysis on polluted sediments, part-II: Evaluation of environmental impact. Environ Technol Lett 1: 506–517.
60. Lambert R, Grant C, Sauvé S (2007) Cadmium and zinc in soil solution extracts following the application of phosphate fertilizers. Sci Total Environ 378: 293–305.
61. Zhang H, Shan BQ (2008) Historical records of heavy metal accumulation in sediments and the relationship with agricultural intensification in the Yangtze–Huaihe region, China. Sci Total Environ 399: 113–120.
62. Alloway BJ (1990) Cadmium. In: Alloway BJ, editor. Heavy metals in soils. Blackie, Glasgow. pp. 100–124.
63. Rath P, Panda UC, Bhatta D, Sahu KC (2009) Use of sequential leaching, mineralogy, morphology and multivariate statistical technique for quantifying metal pollution in highly polluted aquatic sediments – a case study: Brahamani and Nandira rivers, India. J Hazard Mater 163: 632–644.
64. Fytianos K, Lourantou A (2004) Speciation of elements in sediment samples collected at lakes Volvi and Koronia, N. Greece. Environ Int 30: 11–17.
65. Ytreberg E, Karlsson J, Hoppe S, Eklund B, Ndungu K (2008) Effect of organic complexation on copper accumulation and toxicity to the estuarine Red Macroalga *Ceramium tenuicorne*: a test of the free ion activity model. Environ Sci Technol 45: 3145–3153.
66. Jain CK (2004) Metal fractionation study on bed sediments of River Yamuna, India. Water Res 38: 569–578.

67. Zhang J, Liu CL (2002) Riverine composition and estuarine geochemistry of particulate metals in China–weathering features, anthropogenic impact and chemical fluxes. Estuar Coast Shelf S 54: 1051–1070.
68. Duan LQ, Song JM, Xu YY, Li XG, Zhang Y (2010) The distribution, enrichment and source of potential harmful elements in surface sediments of Bohai Bay, North China. J Hazard Mater 183: 155–164.
69. Li XD, Wai OWH, Li YS, Coles BJ, Ramsey MH, et al. (2000) Heavy metal distribution in sediment profiles of the Pearl River estuary, South China. Appl Geochem 15: 567–581.

Acoustic Telemetry Validates a Citizen Science Approach for Monitoring Sharks on Coral Reefs

Gabriel M. S. Vianna[1,2]*, Mark G. Meekan[2], Tova H. Bornovski[3], Jessica J. Meeuwig[1]

1 School of Animal Biology and The UWA Oceans Institute, The University of Western Australia, Perth, Western Australia, Australia, **2** Australian Institute of Marine Science, Perth, Western Australia, Australia, **3** Micronesian Shark Foundation, Koror, Palau

Abstract

Citizen science is promoted as a simple and cost-effective alternative to traditional approaches for the monitoring of populations of marine megafauna. However, the reliability of datasets collected by these initiatives often remains poorly quantified. We compared datasets of shark counts collected by professional dive guides with acoustic telemetry data from tagged sharks collected at the same coral reef sites over a period of five years. There was a strong correlation between the number of grey reef sharks (*Carcharhinus amblyrhynchos*) observed by dive guides and the telemetry data at both daily and monthly intervals, suggesting that variation in relative abundance of sharks was detectable in datasets collected by dive guides in a similar manner to data derived from telemetry at these time scales. There was no correlation between the number or mean depth of sharks recorded by telemetry and the presence of tourist divers, suggesting that the behaviour of sharks was not affected by the presence of divers during our study. Data recorded by dive guides showed that current strength and temperature were important drivers of the relative abundance of sharks at monitored sites. Our study validates the use of datasets of shark abundance collected by professional dive guides in frequently-visited dive sites in Palau, and supports the participation of experienced recreational divers as contributors to long-term monitoring programs of shark populations.

Editor: Danilo Russo, Università degli Studi di Napoli Federico II, Italy

Funding: Financial support for the acoustic telemetry was provided by the Save Our Seas Foundation. The funders had no role in study design, data collection and analysis, decision to publish, or preparation of the manuscript.

Competing Interests: The authors have declared that no competing interests exist.

* E-mail: gabe.vianna@gmail.com

Introduction

Many shark species are experiencing unsustainable rates of mortality due to fishing, a phenomenon that is driving population declines globally [1]. Despite this emerging crisis, our knowledge of the distribution, abundance and ecology of many species is generally poor. In 2014, an assessment of the extinction risk of 1,041 species of elasmobranchs concluded that almost half (487 species) were categorised as "Data Deficient", meaning that a lack of information prevented any firm conclusions being drawn on their population status and trajectories [2]. This has occurred at a time when there is increasing evidence of the importance of sharks as top-down regulators of the structure and function of marine ecosystems [3] and recognition of their current and potential value as a non-consumptive resource that supports local economies through ecotourism [4].

The assessment and monitoring of shark populations through fishery-independent techniques presents considerable challenges due to the naturally low population densities and relatively large home ranges common to most species [5,6]. The large scale (tens to hundreds of km) and long-term (years to decades) monitoring programs that can be required to document the status of populations are thus expensive, particularly if they involve in-water activities such as SCUBA diving. For this reason, implementing these initiatives is often beyond the means of governments of developing countries or organisations with interests in the conservation of sharks. Thus, there is an urgent need for the creation and adoption of simple, standardised and low-cost methods for monitoring shark populations [7].

Data collected by the public can provide a cost-effective means of monitoring populations of wild animals [8]. Such "citizen science" initiatives are growing in popularity as alternatives to conventional scientific sampling as they offer the opportunity to gather large datasets at reduced cost [9,10]. In the marine environment, this approach is particularly useful for the study of conspicuous animals and megafauna inhabiting coastal areas and coral reefs, where data obtained from the public are relatively easy to collate. This approach has been used to describe the distribution and ecology of many species, including green and hawksbill turtles (*Chelonia mydas* and *Eretmochelys imbricata*), minke whales (*Balaenoptera acutorostrata*) and manta rays (*Manta alfredi*) [11–13]. Sharks have also been the target of many of these initiatives, with projects based on data collected by recreational snorkelers and divers used to investigate patterns in distribution, demographics, abundance, habitat use, movement and the effects of environmental and anthropogenic factors [7,14–19]. Structured programs where data are gathered by recreational participants have also been used to provide baseline data and monitor spatial and temporal trends in abundance, which can then be used to design and assess the efficiency of conservation measures [13,15].

Recreational divers and snorkelers have been used to collect data on elasmobranchs in two principal ways: firstly, by recording counts of animals seen during a dive [7,16,20] and secondly, by taking identification photos that can then be used in mark-recapture modelling to estimate trends in abundance and demography [17,21]. The latter approach focuses on those species that have distinctive patterning or scars that allow animals to be identified individually, such as whale sharks (*Rhincodon typus*), manta rays and white sharks (*Carcharodon carcharias*), but is unsuitable for the many reef and pelagic sharks that generally lack any persistent features that might be used to distinguish individuals. For these species, counts by divers provide one of the simplest means to monitor numbers.

Traditional approaches to underwater visual surveys involve standardized techniques that focus on quantifying the area sampled and the abundance and length of individuals within the sample space [5,22]. Such rigorous protocols are a feature of science-based diving programs, but are not necessarily applied during recreational diving. For this reason, datasets of counts collected by recreational divers do not usually generate the data necessary for calculations of total abundance, density and biomass as area-based metrics [5,23]. It is also recognised that other issues may potentially compromise the quality of recreational datasets, such as rounding bias, misidentification and inflation of estimates [7,9,20]. Although simulation and comparative studies have suggested that recreational divers may indeed be able to report shark numbers in an accurate and reliable manner [7,23], there is a need for independent validation of this approach in the field.

Acoustic telemetry can provide a means to address these issues and validate datasets generated by citizen science programmes. Arrays of acoustic receivers are now commonplace in many coastal marine environments that are inhabited by marine megafauna such as sharks [24]. Acoustic tags can be deployed on animals without causing modification of behaviour and will report their presence whenever they are within range of the array [24]. The presence/absence data generated by these tags are commonly used to monitor the attendance of individuals tagged at the monitored sites [24], and provide an index of relative abundance that can be used to identify trends in the populations over time. In places where citizen science initiatives and arrays overlap, acoustic telemetry can be used to assess the validity of citizen science datasets.

In our study, we compared datasets of shark relative abundance collected by professional dive guides with tagging data generated by passive acoustic telemetry at the same sites on coral reefs in Palau, Micronesia. We aimed to determine if the observations reported by dive guides produced comparable estimates of relative abundances and temporal patterns in numbers of sharks as those obtained from presence/absence data derived from acoustic tagging and monitoring. We also used the telemetry data to investigate the effect of the presence of tourist divers (i.e. observer effect) on the relative abundance and depth use by sharks. Finally, we analysed the citizen science data in order to identify environmental correlates of patterns of relative abundance of sharks at dive sites.

Methods

Ethics Statement

This project was conducted under the Republic of Palau Marine Research Permit no. RE-09-26 and the Koror State Marine Research Permit no. 10–204. Shark tagging in 2011 was also conducted under UWA animal ethics permit no. RA/3/100/975, in adherence to provisions contained within the Australian Code of Practice for the Care and Use of Animals for Scientific Purposes. Participants of the shark counts were aware of the use of these data for research and provided written consent for the use of the data collected. This project has been assessed as exempt from ethics review by the Human Research Ethics Office of the University of Western Australia (protocol code: RA/4/1/6457).

Study Area

Palau supports a dive industry consisting of approximately 20 tourism businesses that use mainly small speed boats to transport tourists to local reef sites for two to three dives per day. In 2010, it was estimated that approximately 41,000 tourists visited Palau to engage in dive activities, of which approximately 8,600 visited the country specifically to dive with sharks [4]. Most of the popular dive sites are on the southwest (leeward) area of the barrier reef that surrounds Babeldaob, the main island of Palau (7°N, 134°W). This area encompasses dive sites that vary in topography from relatively sheltered sand flats and coral gardens to steep walls and promontories that project into oceanic waters on the outer reef slope [25]. Diving occurs in a variety of habitats including sandy channels and caverns, however the main drawcard for divers visiting Palau are the "drop-off" dives on the steep reef slope exposed to the open ocean, where there are usually moderate to strong tidal currents. These dives sites are characterised by high visibility (>30 m) and a rich diversity of marine life with high abundances of large pelagic species. Many of these dive sites host aggregations of reef sharks, which are composed mainly of resident grey reef (*Carcharhinus amblyrhynchos*) and whitetip reef sharks (*Triaenodon obesus*) [4,26]. Dives at these sites are typically conducted during periods of relatively high tidal currents when sharks swim just off the edge of the slope. Divers enter the water up-current and, on arrival at the aggregation sites, attach themselves by a hook and line to the reef crest so that they can remain stationary to view sharks and large fish passing along the drop-off [4]. Once safety time limits at these depths (typically between 12–25 m) are attained, the divers release hooks and lines and drift along the reef crest making a slow ascent to the surface. For safety reasons, the routes and durations of these dives are similar through time (GMSV. and MGM. pers. obs.).

Data Collection

Our dataset consisted of counts of sharks sighted during dives by dive guides who worked as employees of a dive tourism business. Standard questionnaires were completed after the day trips by dive guides from October 2007 to November 2012. A total of 62 dive guides recorded information for 2,360 dives at 52 dive sites in Palau. Each questionnaire contained information on the dive site visited, date, species and counts of individual sharks sighted by the guide. Dive guides estimated the depth of the divers during the sightings, current strength (0- no current; 1- weak; 2- moderate; 3- strong current), visibility (in meters), and recorded dive time and number of tourist divers in the group. Each completed questionnaire provided observations from a single dive guide.

Most of the dive guides participating in the study were local residents familiar with the identification reef shark species in Palau. Guides were instructed to report the total number of individual sharks of each species observed during the entire dive. We also instructed the dive guides to be conservative with counts, observing features that could permit individual identification (e.g. pigment patterns, marks and scars) and reduce the potential for repeated counts. An office staff member was responsible for the administration and management of the survey, encouraging dive guides to return questionnaires regularly. This staff member was also trained to enter data and maintain the dataset. To promote engagement and consistency in data collection, we also established an annual event to provide feedback to the participants, where the dive guides who collected data regularly would receive small

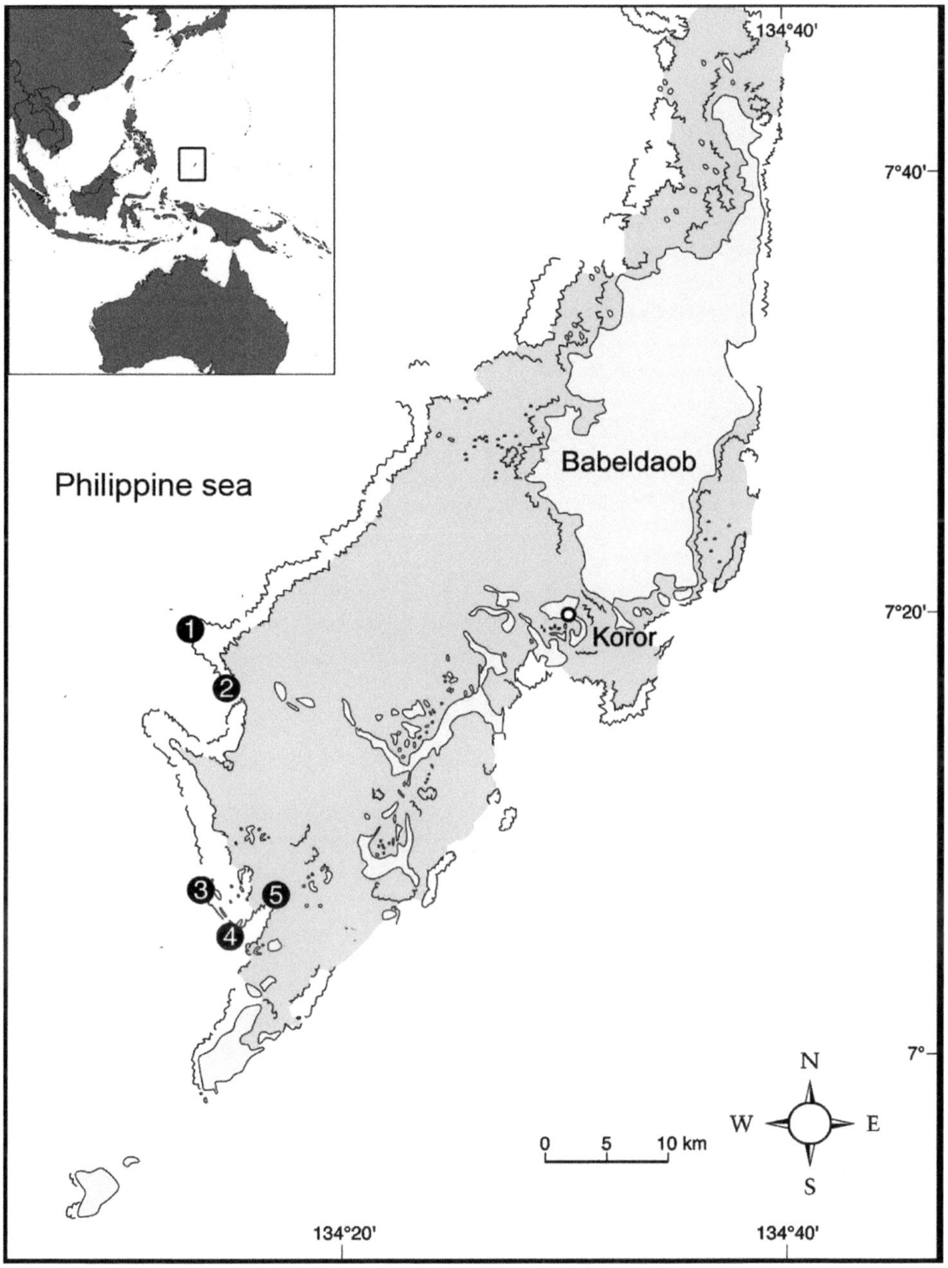

Figure 1. Study area in Palau. Numbers indicate location of dive sites monitored: 1) Siaes Corner, 2) Ulong Channel, 3) Blue Corner, 4) New Drop-off and 5) German Channel. Grey shade represents lagoonal area, light grey indicates islands. Numbers one to four also indicate the location of acoustic receivers.

rewards. During these events, the researchers involved in the project would also provide lectures to the dive community, where updates on relevant issues and results obtained from the dataset were presented.

We used passive acoustic telemetry to monitor the attendance of resident grey reef sharks at four key dive sites (Blue Corner, Siaes Corner, Ulong Channel and New Drop-off), known to host predictable aggregations of sharks (Figure 1). An array of four Vemco VR2w acoustic receivers monitored sharks tagged with acoustic transmitters from November 2008 to December 2012, a period that overlapped the dive guide records of shark counts. The receivers were deployed at depths between 25 and 40 m on the barrier reef drop-off or slope and recorded the presence of tagged sharks within 200 m of the receiver location [26]. We internally tagged 39 grey reef sharks (38 females, 1 male) with Vemco V16 coded tags with battery life ranging between three and a half and ten years. Ten of these tags were also fitted with pressure sensors, which provided a record of depth of the tagged sharks (for full description see [26]). Temperature loggers were deployed near the dive sites at 15 m depth and provided records of daily temperature from January 2009 to March 2012. We used the number of sharks

Table 1. Descriptive metrics of dive guide observations and telemetry data at the monitored sites in Palau.

	Blue Corner					Ulong Channel					Siaes Corner					New Drop-off					German Channel					Total				
Metrics	**Mean**	**Min**	**Max**	**SD**	***n***	**Mean**	**Min**	**Max**	**SD**	***n***	**Mean**	**Min**	**Max**	**SD**	***n***	**Mean**	**Min**	**Max**	**SD**	***n***	**Mean**	**Min**	**Max**	**SD**	**n**	**Mean**	**Min**	**Max**	**SD**	***n***
Number of dives	388	–	–	–	–	223	–	–	–	–	118	–	–	–	–	138	–	–	–	–	385	–	–	–	–	250.4	118	388	130.4	5
Estimated area of dive site (m^2)	17000	–	–	–	–	27000	–	–	–	–	19000	–	–	–	–	22000	–	–	–	–	10000	–	–	–	–	19000	10000	27000	6285	6
Dives per month	7.9	1	31	5.3	48	4.8	1	18	3.4	45	2.8	1	7	1.6	41	3.3	1	13	2.5	40	7.9	1	37	6.0	47	5.7	1	48	4.6	48
Dive time (min)	56	30	70	6.4	354	60	25	88	7.6	206	57	20	68	6.3	108	57	45	85	6.5	123	57	30	80	6.8	336	57.4	20	88	6.9	1127
Divers depth	16.3	5	35	5.1	258	15.4	5	45	4.7	158	18	10	25	5.4	86	15.4	5	30	5.3	87	15.7	5	30	5.1	281	16.2	5	45	5.1	870
Divers per dive	17.5	10	40	7.5	354	16	10	40	6.7	204	15.5	10	30	6.6	110	13.9	10	30	5.2	123	19.4	10	40	8.3	334	17.2	10	40	7.5	1125
Days with telemetry data	185	–	–	–	–	122	–	–	–	–	67	–	–	–	–	32	–	–	–	–	–	–	–	–	–	101.5	32	185	66.9	4
Sharks detected daily (telemetry)	7.2	2	19	2.9	185	6.9	3	13	2.6	122	2.7	1	8	1.3	67	3.2	2	7	1.0	32	–	–	–	–	–	6.0	1	19	3.1	406

SD: standard deviation, *n*: sample size.

Table 2. Frequency of occurrence (FO) and mean daily relative abundance of sharks observed by dive guides (Abund) and standard error (SE) of most common species of sharks at the monitored sites in Palau.

	Blue Corner		Ulong Channel		Siaes Corner		New Drop-off		German Channel		Overall	
Species	**FO (%)**	**Abund**	**FO (%)**	**Abund**	**FO (%)**	**Abund**	**FO (%)**	**Abund**	**FO (%)**	**Abund**	**FO (%)**	**Abund**
All sharks	100	20.3 (0.6)	100	19.0 (0.7)	100	13.8 (0.9)	100	11.5 (0.6)	100	11.1 (0.4)	100	15.6 (0.3)
Grey reef shark	86	12.4 (0.5)	89	13.4 (0.6)	94	10.9 (0.9)	87	6.9 (0.6)	83	6.8 (0.3)	86	10.1 (0.3)
Whitetip reef shark	86	7.7 (0.3)	86	5.5 (0.3)	76	2.7 (0.3)	88	4.3 (0.3)	81	4 (0.2)	83	5.3 (0.1)
Blacktip reef shark	13	0.3 (0.1)	9	0.2 (0.2)	8	0.2 (0.1)	22	0.4 (0.2)	17	0.25 (0.1)	14	0.3 (0.1)
Other species[a]	1	0.01 (0.0)	1	0.01 (0.1)	0	0	1	0.01 (0.0)	4	0.04 (0.1)	2	0.02 (0.0)

[a]Other species include hammerhead, zebra, bull, nurse, and lemon sharks.

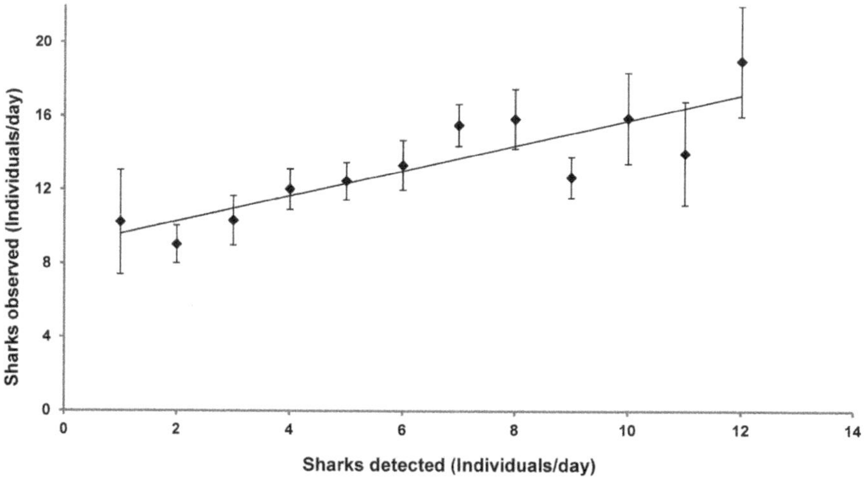

Figure 2. Relationship between daily underwater observations and acoustic telemetry of grey reef sharks in Palau. Mean daily relative abundance of grey reef sharks observed by dive guides as a function of daily number of sharks detected by telemetry. Error bars indicate SE. $y = 0.68x + 8.92$, $R^2 = 0.74$.

detected by the receivers as an index of the relative abundance of the tagged sharks at the monitored sites.

Data Analysis

We limited our analysis to data from experienced dive guides, defined as guides who participated for at least three of the five years during which the study was undertaken, who had returned a minimum of 100 questionnaires. In order to obtain a reasonable representation of questionnaires for each calendar month, our analysis focused on the dive sites that also yielded a minimum of 100 questionnaires. We estimated the daily relative abundance (no. day^{-1}) of sharks observed by dive guides as the mean value of all dives in a given day at the same site. Given daily variation in diving activities, we also calculated a mean daily abundance for each calendar week.

We calculated the frequency of occurrence of each species as the proportion of days reported when a given species was sighted by dive guides. Our statistical analyses focused on grey reef and whitetip reef sharks, the two most abundant species recorded by dive guides at our study sites. We first used linear regression to compare the daily relative abundance of grey reef sharks observed by dive guides and the number of individually tagged sharks recorded by receivers at the same site on the same day. This analysis allowed us to determine whether there were correlations between diver counts and the number of sharks tagged attending the array. Our dataset only included days when sharks were detected by receivers on a minimum of two occasions.

We also used linear regression to investigate the effects of the presence of divers on the behaviour of sharks at the study sites. For this analysis, we regressed the number of tagged grey reef sharks present at a given site and day against the number of tourist divers reported to be in the water during the corresponding day. Multiple dives reported on the same day at a given site were treated as separate samples unless occurring simultaneously (in the same hour), in which case we summed the number of tourist divers reported by each guide as a measure of the total potential influence on the sharks. We also used linear regression to analyse the relationship between the mean depth of the tagged sharks and 1) the number of tourist divers reported to be in the water during a dive and 2) the mean depth of these divers.

We used circular regression [27] to relate monthly patterns in mean daily relative abundance of grey reef sharks observed by dive guides with telemetry records. We used a *t*-test for slopes to compare the regressions of monthly observations by dive guides with the mean number of grey reef sharks detected using telemetry. Monthly estimates of relative abundance observed by dive guides and numbers of sharks detected by telemetry were calculated by averaging the daily values across a given month. For the regression, we included only telemetry values that had two or more corresponding observations by dive guides. The explanatory variable "month" was sine-transformed to account for cyclical variation in abundance of sharks. We also fitted circular regressions to investigate patterns of seasonality in the mean monthly relative abundance of sharks observed by dive guides at the selected study sites. This analysis was performed for all sharks combined but also separately for grey reef and whitetip reef sharks.

We used multiple linear regression to examine the influence of environmental factors on the relative abundance of sharks observed by dive guides. Our response variable was the log-transformed daily abundance of sharks (i.e. all sharks combined, grey reef, and whitetip reef sharks) averaged per week. Explanatory variables included in the models were: year, current strength, temperature, visibility, number of tourist divers and moon phase. We applied the Akaike's Information Criterion (AIC) test statistics including AIC, $_{\Delta}$AIC (difference between AIC of a given model and best fitted model) and $_{w}$AIC (weighted AIC) to select the models with best fit [28]. Since the order of inclusion of variables influences the AIC model selection process, we analysed the correlation coefficients of explanatory and response variables and used the function Regsubsets of the "leaps" R-package [29] to determine the order of inclusion of the variables in building the models. We validated our models by inspecting the residuals for patterns indicating likely violation of assumptions and fitted correlograms to visually inspect our dataset for auto-correlation. This analysis was performed for the data collected at Blue Corner, the site that yielded the largest number of weekly records ($n = 148$) from the dive guides.

All analyses used the R statistical software [30] and all summary metrics were reported as mean values and standard errors (±SE).

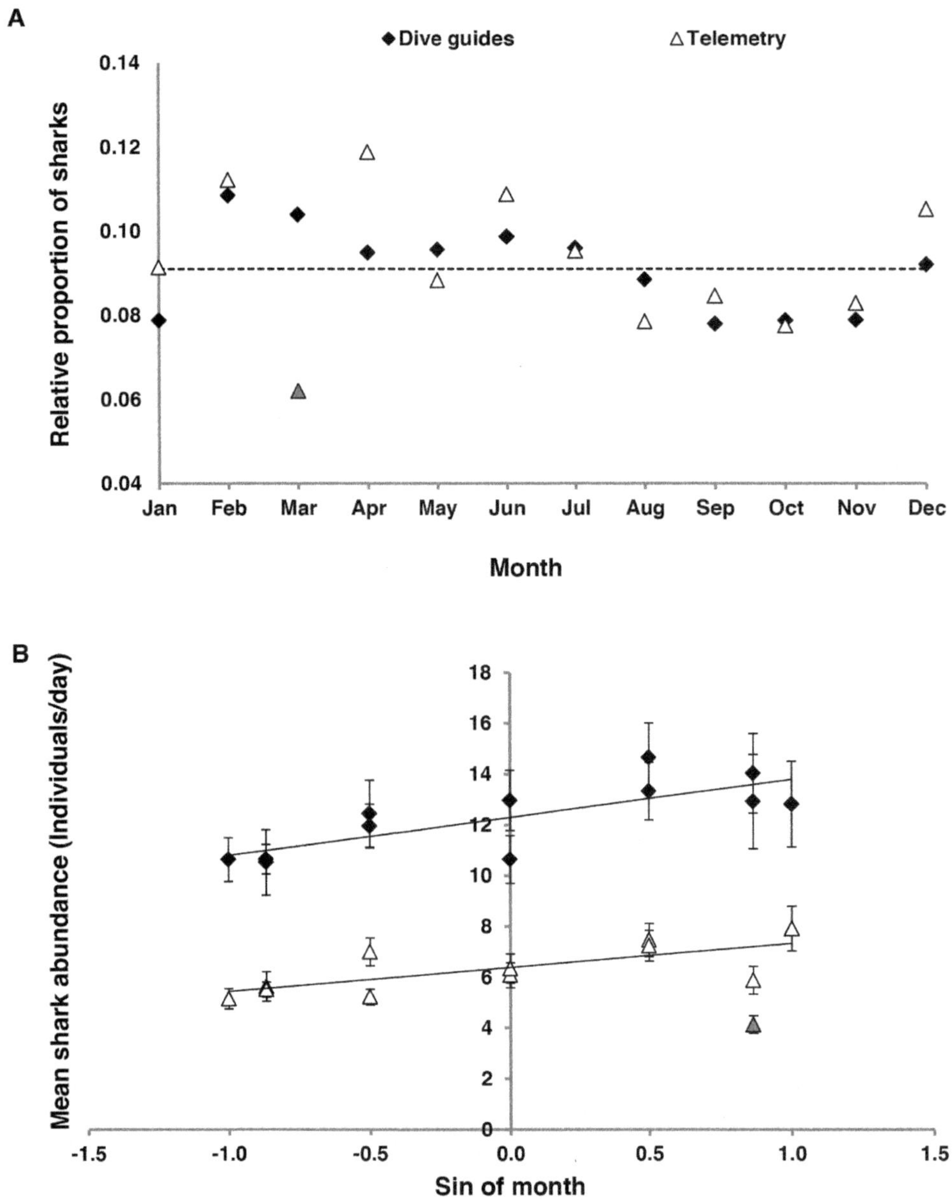

Figure 3. Monthly patterns of abundance of grey reef sharks at monitored sites in Palau. A) Proportion of sharks observed by dive guides and detected by telemetry monthly (calculated as the mean number of sharks observed daily in each month divided by the sum of mean values of all the months). Dashed line indicates the expected values in absence of seasonal variation. B) Mean monthly relative abundance of sharks as function of sine-transformed month. Error bars indicate SE. Dive guides: $y=1.49x+12.23$, $R^2=0.60$. Telemetry: $y=0.94x+6.39$, $R^2=0.51$. *Grey triangle indicates detection during low receiver performance and was not included in the analysis (see discussion).

Results

Dive Guide Datasets

Our final dataset, filtered to only include records of the most frequently visited dive sites and observations of selected guides (those returning more than 100 questionnaires), consisted of data for 1,252 dives (53%) that were collected by 24 dive guides (39%) at five dive sites (10%) (Blue Corner, Siaes Corner, New Drop-off, Ulong Channel and German Channel) over a period of five years. These dive sites are known to be aggregations or hotspots of charismatic megafauna, including reef sharks. The total number of dives at each site varied from 118 at Siaes Corner to 388 at Blue Corner, with an overall mean dive time of 57±0.22 min and a mean dive depth of 16±0.18 m (Table 1).

Dive guides reported seeing sharks during all dives at the selected sites. Grey reef and whitetip reef sharks were the species most commonly observed, with a frequency of occurrence of 86% and 83% and mean relative abundance per dive of 10.1±0.3 and 5.3±0.1 respectively (Table 2). Blacktip reef sharks (*Carcharhinus melanopterus*) were also sighted frequently (14% of dives) but with relative abundance of 0.3±0.1 sharks per dive. The other species of observed sharks were recorded very infrequently (2% of dives) and in low numbers (0.02 sharks per dive) (Table 2). Shark relative abundances at Blue Corner and Ulong Channel were higher than at other sites with mean values of 20.3±0.6 and 19.0±0.7 per dive respectively, while lower relative abundance was recorded at German Channel with mean value of 11.1±0.4 sharks per dive (Table 2).

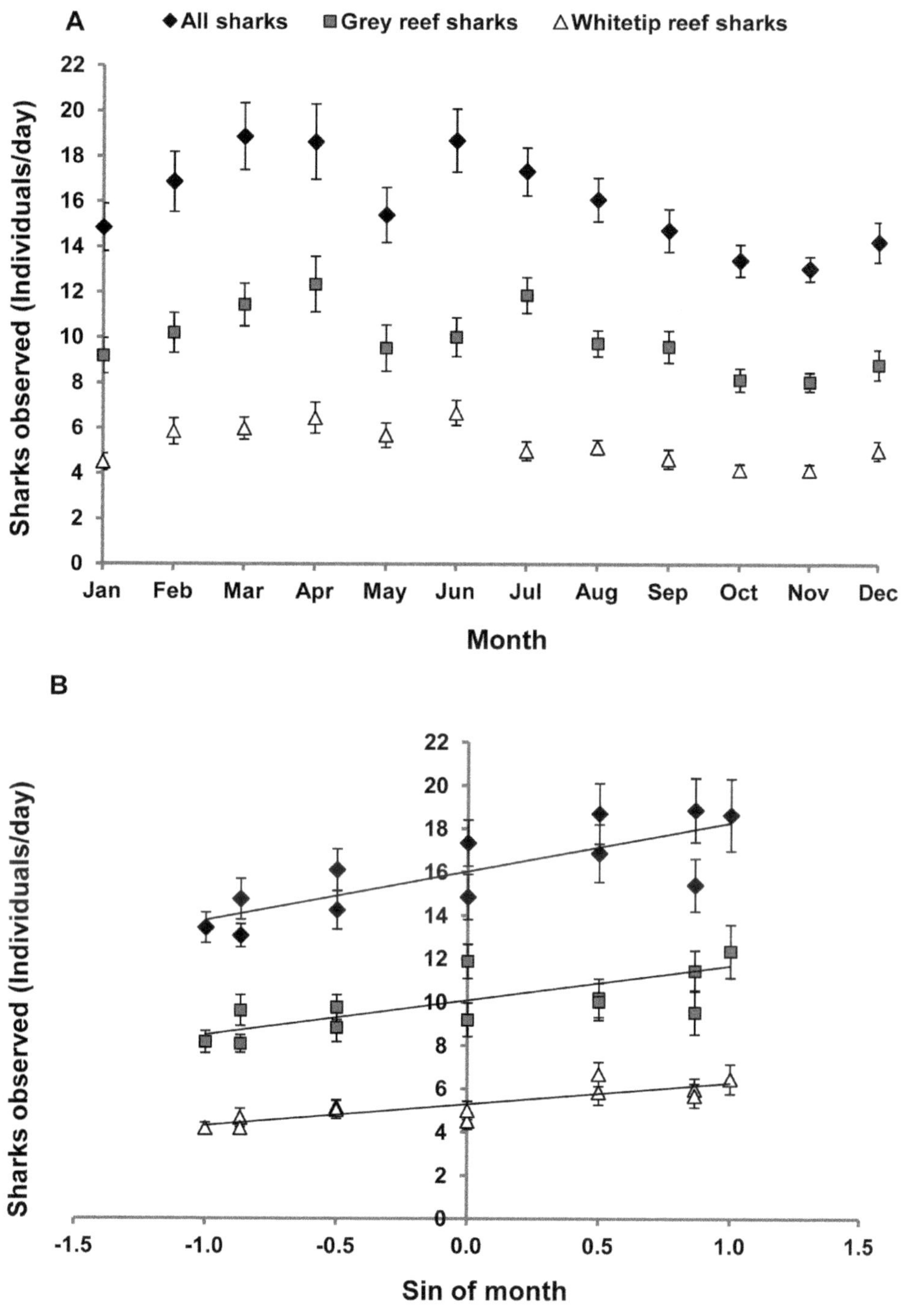

Figure 4. Monthly patterns of abundance of reef sharks observed by dive guides in Palau. A) Relative abundance of common species of reef sharks observed by divers at monitored sites. B) Relative abundance of common species of reef sharks observed by diver guides as a function of sine-transformed months. Error bars indicate SE. All sharks: $y = 2.70x+16.30$, $R^2 = 0.81$, Grey reef: $y = 1.59x+10.08$, $R^2 = 0.64$, Whitetip: $y = 0.99x+5.27$, $R^2 = 0.73$. *May values of abundance of "All sharks" and "Grey reef sharks" not included in regressions.

Integration of Dive Guide and Telemetry Data

Paired sets of abundance estimates provided from dive guides and acoustic detections were available for 406 dives (Table 1). For the dives when telemetry data were available, the number of sharks detected by acoustic receivers at a given site varied from one to 19 with a mean of six grey reef sharks detected per day (±0.02).

The regression analysis indicated a significant and strong relationship ($R^2 = 0.74$, $p<0.001$) between the mean daily relative abundance of grey reef sharks observed by dive guides and the number of tagged individuals detected by telemetry (Figure 2). Lowest relative abundance (10 sharks per dive) was observed on days when the number of sharks detected by telemetry was also low (1–3 sharks per day). An increase in the number of sightings by dive guides corresponded to increased numbers of tagged sharks

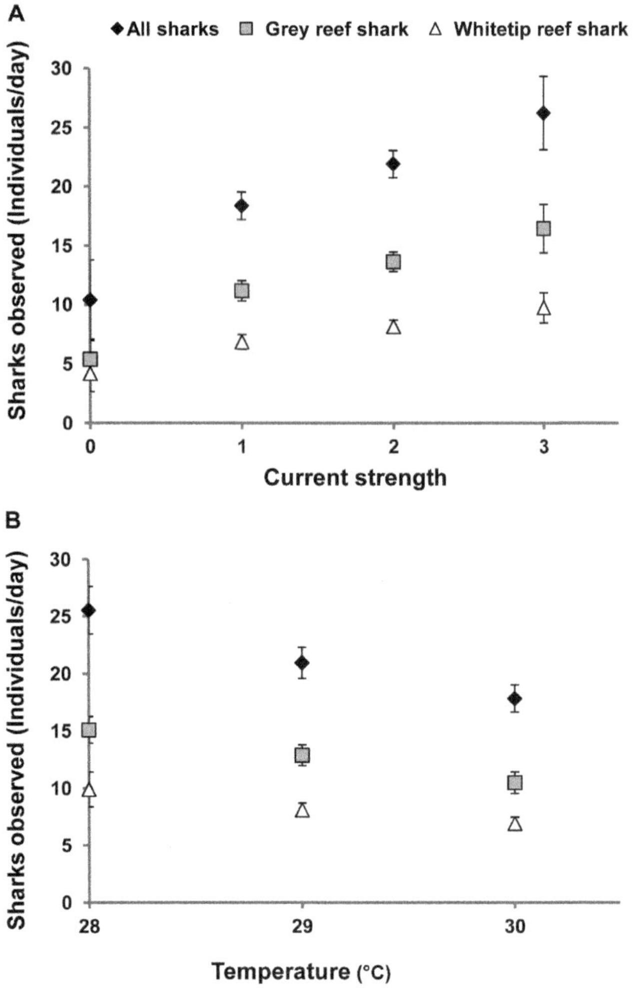

Figure 5. Environmental drivers of relative abundance of sharks at monitored sites in Palau. Relationship of mean daily relative abundance of sharks observed by dive guides and A) Current strength ($n = 143$ dives), and B) Temperature. ($n = 123$ dives). Error bars indicate SE.

detected acoustically. The highest relative abundance of 19 grey reef sharks was observed by guides when a maximum of 12 tagged sharks were detected on acoustic receivers.

Data generated by dive guides showed monthly variation in the number of grey reef sharks (Figures 3 and 4), with peaks of relative abundance occurring from February to June and lower values from August to November. This pattern was generally similar to that observed in the telemetry data (Figure 3). While there was no significant difference between the slopes of the regressions of the mean relative abundance observed by guides and telemetry detections as a function of month (t-test for slopes, $t = -0.76$, $p = 0.47$), there was some divergence between the two data sets during March and April (Figure 3A) when detections appeared proportionally lower than numbers observed by dive guides.

The linear regressions showed no significant relationship between the number of grey reef sharks detected by telemetry and the number of tourist divers present during the dives ($p = 0.48$). Mean values varied from 6.7±0.3 sharks detected when up 10 tourist divers were present to 7.2±1.1 when 40 or more tourist divers were present. Similarly, there were no significant relationships between the mean depth of tagged grey reef sharks and the number of tourist divers present during the dive ($p = 0.31$) or the mean depth of the tourist divers ($p = 0.44$).

Relative Abundance of Sharks Observed by Dive Guides in Relation to Environmental Drivers

Grey reef and whitetip reef sharks were present at the monitored sites throughout the year however, there was seasonal variation in the number of individuals of both species. Lower values of monthly relative abundance occurred in October and November, with means of 8.1±0.4 for grey reef sharks and 4.2±0.2 for whitetip reef sharks. Between March and April, the relative abundance of sharks was higher with monthly means of 12.4±1.2 and 6.5±0.7 for grey reef and whitetip reef sharks respectively (Figure 4). There was a sharp decline in relative abundance of grey reef sharks in May and June to a mean of 9.4±1.2. A similar but less pronounced pattern was also observed for whitetip reef sharks with relative abundance of 5.7±1.0 in May. The circular regressions indicated that the overall seasonal patterns were statistically significant with sine (month) explaining 81% of the variation in relative abundance of all sharks ($p<0.001$), 64% of the variation in relative abundance for grey reef ($p = 0.003$) and 73% for whitetip reef sharks ($p<0.001$).

Multiple regression indicated that current and temperature were the key environmental factors influencing the numbers of sharks recorded by dive guides ($R^2 = 0.18$, $p<0.001$; Table 3). There was a positive linear relationship between current and the relative abundance of all sharks ($R^2 = 0.14$, $p<0.001$), grey reef ($R^2 = 0.13$, $p<0.001$) and whitetip ($R^2 = 0.07$, $p = 0.002$) reef sharks, while temperature displayed a negative linear relationship with relative abundance of all sharks ($R^2 = -0.04$, $p = 0.02$) and grey reef sharks ($R^2 = 0.05$, $p = 0.03$) (Table 3, Figure 5). Year, visibility, moon phase and number of tourist divers in the water had little influence on the number of sharks sighted during the dives (Table 3).

Discussion

Citizen Science as a Means to Monitor Reef Sharks

Our analysis of data generated by dive guides suggests that citizen science initiatives can provide estimates of the relative abundance of reef sharks that are consistent with the estimates from long-term telemetry. Patterns in the relative abundance of grey reef sharks as reported by dive guides and numbers detected by telemetry at the monitored sites followed very similar trends at both daily and monthly scales. The high R^2 value indicated a strong positive relationship between the two metrics, with increases of daily relative abundance of grey reef sharks observed by dive guides matched by a corresponding linear increase in numbers of grey reef sharks detected by telemetry. Data generated by professional dive guides have great potential for providing estimates of relative abundance [15,16], changes in population size over small and large scales [7,19,31] and insights into the ecology and population trends of marine predators in reef and coastal habitats [15,20] however, the biases and limitations of such datasets remain poorly understood. One earlier study found that experience levels of observers made little difference in their abilities to detect sharks [7]. However, another study suggested that observations by dive guides might underestimate site fidelity [32]. Our study is the first to examine the ability of experienced observers to monitor patterns at a variety of temporal scales. For the most part, we found that dive guides produced datasets of shark relative abundance that tightly mirrored patterns generated by acoustic telemetry. Indeed, in some circumstances counts by dive guides may have been more accurate than those obtained by

Table 3. Multiple regression ranking results of the mean relative abundance of sharks observed by divers (Abund, response variable) as a function of the following explanatory variables: current strength (Cur), temperature (Temp), visibility (Vis), year, moon phase (Moon) and number of tourist divers in the water (Divers), $n = 148$ weeks.

Species	Model	df	AIC	ΔAIC	wAIC	R^2	*p*-value	*F*-value
All sharks	Ab_{DG}~Null	2	−32.10	22.78	0.00	0	0.48	0.71
	Ab_{DG}~Cur	3	−50.25	4.64	0.05	0.14	<0.001	4.58
	Ab_{DG}~Temp	3	−36.03	18.86	0.00	0.04	0.02	−2.42
	Ab_{DG}~Vis	3	−30.43	24.46	0.00	−0.01	0.48	0.72
	Ab_{DG}~Year	3	−30.22	24.67	0.00	−0.01	0.54	−0.61
	Ab_{DG}~Moon	5	−30.22	24.67	0.00	0.01	0.28	1.28
	Ab_{DG}~Divers	3	−30.23	24.66	0.00	−0.01	0.98	0.03
	Ab_{DG}~Cur+temp	**4**	**−54.89**	**0.00**	**0.49**	**0.18**	**<0.001**	**14.6**
	Ab_{DG}~Cur+temp+vis	5	−53.73	1.15	0.28	0.18	<0.001	9.9
	Ab_{DG}~Cur+temp+vis+moon	8	−51.98	2.91	0.11	0.19	<0.001	5.72
	Ab_{DG}~Cur+temp+vis+moon+year	9	−50.39	4.50	0.05	0.18	<0.001	4.93
	Ab_{DG}~Cur+temp+vis+moon+divers	10	−49.98	4.90	0.04	0.18	<0.001	4.28
Grey reef shark	Ab_{DG}~Null	2	9.80	22.37	0.00	0	0.54	0.06
	Ab_{DG}~Cur	3	−6.24	6.33	0.01	0.13	<0.001	4.68
	Ab_{DG}~Temp	3	4.30	16.87	0.00	0.05	0.03	−2.26
	Ab_{DG}~Vis	3	11.04	23.61	0.00	−0.01	0.39	0.86
	Ab_{DG}~Year	3	11.79	24.37	0.00	−0.01	0.59	−0.54
	Ab_{DG}~Moon	5	11.90	24.48	0.00	0.01	0.29	1.28
	Ab_{DG}~Divers	3	9.36	21.94	0.00	0.01	0.21	1.26
	Ab_{DG}~Cur+temp	**4**	**−12.58**	**0.00**	**0.32**	**0.18**	**<0.001**	**14.35**
	Ab_{DG}~Cur+temp+moon	7	−12.14	0.43	0.26	0.20	<0.001	6.94
	Ab_{DG}~Cur+temp+moon+divers	8	−11.91	0.66	0.23	0.20	<0.001	6.10
	Ab_{DG}~Cur+temp+moon+divers+vis	9	−10.53	2.05	0.12	0.20	<0.001	5.29
	Ab_{DG}~Cur+temp+moon+divers+vis+year	10	−8.84	3.74	0.05	0.19	<0.001	4.63
Whitetip reef shark	Ab_{DG}~Null	2	41.28	7.63	0.01	0	0.50	0.68
	Ab_{DG}~Cur	**3**	**33.65**	**0.00**	**0.28**	**0.07**	**0.002**	**2.66**
	Ab_{DG}~Temp	3	41.33	7.67	0.01	0.01	0.06	−1.89
	Ab_{DG}~Vis	3	43.01	9.36	0.00	−0.01	0.85	−0.19
	Ab_{DG}~Year	3	42.66	9.01	0.00	−0.01	0.54	−0.61
	Ab_{DG}~Moon	5	42.15	8.50	0.00	0.02	0.17	1.69
	Ab_{DG}~Divers	3	41.86	8.21	0.00	0.01	0.19	−1.33
	Ab_{DG}~Cur+temp	4	33.67	0.02	0.28	0.07	0.003	5.94
	Ab_{DG}~Cur+temp+moon	7	35.34	1.68	0.12	0.08	0.009	3.24
	Ab_{DG}~Cur+temp+moon+divers	8	34.94	1.29	0.15	0.09	0.007	3.11

Table 3. Cont.

Species	Model	df	AIC	ΔAIC	wAIC	R^2	*p*-value	*F*-value
	Ab_{DG}~Cur+temp+moon+divers+year	9	36.46	2.80	0.07	0.09	0.01	2.72
	Ab_{DG}~Cur+temp+moon+divers+year+vis	9	36.46	2.80	0.07	0.09	0.02	2.36

Model with best fit for each analysis highlighted (bold).
AIC: Akaike's Information Criteria, $_{\Delta}$AIC: difference of the AIC of a given model to the model with best fit, $_{w}$AIC: AIC weight, df: degrees of freedom.

telemetry techniques. For example, one of the few discrepancies between dive guide observations and telemetry dataset occurred between March and April, when the relative proportion of sharks detected by telemetry was lower than those recorded by dive guides. This result might be a consequence of the presence of transient sharks during these months or higher attendance of individuals not tagged but that frequently visited the monitoring sites during this period. However, lower values of telemetry more likely reflected a decrease in receiver efficiency [33], since the timing coincided with field work in Palau when new tags were deployed on sharks. In support of this, an analysis of receiver metrics showed that signal collisions resulting from the large number of tags in the vicinity of some receivers at this time undermined the performance of the array (see [26] for more detail).

Although there was a close correlation between dive guide counts and telemetry results, to some extent this may have been a function of the particular circumstances of our study. We used experienced dive guides to gather data and the addition of tourist divers and less experienced guides might have reduced the strength of the relationship. While there is some evidence that diving experience may not necessarily be positively correlated with count accuracy [7], this is likely to depend on the circumstances surrounding the dive and may only be the case under relatively benign conditions of low current, simple topography and with relatively few sharks. During the study, experienced guides leading groups of tourist divers tended to follow predetermined routines, visiting specific landmarks over a time bounded by limits for safe recreational diving. Given the perpetually clear waters on the outer reefs of Palau (visibility typically >30 m), this meant that the sampling area covered by guides was likely to remain relatively constant among dives at a given dive site. The opportunity to view sharks is a focal point of the diving tourist experience in Palau [4], so that dive guides are likely to actively locate sharks during a dive. Additionally, dive guides are familiar with the local fauna at each dive site, which is also likely to reduce misidentifications and search effort. Finally, the sharks in many of the popular dive sites appear to be uninterested in, and relatively unwary of, divers. Indeed, we found no significant correlation between the numbers or depth of sharks recorded by telemetry and the numbers of tourist divers present in the dives. The rapid habituation of sharks to the presence of divers was first noted in some of the early behavioural studies in reef systems [34]. At our study sites, this behaviour meant that sharks were likely to remain in the local area despite the presence of tourist divers, making it relatively easy to obtain reliable counts of numbers. Together, these characteristics of the diving experience in Palau mean that it is ideally suited for a citizen science approach to shark monitoring in partnership with the recreational diving community. While the degree to which such features exist at other recreational diving localities in the wider Indo-Pacific region is unclear, the broad distribution of the species monitored in our study (grey reef and whitetip reef sharks) and the generally favourable diving conditions on coral reefs suggest that similar monitoring programs could be implemented in many locations across the region.

To be successful, monitoring programs need to have well defined objectives and standardised protocols that are effective in collecting accurate data of the target species. In our study, this was possible through data collection by recreational dive guides. However, alternative strategies might be necessary for species of sharks where diving conditions do not allow for underwater visual surveys by recreational divers, such as turbid coastal waters or waters below the limits of recreational diving depths. For such areas, programs designed to collect information from recreational catch and release fishers [35] and visual census conducted from vantage points [36] could provide useful citizen science data for the monitoring of shark populations.

Our study suggests that programs that use dive guides to monitor shark relative abundance in coral reefs could be a cost-effective alternative to traditional science-based surveys. However, this is not to imply that the set up and maintenance of such monitoring programs do not involve considerable logistics. We found that the success of our long-term program relied in part on sufficient resources for personnel, training and management of datasets to ensure data quality. On-ground leadership, encouragement and feedback was required to ensure that participants remained engaged in the program and maintained regular sampling throughout the study, as has been highlighted in other studies [16]. In part, this was done through incentive schemes by the dive tourism operator that involved small rewards (donated by local industry) to those guides that provided the most regular returns of questionnaires on an annual basis.

Environmental Influences on the Abundance Patterns of Sharks

Dive guide data revealed a seasonal cycle in relative abundance of sharks, with peaks from March to June and the lowest values recorded in October and November. Dive guides also recorded relatively low numbers of both grey and whitetip reef sharks in May, a pattern that was more pronounced for the former species. This decrease in relative abundance coincided with an increase in water temperatures from around 25°C in previous months to a peak of 29°C in May [4]. Reproduction in reef sharks is known to be closely tied to temperature variation [37] and it may be that this sudden decline in the numbers of sharks may result from reproductive events occurring elsewhere on the reef that coincided with the peak in water temperatures. Some support for this hypothesis is found in dive guide observations of numerous females with fresh mating scars at this time.

Short-term (daily, weekly) changes in relative abundance of sharks at the monitored sites were correlated with current strength and temperature. Current strength appeared to be the more important driver of variations in relative abundance of whitetip reef sharks, with mean relative abundance steadily increasing with the current flow. For grey reef sharks, our models identified a combination of both current strength and temperature as the principal drivers of relative abundance. These sharks were three times more abundant during dives when currents were strong and relative abundance was also higher when water temperatures decreased from 30°C to 28°C. Earlier studies have noted the association of reef sharks with areas of strong current flow, typically around reef promontories, channels and passes [38], although why this occurs remains unclear. Our previous analyses of telemetry data at the same sites in Palau [26] also show that water temperature strongly influences the vertical movements of reef sharks, so that the mean depths occupied by sharks are greater when the layer of warm water near the surface (<40 m depth) expands to deeper waters (>60 m depth) [26]. Therefore, the reduction in number of sharks observed by dive guides during times of higher water temperatures could be associated with sharks occupying a greater vertical range of habitat, much of which is inaccessible to most recreational divers (i.e. >40 m depth) during these periods.

Conclusions

Some scepticism surrounds the use of citizen science due to potential problems with the quality of data generated by untrained observers [8]. While mindful of the caveats mentioned above, we showed that counts by dive guides in Palau provided an effective method to monitor shark relative abundance. Our approach is relevant to many other citizen science initiatives because the technique of acoustic telemetry that we used to validate data collection by dive guides is one of the most rapidly-growing means of monitoring animals in marine environments. For example, collaborative initiatives such as the Ocean Tracking Network have now deployed acoustic arrays for tracking whales, seals, sharks, penguins, fish and a huge range other species in marine environments worldwide (see: http://oceantrackingnetwork.org/about/ocean). The development of arrays of listening stations by programs such as these offers the opportunity to validate citizen science initiatives, since both often target the charismatic megafauna that inhabit coastal systems. Such comparative studies will be necessary to ensure that the data that citizen science initiatives provide to management and conservation strategies in marine systems is credible, precise and reliable.

Acknowledgments

The authors thank the dive guides, staff and volunteers of Fish 'n' Fins and Micronesian Shark Foundation for data collection and organisation. We acknowledge the Ocean Tracking Network, the Integrated Ocean Observing System and the Australian Animal Tracking and Monitoring System (AATAMS) for logistic contributions regarding the telemetry equipment. The authors thank K. Holland, P. L. Colin and the personnel involved in the shark tagging and receiver downloading: N. Bornovski, I. Field, the staff of Fish 'n' Fins and the Koror State Rangers, in particular M. Moros. We also acknowledge the Coral Reef Research Foundation for access to the temperature data included in the study and C. Pogoreutz and Phil Bouchet for comments on the manuscript.

Author Contributions

Conceived and designed the experiments: GMSV MGM TB. Performed the experiments: GMSV MGM TB. Analyzed the data: GMSV JJM. Contributed reagents/materials/analysis tools: TB MGM. Wrote the paper: GMSV MGM JJM.

References

1. Worm B, Davis B, Kettemer L, Ward-Paige CA, Chapman D, et al. (2013) Global catches, exploitation rates, and rebuilding options for sharks. Marine Policy 40: 194–204.
2. Dulvy NK, Fowler SL, Musick JA, Cavanagh RD, Kyne PM, et al. (2014) Extinction risk and conservation of the world's sharks and rays. eLife 3.
3. Ferretti F, Worm B, Britten GL, Heithaus MR, Lotze HK (2010) Patterns and ecosystem consequences of shark declines in the ocean. Ecology Letters 13: 1055–1071.
4. Vianna GMS, Meekan MG, Pannell DJ, Marsh SP, Meeuwig JJ (2012) Socio-economic value and community benefits from shark-diving tourism in Palau: A sustainable use of reef shark populations. Biological Conservation 145: 267–277.
5. McCauley DJ, McLean KA, Bauer J, Young HS, Micheli F (2012) Evaluating the performance of methods for estimating the abundance of rapidly declining coastal shark populations. Ecological Applications 22: 385–392.
6. Richards BL, Williams ID, Nadon MO, Zgliczynski BJ (2011) A towed-diver survey method for mesoscale fishery-independent assessment of large-bodied reef fishes. Bulletin of Marine Science 87: 55–74.
7. Ward-Paige CA, Lotze HK (2011) Assessing the value of recreational divers for censusing elasmobranchs. PloS One 6: e25609.
8. Goffredo S, Pensa F, Neri P, Orlandi A, Gagliardi MS, et al. (2010) Unite research with what citizens do for fun:"recreational monitoring" of marine biodiversity. Ecological Applications 20: 2170–2187.
9. Bernard A, Götz A, Kerwath S, Wilke C (2013) Observer bias and detection probability in underwater visual census of fish assemblages measured with independent double-observers. Journal of Experimental Marine Biology and Ecology 443: 75–84.
10. Silvertown J (2009) A new dawn for citizen science. Trends in Ecology & Evolution 24: 467–471.

11. Bell CD, Blumenthal JM, Austin TJ, Ebanks-Petrie G, Broderick AC, et al. (2008) Harnessing recreational divers for the collection of sea turtle data around the Cayman Islands. Tourism in Marine Environments 5: 245–257.
12. Higby LK, Stafford R, Bertulli CG (2012) An evaluation of ad hoc presence-only data in explaining patterns of distribution: cetacean sightings from whale-watching vessels. International Journal of Zoology 2012.
13. Jaine FR, Couturier LI, Weeks SJ, Townsend KA, Bennett MB, et al. (2012) When giants turn up: Sighting trends, environmental influences and habitat use of the manta ray *Manta alfredi* at a coral reef. PloS One 7: e46170.
14. Davies TK, Stevens G, Meekan MG, Struve J, Rowcliffe JM (2012) Can citizen science monitor whale-shark aggregations? Investigating bias in mark–recapture modelling using identification photographs sourced from the public. Wildlife Research 39: 696–704.
15. Hussey NE, Stroh N, Klaus R, Chekchak T, Kessel ST (2013) SCUBA diver observations and placard tags to monitor grey reef sharks, *Carcharhinus amblyrhynchos*, at Sha'ab Rumi, The Sudan: assessment and future directions. Journal of the Marine Biological Association of the United Kingdom 93: 299–308.
16. Huveneers C, Luo K, Otway NM, Harcourt RG (2009) Assessing the distribution and relative abundance of wobbegong sharks (Orectolobidae) in New South Wales, Australia, using recreational scuba-divers. Aquatic Living Resources 22: 255–264.
17. Meekan MG, Bradshaw CJA, Press M, McLean C, Richards A, et al. (2006) Population size and structure of whale sharks *Rhincodon typus* at Ningaloo Reef, Western Australia. Marine Ecology Progress Series 319: 275–285.
18. Speed C, Meekan MG, Rowat D, Pierce S, Marshall A, et al. (2008) Scarring patterns and relative mortality rates of Indian Ocean whale sharks. Journal of Fish Biology 72: 1488–1503.
19. Ward-Paige CA, Mora C, Lotze HK, Pattengill-Semmens C, McClenachan L, et al. (2010) Large-scale absence of sharks on reefs in the greater-Caribbean. Plos One 5: e11968.
20. Brunnschweiler JM, Baensch H (2011) Seasonal and long-term changes in relative abundance of bull sharks from a tourist shark feeding site in Fiji. PloS One 6: e16597.
21. Marshall A, Pierce S (2012) The use and abuse of photographic identification in sharks and rays. Journal of Fish Biology 80: 1361–1379.
22. MacNeil M, Graham N, Conroy M, Fonnesbeck C, Polunin N, et al. (2008) Detection heterogeneity in underwater visual-census data. Journal of Fish Biology 73: 1748–1763.
23. Ward-Paige C, Flemming JM, Lotze HK (2010) Overestimating fish counts by non-instantaneous visual censuses: consequences for population and community descriptions. Plos One 5: e11722.
24. Heupel M, Semmens J, Hobday A (2006) Automated acoustic tracking of aquatic animals: scales, design and deployment of listening station arrays. Marine and Freshwater Research 57: 1–13.
25. Colin PL (2009) Marine environments of Palau. San Diego: Indo-Pacific Press. 414 p.
26. Vianna GMS, Meekan MG, Meeuwig JJ, Speed CW (2013) Environmental influences on patterns of vertical movement and site fidelity of grey reef sharks (*Carcharhinus amblyrhynchos*) at aggregation sites. PloS One 8: e60331.
27. deBruyn AMH, Meeuwig JJ (2001) Detecting lunar cycles in marine ecology: periodic regression versus categorical ANOVA. Marine Ecology Progress Series 214: 307–310.
28. Zuur AF, Ieno EN, Smith GM (2007) Analysing ecological data: Springer New York. 672p.
29. Lumley T (2013) Package 'leaps'. http://cran.r-project.org/web/packages/leaps/leaps.pdf.
30. R Development Core Team (2010) R: A language and environment for statistical computing. R Foundation for Statistical Computing, Vienna, Austria. ISBN 3-900051-07-0, URL http://www.R-project.org/.
31. Friedlander AM, Zgliczynski BJ, Ballesteros E, Aburto-Oropeza O, Bolaños A, et al. (2012) The shallow-water fish assemblage of Isla del Coco National Park, Costa Rica: structure and patterns in an isolated, predator-dominated ecosystem. Revista de Biologia Tropical 60: 321–338.
32. Brunnschweiler JM, Barnett A (2013) Opportunistic visitors: Long-term behavioural response of bull sharks to food provisioning in Fiji. PloS One 8: e58522.
33. Simpfendorfer CA, Heupel MR, Collins AB (2008) Variation in the performance of acoustic receivers and its application for positioning algorithms in a riverine setting. Canadian Journal of Fisheries and Aquatic Sciences 65: 482–492.
34. Nelson DR (1977) On the field study of shark behavior. American Zoologist 17: 501–507.
35. Lowry M, Williams D, Metti Y (2007) Lunar landings-Relationship between lunar phase and catch rates for an Australian gamefish-tournament fishery. Fisheries Research 88: 15–23.
36. Speed CW, Meekan MG, Field IC, McMahon CR, Stevens JD, et al. (2011) Spatial and temporal movement patterns of a multi-species coastal reef shark aggregation. Marine Ecology Progress Series 429: 261–275.
37. Speed CW, Meekan MG, Field IC, McMahon CR, Bradshaw CJA (2012) Heat-seeking sharks: support for behavioural thermoregulation in reef sharks. Marine Ecology Progress Series 463: 231–245.
38. Nelson D, Johnson R (1980) Behavior of the reef sharks of Rangiroa, French Polynesia. National Geographic Society Research Reports 12: 479–499.

Radiocarbon Dating and Wood Density Chronologies of Mangrove Trees in Arid Western Australia

Nadia S. Santini[1]*, Quan Hua[2], Nele Schmitz[3], Catherine E. Lovelock[1]

1 The School of Biological Sciences, The University of Queensland, St Lucia, Brisbane, Queensland, Australia, **2** Institute for Environmental Research, Australian Nuclear Science and Technology Organisation, Lucas Heights, New South Wales, Australia, **3** Plant Biology and Nature Management, Vrije Universiteit Brussel, Brussels, Belgium

Abstract

Mangrove trees tend to be larger and mangrove communities more diverse in tropical latitudes, particularly where there is high rainfall. Variation in the structure, growth and productivity of mangrove forests over climatic gradients suggests they are sensitive to variations in climate, but evidence of changes in the structure and growth of mangrove trees in response to climatic variation is scarce. Bomb-pulse radiocarbon dating provides accurate dates of recent wood formation and tree age of tropical and subtropical tree species. Here, we used radiocarbon techniques combined with X-ray densitometry to develop a wood density chronology for the mangrove *Avicennia marina* in the Exmouth Gulf, Western Australia (WA). We tested whether wood density chronologies of *A. marina* were sensitive to variation in the Pacific Decadal Oscillation Index, which reflects temperature fluctuations in the Pacific Ocean and is linked to the instrumental rainfall record in north WA. We also determined growth rates in mangrove trees from the Exmouth Gulf, WA. We found that seaward fringing *A. marina* trees (~10 cm diameter) were 48±1 to 89±23 years old (mean ± 1σ) and that their growth rates ranged from 4.08±2.36 to 5.30±3.33 mm/yr (mean ±1σ). The wood density of our studied mangrove trees decreased with increases in the Pacific Decadal Oscillation Index. Future predicted drying of the region will likely lead to further reductions in wood density and their associated growth rates in mangrove forests in the region.

Editor: Ben Bond-Lamberty, DOE Pacific Northwest National Laboratory, United States of America

Funding: This work was supported by The Mexican Council of Science and Technology (CONACYT), The Secretary of Public Education (SEP, Mexico), the Australian Institute of Nuclear Science and Engineering (AINSE Grant No. 09047 & 11114), ARC Discovery award DP1096749, The School of Biological Sciences at the The University of Queensland (UQ) and the Research Foundation - Flanders (FWO). The funders had no role in study design, data collection and analysis, decision to publish, or preparation of the manuscript.

Competing Interests: The authors have declared that no competing interests exist.

* E-mail: nadiasilvanasantini@gmail.com

Introduction

Mangrove forests are an important component of tropical and subtropical coastal ecosystems. In the arid zone of Australia, the Exmouth Gulf, Western Australia (WA), mangroves support an important fishery and biodiversity [1]. The stature and productivity of mangrove forests is correlated with latitude [2,3] with trees tending to be larger and communities more diverse in tropical latitudes. In arid regions, diversity of mangrove forests is lower than in wetter regions and maximum tree height is also reduced [4,5]. Arid regions are characterized by high evapotranspiration / rainfall ratios (>10) due to low rainfall and high temperatures [6]. Rainfall is likely to be an important factor determining the variation in productivity and diversity of mangrove trees in arid regions because it reduces the salinity of soil porewater (which limits water uptake and growth of mangrove trees) through dilution and recharge of groundwater and increases in soil moisture [7]. Additionally, growth of leaf bearing twigs has been observed to be enhanced after large rain-bearing cyclonic events, possibly due to delivery of nutrients and to increased availability of freshwater that can enhance metabolic rates [8]. Projections of rainfall with climate change indicate an increasing drying with less rainfall and higher temperatures in WA and also in parts of central Australia [9,10]. However, examples of impacts of climate change in estuarine ecosystems are scarce [11,12].

In the arid north west of Australia, at our study site at the Exmouth Gulf, climatic patterns are complex. High rainfall (>600 mm/yr) occurs in the austral summer associated with tropical cyclones, but in some years rainfall can also be associated with southern low-pressure climatic systems during the winter months [13]. The Pacific Decadal Oscillation Index (PDO), which reflects temperature fluctuations in the Pacific Ocean for long periods of time (20 to 30 years), has also been linked to rainfall in central and southern Australia, although the mechanisms underlying the connection between PDO and rainfall in Australia are not clearly resolved [13]. Here we assessed the adequacy of the PDO as a rainfall proxy in the Exmouth Gulf in order to increase the temporal scale of our analysis because reliable rainfall records from the Australian Bureau of Meteorology are relatively recent (from 1966), coincident with economic development of this remote region of WA.

Tree rings have been widely used to infer past climate and establish the response of vegetation to climate and other factors, but the use of ring formation in wood to determine sensitivity of trees to environmental change in mangroves and other tropical tree species can be problematic. In many tropical species annual growth rings are not produced or there is low confidence in the regular timing of tree-ring formation [14]. Examination of wood from the widespread mangrove species *Avicennia marina* (Acanthaceae) and from the Rhizophoraceae, indicates that changes in xylem vessel density throughout the life of trees can reflect changing environmental conditions [15,16,17,18]. These studies suggest that variation in freshwater inputs (rain, rivers, groundwater) may strongly influence the structure of the wood, including its density. Our previous work on *A. marina* suggests that variation in tree growth rates correlates with predictable changes in wood density, and therefore wood density may be used as a proxy for tree growth rates over time [19].

The development of highly accurate and precise radiocarbon dating techniques that use the atomic bomb pulse of radiocarbon in the atmosphere in the 1960s to date recent organic materials including wood [20,21] has offered advances in the examination of the response of tropical and subtropical trees [22,23] to variation in climate. Here, we used radiocarbon techniques combined with X-ray densitometry (which allows analyses of continuous and high-resolution wood density profiles from tree stems over time [24,25]) to develop a wood density chronology. We tested whether wood density chronologies of *A. marina* were sensitive to variation in the Pacific Decadal Oscillation Index, and determined growth rates in mangrove trees from the Exmouth Gulf, WA.

Materials and Methods

Site description and sample collection

The study site was located in Giralia Bay, in the east of the Exmouth Gulf, WA (22.4°S, 114.3°E). The climate of the site is warm and dry, with a mean air temperature of 25°C, a mean minimum temperature of 17°C, and a mean maximum temperature of 32°C, with peaks of up to 47°C in summer. The mean annual rainfall at the site is 250 mm/yr. In the region, cyclones bringing wind gusts in excess of 300 km/h and heavy rainfall (>600 mm/yr) occur every two to three years [26].

In October 2008, four mature stems of the widespread mangrove species *A. marina* were collected. The sampling was 7 months after tropical cyclone *Pancho*, which flooded the region in March 2008 [8]. In the region trees have multiple stems [27], we chose stems of ~10 cm diameter and ~3.3 m tall from the seaward edge of the forest for our study. The reasons were that: 1. Trees of this form are typical for the mangrove seaward fringing zone in the region and 2. The interior of larger stems is damaged by boring insects and bivalves (particularly shipworms), thereby preventing dating of larger girthed trees.

Ethics statement

Our work was conducted under the permit numbers SW014933, SW013522, SW012157, SW010985, Department of Environment and Conservation, Western Australia. *Avicennia marina* is classified as a species of "least concern" according to the International Union for Conservation of Nature and Natural Resources (2013).

Wood density determination

From the tree stems, we cut rectangular blocks of 4 mm ×4 mm x (maximum radius) of the stem with a bandsaw (CarbaTec, Australia), excluding pith and bark. We placed the wood blocks within glass containers with distilled water under vacuum for one day (sufficient time to fully saturate the samples) before scanning them with a compact micro CT scanner 1174 (SkyScan, Belgium). The wood blocks were sealed with plastic film to avoid desiccation and were exposed to X-rays (45 KV, 800 μA) for 30 minutes. We obtained four wood density profiles from each wood block (expressed as 256 grey values) from the CT scanner (resulting in a resolution of *ca.* 22 grey values for each mm of rectangular woody block). The wood density profiles were calibrated with the actual fresh (green) wood density values [28] from the wood blocks using Eqn 1, where A_{ij} is the calculated wood density, B_{ij} is the grey value along the stem (0 – 256) and k is a constant calculated as mean grey value times mean fresh wood density [29].

$$Aij = k\ o(1/Bij) \quad (1)$$

The high resolution of assessed wood density (*ca.* 22 values for each mm of rectangular woody bar) allowed us to perform wood density averages a) per mm of wood, and b) at dates established by radiocarbon dating. Our obtained wood density profiles are available at the International Tree-Ring Data Bank (access numbers *stem 1*: giralia1age, giralia1dist; *stem 2*: giralia2age, giralia2dist; *stem 3*: giralia3age, giralia3dist and *stem 4*: giralia4age, giralia4dist) [30]. To evaluate reliability of the wood density signal of the collected stems, we calculated the expressed population signal (EPS) from the wood density profiles (four wood density profiles per stem, *stem 1 – stem 4*) with the package *detrendeR* from the software R [31,32]. The EPS was determined by calculating the mean correlation between the wood density profiles [33]; a value of 0.85 of the EPS possible range (0 to 1) is considered the minimum to obtain a sufficiently replicated profile [34]. EPS of our wood density profiles was >0.85 (Supplementary material Figure S1, S2). Wood density profiles were detrended by calculating the ratio index of the wood density data to the associated values of the fitted smoothing spline using the package *detrendeR* from the software R [31,32,35]. The arithmetic means and standard deviation of wood density and detrended wood density were calculated for each stem (Supplementary material Figure S1, S2). Given the purpose of detrending data is to remove possible long-term trends due to tree aging, detrended wood density data were used for analysis in which the effect of tree aging had to be removed, while wood density data was used for analysis in which incorporation of tree aging was important. We specified in the Data Analysis section which density data (wood density or detrended wood density) were used for each analysis.

Tree age and growth rate estimation

From the scanned rectangular blocks, we cut cross-sections along the radius of 1 mm length (resulting in 4 mm ×4 mm ×1 mm slices) at intervals of 4 – 20 mm to obtain 6 – 7 samples per stem. These samples were dated using bomb-pulse ^{14}C techniques to estimate the growth history of each stem. The wood samples were pre-treated to extract alpha-cellulose using the method described in [36]. Alpha-cellulose was then combusted to CO_2 and reduced to graphite [37] for ^{14}C analyses using the STAR accelerator mass spectrometry (AMS) facility at the Australian Nuclear Science and Technology Organisation [38]. Measured ^{14}C values were converted to calendar ages using the "Simple Sequence" deposition model of the OxCal calibration program based on chronological ordering (outer samples are younger than inner samples [39]), and a calibration data set for the Southern Hemisphere (SH) for the last 350 years consisting of the updated SH bomb radiocarbon data [20] and the SHCal04 data for the pre-bomb period [40].

Tree growth rates

Growth rates were estimated for the mean calibrated ^{14}C age of each wood slice from the gradient of a cubic spline fitted to the data [41]. Growth rates were determined as increments in stem circumference from the slope of a weighted linear regression through the plotted points and expressed as mm/yr.

Data analyses

We used linear regression analyses to assess the relationship between rainfall and the Pacific Decadal Oscillation Index (PDO, obtained from [42], Supplementary material Figure S3) and sea level pressure and the PDO (obtained from [43], Supplementary material Figure S4). Rainfall data was obtained from the Australian Bureau of Meteorology [26], Learmonth Station (Number 5007, 32 km away from our site). The rainfall record was continuous between 1966 and 2008, but rainfall data were not available or were patchy over periods of time in which growth of our mangrove trees occurred (1919 – 1965). Regression analyses were also used to determine the relationship between wood density and growth rate (using dates established by ^{14}C dating), wood density and distance from pith, wood density and time, detrended wood density and the PDO, detrended wood density and rainfall, growth rate and the PDO and growth rate and rainfall for the time in which rainfall data were available (1961, 1964, 1966 – 2008).

Finally, because of differences in the age of trees and to reflect major changes in the PDO (Supplementary material Figure S3, [44]), we grouped growth rate values within the following periods of time 1952 – 1976 (n = 12), 1977 – 1998 (n = 6) and 1999 – 2008 (n = 6). Differences in growth rates between these three periods of time were tested using Analysis of Variance (ANOVA). These tests were performed with the software Prism ver 5.0 (GraphPad Software, La Jolla, CA, USA).

Results

We found significant but variable trends between rainfall and the Pacific Decadal Oscillation Index (PDO) in the Exmouth Gulf, Western Australia (WA) for the period between 1966 – 2008. Rainfall decreased with increasing values of the PDO ($r^2 = 0.16$, $p = 0.006$, Figure 1A, B).

Our radiocarbon results indicated that *A. marina* mangroves in the Exmouth Gulf (~10 cm diameter, fringe canopy of ~3.3 m tall) were 48±1 to 89±23 years old (mean ±1σ) with an age range of 47 – 112 years old. Determination of age of most of our wood samples had an uncertainty of 1 to 2 years with the exception of stems dated before 1952 (Table 1). Growth rates of *A. marina* expressed as increments in stem circumference per year ranged from 4.08±2.36 mm/yr in the period between 1999 – 2008 to 5.30±3.33 mm/yr in the period between 1952 – 1976, with an overall mean growth rate of 4.84±2.66 mm/yr for the period between 1952 – 2008. Growth rates were not significantly different between the assessed periods of time (Table 2).

Growth rates were positively correlated with wood density values in the fringing *A. marina* trees from the Exmouth Gulf, Western Australia ($r^2 = 0.24$, $p = 0.006$, Figure 2A, B). Results were significant when values of all the four stems were included in the analysis (Figure 2B) but analysis of individual stems of growth rates and wood density were not always significant.

Linear regression analyses indicated that wood density decreased from pith to bark ($r^2 = 0.78 – 0.94$, $p<0.0001$) and with time in *A. marina* trees from the Exmouth Gulf ($r^2 = 0.76 – 0.94$, $p<0.0001$). Slopes of the linear regressions were significantly different among stems ($p<0.0001$) but consistently decreased from pith to bark and with time (1919–2008) for each stem (Figure 3A, B). However, detrended wood density did not significantly decrease from pith to bark or with time, but tended to vary over time (Figure 3C, D).

We found a significant decrease in detrended wood density with increases in the PDO for 1919–2008 ($r^2 = 0.24$, $p = 0.03$, Figure 4A). However, we found unclear trends between detrended wood density and rainfall for 1961 – 2008 (Figure 4B), growth rate and the PDO for 1952 – 2008 and growth rate and rainfall for 1961 – 2008 (Figure 5A, B).

Discussion

Our results indicated that rainfall in the Exmouth region was correlated with the PDO, which in turn links to rainfall patterns in central and southern Australia [13]. Therefore, rainfall in the Exmouth region is associated with rainfall patterns of central and southern Australia.

The age of *A. marina* mangroves from the Exmouth Gulf was 48±1 to 89±23 years old (mean ±1σ) with an age range of 47 – 112 years old (Table 1). Uncertainty associated with age determination of wood samples was 1 – 2 years in recent wood samples (dated after 1952), but the uncertainty increased to 6 – 23 years in older samples (dated before 1952). The accuracy of radiocarbon dating and therefore in growth rates is higher for more recent samples (after 1952) because of significant increases in atmospheric ^{14}C after 1955 resulted from atmospheric nuclear bomb tests [20].

The age of our sampled mangrove trees indicates that they had survived the devastating cyclone in 1998 (Cyclone *Vance*) that directly struck our study site [45] and also previous cyclonic events. Survival of seaward forests of *A. marina* in the Exmouth Gulf may be due to their resistant and flexible wood compared to other mangrove species [46,47]. Some trees at the site reach

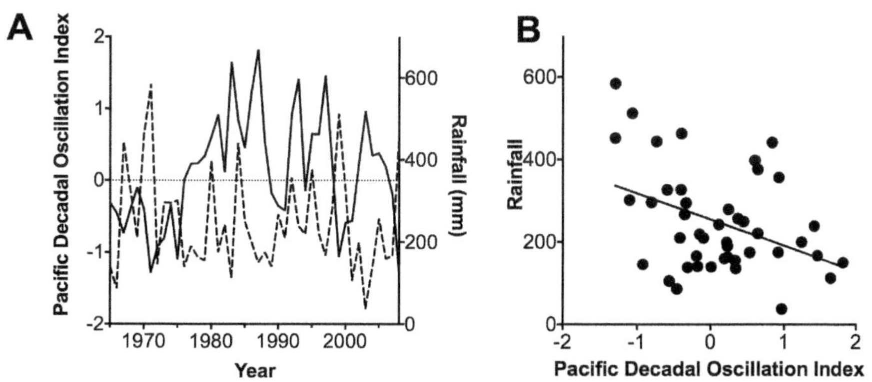

Figure 1. Relationship between annual rainfall and the Pacific Decadal Oscillation Index. Relationship between A) annual rainfall (dashed line) and the Pacific Decadal Oscillation Index (PDO, solid line) in the Exmouth Gulf, Western Australia between 1966 and 2008. B) The line represents the linear regression where Rainfall = −63.2 PDO +255, $r^2 = 0.16$, $p = 0.006$, $n = 44$.

Table 1. Radiocarbon results, estimated tree ages and growth rates of *Avicennia marina*.

Stem	Distance from pith (mm)	^{14}C Mean $\pm 1\sigma$ (pMC)	Modelled calendar age Mean $\pm 1\sigma$ (year AD) [(a)]	Growth rate (mm/yr) [(b)]
1	1	117.20±0.36	1960±1	7.18±0.66
	8	141.12±0.39	1964±1	12.03±4.65
	15	151.92±0.44	1970±1	8.47±2.57
	22	139.72±0.40	1975±1	9.64±1.82
	36	125.53±0.34	1982±1	5.65±2.26
	43	113.36±0.33	1994±1	5.66±0.06
	50	109.37±0.40	2000±2	5.71±0.94
	57(c)	-	2008	7.65±1.16
2	7	98.06±0.69	1936±17	3.42±0.76
	13	101.36±0.31	1947±9	4.85±0.17
	19	97.51±0.28	1951±6	6.69±0.89
	25	104.05±0.54	1957±1	6.38±0.19
	31	119.50±0.51	1961±1	3.49±1.12
	37	133.48±0.53	1976±1	2.36±0.14
	43	110.52±0.50	1999±2	1.83±0.45
	47(c)	-	2008	1.40±1.44
3	1	99.04±0.37	1919±23	1.79±4.82
	11	98.24±0.40	1936±18	1.94±2.15
	16	107.66±0.48	1958±1	2.13±3.62
	21	145.32±0.46	1971±1	2.56±1.08
	26	120.67±0.52	1985±1	3.11±1.51
	36	112.02±0.47	1996±2	3.52±2.20
	41(c)	-	2008	3.62±2.51
4	1	99.67±0.32	1945±10	2.79±3.62
	6	99.01±0.40	1952±2	2.80±0.99
	11	133.72±0.43	1964±1	3.01±1.10
	16	143.48±0.59	1972±1	3.66±1.66
	21	127.98±0.46	1980±1	4.66±1.07
	41	115.22±0.36	1991±2	5.38±1.72
	51(c)	-	2008	4.31±0.10

Values of ^{14}C are shown in percent modern carbon (pMC). [(a)] Modelled calendar age in year AD at 68.2% confidence level. [(b)] Uncertainty associated with mean growth rate represents the maximum difference between growth rate estimates from the spline curve using upper and lower limits of the calibrated ^{14}C age range compared with the mean ^{14}C age [41] [(c)] Outermost samples were collected in 2008 and were not used for ^{14}C analysis.

65 cm in circumference (C.E. Lovelock *unpublished data*), even in the shorter scrub forest (<1.5 m height) which occurs landward of the taller seaward fringe. Given the growth rates of trees at the site, these trees may be ~160 years old (based on circumference increments of 4.08 mm/yr – the minimum growth rate for the last 60 years) or ~120 years old (if circumference increments are 5.30 mm/yr – the maximum growth rate for the last 60 years). The estimated ages of larger girthed (although not taller) trees at this site are comparable to maximum ages of mangroves indicated by Alongi [48].

Table 2. Growth rates of *Avicennia marina* mangroves from the Exmouth Gulf, Western Australia.

Time	Growth rate (mm/yr)	*n*
1952–1976	5.30±3.33	12
1977–1998	4.66±1.11	6
1999–2008	4.08±2.36	6

Values are means $\pm 1\sigma$.

Growth rates of *A. marina* measured as increments in stem circumference per year exhibited an overall mean of 4.84±2.66 mm/yr (Table 2). These growth rates are comparable to stem growth rates of the same species observed in Gazi and Dabaso, Kenya [49]. Growth rates of *A. marina* exhibited higher values than growth rates for *Xylocarpus granatum* (0.62 – 2.51 mm/yr), *Bruguiera gymnorrhiza* (0 – 2.51 mm/yr), *Rhizophora mucronata* (0.94 mm/yr) and *Sonneratia alba* (0.31 – 1.25 mm/yr), all documented from mangrove trees of approximately 15 cm diameter from Micronesia [50].

We found growth rates were positively correlated with wood density values in the fringing *A. marina* trees from the Exmouth Gulf, Western Australia (Figure 2A, B). These results are in

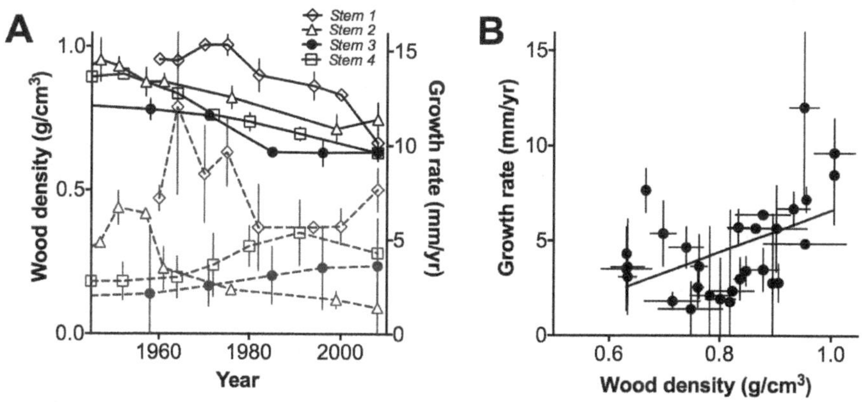

Figure 2. Relationship between growth rate and wood density of *Avicennia marina* collected in Giralia Bay, Western Australia. A) Relationship between growth rate (dashed grey line) and wood density (solid black line) from four stems (*stem 1 – stem 4*) collected in Giralia Bay, Western Australia, points are at dates established using bomb-pulse dating, values are means $\pm 1\sigma$. B) The line represents the linear regression where: Growth rate $= 11.0$ Wood density - 4.32, $r^2 = 0.24$, $p = 0.006$, $n = 30$. Points are means $\pm 1\sigma$ at dates established using bomb-pulse dating.

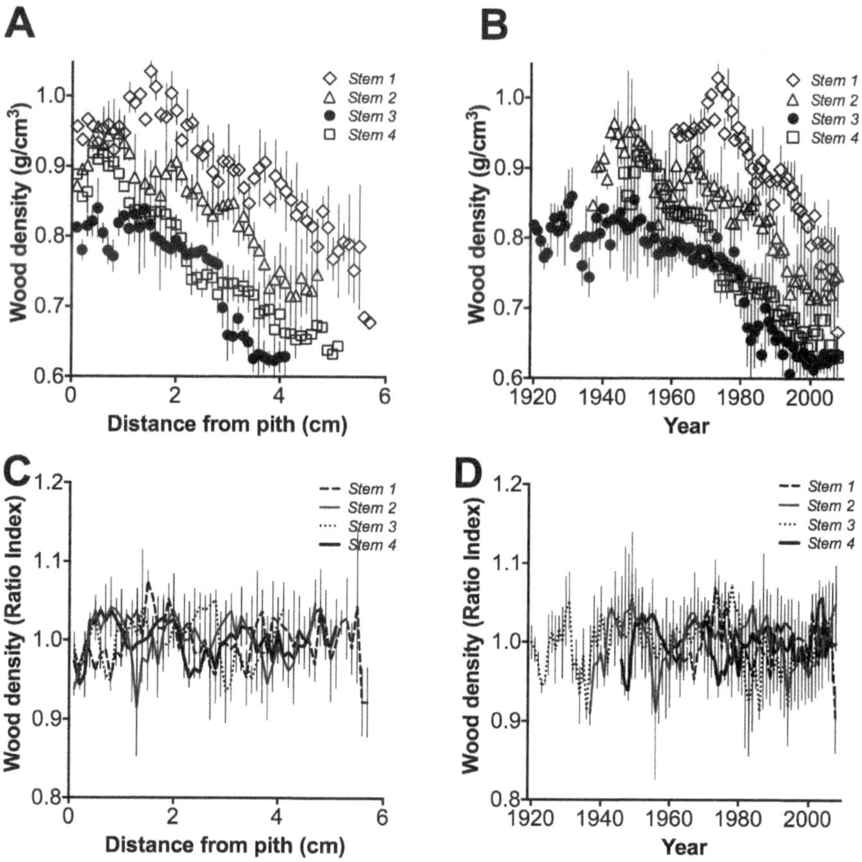

Figure 3. Relationship between wood density and distance from pith and wood density and time in *Avicennia marina*. Relationship between A) wood density and distance from pith, slopes between stems were significantly different (*$p < 0.0001$*), the regression equations were: Wood density (*stem 1*) $= -4.3 \times 10^{-2}$ distance from pith +1.01, $r^2 = 0.78$, $n = 57$; wood density (*stem 2*) $= -4.8 \times 10^{-2}$ distance from pith +0.95, $r^2 = 0.83$, $n = 47$; wood density (*stem 3*) $= -5.5 \times 10^{-2}$ distance from pith +0.86, $r^2 = 0.78$, $n = 41$; wood density (*stem 4*) $= -5.7 \times 10^{-2}$ distance from pith +0.95, $r^2 = 0.94$, $n = 51$; $p < 0.0001$ for all stems (*stem 1 – stem 4*), vertical lines $= 1\sigma$. Relationship between B) wood density and time, slopes between stems were significantly different (*$p < 0.0001$*), the regression equations were: Wood density (*stem 1*) $= -5.07 \times 10^{-2}$ time +10.9, $r^2 = 0.78$, $n = 48$; wood density (*stem 2*) $= -3.1 \times 10^{-2}$ time +7.04, $r^2 = 0.81$, $n = 72$; wood density (*stem 3*) $= -2.5 \times 10^{-3}$ time +5.75, $r^2 = 0.76$, $n = 89$; wood density (*stem 4*) $= -4.7 \times 10^{-3}$ time +9.93, $r^2 = 0.94$, $n = 63$; $p < 0.0001$ for all stems (*stem 1 – stem 4*), vertical lines $= 1\sigma$. C) Detrended wood density and distance from pith, vertical lines $= 1\sigma$ and D) Detrended wood density and time, vertical lines $= 1\sigma$.

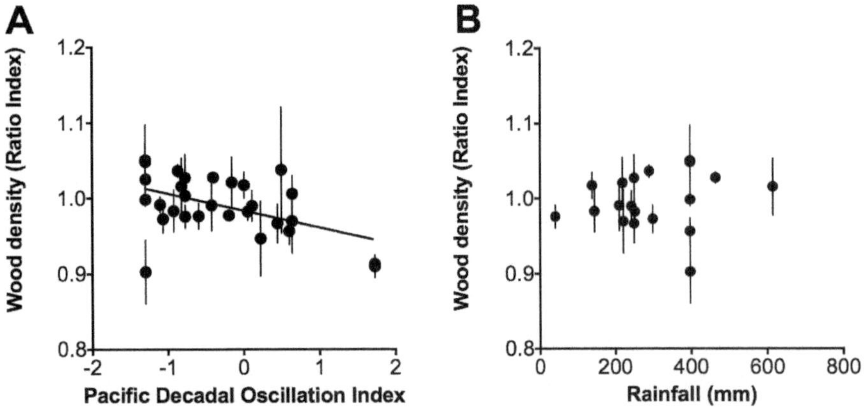

Figure 4. Relationship between detrended wood density and the Pacific Decadal Oscillation Index and detrended wood density and rainfall. Relationships between A) detrended wood density and the Pacific Decadal Oscillation Index (PDO) between 1919 and 2008. The line represents the linear regression where: Wood density (Ratio Index) = −0.02 PDO +0.98, $r^2 = 0.24$, $p = 0.03$, $n = 30$ and B) detrended wood density and rainfall between 1961 – 2008. Points are means $\pm 1\sigma$.

contrast to studies in other tree species where increasing density is often associated with low growth rates [51,52], but are consistent with our previous work with trees of *A. marina* from New Zealand [19]. Increasing growth rates with increasing wood density in *A. marina* is largely due to its unique wood anatomy, which includes phloem and parenchyma tissues (which have low density) within the wood. High wood density in *A. marina* is associated with thick fibre walls and low levels of phloem as well as large vessel diameters, which facilitate water transport, and high rates of photosynthetic carbon gain, while, low-density wood has high amounts of phloem in wood and smaller vessels, traits that are evident in trees growing under higher salinity regimes [18,19]. Therefore, relatively rapid growth rates are achieved with dense wood in *A. marina*.

Our linear regression analyses indicated that wood density decreased from pith to bark (Figure 3A) and with age (Figure 3B) in the *A. marina* mangroves trees from the Exmouth Gulf. Radial variation of wood density in trees can be due to both (1) variation in climate over time and (2) tree aging [53,54,55]. Variation in wood density is responsive to variation in climate because while growth occurs, wood anatomical characteristics (e.g. xylem vessel diameter and fibre wall thickness), which are the basis of wood density variation, vary to respond to changing environmental conditions. In mangroves, changes in wood anatomy have been found to be associated with the availability of freshwater (i.e. rainwater or groundwater) and variation in the salinity of the water in which mangroves occur [16,56]. Reductions in wood density, given that low wood density is associated with low growth rates in *A. marina* (see above), strongly suggest declining growth rates of trees over time.

However, as trees age a pattern of decreased wood density towards the bark can occur because when a woody stem ages, the inner part of the stem (originally functional sapwood) is transformed to heartwood (which lacks functional xylem vessels and parenchyma). Heartwood is denser than sapwood and is formed by polymerization of compounds such as lignin and flavonoids [57,58], although we did not observe this in our wood sections. Additionally, as canopy size usually increases with tree age, wood density is often observed to decrease because the size of xylem vessels increases to accommodate larger water fluxes through the stem [59]. However, increases in canopy size of *A. marina* would be expected to be associated with increasing wood density [19]. Detrended wood density (which does not incorporate tree age effect) was variable over time, but significantly decreased

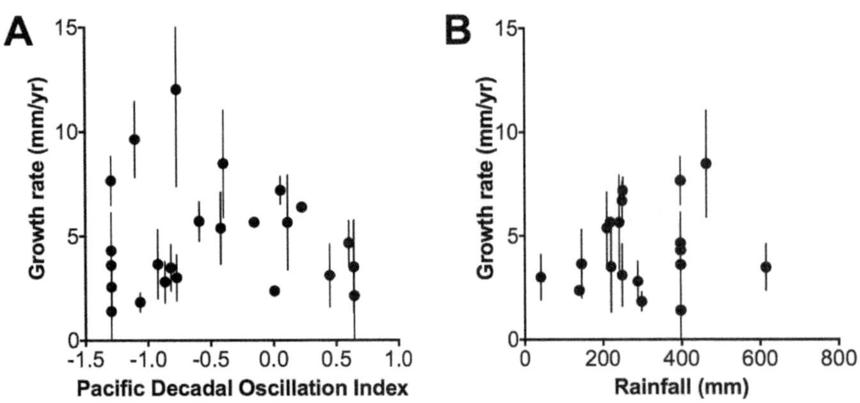

Figure 5. Relationship between growth rate and the Pacific Decadal Oscillation Index and growth rate and rainfall. Relationships between A) growth rate and the Pacific Decadal Oscillation Index (PDO) between 1952 and 2008 and B) growth rate and rainfall between 1961 – 2008. Points are means $\pm 1\sigma$.

with increments in the PDO. The significant decreases in wood density in *A. marina* with increments in the PDO can be partly explained by deteriorating environmental conditions, e.g. increased soil salinity due to reduced rainfall, but wood density was not significantly correlated with rainfall, indicating that rainfall was not the only climatic factor determining wood density variability.

We tested sensitivity of growth rates to the PDO and rainfall, but our results were not significant. Therefore other factors in addition to rainfall, that may interact with variation in rainfall (e.g. variation in sea level pressure or temperature), may also be influencing mangrove growth in the region (Supplementary material Figure S4) [7]. Growth rates tended to be higher during the period between 1952 and 1976, which was an extended negative phase of the PDO, than growth rates from the extended positive PDO phase between 1977 and 1998 (Supplementary material Figure S3). However, the differences in growth rates were not significant (Table 2). Growth rate determinations are variable because of uncertainties associated with dating and sampling wood for dating and because of variation in the performance of individual stems that vary due to spatial variation in growth conditions in the forest. The variation in the relationships between growth and the PDO may also reflect the high variability in the rainfall and PDO ($r^2 = 0.16$) relationship (Figure 1A, B). Increased replication over time would assist in unraveling the effects of variation in rainfall, temperature and sea level on mangrove growth in the region [60].

Conclusions

The Pacific Decadal Oscillation Index is a weak but long-term proxy of rainfall in the Exmouth Gulf, Western Australia. Small, 10 cm diameter seaward fringing stems of *A. marina* in the Exmouth Gulf, Western Australia are relatively old, 48±1 to 89±23 years old (mean ±1σ). The wood density of our studied mangrove trees decreased with increments in the Pacific Decadal Oscillation Index. Future predicted drying of the region will likely lead to further reductions in wood density and their associated growth rates in mangrove forests of the region.

Supporting Information

Figure S1 Wood density profiles displayed as distance from the pith from four stems collected in Giralia Bay, Western Australia. A–D) Wood density profiles of the four stems (*stem 1 – stem 4*) with their EPS value of 0.97. E–H) Wood density profiles showing arithmetic means ± 1σ of the four stems. I–L) Detrended wood density profiles depicting arithmetic means ± 1σ of the four stems.

Figure S2 Wood density profiles over time from four stems collected in Giralia Bay, Western Australia. A–D) Wood density profiles of the four stems (*stem 1 – stem 4*) with their EPS value of 0.92. E–H) Wood density profiles showing arithmetic means ± 1σ of the four stems. I–L) Detrended wood density profiles depicting arithmetic means ± 1σ of the four stems.

Figure S3 Pacific Decadal Oscillation Index. Mean annual Pacific Decadal Oscillation Index (PDO) from 1940 – 2008.

Figure S4 Relationship between sea level pressure and the Pacific Decadal Oscillation Index. Relationship between A) mean annual sea level pressure (dashed line) and the Pacific Decadal Oscillation Index (PDO, solid line) in the Exmouth Gulf, Western Australia between 1951 and 2008. B) The line represents the linear regression where Sea Level Pressure = 0.44 PDO +9.98, $r^2 = 0.32$, $p<0.0001$, $n = 58$.

Acknowledgments

We thank Dr. Ruth Reef and Dr. María Gómez Cabrera from the University of Queensland for technical assistance with the CT Scanner 1174, Dr. Matthew Brookhouse from the Australian National University for providing advice on dendrochronological calculations and two anonymous Reviewers for their valuable comments which improved this manuscript.

Author Contributions

Conceived and designed the experiments: NSS QH NS CEL. Performed the experiments: NSS QH. Analyzed the data: NSS QH CEL. Contributed reagents/materials/analysis tools: NSS QH CEL. Wrote the paper: NSS. Critically revised the manuscript: CEL.

References

1. Penn JW, Caputi N (1986) Spawning stock-recruitment relationships and environmental influences on the tiger prawn *Penaeus esculentus* fishery in Exmouth Gulf, Western Australia. Aus J Mar Fresh Res 37: 491–505.
2. Twilley RR, Chen RH, Hargis T (1992) Carbon sinks in mangroves and their implications to carbon budget of tropical coastal ecosystems. Water Air Soil Poll 64: 265–288.
3. Saenger P, Snedaker SC (1993) Pantropical trends in mangrove aboveground biomass and annual litterfall. Oecologia 96: 293–299.
4. Smith TJ III, Duke NC (1987) Physical determinants of inter-estuary variation in mangrove species richness around the tropical coastline of Australia. J Biogeogr 14: 9–19.
5. Smith TJ III (1992) Forest structure. In, Robertson AI, Alongi DM, editors. Tropical Mangrove Ecosystems. Coastal and Estuarine Studies, vol 41. Washington DC: American Geophysical Union. pp. 101–136.
6. Semeniuk V (1983) Mangrove distribution in Northwestern Australia in relationship to regional and local freshwater seepage Vegetatio 53: 11–31.
7. Ball MC, Farquhar GD (1984) Photosynthetic and stomatal responses of two mangrove species, *Aegiceras corniculatum* and *Avicennia marina*, to long-term salinity and humidity conditions. Plant Physiol 74: 1–6.
8. Lovelock CE, Feller IC, Adame MF, Reef R, Penrose HM, et al. (2011) Intense storms and the delivery of materials that relieve nutrient limitations in mangroves of an arid zone estuary. Func Plant Biol 38: 514–522.
9. Solomon S, Plattner G, Knutti R, Friedlingstein P (2009) Irreversible climate change due to carbon dioxide emissions PNAS 106:1704–1709.
10. Climate Change in Australia (2013) Available: http://www.climatechangeinaustralia.gov.au/. Accessed 2013 April 12.
11. Lovelock CE, Ellison J (2007) Vulnerability of mangroves and associated tidal wetlands of the GBR to climate change. In: Johnson J, Marshall P, editors. Climate Change and Great Barrier Reef. Townsville: Great Barrier Reef Marine Park Authority and Australian Greenhouse Office. pp. 237–269.
12. Richardson A, Brown CJ, Brander K, Bruno JF, Buckley L, et al. (2013) Climate change and marine life. Biol Lett doi:10.1098/rsbl.2012.0530.
13. McGowan H, Marx SK, Denholm J, Soderholm J, Kamber BS (2009) Reconstructing annual inflows to the headwater catchments of the Murray River, Australia, using the Pacific Decadal Oscillation. Geophys Res Lett 36: L06707.
14. Worbes M (2002) One hundred years of tree-ring research in the tropics- a brief history and an outlook to future challenges. Dendrochronologia 20: 217–231.
15. Menezes M, Berger U, Worbes M (2003) Annual growth rings and long-term growth patterns of mangrove trees from the Bragança peninsula, North Brazil. Wet Ecol Manag 11: 233–242.
16. Verheyden A, De Ridder F, Schmitz N, Beeckman H, Koedam N (2005) High-resolution time series of vessel density in Kenyan mangrove trees reveal a link with climate. New Phytol 167: 425–425.
17. Yu KF, Kamber BS, Lawrence MG, Greig A, Zhao JX (2007) High precision analysis on annual variations of heavy metals, lead isotopes and rare earth elements in mangrove tree rings by inductively coupled plasma mass spectrometry. Nucl Instrum Meth B 255:399–408.

18. Robert E, Schmitz N, Okello JA, Boeren I, Beeckman H, et al. (2011) Mangrove growth rings: fact or fiction? Trees 25: 49–58.
19. Santini NS, Schmitz N, Lovelock CE (2012) Variation in wood density and anatomy in a widespread mangrove species. Trees 26:1555 – 1563.
20. Hua Q, Barbetti M (2004) Review of tropospheric bomb ^{14}C data for carbon cycle modelling and age calibration purposes. Radiocarbon 46: 1273–1298.
21. Hua Q (2009) Radiocarbon: A chronological tool for the recent past. Quat Geochronol 4: 378–390.
22. Bowman DMJS, Prior LD, Tng D, Hua Q, Brodribb TG (2011) Continental-scale climatic drivers of growth ring variability in an Australian conifer. Trees 25: 925–934.
23. Pearson S, Hua Q, Allen K, Bowman DMJS (2011) Validating putatively cross-dated *Callitris* tree-ring chronologies using bomb-pulse radiocarbon analysis. Austral J Bot 59: 7–17.
24. Steppe K, Cnudde V, Girard C, Lemeur R, Cnudde JP, et al. (2004) Use of X-ray computed tomography for non-invasive determination of wood anatomical characteristics. J Struct Biol 148: 11–21.
25. Nock C, Geihofer D, Grabner M, Baker P, Bunyavejchewin S, et al. (2009) Wood density and its radial variation in six canopy tree species differing in shade-tolerance in western Thailand. Ann Bot 104: 297–306.
26. Australian Bureau of Meteorology (2013) Available: http://www.bom.gov.au.Accessed 2013 Jan 16.
27. Clough BF, Dixon P, Dalhaus O (1997) Allometric relationships for estimating biomass in multistemmed mangrove trees Aust J Bot 45: 1023–1031.
28. Chave J, Muller-Landau HC, Baker TR, Easdale TA, Steege H, et al. (2006) Regional and phylogenetic variation of wood density across 2456 neotropical tree species. Ecol Appl 16: 2356–2367
29. Shadbolt P (1988) Some aspects of non-destructive testing using computerised tomography. Chisholm Institute of Technology. Melbourne: Monash University. 268 p.
30. International Tree-Ring Data Bank (2013) Available: http://www.ncdc.noaa.gov/cdo-web/. Accessed 2013 July 04.
31. R Development Core Team (2008) R: A language and environment for statistical computing. R Foundation for Statistical Computing, Vienna, Austria. http://www.R-project.org.
32. Campelo F (2012) detrendeR: Start the detrendeR Graphical User Interface (GUI) R package version 1.0.4. http://CRAN.R-project.org/package = detrendeR
33. Haneca K, Boeren I, Acker JV (2006) Dendrochronology in suboptimal conditions: tree rings from medieval oak from Flanders (Belgium) as dating tools and archives of past forest management. Veget Hist Archaeobot 15:137–144.
34. Wigley TML, Briffa KR, Jones PD (1984) On the average values of correlated time series, with applications in dendroclimatology and hydrometeorology. J Clim Appl Meteorol 23: 201–213.
35. Cook ER, Peters K (1981) The smoothing spline: a new approach to standardizing forest interior tree-ring width series for dendroclimatic studies. Tree-Ring Bulletin 41: 45–53.
36. Hua Q, Barbetti M, Zoppi U, Fink D, Watanasak M, et al. (2004) Radiocarbon in tropical tree rings during the Little Ice Age. Nucl Instrum Meth B 223–224: 489–494
37. Hua Q, Jacobsen G, Zoppi U, Lawson E, Williams A, et al. (2001) Progress in radiocarbon target preparation at the ANTARES AMS Centre. Radiocarbon 43: 275–282.
38. Fink D, Hotchkis M, Hua Q, Jacobsen G, Smith AM, et al. (2004) The ANTARES AMS facility at ANSTO. Nucl Instrum Meth B 223–224: 109–115.
39. Ramsey CB (2008) Deposition models for chronological records. Quat Sci Rev 27: 42–60.
40. McCormac G, Hogg AG, Blackwell PG, Buck CE, Higham TFG, et al. (2004) SHCal04 Southern Hemisphere calibration, 0–11.0 cal kyr BP. Radiocarbon 46: 1087–1092.
41. Clarke LJ, Robinson SA, Hua Q, Ayre DJ, Fink D (2012) Radiocarbon bomb spike reveals biological effects of Antartic climate change. Global Change Biol 18: 301–310.
42. University of Washington (2013) Available: jisao.washington.edu/pdo/PDO.-latest. Accessed 2013 March 10.
43. National Climatic Data Center (2013) Available: http://www.cpc.ncep.noaa.gov/data/indices/darwin. Accessed 2013 Aug 19.
44. Peterson WT, Schwing FB (2003) A new climate regime in northeast pacific ecosystems. Geophys Res Lett 30 http://dx.doi.org/10.1029/2003GL017528
45. Paling EI, Kobryn HT, Humphreys G (2008) Assessing the extent of mangrove change caused by Cyclone Vance in the eastern Exmouth Gulf, northwestern Australia. Estuar Coast Shelf Sci 77:603–613.
46. Bardsley KN (1985) The effects of cyclone Kathy on mangrove vegetation. In: Bardsley KN, Davie JDS, Woodroffe CD, editors. Coasts and Tidal Wetlands of the Australian Monsoon Region. Australian National University North Australia Research Unit, Darwin. pp 167–183.
47. Santini NS, Schmitz N, Bennion V, Lovelock CE (2013) The anatomical basis of the link between density and mechanical strength in mangrove branches. Funct Plant Biol 40:400 – 408.
48. Alongi DM (2002) Present state and future of the world's mangrove forests. Environ Conserv 29: 331–349.
49. Schmitz N, Robert EMR, Verheyden A, Kairo JG, Beeckman H, et al. (2008) A patchy growth via successive and simultaneous cambia: Key to success of the most widespread mangrove species *Avicennia marina*? Ann Bot 101: 49–58.
50. Devoe NN, Cole TG (1998) Growth and yield in mangrove forests of the Federal States of Micronesia. Forest Ecol Manag 103: 33 – 48.
51. King DA, Stuart JD, Tan S, Noor NSMD (2006) The role of wood density and stem support costs in the growth and mortality of tropical forests. J Ecol. 94:670–680.
52. Poorter L, McDonald I, Alarcon A, Fichtler E, Licona J, et al. (2009) The importance of wood traits and hydraulic conductance for the performance and life history strategies of 42 rainforest tree species. New Phytol 185:481–492.
53. Woodcock DW, Shier AD (2002). Wood specific gravity and its radial variations: the many ways to make a tree. Trees 16:437–443.
54. Thomas DS, Montagu KD, Conroy JP (2007) Temperature effects on wood anatomy, wood density, photosysnthesis and biomass partitioning of *Eucalyptus grandis* seedlings. Tree Physiol 27:251–260.
55. Chave J, Coomes D, Jansen S, Lewis SL, Swenson NG, et al.(2009) Towards a worldwide economics spectrum. Ecol Lett 12:351–366.
56. Schmitz N, Verheyden A, Beeckman H, Gitundu Kairo J, Koedman N (2006) Influence of a Salinity Gradient on the Vessel Characters of the Mangrove Species *Rhizophora mucronata*. Ann Bot 98: 1321–1330.
57. International Association of Wood Anatomists (IAWA) (2013) Available: http://bio.kuleuven.be/sys/iawa/. Accessed 2013 March 04 March.
58. Schultz T, Harms W, Fisher T, Mcmurtrey K, Minn J, et al. (1995) Durability of angiosperm heartwood: The importance of extractives *Holzforschung* 49:29–34.
59. Enquist B, West G, Charnov E, Brown J (1999) Allometric scaling of production and life-history variation in vascular plants. Nature 401: 909–911.
60. Feng M, Li Y, Meyers G (2004) Multidecadal variations of Fremantle sea level: Footprint of climate variability in the tropical Pacific. Geophys Res Lett doi: 10.1029/2004GL019947.

7

Wintering Habitat Model for the North Atlantic Right Whale (*Eubalaena glacialis*) in the Southeastern United States

Timothy A. Gowan*, Joel G. Ortega-Ortiz¤

Florida Fish and Wildlife Conservation Commission, Fish and Wildlife Research Institute, St. Petersburg, Florida, United States of America

Abstract

The coastal waters off the southeastern United States (SEUS) are a primary wintering ground for the endangered North Atlantic right whale (*Eubalaena glacialis*), used by calving females along with other adult and juvenile whales. Management actions implemented in this area for the recovery of the right whale population rely on accurate habitat characterization and the ability to predict whale distribution over time. We developed a temporally dynamic habitat model to predict wintering right whale distribution in the SEUS using a generalized additive model framework and aerial survey data from 2003/2004 through 2012/2013. We built upon previous habitat models for right whales in the SEUS and include data from new aerial surveys that extend the spatial coverage of the analysis, particularly in the northern portion of this wintering ground. We summarized whale sightings, survey effort corrected for probability of whale detection, and environmental data at a semimonthly resolution. Consistent with previous studies, sea surface temperature (SST), water depth, and survey year were significant predictors of right whale relative abundance. Additionally, distance to shore, distance to the 22°C SST isotherm, and an interaction between time of year and latitude (to account for the latitudinal migration of whales) were also selected in the analysis presented here. Predictions from the model revealed that the location of preferred habitat differs within and between years in correspondence with variation in environmental conditions. Although cow-calf pairs were rarely sighted in the company of other whales, there was minimal evidence that the preferred habitat of cow-calf pairs was different than that of whale groups without calves at the scale of this study. The results of this updated habitat model can be used to inform management decisions for a migratory species in a dynamic oceanic environment.

Editor: Z. Daniel Deng, Pacific Northwest National Laboratory, United States of America

Funding: This research was funded by the National Marine Fisheries Service (NMFS) [grant P.O. Number 6WCA90016; http://www.nmfs.noaa.gov/] and the Save the Manatee Trust Fund [http://myfwc.com/research/manatee/trust-fund/]. Other than a contribution from NMFS to original survey protocols and survey lines design for the right whale early warning system, the funders had no role in the analysis, decision to publish, or preparation of the manuscript.

Competing Interests: The authors have declared that no competing interests exist.

* E-mail: tim.gowan@myfwc.com

¤ Current address: Cooperative Institute for Marine and Atmospheric Studies, Rosenstiel School of Marine and Atmospheric Science, University of Miami, Miami, Florida, United States of America

Introduction

The North Atlantic right whale (*Eubalaena glacialis*) is highly endangered, and although it has received protection from the Endangered Species Act of 1973 and the Marine Mammal Protection Act of 1972, the species remains well below its optimum sustainable population level [1]. Threats to the recovery of this species include entanglement in fishing gear and collisions with ships [2]. Effective management decisions, and assessing risk of injury and mortality from these threats, require knowledge of right whale habitat preferences to identify areas where whales are likely to occur.

Western North Atlantic right whales occupy nearshore habitats of North America, with their primary feeding grounds from spring through autumn off the coast of New England and eastern Canada. The coastal waters off the southeastern United States (SEUS), however, have been identified as their primary calving grounds, with cow-calf pairs and some juveniles and adults occupying the waters off Florida (FL) and Georgia (GA) during winter months (December–March) [3]. Several management measures have been established in the SEUS due to threats from human activities in this region, especially the risk of ship strikes from the high volume of shipping traffic associated with the ports of Jacksonville, FL, Fernandina, FL, Brunswick, GA, and Savannah, GA. Measures designed to reduce the likelihood of ship collisions with right whales in high-risk areas include a mandatory ship reporting system, seasonal management areas with ship speed restrictions, recommended transit lanes for large ships, and aerial surveys during the calving season (Figure 1). In an effort to enhance protection measures for North Atlantic right whales, aerial surveys in the SEUS have been supported since the early 1990s by a collection of agencies, including the National Oceanic and Atmospheric Administration National Marine Fisheries Service (NOAA Fisheries), the US Coast Guard, the US Navy, and the US Army Corps of Engineers. As part of that effort, a multiagency, coordinated survey network, termed the Early Warning System (EWS), has been in place since the 1993/1994 calving season [4]. The level of effort varied in the early surveys, but a core survey area off the Florida-Georgia border was regularly flown [5]. In 2003 EWS survey lines were modified as depicted in Figure 1 to provide consistent coverage of nearshore waters from Butler Beach, FL (29.76° N) to Sapelo Sound, GA (31.56° N). A primary objective of these surveys is to locate right whales in the area and relay this information to mariners; other objectives include monitoring population vital rates and human-related injuries with the aid of photo identification and characterizing whale habitat use and distribution. Additional surveys off South Carolina and northern Georgia (SC-GA) were implemented in the 2004/2005 calving season (Figure 1) with similar objectives, providing coverage from Sapelo Sound, GA (31.56° N) to North Myrtle Beach, SC (33.82° N).

Effective management actions rely on the ability to know where and when right whales are likely to occur, yet the boundaries of management areas are typically static in time and space, while whale distribution varies spatially and temporally in response to dynamic environmental conditions [6,7]. Additionally, right whale distribution changes seasonally as whales migrate between wintering grounds in the SEUS and feeding grounds at higher latitudes. If whale distribution can be modeled and predicted from environmental variables and behavioral patterns, management strategies can be developed that take into account system variability and uncertainty. Several predictive habitat models have been developed for the right whale wintering grounds in the SEUS [5,8,9,10]. However, these models did not analyze the most current survey data, including surveys with more consistent protocols (see Methods) and data from the recent SC-GA surveys, which have significantly expanded coverage in the northern section of the wintering ground. Accurate inferences from species distribution models are frequently limited to the extent (spatial, temporal and environmental) of the input data (*e.g.*, [11]), and development of these models, particularly for rare and endangered species, should be an iterative process validated and informed by newly available data [12,13]. Right whale abundance (sighting rates) observed during more recent surveys in the northern portion of the wintering ground was not consistent with predictions from previous models. Additionally, none of the previous right whale habitat models for the SEUS accounted for latitudinal migration patterns or for differences in the probability of whale detection due to variable platforms (type of plane) and survey conditions, as has been done in other studies (*e.g.*, [14]).

In this study, we characterized right whale habitat in the SEUS using generalized additive models (GAMs) to relate whale sightings from aerial surveys to static and dynamic environmental variables. We modeled the occurrence (presence-absence) and relative abundance of whales using a hurdle model. We analyzed recent survey data (from 2003/2004 through 2012/2013), with spatial coverage ranging from southern Florida through South Carolina. Our model was developed at a two-week temporal resolution, and it included survey effort, adjusted for survey platform and sea state, and inter- and intra-annual effects as covariates. We generated predictions from this model, and the results demonstrate that right whale distribution varies within and among years. Additionally, we compared environmental conditions for sightings with a calf present to those without calves to determine whether cow-calf pairs utilize habitat that is different from other whales that migrate to the SEUS.

Methods

Ethics Statement

Whale surveys were conducted under permits #0594-1467, #594-1759-00 and #15488 issued to the Georgia Department of Natural Resources and permits #655-1652 and #14233 issued to Scott Kraus (New England Aquarium) by NOAA Fisheries. All research protocols were reviewed by NOAA's Office of Protected Resources and complied with the Endangered Species Act and the Marine Mammal Protection Act.

Aerial Surveys

Aerial surveys for right whales have been conducted in the SEUS during the right whale calving season (December–March) since 1991, but effort intensity and spatial coverage have varied considerably among years outside of a core EWS area [5]. We analyzed survey data from 2003/2004 through 2012/2013 for this study because survey methods were standardized and the location of EWS transect lines was consistent during these seasons. Sightings and effort data for these surveys were provided by the North Atlantic Right Whale Consortium [15]. Since 2003/2004, EWS surveys have been flown along fixed track lines running east-west, spaced 3 nautical miles (5.56 km) apart, and extending 24–34 nautical miles (44–63 km) offshore (Figure 1). Surveys were conducted from a Cessna 337 Skymaster or a de Havilland DHC-6 Twin Otter aircraft with a target altitude of 1000 ft (305 m). EWS surveys were designed to be flown daily during the calving season and complete the defined survey track lines, dependent upon weather conditions and aircraft availability. The SC-GA surveys began in 2004/2005 using Skymaster aircraft and the same data-collection protocols as in EWS surveys but with dedicated track lines running east-west and spaced 3 nautical miles (5.56 km) apart in the southern portion of the survey area and running northwest-southeast and spaced approximately 4 nautical miles (7.52 km) apart in the northern portion (Figure 1). Additional coastal ("Florida nearshore," *sensu* [5]) surveys south of these track lines were flown in some years. Most effort during coastal surveys was within 10 nautical miles (18.5 km) off and parallel to the shoreline (generally in a north-south direction).

During aerial surveys, one observer on each side of the aircraft searched for right whales, and a computer program was used to automatically record geographic location (latitude, longitude, and altitude) obtained from the aircraft's GPS every 10–30 seconds. Environmental conditions (sea state, visibility, weather) were recorded at the beginning of the survey and updated as conditions changed. When a right whale was observed, the aircraft deviated from the designated track line to approach the sighting location. Whale locations were recorded as the GPS location when the aircraft flew directly over the whale. After recording the behavior and number of whales and obtaining photographic documentation, the aircraft returned to the designated track line to resume survey.

Survey Data Processing

Data recorded during each flight from all EWS, SC-GA, and coastal surveys were entered into a GIS (ArcGIS version 10.0, Esri Inc., Redlands, CA), and the equidistant Universal Transverse Mercator (UTM) projection was used for data analysis. We filtered survey data to only include portions considered "on-effort": sea state ≤3 (Beaufort scale), altitude ≤365 m, visibility ≥3.7 km, and flying along a designated track line (*i.e.*, not in transit or circling a whale). Whale sightings were considered on-effort if that was the survey status at the time of the initial sighting. We removed all verification survey sightings and their associated effort (*i.e.*, surveys conducted to locate and verify a reported whale sighting) and all duplicate sightings (*i.e.*, whales already sighted on the same survey, as verified by photo identification). Each segment of a flight path was buffered on both sides with an effective search width, according to survey platform and sea state, to estimate the searched area. To determine effective search widths, perpendicular sighting distances were calculated from survey track lines and

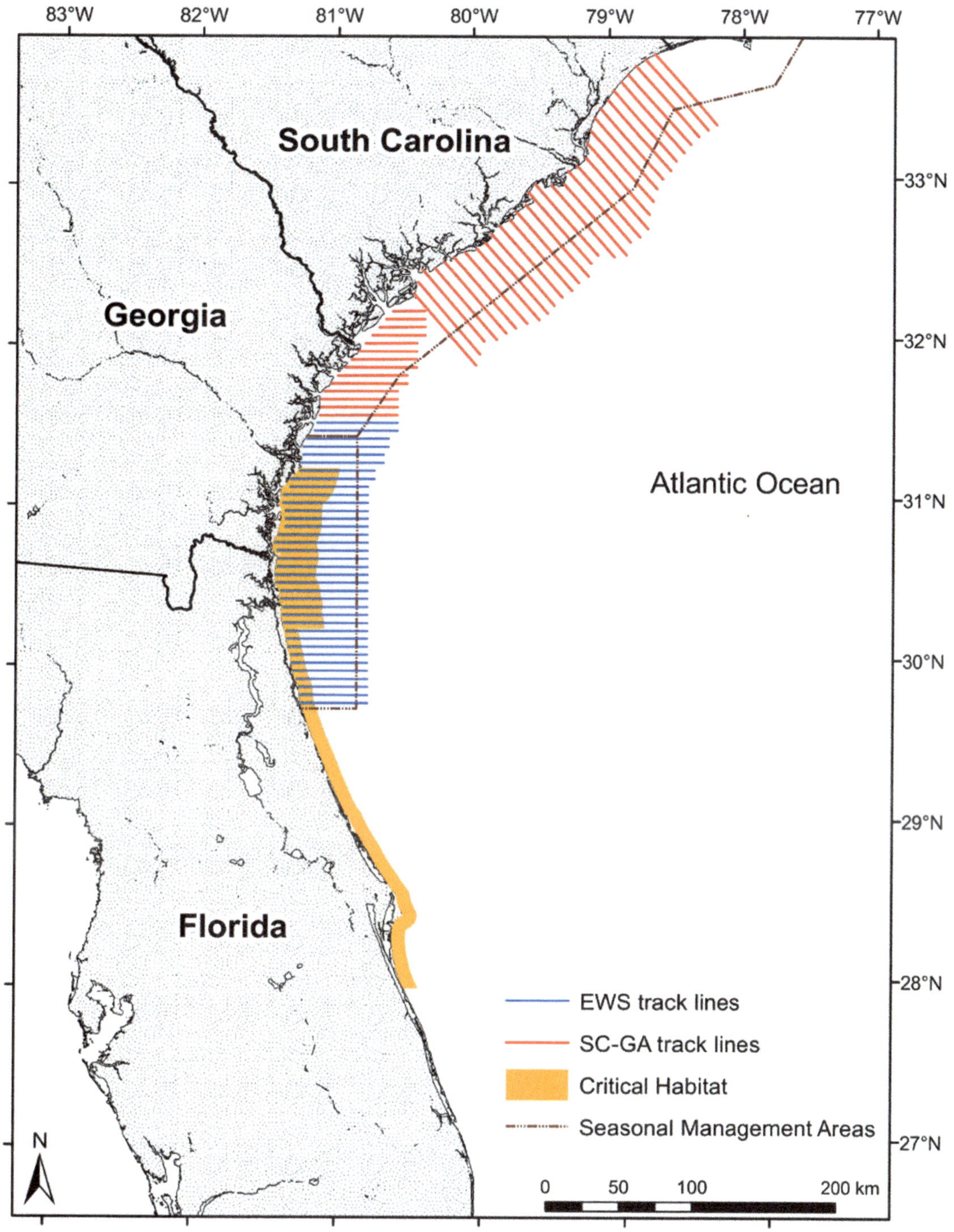

Figure 1. Aerial survey track lines for right whales in the southeastern United States. Transect lines are from the Early Warning System (EWS) and South Carolina/Georgia (SC-GA) survey areas. Critical habitat and seasonal management area boundaries are included for reference.

right whale sighting locations [16], and the multiple-covariate distance sampling (MCDS) engine in Distance 6.0 software was used to model detection probabilities [17]. Separate detection functions were created for each survey platform (Skymaster and Twin Otter aircrafts) using sea state as a covariate in MCDS, resulting in distinct effective search widths (range = 1.3–2.2 km) for each platform/sea state combination. Because the Skymaster has flat windows resulting in a blind spot beneath the aircraft, a section corresponding to 0.186 km (unpublished field data) on both sides of the flight path was removed from the estimated searched area for surveys from this platform. The Twin Otter had no blind spot because it has bubble windows that allowed observers to search directly under the aircraft.

A composite sampling grid for the study area was constructed to accommodate disparities in track line spacing and orientation, consisting of 5.56− ×5.56-km cells oriented east-west in the south and 7.52− ×7.52-km cells oriented northwest-southeast in the north (Figure 2). We overlaid the sampling grid onto the study area so the track lines bisected the grid cells, allowing all survey effort within a cell to be associated with a single track line. Because whale sightings were so rare, we aggregated all survey and environmental data into the grid cells at a semimonthly (approximately 2 week) temporal resolution–either the 1st through the 15th day of a month (A) or the 16th through the last day of a month (B)–to increase the number of records with whale presence, while maintaining a time frame with relatively stable environmental conditions. Thus each survey year consisted of eight semimonthly periods (December A through March B). Each cell summarized over a semimonth is hereafter considered a sampling unit. For each semimonth, the searched area (km^2) from all surveys during that period was summed within each grid cell and used as the measure of survey effort. Two sampling units (survey effort > 340 km^2) were identified as outliers on a Cleveland dotplot [18] and were excluded from analysis.

Environmental Data

Geographic locations (UTM easting and northing) were taken at the center point of each grid cell. Distance to shore was calculated as the distance between the center point of each cell and the high-resolution NOAA composite shoreline GIS layer (http://shoreline.noaa.gov). Bottom depth data were obtained from the 30-arc-second-resolution data set of the General Bathymetric Chart of the Oceans (http://www.gebco.net). Depth was calculated as the mean bottom depth of all values within each grid cell; cells with mean depth above sea level or deeper than 70 m (22 sampling units considered outliers) were excluded from analysis. Bottom slope (degrees) was derived from bathymetric data using Spatial Analyst in ArcGIS and was also summarized as the mean of values within each cell. Sea surface temperature (SST) was derived from Advanced Very High Resolution Radiometer (AVHRR) 1.47-km-resolution imagery from NOAA's CoastWatch data set (http://cwcaribbean.aoml.noaa.gov/data.html) for the East Coast South region. Within each semimonth all available daily images with minimal cloud cover were downloaded (mean = 8.1 images), image pixels with cloud interference were removed, and the semimonthly mean SST was calculated at each pixel. SST for a sampling unit was calculated as the mean of all semimonthly mean SST 1.47 km pixels within each cell. The potential environmental predictors mentioned above were chosen based on previous studies of right whales in the SEUS [5,8,9,10]. Additionally, the semimonthly mean SST data were used to estimate the location of the 22°C SST isotherm (Spatial Analyst in ArcGIS 10.0); this isotherm was chosen because 22°C is at the upper SST range for right whales in this region [5] and can be used as a proxy for the Gulf Stream boundary in winter [19]. Distance from this isotherm was calculated from the center point of each grid cell; distances for sampling units with SST >22°C were set as negative values, indicating that a cell was east or south of the isotherm.

Model Framework

GAMs were used to relate the number of right whale sightings to possible predictor variables. GAMs are extensions of generalized linear models (GLMs) that allow for smooth, nonlinear functions of predictor variables determined by observed data rather than by strict parametric relationships [20,21]. Like GLMs, GAMs use a specified error distribution for the response variable and a link function to relate the response variable to the predictor variables. Because the response variable, number of sighted whales, was overdispersed and zero-inflated due to the large number of sampling units (96%) with no sightings, we used a hurdle model [22]. A hurdle model consists of two steps: modeling presence-absence with a binomial distribution, and then modeling positive abundance, conditional on presence. Hurdle models are therefore useful for modeling zero-inflated data and allow the modeled process which determines presence to differ from the process which determines abundance [22]. We first used a quasibinomial distribution (to deal with excessive number of zeros) with a logit link to model presence-absence from all data. We then used a gamma distribution with a log link to model the number of whales from sampling units with whale sightings [23]. Predicted relative abundance can be calculated by multiplying the probability of occurrence, derived from the first model, by the expected number of whales, derived from the second model. Often, models of count data from wildlife surveys include effort as

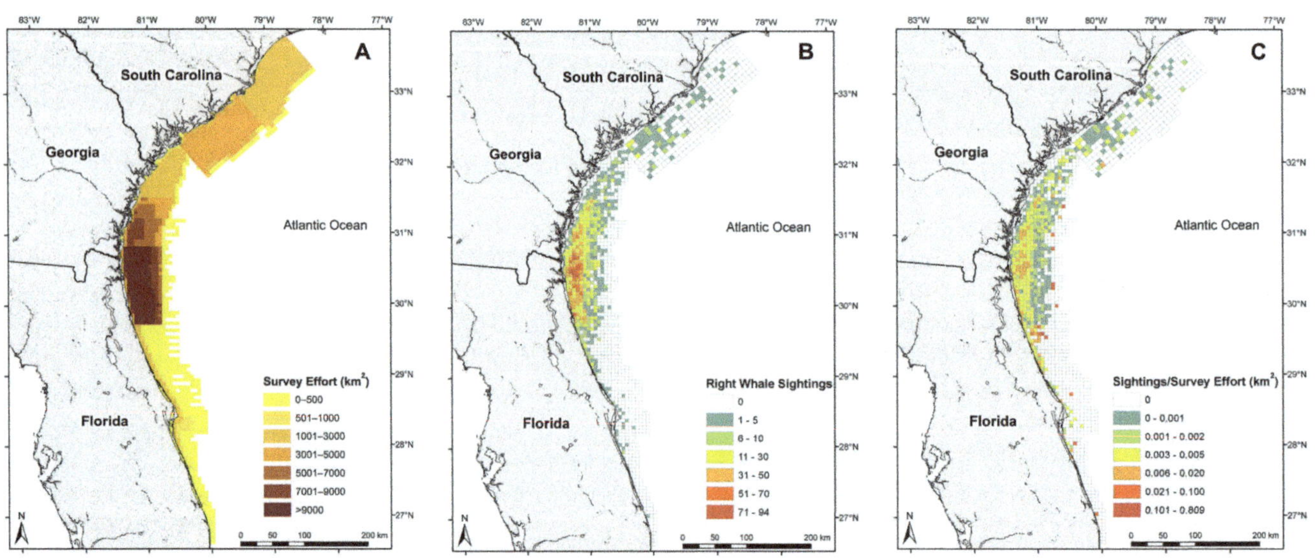

Figure 2. Aerial survey effort and right whale sightings in the southeastern United States. Values represent cumulative area surveyed (A), cumulative number of whales sighted (B), and number of whales sighted divided by area surveyed (C) per grid cell between December 2003 and March 2013 while observers were on-effort.

an offset term, assuming this variable as a linear predictor with a coefficient equal to one [21]; our data, however, did not meet this assumption (see Results), and we included effort as a smoothed covariate. GAMs were constructed using the R statistical software (version 3.0.1) with the mgcv package (version 1.7-24; [24]). The mgcv package determines the appropriate degrees of freedom (df) and smoothing for each predictor variable by minimizing the generalized cross validation (GCV) score [21]. To avoid overfitting by the model and to limit complexities of the smoothing function to an ecologically interpretable relationship, we set the basis dimension parameter to 3, thereby limiting the maximum df for each term to 2 (*e.g.*, [25,26]).

The response variable for the hurdle model was the total number of right whales sighted in each sampling unit. An individual whale could have been sighted multiple times within the same semimonth, and we did not correct for availability bias (*e.g.*, whales submerged at time of survey); therefore, our model is representative of the relative spatial distribution of whales rather than absolute abundance or density. Possible predictor variables included depth, distance to shore, SST, distance to the 22°C isotherm, and survey effort. Survey year was included as a factor variable to account for inter-annual variation in the total number of right whales in the SEUS, and the interaction between northing and semimonthly period was used to account for intra-annual shifts in distribution resulting from the timing of the whale migration. Easting was ultimately excluded as a predictor variable in favor of more informative environmental variables with which it was correlated. Slope was also excluded because all survey effort occurred on the continental shelf within a narrow range of values (0.01–1.29°), and it was consistently found to be the least significant term in a stepwise selection (see Table S1). Although collinearity was present among some of our explanatory variables, its impact was minimized by removing easting and slope to reduce variance inflation factors for the remaining predictors to <4.8 [27], employing penalized regression splines with shrinkage in the mgcv package to estimate smoothing functions [28,29], and limiting model predictions to the range of sampled data [29].

Model Selection and Validation

Model selection was accomplished with a forward stepwise selection procedure, using the following evaluation criteria: model GCV scores, percentage of deviance explained, and analysis of deviance tests. Starting with a null model, each term was added individually; the term resulting in the lowest GCV score was included in the next step. At each step, analysis of deviance was used to determine whether increasing model complexity with the addition of the selected term significantly improved the model. A term was not included in the final selected model if the confidence interval for the fitted response included zero for all observed values of the term or if the model at the previous step (which did not contain the term) had a lower model GCV score [24].

A five-fold cross-validation was used to evaluate each candidate model's performance in predicting novel data [30]. Model training was based on a random subset of 80% of the data and used to predict the validation subset (remaining 20%). Average squared prediction error (ASPE) was calculated from the predicted and observed number of whales sighted in the validation subset. This cross-validation was run five times, and the mean ASPE was calculated for each model to assist in model selection. Final specification of the selected best model used to estimate smoothing functions and create prediction maps was based on the complete data set.

Spatial autocorrelation in species distribution models can be problematic because it violates the assumption that residuals are independently distributed. We examined whether spatial autocorrelation was present by calculating Moran's I of deviance residuals of predicted relative abundance for all 80 semimonths (8 semimonths/year × 10 years) using the R-package spdep (version 0.5-56; [31]). Moran's I was calculated as a global statistic, considering all grid cells as neighbors, with an inverse distance weighting such that nearby grid cells exerted greater influence. Moran's I ranges from -1 (perfect negative correlation) to $+1$ (perfect positive correlation), with values near 0 indicating no spatial autocorrelation present.

Segregation of Calves

Studies of baleen whales, including southern right whales (*Eubalaena australis*), indicate that cows with calves may prefer environmental features different from those used by other demographic groups on wintering grounds [32,33], and it has been suggested that cow-calf pairs receive fitness benefits by segregating themselves from the harassment of other whales [34] or by avoiding areas with more predators [8]. We therefore compared environmental variables (depth, distance to shore, SST, distance to the 22°C isotherm, and UTM northing) at locations of sightings with calves present to those without calves. For this analysis, we overlaid exact sighting locations on the original environmental GIS layers (not the sampling grid) to maximize resolution for each variable. SST values were estimated only for sightings that occurred on days for which SST images were available and at locations free of cloud interference. Isotherms were constructed as stated above, based on semimonthly mean SST data. We excluded all sightings that were not on-effort or those with an estimated depth above sea level due to data precision. Sightings were classified based on the presence of a calf, pooled across years, semimonths, and levels of survey effort, and compared using a Mann-Whitney test. Due to the potential effect of migration timing, we tested for differences in UTM northing in each semimonthly period using a series of pair-wise comparisons with sequential Bonferroni correction [35].

Results

A total of 2191 surveys was flown in the EWS and SC-GA areas from December 2003 through March 2013, resulting in sightings of 3286 right whale groups and 7369 whales (not unique individuals, as some whales were resighted within and between seasons) while on-effort. The greatest concentration of both survey effort and sightings occurred in the EWS area near the coast of northern Florida and southern Georgia (Figure 2). A total of 56143 sampling units, containing 3104 sightings of 6953 whales, was retained for analysis after removing statistical outliers and those with missing SST values due to cloud interference.

Occurrence Model

For the presence-absence model, the GAM stepwise selection procedure yielded higher explained deviances, lower GCV scores, and lower ASPE values at each step, signifying that models had a better fit as complexity increased (Table 1, Table S1A). This finding was supported by analysis of deviance tests at each step; the more complex models were consistently identified as the most parsimonious despite having more parameters (Table S2A). The selected best model included seven predictor variables as significant (survey effort, SST, distance to the shoreline, bottom depth, interaction between semimonthly period and UTM northing, survey year, and distance to the 22°C SST isotherm) and explained 22.8% of the total deviance (Table 1). This model had the lowest ASPE values for all validation data sets,

Table 1. Summary of stepwise selection procedure for presence-absence model of right whales in the southeastern US.

Model	% Deviance	GCV	mean ASPE
null	0	0.3417	0.0392
s(SemiMonth:Northing)	10.3	0.3065	0.0377
s(SemiMonth:Northing)+s(DistToShore)	15.7	0.2882	0.0364
s(SemiMonth:Northing)+s(DistToShore)+Year	18.8	0.2777	0.0352
s(SemiMonth:Northing)+s(DistToShore)+Year+s(Effort)	20.9	0.2706	0.0345
s(SemiMonth:Northing)+s(DistToShore)+Year+s(Effort)+s(SST)	22.3	0.2658	0.0342
s(SemiMonth:Northing)+s(DistToShore)+Year+s(Effort)+s(SST)+s(Depth)	22.6	0.2647	0.0340
s(SemiMonth:Northing)+s(DistToShore)+Year+s(Effort)+s(SST)+s(Depth)+s(DistTo22Iso)	22.8	0.2642	0.0340

Predictor variables include interaction between semimonthly period and UTM northing, distance to the shoreline (DistToShore), survey year, survey effort, sea surface temperature (SST), bottom depth, and distance to the 22°C SST isotherm (DistTo22Iso). Smoothed covariates indentified by "s()". Evaluation criteria include the proportion of deviance explained, generalized cross validation score (GCV), and mean average squared prediction error (ASPE) from a five-fold cross-validation.

demonstrating its superior ability to predict the spatial distribution of right whales from novel data in the study area.

Smoothing functions for all terms in this model indicated nonlinear relationships and were consistent with the locations of sightings from the surveys. As expected, the probability of a right whale sighting increased with increasing survey effort but leveled off when effort was >250 km^2 (Figure 3A); this result supports our decision to model survey effort as a smoothed covariate rather than as an offset term. The SST for sampling units with a right whale present ranged from 9.0 to 22.6°C, with a mean (± SE) of 14.7°C (±0.04). The GAM predicted the highest probabilities between 12 and 16°C (Figure 3B). Right whales were more likely to be sighted close to shore, particularly within 25 km of the coast (Figure 3C), and at intermediate depths, between 10 and 25 m (Figure 3D). Generally, sightings occurred in cooler waters far from the 22°C isotherm, with only one sighting made south or east of this isotherm. However due to the proximity of the Gulf Stream to shore, especially in the southern portion of the study area, whales were limited by land to a maximum distance from the 22°C isotherm, and the response to this variable reached a plateau near this distance (Figure 3E). The interaction term between semimonth and northing was remarkably useful for capturing temporal shifts in distribution due to right whale migration in and out of the SEUS. Consistent with sighting data, the GAM predicted the highest probability of a whale sighting in January and February. Across all semimonthly periods, sighting rates were highest in intermediate northings of our study area, between latitudes of approximately 29.5 and 31.0°N. The interaction of these variables, however, demonstrated that whales occur in more northerly areas as they arrive at the beginning of the calving season in December, shift farther south in the middle of the season in January and February, and return north as they depart from the SEUS in March (Figure 3F). Survey year was also a significant predictor variable, as some years (*e.g.*, 2007/2008 and 2008/2009) had higher sighting rates than others (Figure 3G, Table S3).

Abundance Model

For the positive abundance model, the selected best model included six predictor variables as significant (SST, distance to the shoreline, distance to the 22°C SST isotherm, bottom depth, interaction between semimonthly period and UTM northing, and survey year) and explained 12.2% of the deviance for the presence-only data (Table 2). Adding survey effort increased the GCV score and did not significantly decrease the explained deviance for modeling the number of whales (Table S1B, Table S2B), so this term was excluded from the model.

The range of training data for the positive abundance model included only data from sampling units with whale sightings present, and the response for this model is the expected number of sighted whales per sampling unit, given whale presence. In general, smoothing functions in this model were similar to those in the presence-absence model, with more whales likely in sampling units with intermediate SST (Figure 4A), close to shore (Figure 4B), and in cooler waters far from the 22°C isotherm (Figure 4C). More whales were predicted, although with high uncertainty, near the upper limit of depth values with whales present (Figure 4D). Survey year was again significant, although the likelihood of multiple whales sighted in a sampling unit for a given year did not necessarily correspond with the overall sighting rate or probability of occurrence (Figure 4E, Table S3). Of the sampling units with whales present, multiple whales were more likely in sampling units at southerly northings and when whale densities were greatest, from late January through late February (Figure 4F). Smoothing functions for the positive abundance model generally had a higher standard error than those for the presence-absence model due in part to the smaller sample size.

Model Predictions

The final, combined hurdle model was successful at predicting the number of observed whales, as deviance residuals were overall close to zero (mean ± SE = −0.219±0.003). However, examination of the model residuals revealed that, in general, the model tended to overpredict when whales were not sighted (*i.e.*, predicting whale occurrence in sampling units with no whales observed) and underpredict when whales were sighted. For sampling units where no whales were sighted, the mean observed number of whales – the mean predicted number of whales = 0–0.106. For sampling units where whales were sighted, the mean observed number of whales – the mean predicted number of whales = 3.027–0.524. Excluding the last semimonthly period from March 2013, we found no evidence for spatial autocorrelation of the residuals during any of the semimonthly periods, as all Moran's I values approximated zero (mean = 0.030, range = −0.002 to 0.158). Moderate spatial autocorrelation was observed for late March 2013 (Moran's I = 0.376), although no sightings were made during this semimonth and all residuals were thus negative.

Using the selected hurdle model, we created hindcasts predicting the spatial distribution of whales for all semimonthly periods in the study. With the predict.gam function in the mgcv

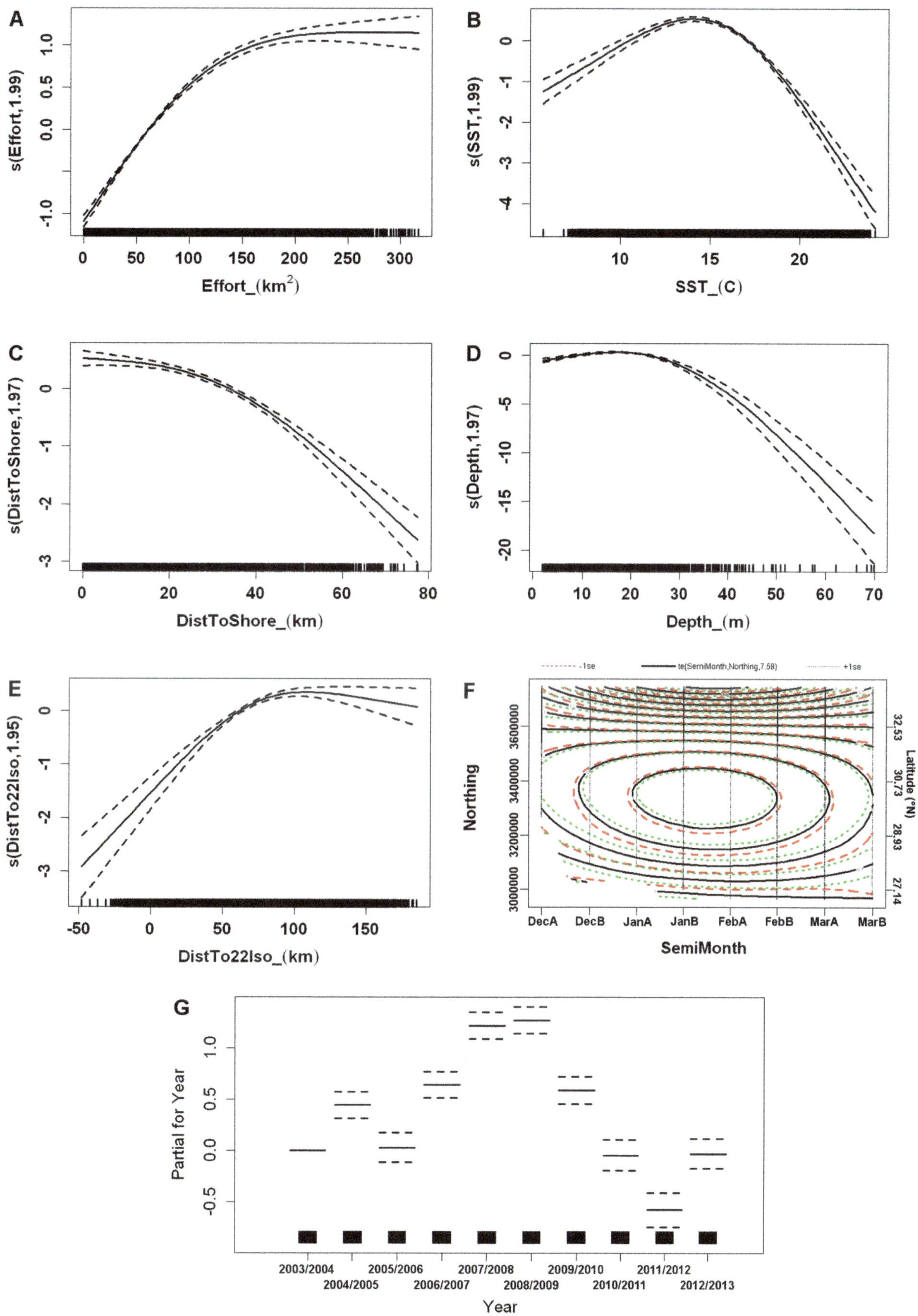
A
s(Effort,1.99)
Effort_(km2)
B
s(SST,1.99)
SST_(C)
C
s(DistToShore,1.97)
DistToShore_(km)
D
s(Depth,1.97)
Depth_(m)
E
s(DistTo22Iso,1.95)
DistTo22Iso_(km)
F
-1se
te(SemiMonth,Northing,7.58)
+1se
Northing
Latitude (°N)
DecA DecB JanA JanB FebA FebB MarA MarB
SemiMonth
G
Partial for Year
2003/2004 2004/2005 2005/2006 2006/2007 2007/2008 2008/2009 2009/2010 2010/2011 2011/2012 2012/2013
Year

Figure 3. Smooth functions of predictor variables for presence-absence model of right whale sightings. Predictor variables include survey effort (A), sea surface temperature (B), distance to the shoreline (C), bottom depth (D), distance to the 22°C isotherm (E), interaction between semimonthly period and UTM northing (F), and partial effects for survey year (G). The y-axis is on the scale of the linear predictor, and dashed lines indicate ±2 standard errors. Estimated degrees of freedom for each smoothed variable are in parentheses on the y-axis; tick marks on the x-axis indicate all sampled values.

package, the observed values for predictor variables (except survey effort) in all grid cells were used to generate the predicted and standard error estimates for probability of a whale sighting and number of whales sighted from the first and second steps of the hurdle model, respectively. Predicted relative abundance was then calculated by multiplying the probability of occurrence by the expected number of sighted whales. For these predictions, survey effort was set constant for all grid cells (at 250 km^2, near which the smoothing function for the presence-absence model reached a plateau) to eliminate the confounding effect of variable survey effort on number of sightings. We did not extrapolate our results or predictions outside the range of our sampled data (*e.g.*, where depth was >70 m, the maximum value in our training data), nor did we make predictions for cells with missing data (*e.g.*, where SST data were lacking due to cloud interference). For illustrative purposes, we have included prediction maps for the first 15 days of each month for the 2009/2010 and 2011/2012 seasons (Figure 5). In general, sea surface temperatures were near average in December 2009 but colder than average for the remainder of the 2009/2010 season (Figure S1), and this season had intermediate sighting rates relative to other years in our data set (Figure 3G, Table S3). In contrast, the 2011/2012 season was warmer than average (Figure S1) and had low observed sighting rates relative to other years in our data set (Figure 3G, Table S3). Consistent with the smoothing functions and observed sightings from surveys, the greatest concentrations of predicted whale sightings for all semimonths occurred close to shore, in the relatively shallow and cooler waters west of the Gulf Stream (Figure 5). High predicted sighting rates were more common and distributed farther south in January and February than in December and March (Figure 5). Because 2009/2010 had a greater partial effect for the year term (Figure 3G), this season had higher predicted sighting rates overall compared to 2011/2012 (Figure 5). Notably, areas within the SEUS with the highest predicted number of whales were located farther south in a cold year (2009/2010; *e.g.*, Figure 5B) compared to a warm year (2011/2012; *e.g.*, Figure 5F). The presence-absence model seemed to drive the pattern for predicted relative abundance (Pearson correlation coefficient between predicted probability of presence and predicted relative abundance = 0.98); standard errors were generally larger for cells with high predicted probability of occurrence.

Segregation of Calves

Of the total 3286 sightings, 1344 (41%) were of unaccompanied cow-calf pairs and 1919 (58%) were sightings without calves (*i.e.*, juveniles, adult males, non-calving adult females, pregnant females, and females that had lost a calf that year). The remaining 1% of sightings was composed of a calf by itself (6), two cow-calf pairs in close association (4), or a cow-calf pair accompanied by other individuals (13). We were able to extract SST data for 459 sightings with a calf present and 675 sightings with a calf absent, and all other environmental data for 1349 sightings with a calf present and 1907 sightings with a calf absent. Sightings with a calf present were made in locations significantly shallower and closer to shore and tended to occur in warmer water than sightings without calves, although the differences were small and there was extensive overlap in the range of values between group types (Table 3). Distance to the 22°C isotherm did not differ significantly among groups (Table 3). UTM northing did not differ significantly between groups with or without calves for any semimonthly period except the first period in December, when sightings without calves occurred farther north ($p<0.01$; Figure 6).

Discussion

In agreement with previous studies, our results indicate that GAMs are a useful tool for relating cetacean distribution to environmental variables and predicting cetacean occurrence and relative density based upon those variables [26,36]. The development of a GAM, along with the ability to collect environmental data via remote sensing and to apply GIS processing techniques, permitted us to interpolate our results and estimate relative abundance of right whales, with associated standard error, in regions of our study area not sampled by aerial surveys. The flexibility of GAMs allowed right whale encounter rates to be modeled as a complex, nonlinear response to predictor variables, and a hurdle model approach allowed us to deal with zero-inflated data from surveys of a rare species summarized at a high spatial resolution. By using dynamic predictor variables and accounting

Table 2. Summary of stepwise selection procedure for positive abundance model of right whales in the southeastern US.

Model	% Deviance	GCV	mean ASPE
null	0	0.4534	8.284
s(SemiMonth:Northing)	6.5	0.4261	7.981
s(SemiMonth:Northing)+Year	10.2	0.4126	7.842
s(SemiMonth:Northing)+Year+s(SST)	11.2	0.4082	7.822
s(SemiMonth:Northing)+Year+s(SST)+s(DistTo22Iso)	11.7	0.4066	7.819
s(SemiMonth:Northing)+Year+s(SST)+s(DistTo22Iso)+s(DistToShore)	11.8	0.4065	7.820
s(SemiMonth:Northing)+Year+s(SST)+s(DistTo22Iso)+s(DistToShore)+s(Depth)	12.2	0.4054	7.807

Predictor variables and abbreviations same as in Table 1.

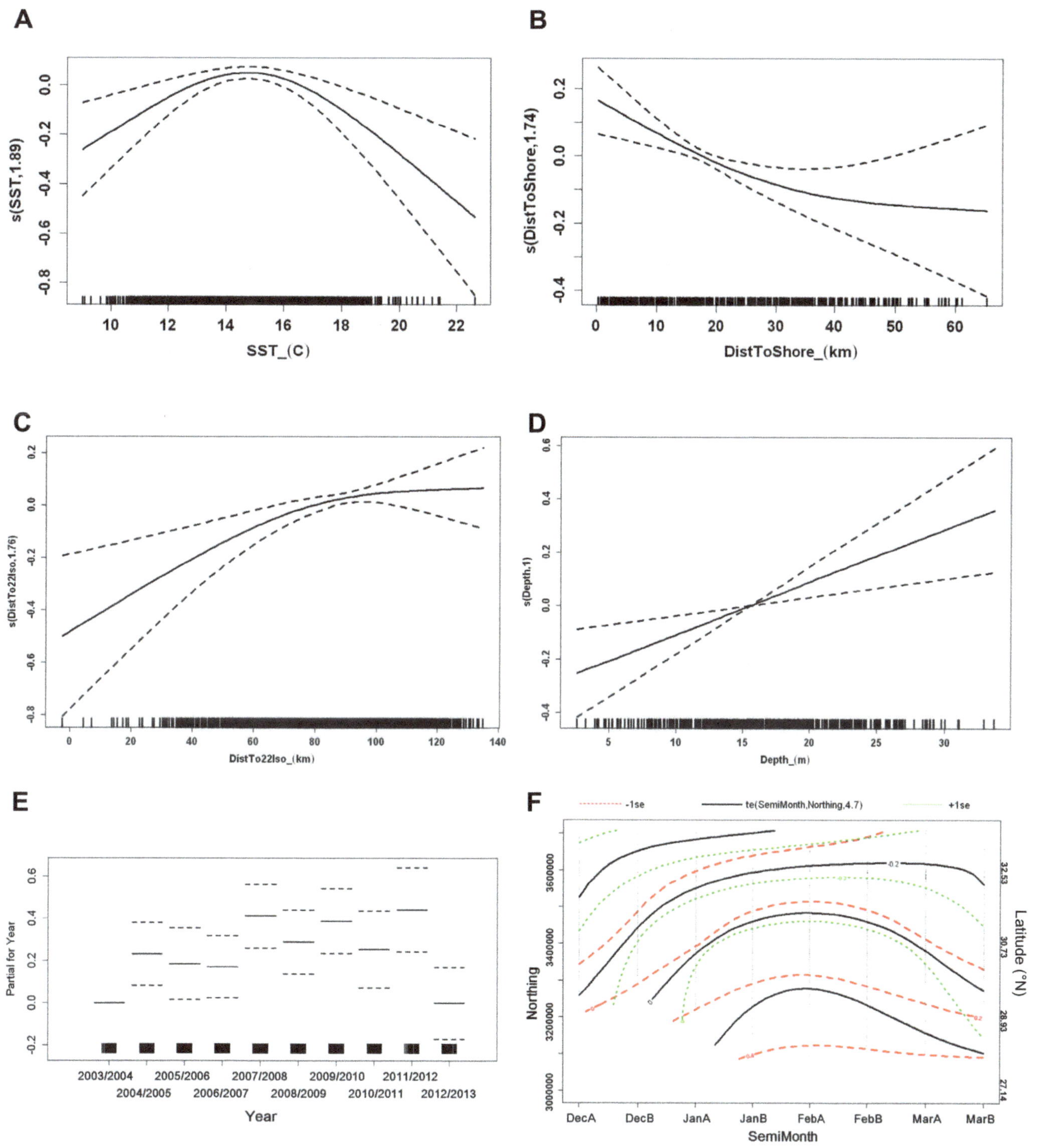

Figure 4. Smooth functions of predictor variables for positive abundance model of right whale sightings. Predictor variables and display same as in Figure 3.

for migration patterns, we were able to model changes in whale distribution through time.

Concordant with other studies in the SEUS [8,9,10], bottom depth and SST were significant predictors of right whale distribution; right whales were more likely to be sighted in waters 10 to 25 m deep and 12 to 16°C SST. Fonnesbeck *et al.* [9] and Keller *et al.* [10] also found year to be a significant factor variable related to differences in the total number of whales in the SEUS. Based on SST at sighting locations, Keller *et al.* [5,10] suggested that right whales likely avoid the Gulf Stream, although they did not include a term to represent Gulf Stream variability in their model (such as distance to the 22°C SST isotherm). Distance to

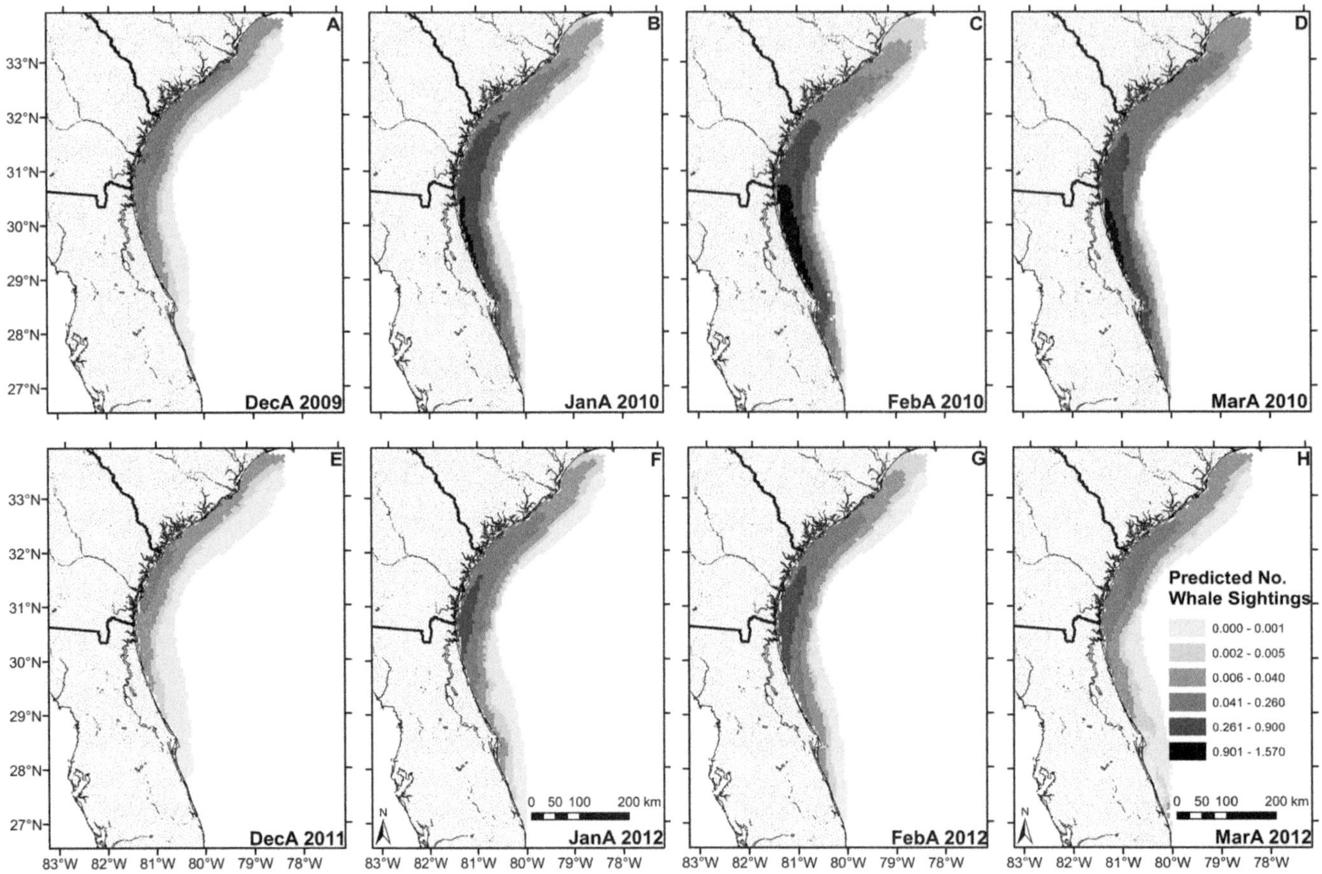

Figure 5. Predicted right whale relative abundance. Values represent predicted number of sighted right whales per grid cell (assuming uniform survey effort) during the 2009/2010 calving season (a relatively cold season with high sighting rates) for December 1–15 (A), January 1–15 (B), February 1–15 (C), and March 1–15 (D); and during the 2011/2012 calving season (a relatively warm season with low sighting rates) for December 1–15 (E), January 1–15 (F), February 1–15 (G), and March 1–15 (H).

shore was a significant predictor in our model, in contrast to these previous studies which either did not consider this term [8,9] or limited model complexity to four terms [10]. The range of preferred SST and avoidance of waters >22°C SST may be related to physiological optima and constraints on thermoregulation and growth, particularly in pregnant or nursing females and newborn calves [5,37,38]. It is less clear why whales were more common in shallow water close to shore, but it has been hypothesized that predator avoidance, weaker currents, and calm waters provided by these physical factors may be advantageous to calves and juveniles that are still developing swimming skills or more susceptible to predation [8,34]. Unfortunately, it will be difficult to adequately address these hypotheses unless predator abundance and wave/surface roughness data become available at finer temporal and spatial resolutions.

Our model also included the interaction between northing and semimonthly period to account for whale migration during the calving season. Both Good [8] and Keller *et al.* [10] predicted suitable habitat off northern Georgia and South Carolina throughout the winter. We, however, predict that these areas have relatively lower encounter rates compared to core-use areas off northern Florida and southern Georgia, particularly during the middle of the winter (Figure 5), congruent with sighting data (Figure 2). We believe that this difference in predictions is driven by the additional data with low sighting rates from the SC-GA surveys and by the inclusion of the latitude/semimonth interaction term in our model, which accounted for latitudinal shifts in whale distribution that were independent of SST variability. This is the first habitat model, to our knowledge, that explicitly considers a migration process in an area outside of the species' feeding grounds. Although the latitude/semimonth interaction term is only a proxy for the inherent processes that truly drive the migration timing, and it limits the application of the model to the area analyzed, it nevertheless improved the predictive capability of our model and is therefore useful for identifying areas in the SEUS with high whale encounter rates and predicting how these encounter rates change throughout a winter season. Moreover, the ecological factors that may influence the timing of whale migration into and within the SEUS (e.g., predation risk, calm seas, and historical distribution of con-specifics) are difficult to quantify and remain unknown. Although our model predicts that sightings are less common and whales likely spend less time in the area off northern Georgia and South Carolina, this does not mean that right whales do not use (and possibly give birth in) these areas. We hypothesize that, in most years, these more northerly areas are part of the right whale migration corridor [39,40]. Whales may display behaviors in these corridors that affect detectability (if they spend more time below the surface) and may have shorter residence times there than in the core wintering area. We therefore suggest that a separate model be created to characterize the right whale migration corridor when more data on migrating whales become available. Variability in sea surface roughness (not

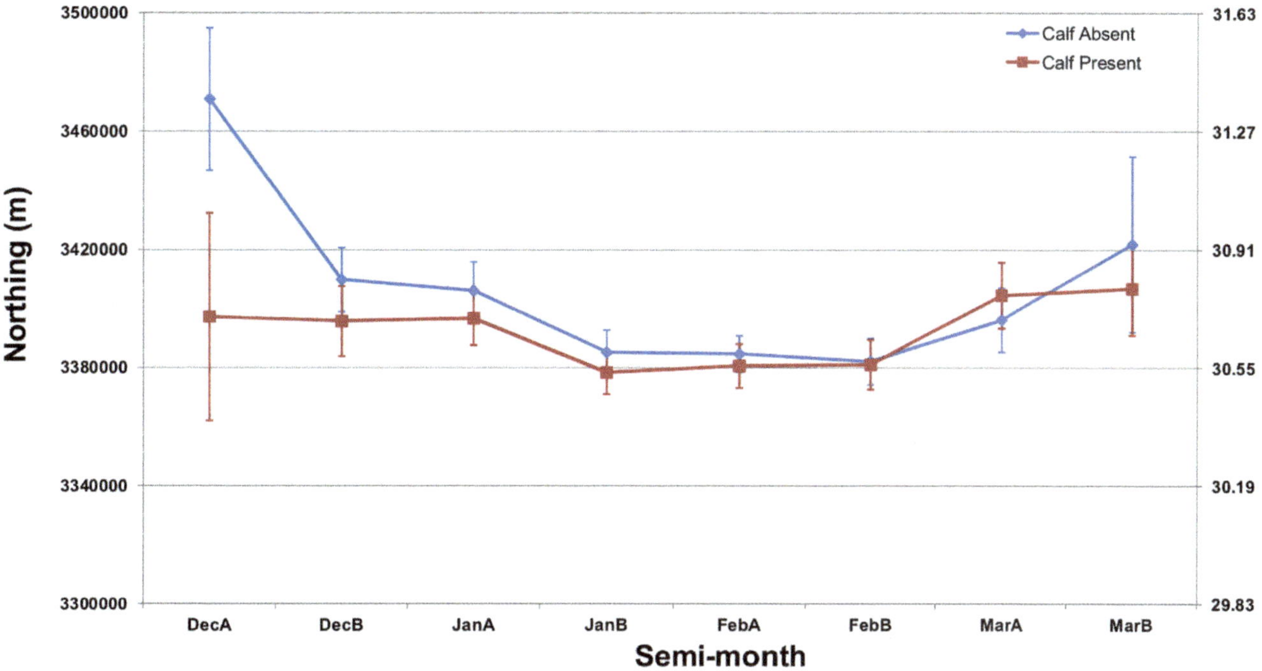

Figure 6. UTM northings for right whale sightings in the southeastern United States. Values indicate mean (±2 standard errors) UTM northing of sightings for each semimonthly period during all calving seasons. Sightings with a calf present displayed as red squares; sightings without a calf displayed as blue diamonds.

addressed in this study) and the occurrence of Gulf Stream meanders north of Charleston, SC, which generate onshore movements of unsuitably warm water [19,41], may yield a less stable habitat for whales in the northern areas of the SEUS. Nevertheless, we agree with Good [8] and Keller *et al.* [10] that in some years the core wintering area extends farther north than the current critical habitat boundary (*e.g.*, Figure 5G). Although extrapolating model results can help identify potential areas for future surveys [10,12], the different predictions among studies here highlight the importance of validating and updating models with additional data and may emphasize that caution is needed when applying model results beyond the spatial and temporal extent of training data [42].

Environmental variability among years leads to differences in both the total number of whales that migrate to the SEUS and the spatial distribution of whales once they arrive there (Figure 5, Keller *et al.* [10]). Variability in right whale spatial distribution among years seems to be driven by local conditions such as water temperature and location of the Gulf Stream, which vary in the SEUS among years and within a calving season (Figure S1). For example, whales are distributed farther south in relatively cold years (Figure 5), presumably to occupy areas of preferred SST. However, the number of whales that migrate to the SEUS in a given year and the amount of time they spend there are likely influenced by factors outside the SEUS [43,44]. Greene *et al.* [43] and Kenney [44] suggest that female right whale calving rates (and consequently the number of calving females that migrate to the SEUS each year) are determined by the number of females available to calve and by recovery times for reproductive females, ultimately influenced by food availability in summer foraging grounds. Encounter rates of whales will also be affected by the duration of whale residence in the SEUS, which may be a consequence of demographics (with calving females possibly having longer residence times than other whales [45]), energy reserves, changes in weather, or other, unknown factors. Although including year as a factor variable significantly improved our model fit and the ability to hindcast whale relative abundance for the years analyzed in our study, it precluded the ability to predict the magnitude of whale densities for years beyond our data set. Replacement of this term with a variable that describes the

Table 3. Mean (range) environmental conditions at locations of right whale sightings with a calf present or absent.

Variable	Calf Present	Calf Absent	z-score	*p*
Depth (m)	14.9 (3.0–34.0)	15.5 (2.0–33.0)	−2.95	0.003
DistToShore (km)	17.6 (0.2–62.1)	18.5 (0.1–66.2)	−2.11	0.035
SST (°C)	14.9 (9.9–21.3)	14.7 (8.9–22.9)	−1.84	0.066
DistTo22Iso (km)	86.2 (11.7–135.3)	85.7 (−3.9–132.3)	−0.44	0.664

Environmental variables include bottom depth, distance to the shoreline (DistToShore), sea surface temperature (SST), and distance to the 22°C SST isotherm (DistTo22Iso).

number and demography of right whales predicted to migrate to the wintering grounds each year (*e.g.*, [43]) would greatly improve the predictive capabilities of our model, particularly for forecasting the true value of whale encounter rates. Overall, the results of this study demonstrate that whale abundance and distribution vary within and between years. Recognizing and understanding this temporal variability can better inform the location and timing of management actions as whale distribution changes throughout a season, which would result in more effective risk mitigation and population monitoring of right whales.

While the final model explained the most variation in our dataset, overdispersion (partly caused by the high spatio-temporal resolution of our sampling units, the rarity of right whale sightings, and the potentially high number of false absences generated by aerial survey data) created challenges to achieving a good model fit. Indeed, it is common for GAMs modeling cetacean occurrence at a high spatial resolution, and non-normal models in general, to explain only a small proportion of the deviance in a dataset [26,36,46]. For example, our model frequently under-estimated the number of whales in grid cells with observed sightings, and this may be related to how the data were aggregated. Rather than predicting a high sighting probability in a single grid cell, the model smoothed predicted whale occurrence over several nearby cells with similar, preferred habitat conditions; in this regard, the model predictions may describe more accurately where a whale may be located over the span of two weeks (*i.e.*, temporarily occupying multiple nearby cells) than would an aerial survey sighting fixed in time and location. Conversely, the model tended to overpredict whale occurrence in areas where no whales were sighted. The small population size of right whales could have resulted in the absence of whales from suitable habitat [47]. Additionally, the lack of a sighting does not necessarily mean that whales were truly absent. Whales might not have been detected due to availability bias (*e.g.*, whales were submerged [48]), perception bias (*e.g.*, observer error [49]), or incomplete temporal coverage by surveys (*i.e.*, not flying at night, not flying during poor weather). In consideration of these limitations of aerial surveys and the confounding effect of unequal survey effort on the number of sightings, we believe that predictive habitat models can better characterize the distribution of whales than can sightings-per-unit-effort data (*e.g.*, Figure 2C). Rather than relying solely on costly surveys with imperfect detection probabilities and limited coverage, predictions from these models can be particularly helpful in making management decisions that require risk assessment based on expected whale distribution [50].

Whale groups with a calf present were sighted in waters that were shallower and closer to shore than were other whale groups. Even though shallow, nearshore locations in our study area generally had the coldest water, there was also a trend for groups with calves to be sighted in slightly warmer water. Sightings with a calf present occurred farther south during early December, although this finding is confounded by few sightings of calves early in the calving season and the fact that calves-of-the-year are not likely to be migrating from the north at this time. Studies of baleen whales on their wintering grounds have also found cow-calf pairs closer to shore and in shallower water than other demographic groups [32,33,51]. Differences in habitat preference may be a result of different energetic constraints or predation risks between demographic groups. Although our results were statistically significant and supported by other studies, the differences between groups were minimal, the range of values for environmental variables were very similar, and sightings of groups with calves compared with those without calves were not spatially segregated at the scale and spatial extent of our model. We therefore conclude that separate habitat models for cow-calf pairs and other demographic groups in the SEUS are not warranted and are impracticable at the scale used in this study, and management areas for one group will likely be equally effective for the other. Despite the large number of sightings in our data, however, it was extremely rare for a cow-calf pair to be sighted with another whale. Thus, at a fine scale, it appears that cow-calf pairs are segregated from other whales, but this pattern may be influenced more by social dynamics (*e.g.*, harassment avoidance; [34,52]) than by environmental features.

Our analysis presents an example of the use of static and dynamic environmental variables to predict species distribution and characterize habitat preferences of a migratory species in a dynamic seascape. Using temporally dynamic variables, such as SST, isotherms, and a seasonal migration index, we were able to predict semimonthly right whale distribution. Although our framework does not estimate absolute densities, it provides additional insights into how whale distribution changes over time compared to a single depiction of instantaneous density. This framework could be applied to other highly mobile migratory species in variable habitats that may benefit from dynamic management actions [6,53]. The model results corroborate that right whale distribution varies within and among years. We believe that the addition of data from a broader survey area relative to previous studies, and accounting for migration patterns, improved the model by predicting how whale distribution changes within a calving season and by separating core-use areas from short-term-use areas in the migration corridor. Hindcast predictions from this model can be used to assess risks to whales and effectiveness of management actions. Improvements of the model, including the addition of parameters for predicting the total number of whales that migrate to the wintering grounds [43], would allow for near-real-time forecasts of whale distribution that could be used to better inform management decisions. We encourage future evaluations of this and other habitat models as additional data, particularly from other predictors and from surveys in areas previously receiving little coverage, become available.

Supporting Information

Figure S1 Average daily sea surface temperatures for December through March. Data obtained from the National Data Buoy Center (http://www.ndbc.noaa.gov) at Gray's Reef (top panel, station 41008, 31.400°N 80.868°W) and at St. Augustine (bottom panel, station SAUF1, 29.857°N 81.265°W). Data include 2009/2010 (blue line), 2011/2012 (red line), long-term historical average with 95% confidence intervals (1991/1992 and 1997/1998–2012/2013 at Gray's Reef; 1988/1989–2001/2002 and 2004/2005–2011/2012 at St. Augustine), and long-term historical range.

Table S1 A. Summary of all models tested in stepwise selection procedure for presence-absence models of right whales in the southeastern United States. Predictor variables include interaction between semimonthly period and UTM northing, distance to the shoreline (DistToShore), survey year, survey effort, sea surface temperature (SST), bottom depth, distance to the 22°C SST isotherm (DistTo22Iso), and slope. Smoothed covariates indentified by "s()". Evaluation criteria include the proportion of deviance explained, generalized cross validation score (GCV), and mean average squared prediction error (ASPE) from a five-fold cross-validation. The best model at each step is in bold. B. Summary of all models tested in stepwise selection procedure for positive abundance models of right whales in the southeastern United

States. Predictor variables and abbreviations same as in Table S1A. The best model at each step is in bold.

Table S2 A. Results of analysis of deviance tests comparing presence-absence models at each step of the stepwise selection procedure. Models at each step refer to the respective best model (in bold) from Table S1A. Reductions in deviance, *F*-statistics, and *p*-values compare each model to the model in the previous step. B. Results of analysis of deviance tests comparing positive abundance models at each step of the stepwise selection procedure. Models at each step refer to the respective best model (in bold) from Table S1B. Reductions in deviance, *F*-statistics, and *p*-values compare each model to the model in the previous step.

Table S3 Number of observed sightings, number of observed whales (not unique individuals), total survey effort, and overall sighting rate (sightings/1000 km^2 surveyed) for each calving season from our data set.

Acknowledgments

We are grateful to the dedicated observers who conducted many hours of aerial surveys throughout the years and collected the data used in the analysis presented here. The North Atlantic Right Whale Consortium curates the aerial survey data provided to us by R. Kenney of the University of Rhode Island. We thank D. Dick, S. Cain, and A. Krzystan for their assistance with aerial survey data processing and GIS analyses. We also thank R. Muller, L. McEachron, L. Ward, B. Crowder, and two anonymous reviewers for their comments on this manuscript.

Author Contributions

Conceived and designed the experiments: JGO TAG. Performed the experiments: TAG JGO. Analyzed the data: TAG JGO. Contributed reagents/materials/analysis tools: TAG JGO. Wrote the paper: TAG JGO.

References

1. Waring GT, Josephson E, Maze-Foley K, Rosel PE, editors (2013) U.S. Atlantic and Gulf of Mexico marine mammal stock assessments - 2012. NOAA Technical Memorandum NMFS-NE-223: Available from: National Marine Fisheries Service, 166 Water Street, Woods Hole, MA 02543–1026. Available: http://www.nefsc.noaa.gov/nefsc/publications/. 319 p.
2. Knowlton AR, Kraus SD (2001) Mortality and serious injury of northern right whales (*Eubalaena glacialis*) in the western North Atlantic Ocean. J Cetacean Res Manag (Special Issue 2): 193–208.
3. Kraus SD, Rolland RM (2007) Right whales in the urban ocean. In: Kraus SD, Rolland RM, editors. The urban whale: North Atlantic right whales at the crossroads. Cambridge, MA: Harvard University Press. 1–38.
4. National Marine Fisheries Service (2005) Recovery plan for the North Atlantic right whale (*Eubalaena glacialis*). Silver Spring, MD: National Marine Fisheries Service, Office of Protected Resources. 137p.
5. Keller CA, Ward-Geiger LI, Brooks WB, Slay CK, Taylor CR, et al. (2006) North Atlantic right whale distribution in relation to sea-surface temperature in the southeastern United States calving grounds. Mar Mamm Sci 22: 426–445.
6. Redfern JV, Ferguson MC, Becker EA, Hyrenbach KD, Good C, et al. (2006) Techniques for cetacean-habitat modeling. Mar Ecol Prog Ser 310: 271–295.
7. Pendleton DE, Sullivan PJ, Brown MW, Cole TVN, Good CP, et al. (2012) Weekly predictions of North Atlantic right whale *Eubalaena glacialis* habitat reveal influence of prey abundance and seasonality of habitat preferences. Endanger Species Res 18: 147–161.
8. Good CP (2008) Spatial ecology of the North Atlantic right whale (*Eubalaena glacialis*) [Ph.D. dissertation]. Durham, NC: Duke University. 135 p.
9. Fonnesbeck CJ, Garrison LP, Ward-Geiger LI, Baumstark RD (2008) Bayesian hierarchichal model for evaluating the risk of vessel strikes on North Atlantic right whales in the SE United States. Endanger Species Res 6: 87–94.
10. Keller CA, Garrison L, Baumstark R, Ward-Geiger LI, Hines E (2012) Application of a habitat model to define calving habitat of the North Atlantic right whale in the southeastern United States. Endanger Species Res 18: 73–87.
11. Hamazaki T (2002) Spatiotemporal prediction models of cetacean habitats in the mid-western North Atlantic Ocean (from Cape Hatteras, North Carolina, U.S.A. to Nova Scotia, Canada). Mar Mamm Sci 18: 920–939.
12. Guisan A, Broennimann O, Engler R, Vust M, Yoccoz NG, et al. (2006) Using niche-based models to improve the sampling of rare species. Conserv Biol 20: 501–511.
13. Guisan A, Zimmermann NE (2000) Predictive habitat distribution models in ecology. Ecol Modell 135: 147–186.
14. Hedley S, Buckland S (2004) Spatial models for line transect sampling. J Agric Biol Environ Stat 9: 181–199.
15. North Atlantic Right Whale Consortium (2012) North Atlantic right whale consortium sightings database (New England Aquarium, Boston, MA, U.S.A.).
16. Buckland ST, Anderson DR, Burnham KP, Laake JL, Borchers DL, et al. (2001) Introduction to distance sampling: Estimating abundance of biological populations. Oxford: Oxford University Press. 432 p.
17. Thomas L, Buckland ST, Rexstad EA, Laake JL, Strindberg S, et al. (2010) Distance software: Design and analysis of distance sampling surveys for estimating population size. J Appl Ecol 47: 5–14.
18. Zuur AF, Ieno EN, Elphick CS (2010) A protocol for data exploration to avoid common statistical problems. Methods Ecol and Evol 1: 3–14.
19. Stegmann PM, Yoder JA (1996) Variability of sea-surface temperature in the South Atlantic Bight as observed from satellite: Implications for offshore-spawning fish. Cont Shelf Res 16: 843–861.
20. Hastie TJ, Tibshirani RJ (1990) Generalized additive models. Monographs on statistics and applied probability 43. London: Chapman & Hall. 335 p.
21. Wood SN (2006) Generalized additive models: An introduction with R. Boca Raton, FL: CRC Press.
22. Potts JM, Elith J (2006) Comparing species abundance models. Ecol Modell 199: 153–163.
23. Fauchald P, Skov H, Skern-Mauritzen M, Hausner VH, Johns D, et al. (2011) Scale-dependent response diversity of seabirds to prey in the North Sea. Ecology 92: 228–239.
24. Wood SN (2001) mgcv: GAMs and generalized ridge regression for R. R News 1: 20–25.
25. Austin M (2007) Species distribution models and ecological theory: A critical assessment and some possible new approaches. Ecol Modell 200: 1–19.
26. Best BD, Halpin PN, Read AJ, Fujioka E, Good CP, et al. (2012) Online cetacean habitat modeling system for the US East Coast and Gulf of Mexico. Endanger Species Res 18: 1–15.
27. Montgomery DC, Peck EA (1992) Introduction to linear regression analysis. New York: John Wiley. 527p.
28. Wood SN (2003) Thin plate regression splines. J R Stat Soc Series B Stat Methodol 65: 95–114.
29. Dormann CF, Elith J, Bacher S, Buchmann C, Carl G, et al. (2013) Collinearity: A review of methods to deal with it and a simulation study evaluating their performance. Ecography 36: 027–046.
30. Dormann CF, Purschke O, García Márquez JR, Lautenbach S, Schröder B (2008) Components of uncertainty in species distribution analysis: A case study of the great grey shrike. Ecology 89: 3371–3386.
31. Bivand R (2007) spdep. Spatial dependence: Weighting schemes, statistics and models. R Package Version 0.5–56.
32. Elwen SH, Best PB (2004) Environmental factors influencing the distribution of southern right whales (*Eubalaena australis*) on the south coast of South Africa I: Broad scale patterns. Mar Mamm Sci 20: 567–582.
33. Ersts PJ, Rosenbaum HC (2003) Habitat preference reflects social organization of humpback whales (*Megaptera novaeangliae*) on a wintering ground. J Zool 260: 337–345.
34. Elwen SH, Best PB (2004) Female southern right whales *Eubalanena australis*: Are there reproductive benefits associated with their coastal distribution off South Africa? Mar Ecol Prog Ser 269: 289–295.
35. Rice WR (1989) Analyzing tables of statistical tests. Evolution 43: 223–225.
36. Forney KA, Ferguson MC, Becker EA, Fiedler PC, Redfern JV, et al. (2012) Habitat-based spatial models of cetacean density in the eastern Pacific Ocean. Endanger Species Res 16: 113–133.
37. Brodie PF (1975) Cetacean energetics, an overview of intraspecific size variation. Ecology 56: 152–161.
38. Kshatriya M, Blake RW (1988) Theoretical model of migration energetics in the blue whale, *Balaenoptera musculus*. J Theor Biol 133: 479–498.
39. Knowlton A, Ring J, Russell B (2002) Right whale sightings and survey effort in the Mid-Atlantic region: Migratory corridor, time frame, and proximity to port entrances. A report submitted to the NMFS Ship Strike Working Group. Available: www.nero.noaa.gov/shipstrike/ssr/midatanticreportrFINAL.pdf.
40. Schick RS, Halpin PN, Read AJ, Slay CK, Kraus SD, et al. (2009) Striking the right balance in right whale conservation. Can J Fish Aquat Sci 66: 1399–1403.
41. Bane JM, Brooks DA (1979) Gulf Stream meanders along the continental margin from the Florida Straits to Cape Hatteras. Geophys Res Lett 6: 280–282.
42. Randin CF, Dirnböck T, Dullinger S, Zimmermann NE, Zappa M, et al. (2006) Are niche-based species distribution models transferable in space? J Biogeogr 33: 1689–1703.

43. Greene CH, Pershing AJ, Kenney RD, Jossi JW (2003) Impact of climate variability on the recovery of endangered North Atlantic right whales. Oceanography 16: 98–103.
44. Kenney RD (2007) Right whales and climate change: Facing the prospect of a greenhouse future. In: Kraus SD, Rolland RM, editors. The urban whale: North Atlantic right whales at the crossroads. Cambridge, MA: Harvard University Press. 436–459.
45. Burnell SR, Bryden MM (1997) Coastal residence periods and reproductive timing in southern right whales, *Eubalaena australis*. J Zool 241: 613–621.
46. Hosmer DW, Lemeshow S, Sturdivant RX (2013) Applied logistic regression. New York: John Wiley. 528p.
47. Pulliam HR (2000) On the relationship between niche and distribution. Ecol Lett 3: 349–361.
48. Hain JHW, Ellis SL, Kenney RD, Slay CK (1999) Sightability of right whales in coastal waters of the southeastern United States with implications for the aerial monitoring program. In: Garner GW, Amstrup SC, Manly BFJ, Laake JL, McDonald LL et al., editors. Marine mammal survey and assessment methods. Rotterdam, Netherlands: A.A. Balkema. 191–207.
49. Marsh H, Sinclair DF (1989) Correcting for visibility bias in strip transect aerial surveys of aquatic fauna. J Wildl Manage 53: 1017–1024.
50. Redfern JV, McKenna MF, Moore TJ, Calambokidis J, Deangelis ML, et al. (2013) Assessing the risk of ships striking large whales in marine spatial planning. Conserv Biol 27: 292–302.
51. Felix F, Haase B (2005) Distribution of humpback whales along the coast of Ecuador and management implications. J Cetacean Res Manag 7: 21–31.
52. Cartwright R, Sullivan M (2009) Associations with multiple male groups increase the energy expenditure of humpback whale (*Megaptera novaeangliae*) female and calf pairs on the breeding grounds. Behaviour 146: 1573–1600.
53. Bull JW, Suttle KB, Singh NJ, Milner-Gulland EJ (2013) Conservation when nothing stands still: Moving targets and biodiversity offsets. Front Ecol Environ 11: 203–210.

8

The Role of the Sedimentary Regime in Shaping the Distribution of Subtidal Sandbank Environments and the Associated Meiofaunal Nematode Communities: An Example from the Southern North Sea

Michaela Schratzberger[1]*, Piers Larcombe[2]

1 Centre for Environment, Fisheries and Aquaculture Science, Lowestoft Laboratory, Lowestoft, Suffolk, United Kingdom, **2** RPS MetOcean, Subiaco, Western Australia, Australia

Abstract

We combined sediment and faunal data to explore the role of the sedimentary regime in shaping the distribution of subtidal sandbank environments and the associated meiofaunal nematode communities at Broken Bank and Swarte Bank, in the southern North Sea. A variety of sediment transport processes occur in the area, differing in the frequency and magnitude of sediment mobility, and the continuum between erosion, translation and sediment accumulation. The seabed contained a variety of bedforms, including longitudinal furrows, and small to very large sandwaves. The bed sediments were dominated by fine and medium sands, with admixtures of silt and gravel. Based on sedimentary bedforms and grain size analysis, a total of 11 sedimentary facies were delineated, of which 8 were analysed in detail for their relationships with the meiofauna. The sedimentary facies fell clearly into groups of facies, respectively representing high, high-moderate and moderate, and episodic sediment mobility. For those sedimentary facies where daily movement of sediments and bedforms occurred ('high' sediment mobility), the resulting spatially homogeneous environments were dominated by an impoverished nematode community comprising small deposit feeders and large predators. Resistance to sediment movement and the ability to exploit alternative food sources were prominent functional features of the successful colonisers. Those facies characterised by relatively infrequent sediment mobility ('episodic' and 'high-moderate and moderate' sediment mobility) comprised a heterogeneous suite of benthic habitats, containing taxonomically and functionally diverse assemblages of nematodes of various sizes, feeding types and reproductive potential. Faunal distribution patterns here indicated trade-offs between the resistance to sediment movement, environmental tolerance and competitive abilities. Our focus on diverse assemblages of organisms with high turnover times, inhabiting highly dynamic sedimentary environments, has revealed new animal-sediment relationships of relevance to pure and applied science.

Editor: David William Pond, Scottish Association for Marine Science, United Kingdom

Funding: The work was supported by the Department for Environment, Food and Rural Affairs (Defra, research project AE1148), by EDF Energy and by RPS MetOcean. Early work on the manuscript was supported by the Marine Environment Protection Fund (MEPF), a delivery partner of Defra's Aggregate Levy Sustainability Fund (ALSF). The funders had no role in study design, data collection and analysis, decision to publish, or preparation of the manuscript.

* Email: michaela.schratzberger@cefas.co.uk

Introduction

Scientists, socio-economists, environmental managers, policy-makers and the public all have interests in marine environmental studies and their outcomes. A common motivation of marine studies is the implicit or explicit search for indicators of natural and anthropogenic change, with an aim of identifying and assessing human impacts and the effects of management actions. As an example policy driver, the EU Habitats Directive for the protection of species and habitats [1] requires that management plans be created in order to detect and respond to human-induced change. In particular, such plans are to be prepared for the candidate Special Areas of Conservation (SAC), including subtidal mobile sandbanks, on which we expand below. Thus, it is a requirement that plans are able to detect human-induced change as distinct from the various types and timescales of natural variability, and that actions are identified which are able to be enacted to counter such (assumed) detrimental change. In most environments this is not a simple task, but in the dynamic marine sedimentary environment of much of the western European shelf it appears to be a major challenge indeed.

Effective conservation of marine habitats in European seas requires specific conservation objectives. These need to be achieved by closely linked management measures that regulate the activities of those marine industries which may threaten the integrity of the habitat, or the viability of the associated populations [2]. Conservation objectives for subtidal sandbanks generally demand that, subject to natural processes, the extent and distribution of sandbank habitats, their faunal structure and function, their gross morphology (e.g. depth, distribution and profile) and the supporting processes on which sandbank habitats rely, are maintained or recovered [3]. In order to be able to measure the success or otherwise of such management goals, detailed investigations of the links between the sedimentary habitat and the biota are essential.

Reviewing the literature on animal-sediment relationships, [4] found little evidence that animal distributions are determined by any of the sediment variables derived from grain size alone. Indeed, [5] summarised the results of nine recent studies, which used abiotic variables such as percentage mud, percentage calcium carbonate, percentage gravel, total organic carbon content, a bed disturbance factor (named seabed exposure by [6]), water depth, latitude, longitude, slope, turbidity and distance to ocean. In these studies, between 25 and 75% of variability in faunal distribution patterns was accounted for; most of the stronger relationships involved using physical measures of the sea floor and sediments. Hence, meaningful and predictive explanations for faunal distributions are likely to emerge if these are evaluated relative to the suite of hydrodynamic and sediment transport processes that are responsible for sediment distributions. It is also important to understand the physical form of the sediments (bathymetry and the nature of sedimentary bedforms) at a range of spatio-temporal scales, and the details of the grain size distribution.

A general appreciation of the highly heterogeneous nature of continental shelf sediments has progressively emerged in recent decades. Multi-beam bathymetric surveys have revealed numerous types of sedimentary and geomorphological features on the sea floor that occur at a variety of spatial scales, ranging from a few meters to several kilometres [7,8,9,10]. Ground-truthing acoustic data to confirm the nature of the seabed is an important process, but the next step of improving understanding of the relationships with the fauna is hampered by gaps in our knowledge. These gaps include the difficulty of identifying organisms to fine taxonomic levels [11] and the comparative shortage of information on those organisms with a range of life-history characteristics which tie them closely to the sediment in which they live, such as meiofauna (comprising invertebrates that are retained on a 500–1000 μm mesh; [12]).

There are obvious possible dependencies of meiofauna upon the distribution and mobility of sediments [13,14,15]. When patches of suitable habitat are spatially or temporally discontinuous, an organism's dispersal capability will greatly affect its ability to colonise available space. Where the dispersal mechanism is closely related to the transport mechanisms of the associated sediment (e.g. meiofaunal nematodes) we would expect that the distribution and location of the sediment to be a key factor in controlling faunal distribution patterns. Mobile sediments at subtidal sandbanks could affect meiofaunal assemblages by altering the composition of sedimentary environments, thereby changing their suitability for settling organisms and/or by increasing the resuspension of organisms as a result of sediment deposition and/or sediment instability.

A detailed description of the benthic fauna associated with two linear sandbanks off the North Norfolk coast in the southern North Sea by [16] provided a general overview of the benthic ecology of the area. Here, we further explore these data, taking account of the topographic complexity of the surveyed sandbank habitats, to improve our understanding of the dynamics of the sandbank ecosystem. As such, this scoping study has two objectives: 1) to qualitatively describe the nature and origins of several kinds of sedimentological features and processes at two subtidal sandbanks in the southern North Sea, explicitly taking account of hydrodynamics and sediment transport processes, the bathymetry and nature of sedimentary bedforms, and grain size distributions and 2) to depict the taxonomic and functional characteristics of meiofaunal nematode communities which occupy these environments. We use our findings to better understand the way in which the physical dynamics of the seabed and its spatial heterogeneity might combine to determine the number and type of available niches and hence influence the diversity and composition of meiofaunal nematode communities.

Field Surveys and Sample Processing

All data reported in this article were collected by the Centre for Environment, Fisheries and Aquaculture Science (Cefas), as part of a government-funded research project. As an executive agency of the UK Government, Cefas has an exemption from The Crown Estate (i.e. land owner of our study site) to collect sediment and other scientific material (except vertebrates) in UK waters. The activities involved in the survey of the seabed and the collection of benthic invertebrates were not subject to specific regulation and did not involve endangered or protected species. Throughout the survey, scientists and ship crew acted in a responsible and environmentally friendly manner.

We analysed the bathymetry, sedimentary bedforms, sediments and meiofaunal nematodes at and near Broken Bank (BB) and Swarte Bank (SB), in the southern North Sea, to examine the association of faunal assemblages and their sedimentary habitat. The sampling strategy was not designed to fully characterise nematode-sediment relationships at the two sandbanks, but rather to enable us, with minimal cost and effort, to make comparisons between nematode communities inhabiting a diverse range of sedimentary environments.

Study site

Broken Bank and Swarte Bank are the outermost of the Norfolk Banks in the southern North Sea. Broken Bank lies 67 km offshore, and has dimensions typical of the tidally-controlled open-shelf linear sandbanks [17] of the outer Norfolk Banks area, at about 30 km long, 1 km wide and with relief of around 30 m above the surrounding sea floor (Fig. 1). Both banks are asymmetric, with a gently sloping SW face and a relatively steep NE face. For the Norfolk Banks in general, based on analyses of past charts and repeat bathymetric surveys, bank migration over decadal and century timescales is generally to the NE, at rates up to 11–15 m yr^{-1} [18,19], although rates of 40 m yr^{-1} have been recorded [20]. Banks may repeatedly shorten and lengthen, and may migrate SW then NE. For all banks, the bathymetric changes at the head and tail appear high compared to the sideways migration rates [19]. The surface tidal currents in the area of Broken Bank and Swarte Bank are strong, with speeds of 1.6 m s^{-1} and 1.0 m s^{-1} on spring and neap tides, respectively [21]. The bed sediments on and around the banks are loosely-consolidated, well-sorted, fine-grained sands which are transported in an overall clockwise rotation [22]. This transport is thought to take place

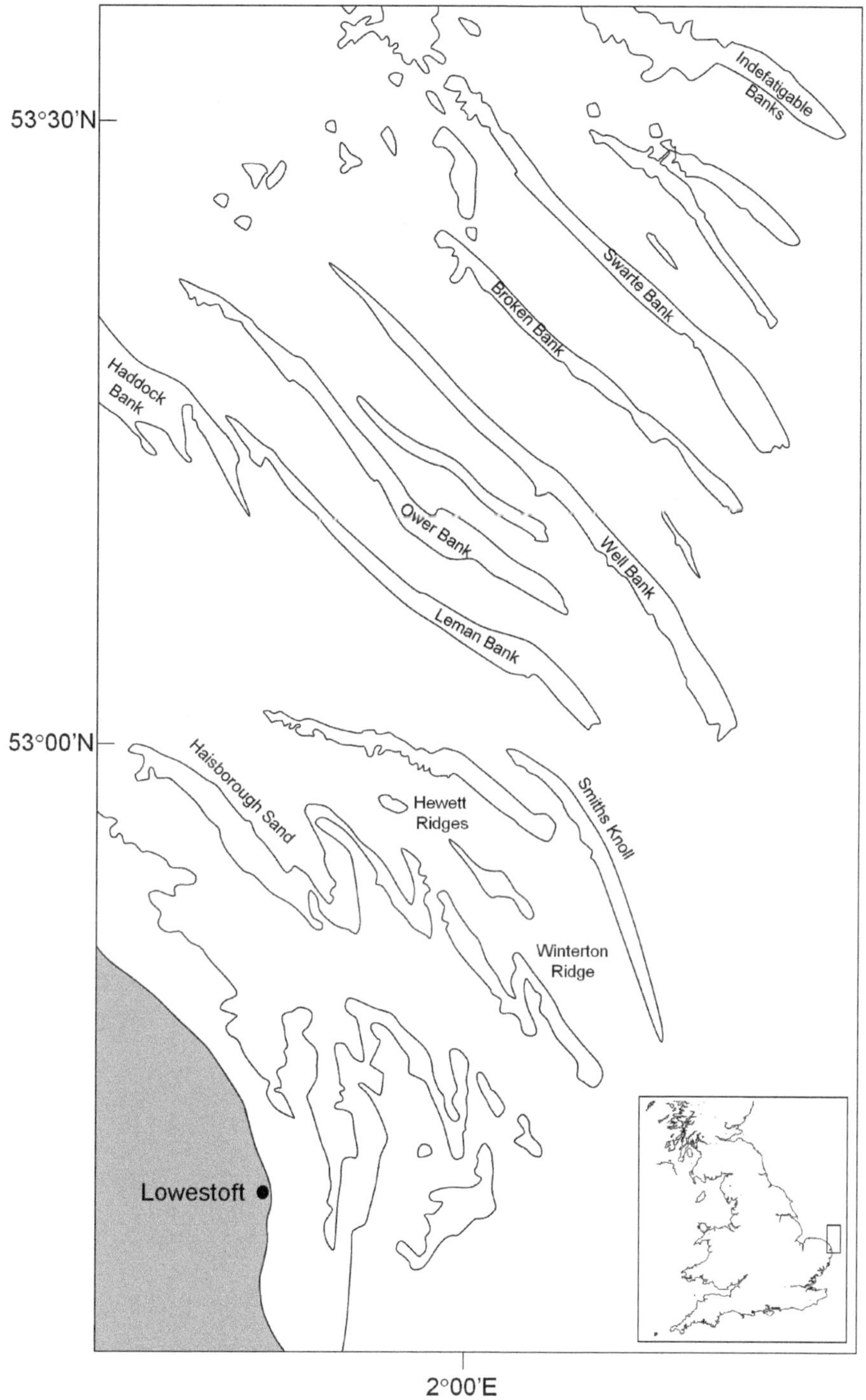

Figure 1. Location of Broken Bank and Swarte Bank in the southern North Sea. Figure adapted from [16].

nominally as bedload, under the influence of tidal currents alone (see also [23]). The addition of wave-induced bed stresses or storm surges can act to increase the transport towards the bank crests and lead to some sand being moved into suspension [22].

Geophysics

In April 2006, the southern portions of Broken Bank and Swarte Bank were surveyed using RV Cefas Endeavour to run a Kongsberg EM3000D multi-beam echo-sounder to provide high-resolution bathymetric data, and a Benthos 1500 digital side-scan sonar to measure seabed acoustic reflectance, which can indicate substrate hardness [11]. The acoustic survey was carried out in parallel to a multi-disciplinary sampling programme designed to provide a general description of the benthic fauna (including meiofauna, macrofauna, epifauna and fish) associated with Broken Bank and Swarte Bank [16]. Individual swath widths covered approximately 150 m on the seabed, and survey lines were designed to run parallel and perpendicular to the overall NW-SE alignment of the bathymetric features. More complete coverage was performed in selected areas of sedimentary interest, including a N-S strip 3×1 km across the SW face, crest and NE face of Broken Bank, and an area 500–800 m wide extending for 7 km along the NE flank of Swarte Bank (Fig. 2 and 3). The survey included coverage of the swale between the banks and an extended swath line off Broken Bank to the SSE. The data were processed using Fledermaus to drape the side-scan images onto the measured bathymetry, with tidal corrections applied. The morphology of the banks and their associated sedimentary bedforms were measured on screen using the Fledermaus viewing software ('iview4D'), leading to the delineation of a series of areas defined by their sedimentary morphology (bedform groups).

Sediments

Sediments from both Broken Bank and Swarte Bank were sampled by means of a 1 m^2 Day grab (Fig. 2; Table 1). Four samples each were collected at five stations along the crest of each sandbank (X, A, B, C, D) and at single sites either side of the sandbanks (F, I). Single samples were collected at four other stations located on the deeper surrounding seabed of each sandbank (E, G, H, J). From each grab sample, two sub-samples were collected with a Perspex corer (3.6 cm diameter, 10 cm^2 surface area) to a depth of 5 cm, one for the analysis of grain size distribution and one for the study of meiofaunal nematodes. The location of sampling stations was largely dictated by the seabed topography because sufficient draught was necessary under the vessel to allow the collection of sediment samples.

Sub-samples for grain size analysis were combined by station and frozen at −20°C prior to analysis. After thawing, the 22 sediment samples were wet sieved through a 500 µm sieve, and the fraction> 500 µm was oven dried at 90°C for 24 h. This fraction was then dry sieved at 0.5 phi intervals, down to 1 phi (500 µm). The fraction <500 µm was freeze dried and analysed on a Malvern Mastersizer S laser particle sizer. The sieve and laser sizer data were merged to form one uniform distribution spanning the size range 0.49 µm −22 mm. It is recognised that freezing samples can alter the grain size distribution, especially of clay-sized grains, but microscope and size analysis both indicated that no clay-sized grains were present in any samples, consistent with previous findings [22], and all samples were treated identically.

Grain size data were grouped using the computer programme ENTROPY [24]. Entropy analysis has been widely used to analyse grain size distributions, avoiding the use of simplified and often inaccurate measures of the shape of the distribution (such as the mean, skewness and standard deviation). Details of the data processing technique have been described in numerous previous studies [24,25,26,27,28,29] and are not repeated here. The analysis was used to group the data into a number of grain size groups, ranging in number from two (over-simplified) to 12 (over-complicated) inclusive, the spatial distributions of which were assessed visually. The choice of the number of groups to use in subsequent interpretations is a compromise between achieving a high R(s) statistic (i.e. statistically explaining more of the data; [30]) and accommodating empirical observations within a sedimentologically reasonable model of sediment distribution (e.g. [24]).

Distribution patterns of the various sedimentary bedforms and the grain size data were used together to delineate a series of sedimentary facies for the study area. In the geological literature, there are a number of ways that the term facies has been used [31], including regarding a rock's measured distinctive lithological and/or biological characteristics, as well as the interpreted sedimentary processes and/or sedimentary environment. As used here, the term sedimentary facies means those distinctive physical characteristics of the sediments that occur under particular conditions of sedimentation, reflecting the sum of sedimentary and depositional processes and the environment (e.g. [31]) and which act to distinguish them from adjacent deposits. Geologists have long used fossil evidence of the associated biology as an integral part of defining sedimentary facies.

Meiofauna

All faunal samples were fixed in 5% formaldehyde in filtered (63 µm) sea water. After washing the meiofauna samples onto a 63 µm sieve, the fauna from all replicate samples were extracted by floatation in Ludox 40 [32]. The extraction was repeated three times before the extracted material was evaporated slowly in anhydrous glycerol and mounted on slides for identification and counting.

Nematodes were identified to genus or species level, using taxonomic keys based on a phylogenetic classification scheme [33,34,35,36]. Species and genera were then assigned to a number of biological traits, i.e. a number of morphological and life-history characteristics, thought or known to represent an important ecological function [37]. Here, six traits were included: mouth morphology, tail shape, pattern of the body wall (cuticle), body size, body shape and life-history strategy (Table 2). There were a total of 25 categories of these traits. A matrix of these biological traits was constructed by assigning to each nematode species/genus its affinity to each trait category.

Species and trait diversity and dominance in each sample were expressed as Hill numbers N_1 and N_2 [50], respectively. These evenness measures describe different aspects of the community and differ only in their tendency to include or ignore the relatively rare species or traits. Bartlett's and Cochran's tests were used to test for homogeneity of variance before facies-specific differences in species and trait diversity, as well as the relative abundance of functional groups (Table 2) were explored using a one-way analysis of variance (ANOVA). Following the detection of significant differences ($P \leq 0.05$) between facies, the Tukey HSD multiple comparisons test was used.

Non-metric multi-dimensional scaling (MDS) ordination using the Bray-Curtis similarity measure was applied to relative abundance data to compare spatial patterns in the taxonomic (i.e. species) and functional (i.e. traits) composition of nematode communities amongst the putative sedimentary facies. One-way analysis of similarities (ANOSIM), the multivariate analogue of analysis of variance, was used to test for significant taxonomic and functional differences between sedimentary facies. Similarity percentages (SIMPER) were calculated to identify the species and traits which both characterised *a priori* sedimentary facies and distinguished between them. The dissimilarity between faunal distribution patterns recorded in the sedimentary facies was computed and an average calculated. We then calculated the contribution of each species and trait to the average similarity within each sedimentary facies and the average dissimilarity between facies. All statistical analyses were performed using the software packages Statgraphics Plus version 3.3 and Primer version 6.1.5 [51].

Disturbance versus mobility

A variety of publications have used the term disturbance to cover a wide range of anthropogenic and natural sediment transport and mixing processes including dredging, dredge material emplacement, trawling and natural sediment mobility [52,53,54]. In our study, we are dealing specifically and purely with the natural movement of sediment, and hence use sediment mobility for maximum clarity. Where we use the term disturbance,

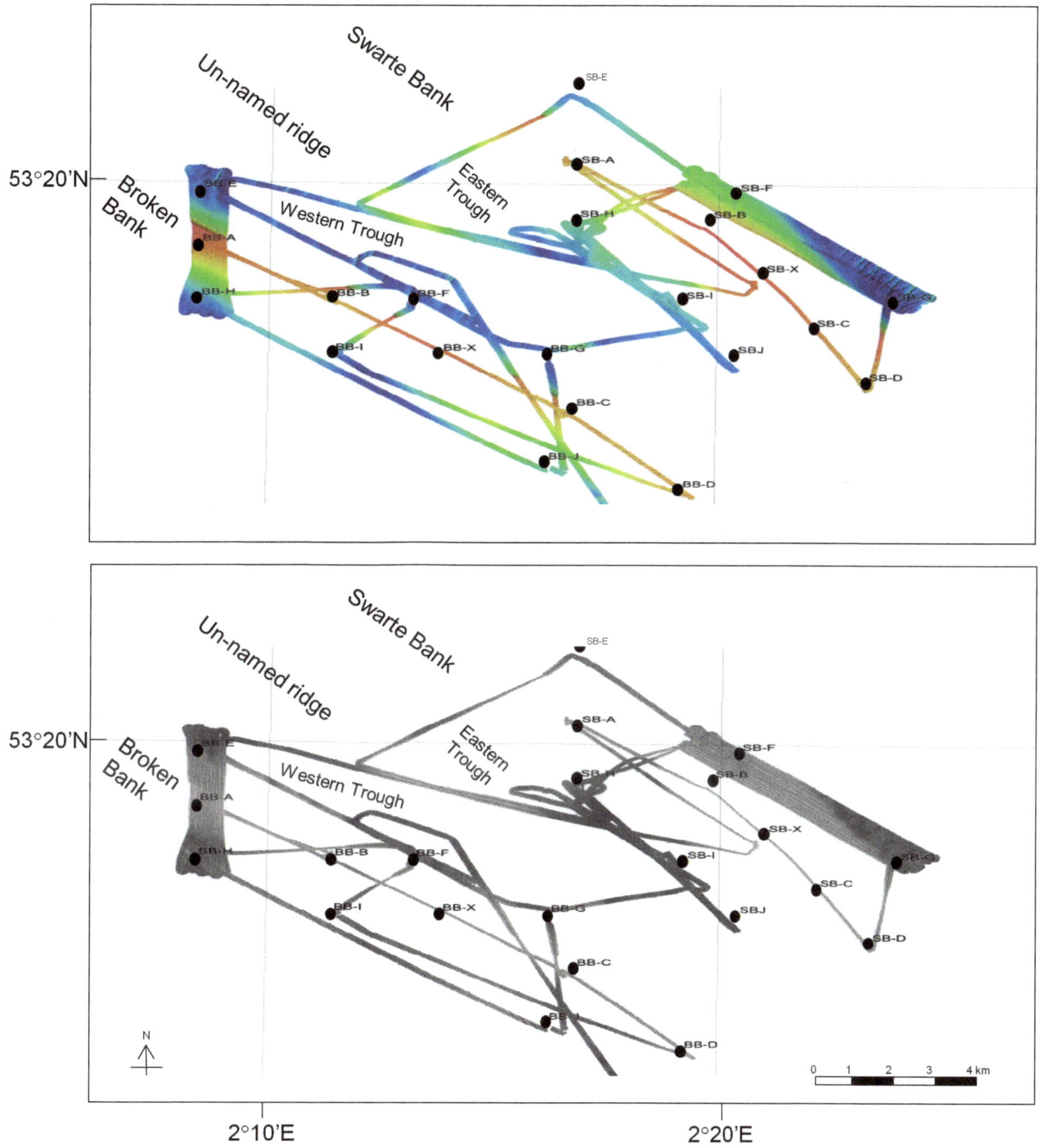

Figure 2. Acoustic survey coverage. The acoustic survey coverage, showing (upper) recorded depth (red = shallow, dark blue = deep) and (lower) acoustic backscatter, where darker areas indicate greater bed roughness. Sampling stations are also shown. BB = Broken Bank, SB = Swarte Bank.

it generally refers to the disturbance experienced by organisms during a period of seabed mobility.

Results

Sandbank form and superposed bedforms

The banks' gross morphologies include measured lee slopes of nearly 5° (Table 3). Superposed (flow-normal) bedforms include a wide variety of megaripples and sandwaves, of different planform, full-beddedness, size and symmetry (orientation). Some bedforms occur in extensive bedform fields and others in isolation (see [55] for explanation of terminology). Superposed in many areas on both sandbanks are smaller megaripples, which, where resolved, have a steep morphology, indicating that they are probably active. In contrast, some areas between Broken Bank and Swarte Bank and areas associated with their lower stoss flanks contain fields of longitudinal furrows, whose formative processes [56] indicate that these will only be active episodically.

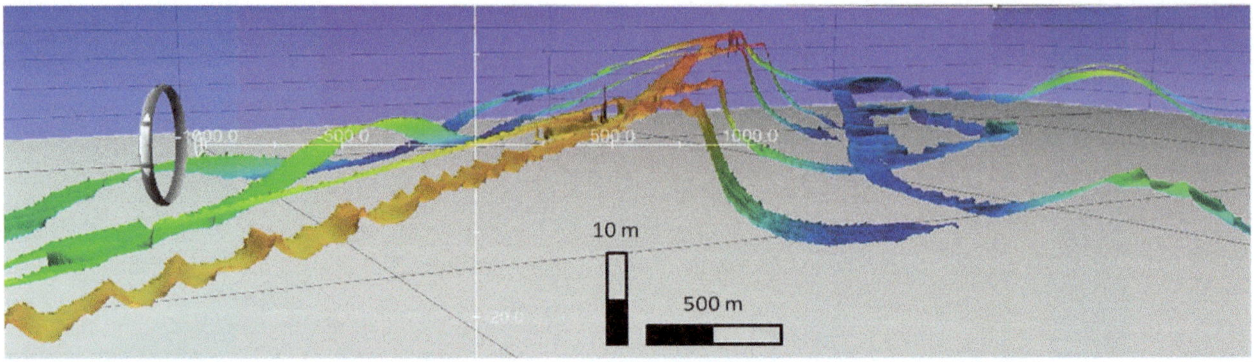

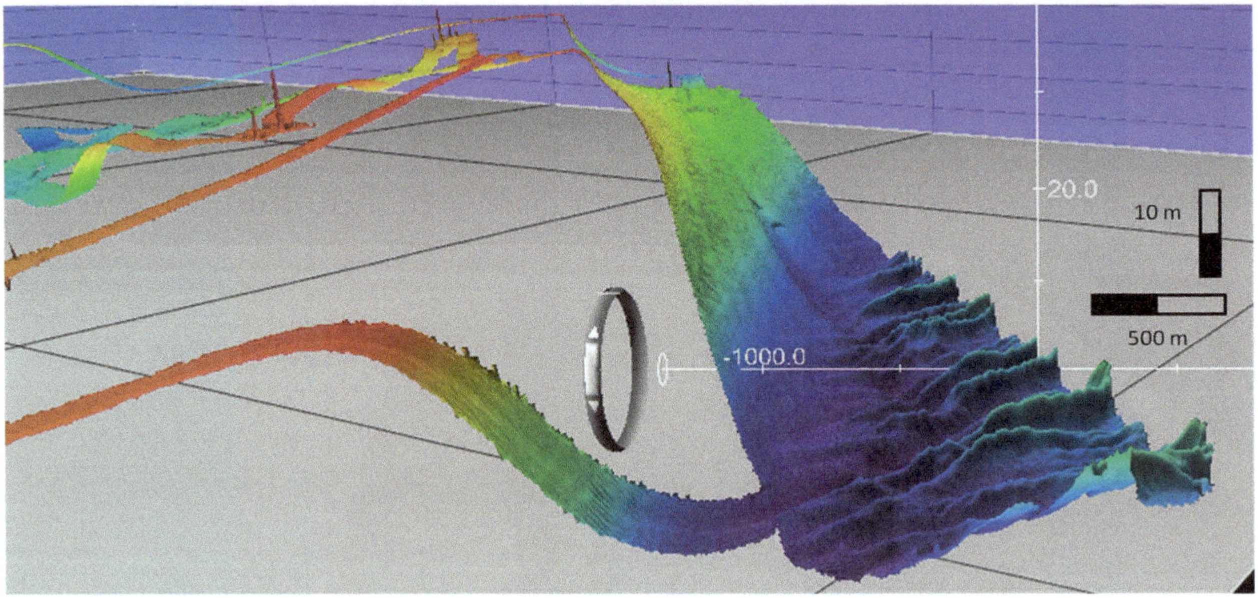

Figure 3. Bathymetry of the surveyed area. Oblique colour-contoured bathymetry of the surveyed area. Vertical exaggeration×35. Black grid squares 5 km apart. Scale bars refer to the intersection point of the white axes. Upper - View towards NW of the asymmetrical cross-section of Broken Bank (sharp crestline in centre) and the low rounded crest of the Un-named ridge (right). Note large asymmetric sandwaves migrating up the SE stoss of Broken Bank (bottom left; details in Table 4). Lower - View towards the NNW of the megarippled lee slope (bedform group v; centre) of Swarte Bank, with field of large asymmetric sandwaves (bedform group i) in the trough migrating to the SE (right; details in Table 4).

Table 1. Positions of the 22 stations sampled in the southern North Sea.

Swarte Bank			Broken Bank		
Station	**Latitude**	**Longitude**	**Station**	**Latitude**	**Longitude**
SB-X	53.30 N	2.35 E	BB-X	53.27 N	2.23 E
SB-A	53.34 N	2.28 E	BB-A	53.31 N	2.14 E
SB-B	53.32 N	2.33 E	BB-B	53.29 N	2.19 E
SB-C	53.28 N	2.37 E	BB-C	53.25 N	2.28 E
SB-D	53.26 N	2.39 E	BB-D	53.22 N	2.32 E
SB-E	53.37 N	2.28 E	BB-E	53.33 N	2.14 E
SB-F	53.33 N	2.34 E	BB-F	53.29 N	2.22 E
SB-G	53.29 N	2.40 E	BB-G	53.27 N	2.27 E
SB-H	53.32 N	2.28 E	BB-H	53.29 N	2.14 E
SB-I	53.29 N	2.32 E	BB-I	53.27 N	2.19 E
SBJ	53.27 N	2.34 E	BB-J	53.23 N	2.27 E

Table 2. Morphological characteristics and life-history of nematode species and genera.

Functional group	Functional importance	Trait categories
Mouth morphology [38]	The mouth cavity shows great diversity in form and reflects the food ingested by the nematodes. It can thus be a good indicator of a species' feeding strategy. Feeding type groupings are fundamental to carbon and energy fluxes through ecosystems, and are linked to nutrient cycling.	Selective deposit feeder, non-selective deposit feeder, epistrate feeder, predator
Tail shape [39,40,41]	The tail shape is important in locomotion and reproduction. Long, filiform tails, for example, are considered a special adaptation to fine sand and muddy sediments where only an incomplete interstitial system exists. In these sediments, the tail enables animals to retract from dead-end interstitial passageways that are too narrow to allow the worm to turn around and escape.	Short/round, conical, clavate, elongated/ filiform
Pattern of the body wall [42,43]	The body wall (cuticle) is used as a support and leverage point for movement. It protects the body from drying out and other harsh environments. Nematodes with elaborate cuticular ornamentation tend to be associated with coarser, silt-free sediments, which may be correlated both with their mode of locomotion and with the need for mechanical protection in unstable substrata.	Smooth, annulated, striated, rows of dots/ punctuated, rows of structures, with desmen
Body size and shape [44,45,46]	Body size and shape affect physiological and ecological features of populations, including metabolic rates, tolerance to chemicals and anoxia, ability to move or migrate, vulnerability to predation etc. Slender nematodes are able to move swiftly through the sediment, but are vulnerable to predation. Predation pressure on stout species may be reduced but so is mobility.	Body size: <1 mm, 1 - 2 mm, 2 - 4 mm,> 4 mm. Body shape: Long/thin, slender, stout
Life-history strategy [47,48,49]	Many small species have short generation times of usually about one month or less with high reproduction rates. This largely r-selected life-history strategy is in contrast to the longer life-cycles and fewer offspring of more K-selected species. Some of these often larger species have an annual reproductive cycle.	Coloniser (largely r-selected), intermediate (2 categories), persister (largely K-selected)

Functional groups were chosen because they have been shown or inferred to be functionally important and are thus potentially ecologically meaningful (detailed in [37]). Terminology adapted from [34].

The bedforms (numbered i – viii) superposed on and between the sandbanks vary across the survey area in a relatively ordered fashion (Fig. 4; detail in Table 4). On the upper stoss flanks (i.e. the SW side), the banks themselves have few clear sedimentary bedforms and show only low relief (<0.5 m; bedform group iii), with, in places, trains of solitary or isolated large sandwaves (bedform group iv). The sandbank lee slopes exhibit very extensive fields of megaripples (bedform group v), with bedform asymmetries indicating bedload sediment transport to the SE. There are also smaller patches of large sandwaves, which are either symmetrical or indicative of net bedload transport to the SE (bedform groups i, vi and vii). The swales between the banks, and areas low on the stoss flanks have a relatively rough topography, which includes areas of longitudinal furrows (bedform group ii) and fields of undulating megaripples (bedform group viii) which occur superposed on the furrows in places. Here, the net bedload transport inferred from the megaripples is generally to the NW.

The key physical features are therefore that (Fig. 4; Table 4):

(a) For the positive relief features of Broken Bank and Swarte Bank, bedform transport is relatively active, with transport of sand to the NW across the stoss slope and along the crest, and to the SE along the lee slope, i.e. in an apparent clockwise circulation (agreeing with [22]).

(b) Between the two sandbanks, including the Western Trough, the Un-named ridge and the Eastern Trough, the bed contains a variety of bedforms and a complex distribution, closely associated with the major bathymetric relief.

(c) The flat seabed away from the sandbanks is relatively starved of sand, and here fast storm-associated shelf flows have caused seabed erosion in the form of longitudinal furrows. Where superposed by megaripples, the furrows have not recently been active.

(d) Overall, the troughs and lower stoss slopes of Broken Bank and Swarte Bank contain coarser sediments and have a rougher seabed, together with bedforms (longitudinal furrows) which indicate episodically fast flows. This is in contrast to the upper stoss where the sediment is finer and where the mostly smoother seabed is inferred to be more continuously mobile.

Sediment granulometry

The seabed sediments are mostly fine and medium sands, with a modal grain size of 180 – 250 μm, and with a minor gravel component in some samples. Based on their entire size distributions, the samples fall into four broad grain size groups (numbered a–d; Fig. 5A, B). In relative terms, the slightly gravelly sediments of groups a and b are poorly sorted and tend to occur off the main banks in areas with uneven topography and high acoustic roughness, including in areas close to the northern termini of longitudinal furrows. The groups' characteristics are relatively simple and, in combination with the bathymetric and side-scan datasets, relationships between them are readily interpretable (Table 5). The data are consistent with the finding of [22] that sand transport on the banks themselves will generally occur as bedload, but with suspension of sand likely at times of superposed waves; this may help explain the relatively poorly defined bedforms on the upper stoss of the banks. During severe storms, mobilisation of all the sampled grain sizes is possible.

Sedimentary facies

The survey area comprised a total of 11 sedimentary facies, eight of which (numbered 1 – 8) were included in our statistical analyses (Fig. 6). Facies are primarily based on the physical evidence for the frequency, magnitude and inferred direction of sediment movement, based mostly on bedforms, and the tendency

Table 3. Morphological characteristics of the positive relief features forming Broken Bank and Swarte Bank.

	Length	Max. width	Max. relief above surrounding seafloor	Water depth at sandbank crest	Max. stoss[a] slope (in surveyed area)		Max. lee slope (in surveyed area)	
	km	km	m	m	degrees	m/100m	degrees	m/100 m
Broken Bank	33	1.1	30	13–16	1.3	2.2	4.9	8.5
Swarte Bank	37	1.4	28	12–15	0.6	1.1	4.0	7.0

[a]Many mobile sandy sedimentary bedforms are asymmetrical in vertical cross-section. For tidal sandbanks, sand may be transported over the bedforms in different directions at different times, and may circulate around it in the long term. Nonetheless, there is often a long-term net direction of movement of the bedform, indicated by its asymmetry. Sand is generally transported (obliquely) up the longer shallower stoss slope, over the crest and down the shorter steeper lee slope, which faces the overall direction of movement of the sandbank.

for long-term erosion or accumulation, based on the evidence of exposed coarse substrates. Facies with less than three observations were omitted from our statistical analyses, including 'Active large full-bedded sandwaves' at stations BB-G and SB-G (total of two observations), 'Washed-out sandwaves' at the lower lee to trough of Broken Bank (single observation at station BB-E) and the 'Washed-out megarippled stoss' of Broken Bank (single observation at station BB-H).

The facies names highlight the major features which define each facies and which distinguish them from adjacent deposits (Table 6; Fig. 6). The sedimentary facies delineated for the study area differed with regard to 1) their location across the study area, 2) their observed grain size distribution and seabed topography, and 3) their inferred history, intensity and frequency of sediment mobility. The concept of mobility, as used here, can thus include factors such as duration, magnitude, period since previous mobilisation, time of year, depth, and physical nature (i.e. erosion, transport, accumulation, mixing and unmixing, covering and uncovering of lag surfaces etc.) as well as associated changes in oxidation state and other biogeochemical parameters. With a view towards timescales of around a month, in line with meiofauna turnover times, our eight facies fall into four groups with respect to their relative level of sediment mobility. Facies 1 is termed high, facies 2, 3 and 4 are high-moderate, facies 5 is moderate and facies 6, 7 and 8 are episodic (Table 6).

Distribution of nematode species and functional traits relative to sedimentary facies

At all stations, nematodes consistently comprised> 81% of the total meiofauna abundance. A total of 9171 nematodes was identified, with abundances per sample ranging from 47 in washed-out megaripples at the stoss off Broken Bank (station BB-D) to 434 in active megaripples at Swarte Bank (station SB-F). Nematode assemblages in the study area were taxonomically and functionally diverse. A total of 118 species was recorded which fell into 77 trait groups.

Results from the one-way ANOVA revealed significant differences between sedimentary facies in terms of taxonomic and trait diversity ($N_{1(species)}$: $F=6.71$, $P<0.01$; $N_{1(traits)}$: $F=4.66$, $P<0.01$) and dominance ($N_{2(species)}$: $F=6.96$, $P<0.01$; $N_{2(traits)}$: $F=3.72$, $P=0.02$). Taxonomic and trait diversity (N_1) were significantly lower in facies 1 and 8 ('Active megaripples at Swarte Bank' and 'Sandwaves over longitudinal furrows off the banks') than in most of the others. High numbers of some dominant species and traits in facies 1 resulted in a low N_2 value, whereas the low N_2 value observed in facies 8 was a consequence of the distribution of nematodes across a limited number of species and traits (Fig. 7).

Functionally, nematode assemblages were dominated by epistrate-feeding species that combined fast turnover times and an ornamented, relatively small, slender body (Table 7). A total of five chromadorid and cyatholaimid species fell into this category. Another multi-species trait group, differing from the dominant trait group only in the type of body ornamentation, as well as two single-species trait groups (comprising *Xyala striata* and *Metadesmolaimus pandus*, respectively) were also highly influential in determining taxonomic and functional distribution patterns across sedimentary facies.

Results from SIMPER and ANOSIM analyses (Tables 6 and 8) indicate that there were differences in nematode community structure between facies. Nematode communities assigned to facies 1 ('Active megaripples at Swarte Bank') exhibited relatively high average similarity in taxonomic and functional composition, probably reflecting the relative homogeneity of the sedimentary

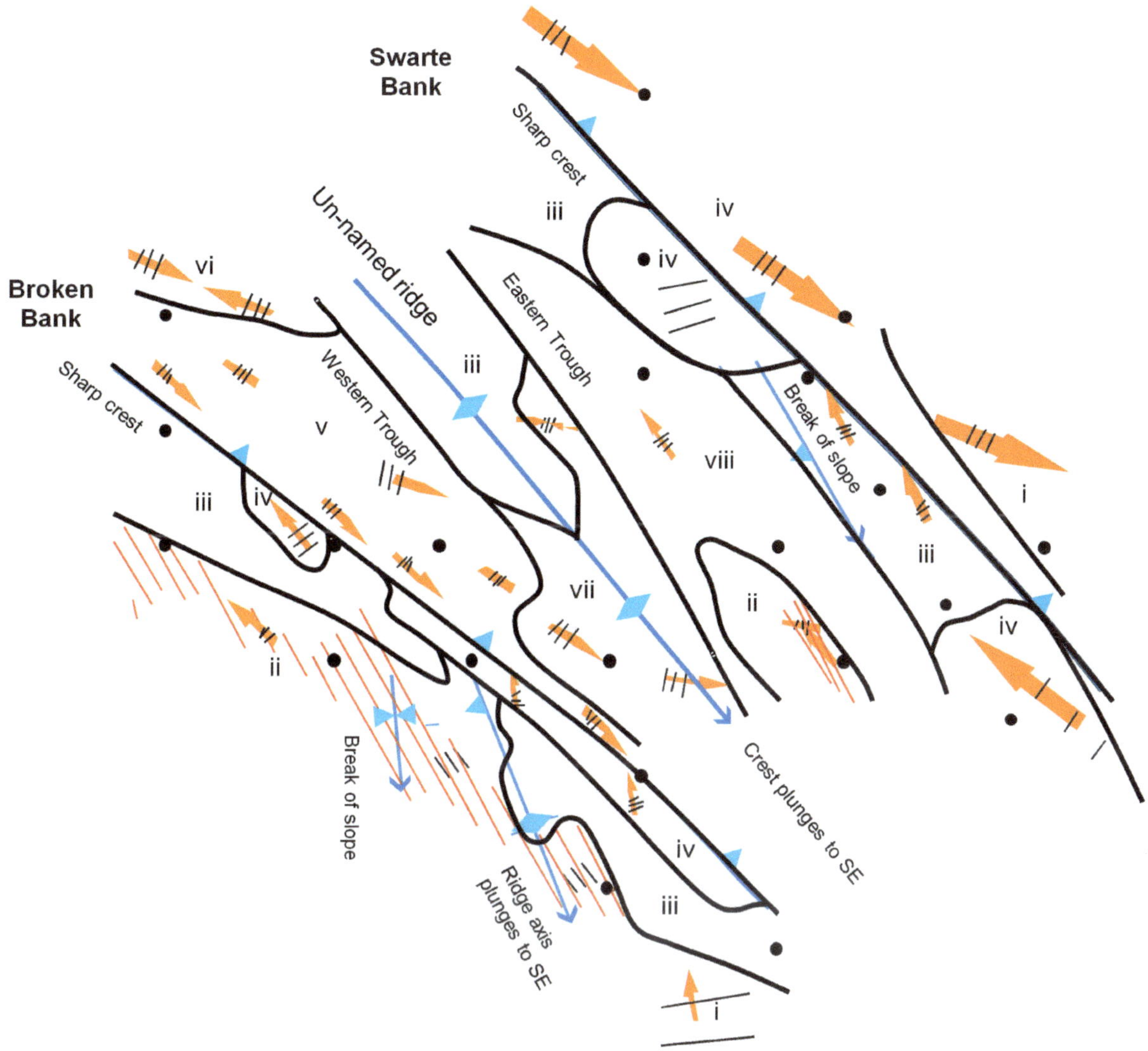

Figure 4. Bedforms in the study area. The 22 stations (black dots) fall into eight broad bedform groups (numbered i – viii; see Table 4), differing in the nature of contained grain size groups (see Table 5). Note - all blue arrows point downhill, indicating the orientation of a significant bed slope. Orange arrows indicate inferred bedform transport direction, with superposed black lines indicating orientation of sandwave crests and sandwave spacing: close lines = <100 m; medium = 100 – 300 m; far apart => 300 m. Narrow parallel orange lines indicate areas and orientation of longitudinal furrows.

environment. In contrast to other facies, the communities of facies 1 were taxonomically more similar than functionally, indicating that those species that were unique to individual samples were mostly single-species trait groups. The samples were highly dominated by contrasting trait groups, comprising *Metadesmolaimus pandus* and large predators, respectively. The trait groups differed not only with regard to feeding habit, but also in cuticle pattern (striated versus smooth) and body size (<1 mm versus> 2 mm). The relative abundance of non-selective deposit feeders was significantly higher ($F = 4.66$, $P < 0.01$) in this facies and that of epistrate feeders significantly lower ($F = 7.07$, $P < 0.01$) compared to other sedimentary environments (Fig. 8). This was the only sedimentary facies where larger predatory nematodes were present in notable numbers ($F = 12.14$, $P < 0.01$; bottom of Fig. 9).

A high proportion of nematodes with a heavily cuticularised body wall was a common and significant functional feature of facies 2 ('Active full-bedded sandwaves at the crest of Broken Bank'; top of Fig. 9) as well as other hydrodynamically active facies containing either megaripples (e.g. facies 3 and 4) or sandwaves (e.g. facies 5). In this group of facies, the proportion of epistrate feeders tended to be higher than the relative abundance of non-selective deposit feeders (Fig. 8; Table 6). The species- and trait-rich facies 5 ('Active large solitary sandwaves at the stoss/apron of Swarte Bank') was associated with the most taxonomically and functionally varied assemblages. The dominant trait groups differed little from those encountered in facies 8, but their relative abundance was notably different. In facies 5, the proportion of *Metadesmolaimus pandus* was lower, with associated higher relative abundance of heavily ornamented nematodes such as *Xyala striata* (top of Fig. 9). This facies comprises a range of sedimentary environments, which are correspondingly colonised by functionally diverse assemblages (Table 6). These assemblages encompassed deposit and epistrate feeders of various sizes (<1 – 2 mm), body ornamentations, ranging from moderately (i.e. striated or punctuated) to heavily (i.e. annulated or structured)

Table 4. Summary characteristics of main bedform groups.

Bedform group	Location	Principal bedforms	Orientation/Asymmetry (degrees)	Size			Description
				Length or spacing	Height or depth	Width or crest length	
Broken Bank							
i	South of Broken Bank	Very large sandwaves	080 – 090, lee to N	~ 600 m	<5 m	> 250 m	Two full-bedded, straight-crested very large sandwaves, with sandwaves in their troughs.
ii	At and beyond the western edge of Broken Bank	Longitudinal furrows overprinted by full-bedded sandwaves	Furrows – 325–330 Sandwaves – 040, lee to NW	Spacing 80 – 180 m	0.3 – 0.5 m	> 2 km long	Longitudinal furrows with superposed sandwaves.
iii	Upper stoss and crest						Bedforms poorly defined, low relief (<0.5 m).
iv	Central and southern crest	Sandwaves	060 – 090, lee to NW	80 – 100 m	2 – 3 m	> 500 m	Highly asymmetric, full-bedded straight crested sandwaves.
v	Lee slope	Sandwaves	Lee to SE	<40 m	<1.5 m on upper slope		Extensive and uniform field of full-bedded straight-wavy crested sandwaves.
Western Trough							
vi	Lee slope of Broken Bank	Large sandwaves	020 (convergence in the north)	100 – 300 m	> 5 m	Unclear	Complex large relief, including asymmetric straight-wavy crested sandwaves.
v	Western Trough	Small sandwaves	020 – 050	10 – 20 m	<0.3 m	Unclear	Symmetry unclear.
vii	Northern part of Western Trough	Sandwaves	Lee to SE	120 – 150 m	<0.5 m	> 200 m	Wavy-crested.
Un-named ridge							
vii	Trough and southern lee slope	Large sandwaves	350 – 020, lee to E	150 m	<2 m	> 250 m	Asymmetric straight-crested sandwaves.
vii	Lower E slope of un-named bank	Sandwaves	005 – 015 (convergence)	80 – 110 m	<1 m	> 300 m	Asymmetric straight-wavy crested sandwaves.
Eastern Trough							
viii	Broad swale west of Swarte Bank and central stoss slope of Broken Bank	Small sandwaves	050 – 070, some have lee to N	~ 20 m	<0.5 m	> 80> 150 m (some straight crested> 400 m)	Undulating to rough relief, patchy. Centre and S has topography parallel to longitudinal furrows. Discrete fields of straight-wavy crested sandwaves.
Swarte Bank							
iv	NW part of crest (and southern apron)	Large sandwaves	060 – 090, lee to NW	200 – 600 m	2 – 3 m (0.5 – 2 m in E)	> 500 m	Train of five solitary highly asymmetric sandwaves.

Table 4. Cont.

Bedform group	Location	Principal bedforms	Orientation/Asymmetry (degrees)	Size			Description
				Length or spacing	Height or depth	Width or crest length	
v	Lee slope	Sandwaves	Crests 030 – 050, lee to SE	30 – 40 m (smaller at boundary with i)	<1 m on upper slope, <0.4 m on lower slope	> 200 m	Very extensive and uniform field of full-bedded straight-crested sandwaves along the bank's lee slope.
i	South-eastern edge of lee slope	Large sandwaves	030 – 040, lee to SE	50 – 200 m	2 – 6 m	> 800 m	Asymmetric full-bedded, wavy-straight crested (some double-crested). Height increases to the E.
iii	Upper stoss of Swarte Bank				0.3 – 0.5 m		Bedforms poorly defined, low relief (<0.5 m).

cuticularised, and varied reproductive potential (coloniser to intermediate).

Metadesmolaimus pandus (23% of all nematodes) and the dominant multi-species trait group (24% of all nematodes) were abundant in facies 8, ('Sandwaves over longitudinal furrows') as was a limited range of other trait groups which combined a deposit- or epistrate-feeding habit with a striated, medium-sized (1 – 2 mm), slender body. Nematode assemblages in this facies exhibited a relatively high similarity with those recorded in facies 6 and 7 (Tables 6 and 8).

From the above, the key faunal features are that:

(a) Taxonomically and functionally distinct nematode communities were present in facies 1 and 5, characterised by low and high faunal diversity, respectively (Table 6).

(b) There was a considerable overlap in the species and trait composition of nematode assemblages collected in facies 2, 3 and 4. Heavily ornamented nematodes were common here and numbers of epistrate feeders generally exceeded that of other feeding types.

(c) Some sedimentary environments located in the Western Trough and Eastern Trough and off the banks (i.e. facies 6, 7 and 8) differed little with regard to their taxonomic and functional structure. Small and medium-sized deposit- or epistrate-feeding species dominated nematode assemblages.

These key features tend to indicate a range of faunal similarities between sets of facies, which led to readily interpretable patterns in the MDS ordination (Fig. 10). Distinct nematode communities were present in active megaripples at Swarte Bank (facies 1, termed high mobility), where spatially homogeneous bedforms are mobilised daily by tides. These communities form a tight cluster at the bottom right hand side of the ordination plots. Nematode assemblages from environments where sediment mobility was more intense yet infrequent or absent on timescales of a month (facies 6 – 8, termed episodic mobility) form a broad cluster across the centre and upper part of the ordination plots. This is consistent with the variable suite of sedimentary environments. Data from facies 8 are particularly widely spaced on all plots. Sampling stations where nematodes experience high to moderate intensities and frequencies of sediment mobility (facies 2 – 5, termed high-moderate and moderate mobility) form a medium-sized cluster, within which most facies overlap except for facies 3, representing relatively active megaripples, which forms a separate cluster.

Discussion

Interstitial habitats such as those occupied by meiofauna are broadly distributed across the planet, and the complexity of the associated physical sedimentary environments has been present throughout the evolutionary history of meiofauna (e.g. [57]). Hence, it is clear that processes operating in the geological past have shaped what we sample and see today. A key relevant timescale is the last few thousand years, when the southern North Sea was inundated by the rising Post-Glacial sea [58] and became subject to marine processes [59]. At shorter timescales, the periods between major storms and the various tidal cycles became relevant to the benthic biology. Put simply, the benthic fauna we sample depends in part on the day-to-day modern processes but also on the episodic storm-associated processes, and the history of both.

A variety of sediment transport processes in our survey area, operating at different magnitudes and frequencies, created a wide range of sedimentary environments for meiofaunal nematodes (Table 6). The presence of active sedimentary bedforms, such as

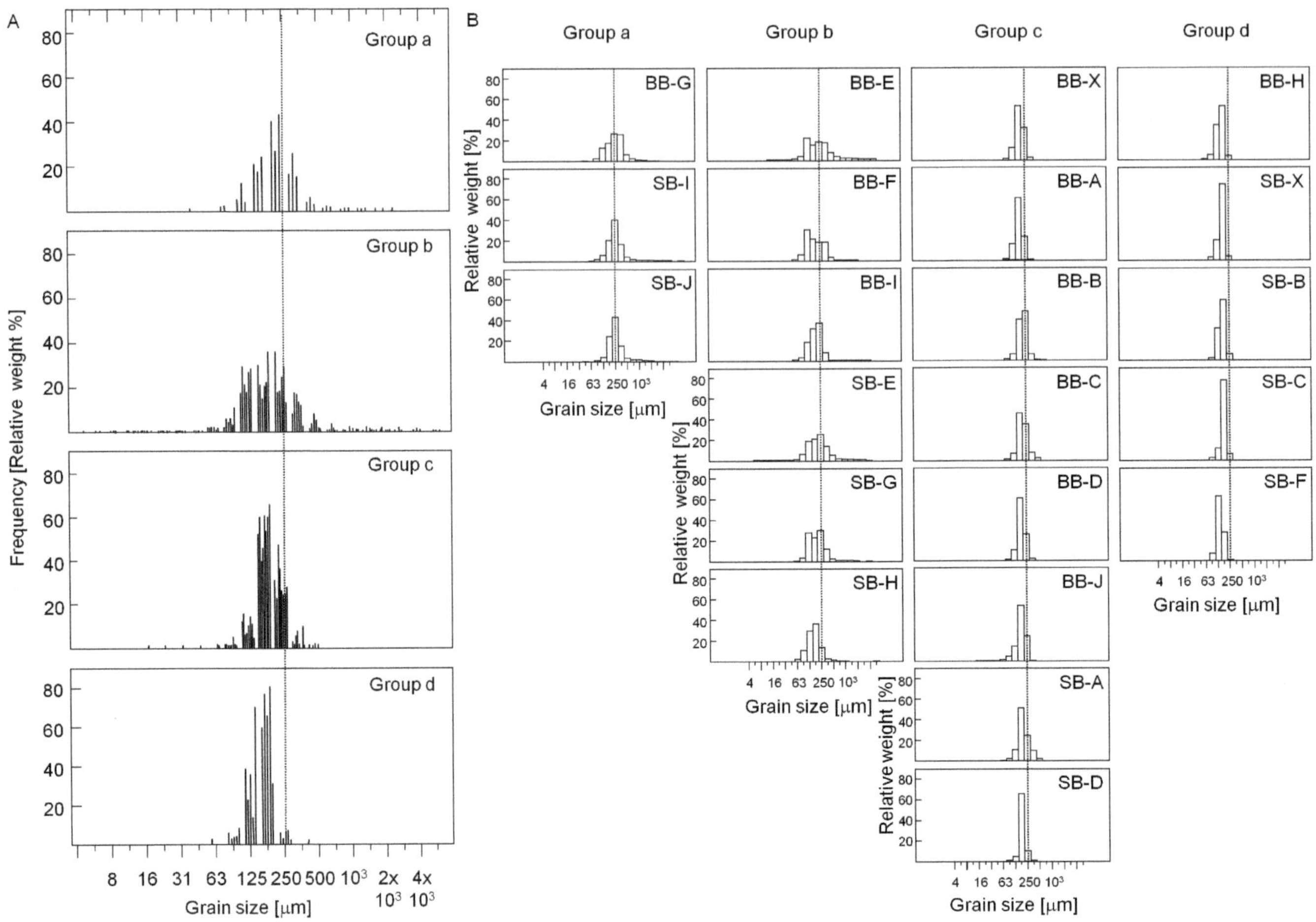

Figure 5. Grain size distribution of all 22 sediment samples. A. All samples of each group, plotted together to illustrate the key differences between the groups. B. All samples within each group. The vertical dotted line at 250 μm helps illustrate the between-group shift in the peak of the grain size distribution. Note that grain size groups a and b have a minor gravel component.

Table 5. Summary characteristics of grain size groups.

Grain size group	Sediment	Main characteristics and water depth	Spatial distribution (i.e. samples forming the group)	Comparison between groups	Interpreted sedimentological processes
a	Slightly gravelly unimodal medium sand	Modal size ~ 250 μm, with up to 25% fine sand. 34 - 39 m	Samples in the Western Trough (BB-G) and the southern part of the Eastern Trough (SB-I, SB-J).	Coarsest sediments, with less sand than other groups. Sand is medium-grained.	Sand fraction probably mobile on regular basis, over a gravel lag and/or medium sand is being eroded from the seabed.
b	Slightly gravelly fine and medium sand	Poorly sorted fine-medium sand with a minor gravel component. 34 - 40 m	In the troughs between the main banks and low on the main banks' stoss slopes (BB-E, BB-F, BB-I, SB-E, SB-G, SB-H).	Equivalent to group a plus fine sand. All sizes present across all groups.	Mixture of other groups; likely to represent a thin layer of mobile fine sands over surface of group a material.
c	Unimodal fine to medium sand	Modal size ~ 180 μm, with up to 30% medium sand. 17 - 33 m	Along the crest of Broken Bank (BB-A, BB-B, BB-X, BB-C, BB-D) and Swarte Bank (SB-A), and on the stoss slope of SE Broken Bank (BB-J) and Swarte Bank (SB-D).	Relatively well sorted.	Regular movement of sand with, in calm seas, perhaps only a few tides per fortnight with no movement. Long-term accumulation.
d	Unimodal fine sand	15 - 34 m	On the crest and NE flank of Swarte Bank (SB-F, SB-B, SB-X, SB-C) and one sample on the northern stoss of Broken Bank (BB-H).	Relatively well sorted.	Daily movement of sand within megaripple fields. Long-term accumulation.

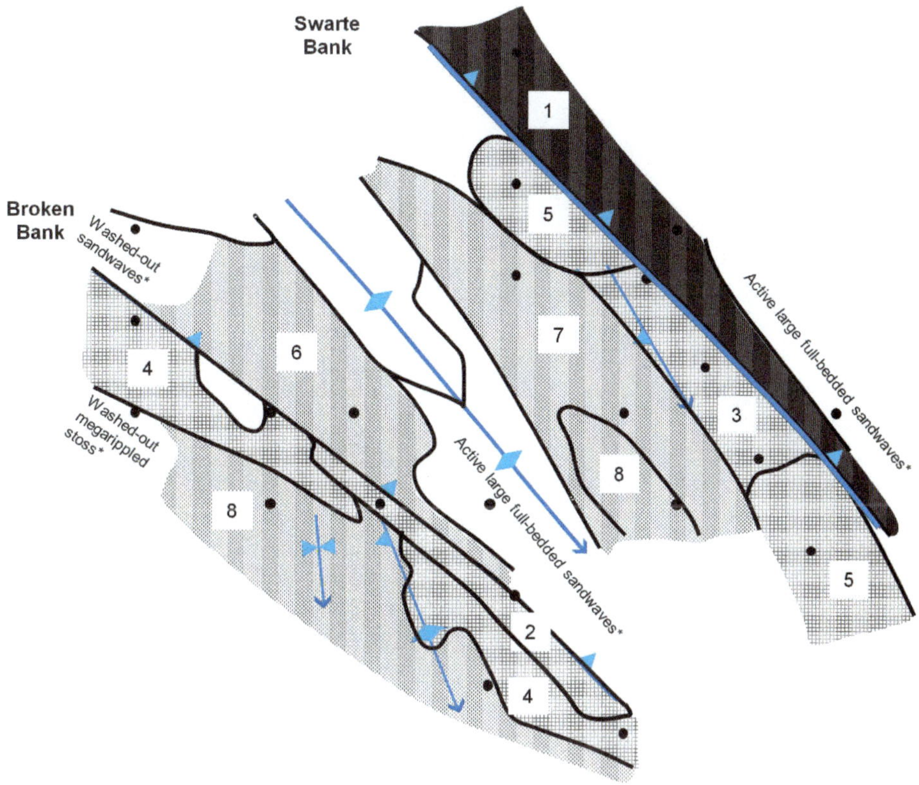

Figure 6. Distribution of sedimentary facies groups in the study area. Facies shadings are grouped according to their relative level of sediment mobility (Table 6). The 22 stations (black dots) fall across a range of sedimentary facies, numbered 1 – 8 (details in the text and in Table 6). Facies with less than three observations (*) were excluded from statistical analyses. For explanation of annotations see Fig. 4.

sandwaves (i.e. crest spacing of 10s – 100s m) and megaripples (i.e. crest spacing of 1s – 10s m), for example, have important consequences for the benthic ecology of sandbanks. Sandwave and megaripple crests tend to have different sediments, sometimes subtly, than their troughs [60,61,62,63], providing a variety of different niches for colonising meiofauna. Previously published review articles on the effects of physical seabed disturbance on benthic ecosystems [64,65] suggest that species-specific properties such as population growth rates and tolerance to environmental change affect a species' susceptibility to disturbance as well as its recovery from it. However, only recently have scientists begun to consider the key role of the sedimentary regime in determining the pre-disturbance faunal distribution patterns and as a factor in the post-disturbance phase of 'recovery' [66,67].

Benefits of the sedimentary facies approach

Delineating and using sedimentary facies can help to better understand the way in which the local complexity of the seabed might determine the number and type of available niches for meiofaunal nematode communities. In our context, sedimentary facies inherently include the key features of the bedforms and sediments themselves (e.g. [31]), thus incorporating many aspects of sediment mobility and the tendency for erosion or accumulation, which vary across environments. Therefore, sedimentary facies tend to group various combinations of physical and time-variable factors, improving the likelihood of forming meaningful and readily interpretable relationships with the infaunal data.

With an emphasis on the sandy facies appropriate to our study area, the general relationships between the groups of sedimentary facies representing high, high-moderate and moderate, and episodic sediment mobility can be illustrated by using three key parameters: the frequency of sediment mobility, its magnitude, and the continuum between erosion, translation (i.e. sediment throughput) and sediment accumulation (Fig. 11). Our facies groups fall into different areas of this diagram. This diagram helps visualise some of the concepts of our data and its interpretation, and we hope helps others in designing measurement and monitoring programmes, and interpreting collected data.

High mobility

A common assumption of ecological theories addressing the causes of faunal variability in space and time is that species or functional groups with different morphology, physiology or behaviour dominate under different environmental regimes [68,69,70,71] and this is consistent with our findings. In facies 1 ('Active megaripples at Swarte Bank'), for example, daily movement of sediments and bedforms will most likely have led to the resuspension, burial and/or mechanical damage of microalgal cells [72]. It is thus logical to find that epistrate feeders, whilst abundant elsewhere on the sandbanks, were less numerous here. We hypothesise that organic material accumulates preferentially in megaripple troughs at slack tide, some of which is subsequently buried by migrating ripple crests during succeeding tidal flows, a recognised sedimentary process [60] which has been postulated as important for meiofauna on intertidal sandflats (e.g. [73]). This subsequently attracts deposit-feeding nematodes from surrounding areas such as the small-sized *Metadesmolaimus pandus* who, together with a contrasting trait group comprising large predatory nematodes (including *Viscosia*), dominated colonist

Table 6. Physical features and key meiofaunal characteristics of sedimentary facies.

	Active megaripples at Swarte Bank (1)	Active full-bedded sandwaves at the crest of Broken Bank (2)	Megarippled stoss of Swarte Bank (3)	Washed-out megaripples at the stoss of Broken Bank (4)	Active large solitary sandwaves at the stoss/apron of Swarte Bank (5)	Active large solitary sandwaves in the Western Trough (6)	Washed-out undu-lating bed with sand-wave fields in the Eastern Trough (7)	Sandwaves over longitudinal furrows off the banks (8)
Station(s)	SB-E, SB-F	BB-X, BB-C	SB-X, SB-B, SB-C	BB-A, BB-B, BB-D	SB-A, SB-D	BB-F	SB-H, SB-I	BB-I, BB-J, SB-J
No. of obs.	5	8	12	12	8	4	5	6
GS groups	b, d	c	d	c	c	b	a, b	a, b, c
Bedf. groups	v	iv	iii	iii	iv	v	viii	ii
Rel. level of sediment mobility	High	High-moderate	High-moderate	High-moderate	Moderate	Episodic	Episodic	Episodic
Description (partly interpretative)	Sediments and megaripples mobile on a daily basis, except possibly at neap tides.	Large sandwaves migrate frequently.	Sediments mobile on a daily basis, but megaripples less frequently so.	Sediments mobile on a daily basis, but megaripples less frequently so.	Large sandwaves migrate frequently.	Large sandwaves migrate, superposed by small sandwaves and megaripples. Gravel grains widespread.	Sandy grains regularly mobile across and formed into sandwaves and megaripples at times, perhaps during storms, when widespread gravel grains also mobile.	Furrows formed and reactivated during severe storms, then become inactive for long periods, during which sandwaves become superposed and finer sediment accumulates in furrows.
Sedimentary environment (partly interpretative)	Little spatial variation in surface sediment, and probably also homogenous to a depth of a few decimetres.	Minor gradients in the nature of benthic habitats across the sandwaves. Sediments probably homogenous to a depth of a few decimetres.	Little spatial variation in surface sediment, and homogenous to a depth of a few decimetres.	Little spatial variation in surface sediment, and probably also homogenous to a depth of a few decimetres.	Moderate gradients in the nature of benthic habitats across the sand-waves, although sediments relatively similar; different ha-bitats in the troughs. Variation with depth, especially across the troughs.	Moderate gradients in the nature of benthic habitats across sandwaves; different habitats in the troughs. Patchy degree of variation with depth, especi-ally across the troughs between sandwaves.	Moderate gradients in the nature of benthic habitats. Potential rapid accumulation after storms.	Range of benthic habitats in and between furrows and sandwaves, affected by the time since creation or re-activation and rate of infilling. Significant spatial and depth-related differences in sediment type.
Nematode assemblages	Highly similar nematode communities dominated by contrasting trait groups, comprising small deposit feeders and large predators.	High proportions of nematodes with a heavily cuticularised body wall; a ratio of epistrate feeders vs. non- selective deposit feeders> 1.	Considerable faunal similarity to facies 2; subtle differences in the relative abundance of dominant species and traits.		Varied assemblages encompassing deposit and epistrate feeders of various sizes, body ornamentations and reproductive potential.	Considerable overlap in the species and trait composition with facies 8. Trait similarity between facies 6 and 7 exceeds similarity within the individual facies.		High abundance of the dominant multi-species trait group and a limited range of other trait groups combining a deposit or epistrate-feeding habit with a medium-sized, slender body.

Facies number in brackets. GS = grain size.

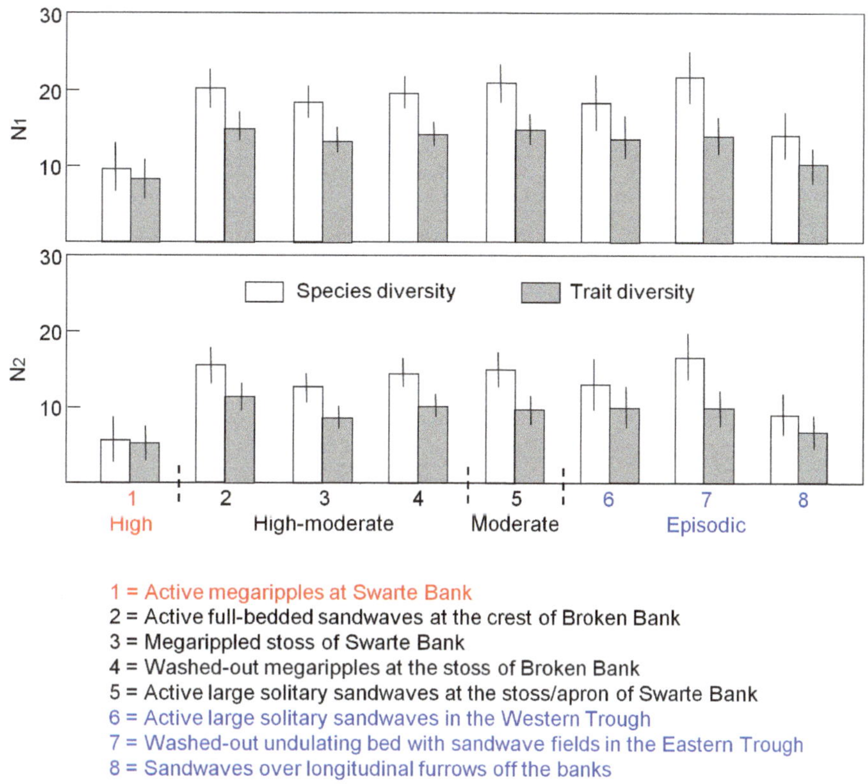

Figure 7. Species and trait diversity of nematode assemblages. Mean (± 95% pooled confidence interval) species and trait diversity of nematode assemblages recorded in eight sedimentary facies coded by relative level of sediment mobility. Diversity is expressed as Hill numbers N_1 and N_2.

communities. These disturbance-resistant nematodes [74] were most efficient in exploiting available food sources.

High-moderate mobility

Larger sedimentary bedforms comprising small sandwaves and washed-out megaripples are located along the crests of both banks (facies 2 – 4) and indicate relatively high magnitudes and frequencies of bed sediment transport. However, in contrast to facies 1, the bedforms may persist for longer (i.e. weeks to years as opposed to hours to days) and be more spatially varied. Accordingly, we recorded a range of coexisting epistrate feeders and sub-dominant deposit feeders. As has been shown for other dynamic sedimentary environments [75,76], much of the available space was filled by competitors most resistant to physical damage, in our case species with a heavily cuticularised body wall.

Moderate mobility

Solitary sandwaves on the lower stoss and apron of Swarte Bank (facies 5) were characterised by moderate spatial gradients in the benthic habitat across and along the bedform (Tables 4 and 6). Compared to smaller, more frequently active bedforms (i.e. megaripples and smaller sandwaves, see above), the physical

Table 7. Taxonomic affiliation and morphological/functional characteristics of nematode trait groups which discriminate between sedimentary facies.

	Dominant multi-species trait group	Sub-dominant multi-species trait group	Single-species trait group	Single-species trait group
No. of species	5	3	1 (*Xyala striata*)	1 (*Metadesmolaimus pandus*)
Taxonomic affiliation	Chromadoridae, Cyatholaimidae	Microlaimidae	Xyalidae	Xyalidae
Buccal morphology	Epistrate feeder	Epistrate feeder	Non-selective deposit feeder	Non-selective deposit feeder
Tail shape	Conical	Conical	Conical	Clavate
Cuticle pattern	Rows of dots/punctuated	Striated	Structures	Striated
Body size	1–2 mm	1–2 mm	1 - 2 mm	<1 mm
Body shape	Slender	Slender	Slender	Slender
Life-history strategy	Coloniser	Coloniser	Intermediate	Coloniser

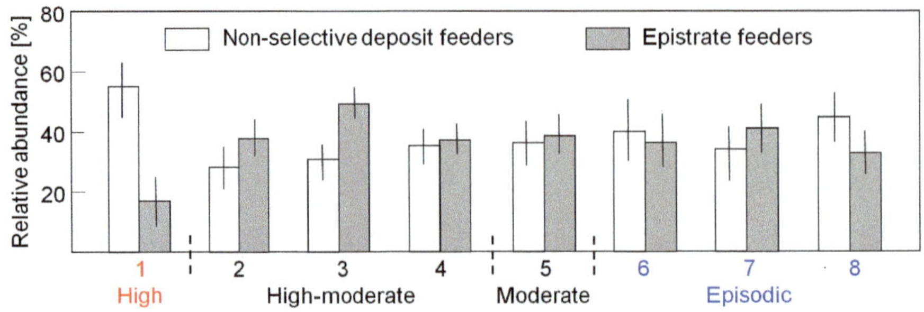

Figure 8. Relative abundance of non-selective deposit feeders and epistrate feeders. Mean (± 95% pooled confidence interval) relative abundance of non-selective deposit feeders and epistrate feeders recorded in eight sedimentary facies coded by relative level of sediment mobility. Facies numbers as in Fig. 7.

conditions here appear less critical in controlling the success or failure of species and functional groups. Biological stress such as intense competition for food and space is likely to be gradually mediated through biological interactions (see review by [77], [78]). Because different functional types use different resources, use the same resources in different ways or use the environment in a way that allows them to reduce or avoid competition with other functional types [79], the relatively high environmental heterogeneity increased the number of microhabitats for nematodes. The functional types recorded in our samples showed responses indicative of the trade-offs between resistance to disturbance, environmental tolerance and competitive abilities. Deposit and epistrate feeders of various sizes, body ornamentations and reproductive potential coexisted in taxonomically and functionally diverse colonist communities.

Episodic mobility

The relatively low elevation seabed between Broken Bank and Swarte Bank comprised sandwave fields of varying topography and extent (e.g. facies 6 and 7). In common with facies 8, at the steep lee slope of Swarte Bank, these areas of seabed probably undergo repeated and complex ecological transitions, associated with the irregular timing and nature of seabed mobilisation (e.g. as a result of storms in the study area). The initiation and episodic regeneration of bedforms (including longitudinal furrows) by strong storm-associated flows is related to very high rates of sediment transport, involving largely passive transport of meiofauna, and at these times, much of the study area would constitute a solely physically driven benthic system (sensu [80]). However, between such events, the area occupied by facies 6 – 8 would experience longer periods of being a more biologically influenced system, geologically characterised by the inter-storm formation of superposed sandwaves and infilling of the furrows, and where nematode communities respond to both the varied sedimentary environment but also to various biological interactions.

Such an episodically influenced change in seabed character probably also extends to those facies located North and West of the furrowed areas, away from the SE tips of the banks. It is likely that, especially during periods of decreasing sediment transport associated with the decaying phase of storms, a phase of relatively high net sediment accumulation occurs here over a few hours, and perhaps up to a day or so, depending upon the particular storm's characteristics. This phase of accumulation would be followed by an extended period where colonisation and assemblage development will take place, associated with the ambient, strongly tidal,

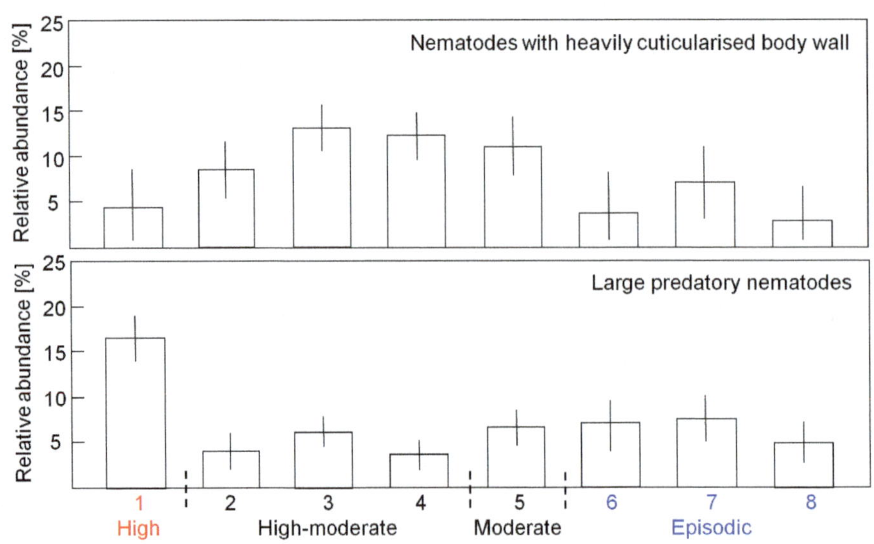

Figure 9. Relative abundance of nematodes exhibiting specific functional traits. Mean (± 95% pooled confidence interval) relative abundance of nematodes exhibiting a heavily cuticularised body wall (top) and mean (± 95% pooled confidence interval) relative abundance of large predators (bottom) recorded in eight sedimentary facies coded by relative level of sediment mobility. Facies numbers as in Fig. 7.

Table 8. Faunal differences (R-values) between sedimentary facies.

	Active megaripples at Swarte Bank (1)	Active full-bedded sandwaves at the crest of Broken Bank (2)	Megarippled stoss of Swarte Bank (3)	Washed-out mega-ripples at the stoss of Broken Bank (4)	Active large solitary sandwaves at the stoss/apron of Swarte Bank (5)	Active large solitary sandwaves in the Western Trough (6)	Washed-out undulating bed with sandwave fields in the Eastern Trough (7)
Active full-bedded sandwaves at the crest of Broken Bank (2)	**1/0.99**						
Megarippled stoss of Swarte Bank (3)	**1/0.99**	**0.83/0.80**					
Washed-out megaripples at the stoss of Broken Bank (4)	**0.99/0.94**	**1**	**0.54/0.43**				
Active large solitary sandwaves at the stoss/apron of Swarte Bank (5)	**0.99/0.90**	**0.51/0.35**	**0.40/0.47**	0.22/0.03			
Active large solitary sandwaves in the Western Trough (6)	**0.89/0.73**	**0.99/0.90**	**1/0.93**	**0.97/0.75**	**0.97/0.69**		
Washed-out undulating bed with sandwave fields in the Eastern Trough (7)	**0.97/0.88**	**0.78/0.77**	**0.93/0.85**	**0.65/0.51**	**0.70/0.46**	0.68/0.28	
Sandwaves over longitudinal furrows off the banks (8)	**0.50/0.34**	**0.74/0.66**	**0.92/0.87**	**0.78/0.65**	**0.73/0.57**	0.14/−0.18	0.44/0.34

R-values derived from pair-wise analysis of similarities (ANOSIM) tests for differences between sedimentary facies based on relative abundance data of species/functional traits. Bold indicates significant difference at $P < 0.01$. Facies number in brackets.

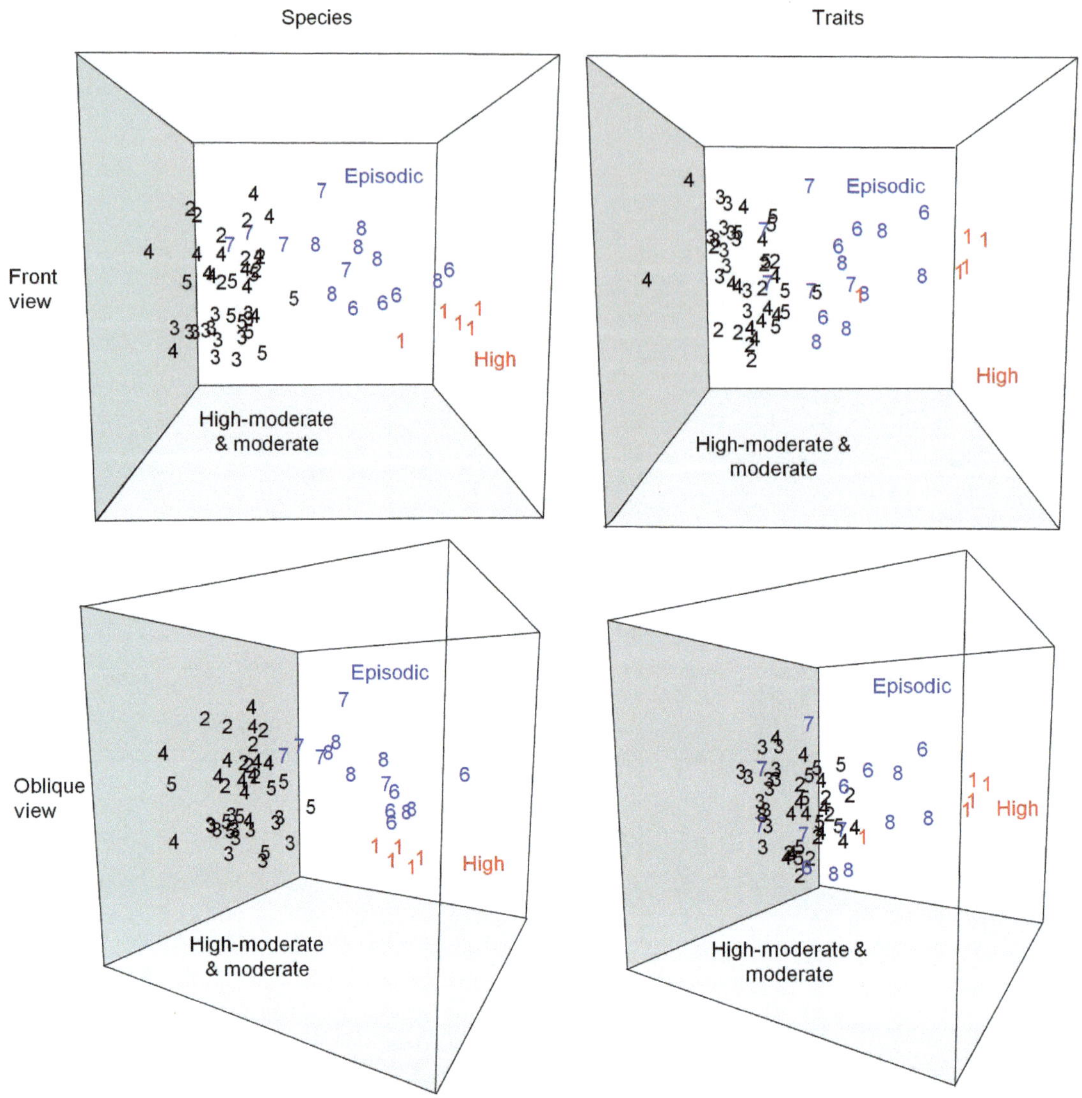

Figure 10. Non-parametric multi-dimensional scaling (MDS) ordination of nematode assemblages. 3D-view of non-parametric MDS ordination of nematode assemblages coded by facies (numbered as in Table 6 and Fig. 7) and relative level of sediment mobility (see text for details). The plots are derived from a similarity matrix of relative abundance of species (left) and traits (right).

oceanographic regime. Resident nematode communities appear to reflect such sedimentary complexity. In our particular case, the dominance of small- to medium-sized deposit- and epistrate-feeders of high reproductive potential most likely represents communities at a relatively early post-mobility stage.

Implications for future studies

Taxonomically and functionally disparate communities inhabiting the sedimentary facies defined in our survey area can be considered a complex product of the various sedimentary contrasts inherent within different sedimentary facies and the general sedimentary nature of the area being considered. Regarding variation within facies, the sedimentary environments experienced by meiofauna will be distinctly different across the variety of sub-environments across a sandwave-dominated facies, where troughs, stoss slopes, crests and lee slopes are dynamically and texturally different. There will also be great differences in those facies where longitudinal furrows occur with differing levels of infill. Further, sandwaves are much larger physical features than megaripples, and therefore their rates of horizontal migration and morphological change tend to be much slower. In our study area, sandwaves also tend to be much more patchy in their distribution than megaripples. Therefore, we might expect that 1) a similar number of sediment grabs in sandwave facies is more likely to include a greater variety of seabed types (and meiofauna) than in megarippled facies and that 2) the more frequent and regular mobility of megaripples tends to favour the presence of fewer meiofauna species and functional groups.

The long-term (i.e. annual to decadal) tendencies for seabed erosion, sediment throughput or sediment accumulation are important because they exert key controls upon habitat type and sediment mobility (Fig. 11). For example, in an environment where sandwaves occur, a long-term tendency towards erosion might increase habitat complexity by producing a series of solitary sandwaves [55] with layers of less mobile sediment between individual bedforms. In contrast, a long-term tendency towards accumulation might be associated with full-bedded sandwaves and lower habitat complexity.

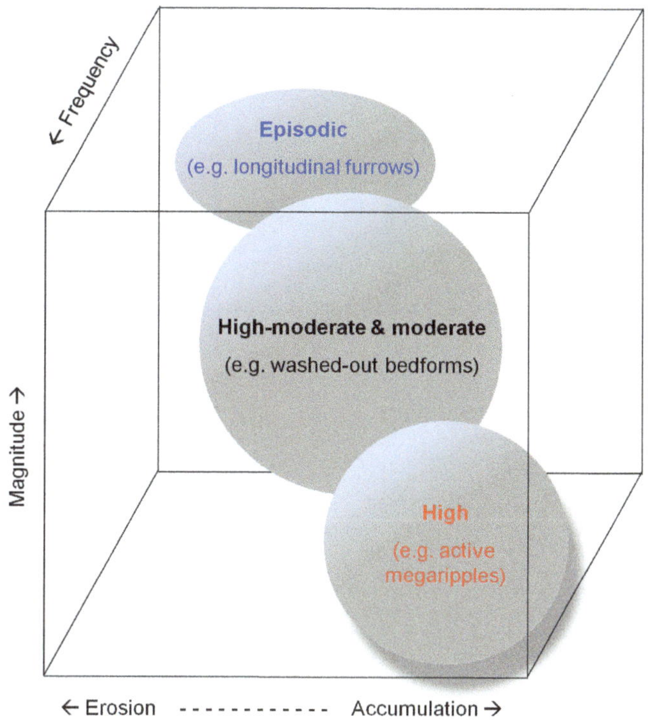

Figure 11. Conceptual diagram of three key controls on sedimentation in the study area. General relative locations of groups of sedimentary facies representing high, high-moderate and moderate, and episodic sediment mobility, plotted against the frequency of sediment mobility, its magnitude, and the continuum between erosion, translation (i.e. sediment throughput) and sediment accumulation.

Two main points emerge from our data. Firstly, a broad-scale characterisation of sedimentary environments is unable to account for the microhabitat heterogeneity to which the associated meiofauna will tend to respond. A key aim of future studies should be to incorporate fine-scale spatial information on sedimentary environments into the analysis of broader-scale animal-sediment relationships. For deeper insights, such studies should build on a good conceptual understanding of sedimentary processes operating at a range of spatial and temporal scales, including the movement of sediment across the full range of marine sedimentary bedforms [56], each of which will have important ecological associations (e.g. [81] and our study).

Secondly, small-scale faunal distribution patterns do not always reflect only those marine processes occurring concurrently with sample collection. Whilst this is not a new finding, our results emphasise how important past events can be in affecting animal distributions. Caution must be exercised when attempts are made to explain spatial distributions of animals from physical or biological conditions measured at a single point in time. For example, analyses conducted of pre-storm samples from our facies 8 ('Sandwaves over longitudinal furrows') would inevitably generate different faunal results, and hence differing expectations regarding state, than analyses of samples collected shortly after major storms. Further, for those areas where longitudinal furrows occur, there would be relatively high variation in sediment types after storm-associated currents which activate the furrows, but the degree of variation would change through time. Within the furrows themselves, sediments would accumulate relatively rapidly and become finer through time, favouring mud-dwelling meiofauna whereas immediately outside the furrows, the sediments would change in completely different fashion in response to the re-development of ripples and megaripples, with associated different meiofaunal responses.

Conclusions

The variety of sedimentary regimes at and around Broken Bank and Swarte Bank clearly shapes the meiofaunal nematode communities, and the interaction of local hydrodynamics and sediment transport is clearly a key driver of the processes involved. Their influence occurs in a variety of different ways, and, importantly, in ways that would not have been predictable from using 1) existing generalised theories of seabed disturbance based on frequency or intensity alone [64,65,82], as opposed to including duration, period since previous event, etc., 2) sediment summary descriptors alone, as opposed to sedimentary facies, nor using 3) faunal analyses based on taxonomic relationships alone, as opposed to including functional traits.

Our focus on diverse assemblages of organisms with high turnover times (i.e. meiofaunal nematodes), inhabiting highly dynamic sedimentary environments (i.e. subtidal sandbanks and inter-bank swales), has revealed interesting animal-sediment relationships worthy of investigation by future studies of these and a broader range of marine environments. Results from such studies should facilitate a new generation of hypotheses concerning the relative importance of different environmental drivers in structuring benthic communities, and movement towards their rigorous testing. This will provide better indications of the stability of the sediments that benthic communities inhabit and allow better insights into how such communities may respond to natural sediment mobility and to anthropogenic disturbance. Taking account of the past evolution and modern processes of the sedimentary environment is paramount. Contemporary faunal distribution patterns reflect the integrated response of species and functional groups to a wide suite of sedimentary processes, so that effective descriptors of those sedimentary processes will help reveal meaningful animal-sediment relationships.

As noted in the introduction, the EU Habitats Directive requires that "management plans be created in order to detect human-induced change"; such words are easy to write but are likely to be very difficult to achieve. In the case of the UK's candidate Special Areas of Conservation (SAC), which include subtidal mobile sandbanks, there is much to be learnt before such plans can realistically be formed. Our scoping study has indicated some physically and ecologically meaningful measures of seabed physical texture that will help improve seabed characterisation and associations with benthic assemblages.

Acknowledgments

We thank the scientists, officers and crew of RV Cefas Endeavour for practical assistance in the collection of samples and the conduct of surveys. We are grateful to colleagues Claire Mason for carrying out grain size analyses, Karema Warr for processing meiofauna samples and Annie Brown, Jon Rees and Christian Wilson for processing side-scan and multi-beam data.

Author Contributions

Conceived and designed the experiments: MS. Performed the experiments: MS. Analyzed the data: MS PL. Contributed reagents/materials/analysis tools: MS PL. Wrote the paper: MS PL.

References

1. European Commission (1992) Council Directive 92/43/EEC of 21 May 1992 on the conservation of natural habitats and of wild fauna and flora. Available: http://ec.europa.eu/environment/nature/legislation/habitatsdirective/index_en.htm.
2. Pedersen SA, Fock H, Krause J, Pusch C, Sell AL, et al. (2009) Natura 2000 sites and fisheries in German offshore waters. ICES Journal of Marine Science 66: 155–169.
3. Joint Nature Conservation Committee (2012) Offshore Special Area of Conservation: North Norfolk sandbanks and Saturn Reef. Conservation objectives and advice on operations. Available: http://jncc.defra.gov.uk/pdf/NNSandbanksandSaturnReef_ConservationObjectives_AdviceonOperations_6.0.pdf)
4. Snelgrove PVR, Butman CA (1994) Animal-sediment relationships revisited: cause versus effect. Oceanography and Marine Biology: an Annual Review 32: 111–177.
5. Harris PT (2012) Surrogacy. In: Harris PT, Baker EK, editors. Seafloor geomorphology as benthic habitat. GeoHAB atlas of seafloor geomorphic features and benthic habitats. UK: Elsevier. pp.93–108.
6. Post AL, Wassenberg TJ, Passlow V (2006) Physical surrogates for macrofaunal distributions and abundance in a tropical gulf. Marine and Freshwater Research 57: 469–483.
7. Belderson RH, Kenyon NH, Stride AH, Stubb AR (1972) Sonographs of the sea floor - a picture atlas. Amsterdam: Elsevier. 185 p.
8. Baptist MJ, van Dalfsen J, Weber A, Passchier S, van Heteren S (2006) The distribution of macrozoobenthos in the southern North Sea in relation to meso-scale bedforms. Estuarine, Coastal and Shelf Science 68: 538–546.
9. Clements AJ, Strong JA, Flanagan C, Service M (2009) Objective stratification and sampling-effort allocation of ground-truthing in benthic-mapping surveys. ICES Journal of Marine Science 67: 628–637.
10. Galparsoro I, Borja Á, Legorburu I, Hernández C, Chust G, et al. (2010) Morphological characteristics of the Basque continental shelf (Bay of Biscay, northern Spain); their implications for Integrated Coastal Zone Management. Geomorphology 118: 314–329.
11. McArthur M, Brooke B, Przeslawski R, Ryan DA, Lucieer V, et al. (2010) On the use of abiotic surrogates to describe marine benthic biodiversity. Estuarine, Coastal and Shelf Science 88: 21–32.
12. Giere O (2009) Meiobenthology. The microscopic motile fauna of aquatic sediments. Second edition. Berlin: Springer Verlag. 527 p.
13. Palmer MA, Gust G (1985) Dispersal of meiofauna in a turbulent tidal creek. Journal of Marine Research 43: 179–210.
14. Palmer MA (1988) Dispersal of marine meiofauna: a review and conceptual model explaining passive transport and active emergence with implications for recruitment. Marine Ecology Progress Series 48: 81–91.
15. Boeckner MJ, Sharma J, Proctor HC (2009) Revisiting the meiofauna paradox: dispersal and colonisation of nematodes and other meiofaunal organisms in low- and high-energy environments. Hydrobiologia 624: 91–106.
16. Ellis JR, Maxwell TAD, Schratzberger M, Rogers SI (2011) The benthos and fish of offshore sandbank habitats. Journal of the Marine Biological Association of the United Kingdom 91: 1319–1335.
17. Dyer KR, Huntley DA (1999) The origin, classification and modelling of sand banks and ridges. Continental Shelf Research 19: 1285–1330.
18. Caston VND (1972) Linear sand banks in the southern North Sea. Sedimentology 18: 63–78.
19. Pravotorov IA (1983) Application of the cartographic method in the study of submarine sand ridges for navigational purposes. Geomorphologiya 2: 39–43.
20. Johnson MA, Caston GF (1984) Assessment of Ower Bank stability. Institute of Oceanographic Sciences Internal Report. 51pp.
21. Balson PS, Harrison DJ (1988) Marine Aggregate Survey Phase 1: Southern North Sea. British Geological Society Marine Report No. 86/38. 20 pp. and maps.
22. Collins MB, Shimwell SJ, Gao S, Powell H, Hewitson C, et al. (1995) Water and sediment movement in the vicinity of linear sandbanks: the Norfolk Banks, southern North Sea. Marine Geology 123: 125–142.
23. Huntley DA, Amos CL, Williams JJ, Humphery JD (1991) Estimating bedload transport on continental shelves by observations of ripple migration - an assessment. In: Soulsby RL, Bettess R, editors. Proceedings of EUROMECH 262 - Sand transport in rivers, estuaries and the sea, Balkema, Rotterdam, Netherlands, pp.17–22.
24. Woolfe KJ, Michibayashi K (1995) 'Basic' entropy grouping of laser-derived grain size data: an example from the Great Barrier Reef. Computers and Geosciences 21: 447–462.
25. Forrest J, Clark NR (1989) Characterizing grain size distribution: evaluation of a new approach using a multivariate extension of entropy analysis. Sedimentology 36: 711–722.
26. Woolfe KJ, Larcombe P, Stewart LK (2000) Shelf sediments adjacent to the Herbert River delta, Great Barrier Reef, Australia. Australian Journal of Earth Sciences 47: 301–308.
27. Orpin AR, Kostylev VE (2006) Towards a statistically valid method of textural sea floor characterization of benthic habitats. Marine Geology 225: 209–222.
28. Mikkelsen OA, Curran KJ, Hill PS, Milligan TG (2007) Entropy analysis of in-situ particle size spectra. Estuarine, Coastal and Shelf Science 72: 615–625.
29. Okada T, Larcombe P, Mason C (2009) Estimation of the spatial distribution of dredged material disposed of at sea by using particle-size distributions and metal concentrations. Marine Pollution Bulletin 58: 1164–1177.
30. Semple RK, Youngman CE, Zeller RE (1972) Economic regionalisation and information theory: an example. Discussion paper 28, Department of Geography, Ohio State University, Columbus, USA. 64 p.
31. Reading HG, Levell BK (1996) Controls on the sedimentary rock record, in: Reading HG, editor. Sedimentary environments: processes, facies and stratigraphy (Third edition). Cambridge: Cambridge University Press. pp.5–35.
32. Somerfield PJ, Warwick RM (1996) Meiofauna in marine pollution monitoring programmes. A laboratory manual. Lowestoft: Ministry of Agriculture, Fisheries and Food, Directorate of Fisheries Research. 71 p.
33. Platt HM, Warwick RM (1983) Free-living marine nematodes. Part I. British enoploids. Cambridge: Cambridge University Press. 307 p.
34. Platt HM, Warwick RM (1988) Free-living marine nematodes. Part II. British chromadorids. Cambridge: Cambridge University Press. 502 p.
35. Warwick RM, Platt HM, Somerfield PJ (1998) Free-living marine nematodes. Part III. British monhysterids. Shrewsbury: Field Studies Council. 296 p.
36. Deprez T (2005) The NeMys database. World Wide Web electronic publication, Available: www.nemys.ugent.be, version (2/2008).
37. Schratzberger M, Warr K, Rogers SI (2007) Functional diversity of nematode communities in the southwestern North Sea. Marine Environmental Research 63: 368–389.
38. Wieser W (1953) Die Beziehung zwischen Mundhöhlengestalt, Ernährungsweise und Vorkommen bei freilebenden marinen Nematoden. Arkiv för Zoologi 4: 439–484.
39. Riemann F (1974) On hemisessile nematodes with flagilliform tails living in marine soft bottoms and on microtubes found in deep sea sediments. Mikrofauna des Meeresbodens 40: 1–15.
40. Thistle D, Sherman KM (1985) The nematode fauna of a deep-sea site exposed to strong near-bottom currents. Deep-Sea Research 32: 1077–1088.
41. Thistle D, Lambshead PJD, Sherman KM (1995) Nematode tail-shape groups respond to environmental differences in the deep sea. Vie et Milieu 45: 107–115.
42. Ward AR (1975) Studies on the sublittoral free-living nematodes of Liverpool Bay. II. Influence of sediment composition on the distribution of marine nematodes. Marine Biology 30: 217–225.
43. Wright KA (1987) The nematode's cuticle: its surface and the epidermis: function, homology, analogy: a current consensus. Journal of Parasitology 73: 1077–1083.
44. Jensen P (1987) Differences in microhabitat, abundance, biomass and body size between oxybiotic and thiobiotic free-living marine nematodes. Oecologia 71: 564–567.
45. Soetaert K, Muthumbi A, Heip C (2002) Size and shape of ocean margin nematodes: morphological diversity and depth-related patterns. Marine Ecology Progress Series 242: 179–193.
46. Vanaverbeke J, Soetaert K, Vincx M (2004) Changes in morphometric characteristics of nematode communities during a spring phytoplankton bloom deposition. Marine Ecology Progress Series 273: 139–146.
47. Bongers T (1990) The maturity index: an ecological measure of environmental disturbance based on nematode species composition. Oecologia 83: 14–19.
48. Bongers T, Alkemade R, Yeates GW (1991) Interpretation of disturbance-induced maturity decrease in marine nematode assemblages by means of the Maturity Index. Marine Ecology Progress Series 76: 135–142.
49. Bongers T, de Goede RGM, Korthals GW, Yeates GW (1995) Proposed changes of c-p classification for nematodes. Russian Journal of Nematology 3: 61–62.
50. Hill MO (1973) Diversity and evenness: a unifying notation and its consequences. Ecology 54: 427–432
51. Clarke KR, Gorley RN (2006) PRIMER v6. User manual. Plymouth: Primer-E Ltd. 190 p.
52. Jennings S, Pinnegar JK, Polunin NVC, Warr KJ (2001) Impacts of trawling disturbance on the trophic structure of benthic invertebrate communities. Marine Ecology Progress Series 213: 127–142.
53. Wilber DH, Ray GL, Clarke DG, Diaz RJ (2008) Responses of benthic infauna to large-scale sediment disturbance in Corpus Christi Bay, Texas. Journal of Experimental Marine Biology and Ecology 365: 13–22.
54. Foden J, Rogers SI, Jones AP (2009) Recovery rates of UK seabed habitats after cessation of aggregate extraction. Marine Ecology Progress Series, 390: 15–26

55. Ashley GM (1990) Classification of large-scale bedforms: a new look at an old problem. Journal of Sedimentary Petrology 60: 160–172.
56. Belderson RH, Johnson MA, Kenyon NH (1982) Bedforms, in: Stride AH, editor. Offshore tidal sands - processes and deposits. London: Chapman and Hall. pp.27–57.
57. Rundell RJ, Leander BS (2010) Masters of miniaturisation: convergent evolution among interstitial eukaryotes. Bioassays 32: 430–437.
58. Shennan I, Lambeck K, Flather F, Horton B, McArthur J (2000) Modelling western North Sea palaeogeographies and tidal changes during the Holocene. In: Shennan I, Andrews J, editors. Holocene Land-Ocean Interaction and Environmental Change around the North Sea (Geological Society Special Publications 166). London: Geological Society. pp.299–319.
59. Ward I, Larcombe P (2008) Determining the preservation rating of submerged archaeology in the post-glacial southern North Sea: a first-order geomorphological approach. Environmental Archaeology 13: 59–83.
60. Reineck H-E, Singh IB (1980) Depositional sedimentary environments. Berlin: Springer-Verlag. 549 p.
61. Visser MJ (1980) Neap-spring cycles reflected in Holocene subtidal large-scale bedform deposits: a preliminary note. Geology 8: 543–546.
62. Langhorne DN (1982) A study of the dynamics of a marine sandwave. Sedimentology 29: 571–594
63. Nio SD, Yang C (1991) Diagnostic attributes of clastic tidal deposits: A review. In: Smith DG, Zaitlin BA, Reinson GE, Rahmani RA, editors. Clastic tidal sedimentology, Canadian Society of Petroleum Geologists Memoir 16: 3–28.
64. Hall SJ (1994) Physical disturbance and marine benthic communities: life in unconsolidated sediments. Oceanography and Marine Biology: an Annual Review 32: 179–239
65. Jennings S, Kaiser MJ (1998) The effects of fishing on marine ecosystems. Advances in Marine Biology 34: 201–352
66. Harris P, Hughes M (2012) Predicted benthic disturbance regimes on the Australian continental shelf: A modelling approach. Marine Ecology Progress Series 449: 13–25
67. Harris PT (2014) Shelf and deep-sea sedimentary environments and physical benthic disturbance regimes: A review and synthesis. Marine Geology 353: 169–184
68. Petraitis PS, Latham RE, Niesenbaum RA (1989) The maintenance of species diversity by disturbance. Quarterly Review of Biology 64: 393–418.
69. Tilman D (1990) Constraints and trade-offs: toward a predictive theory of competition and succession. Oikos 58: 3–15.
70. Steneck RS, Dethier MN (1994) A functional group approach to the structure of algal-dominated communities. Oikos 69: 476–498.
71. Wellborn GA, Skelly DK, Werner EE (1996) Mechanisms creating community structure across a freshwater habitat gradient. Annual Review of Ecology and Systematics 27: 337–363.
72. Delgado M, de Jonge VN, Peletier H (1991) Effect of sand movement on the growth of benthic diatoms. Journal of Experimental Marine Biology and Ecology 145: 221– 231.
73. Hogue EW, Miller CB (1981) Effects of sediment microtopography on small-scale spatial distributions of meiobenthic nematodes. Journal of Experimental Marine Biology and Ecology 53: 181–191.
74. Vanaverbeke J, Vincx M (2008) Short-term changes in nematode communities from an abandoned intense sand extraction site on the Kwintebank (Belgian Continental Shelf) two years post-cessation. Marine Environmental Research 66: 240–248.
75. Connell JH (1978) Diversity in tropical rain forests and coral reefs. Science 199: 1302–1310.
76. Raes M, Vanreusel A (2006) Microhabitat type determines the composition of nematode communities associated with sediment clogged cold-water coral framework in the Porcupine Seabight (NE Atlantic). Deep-Sea Research I 53: 1880–1894.
77. Branch GM (1984) Competition between marine organisms: ecological and evolutionary implications. Oceanography and Marine Biology: an Annual Review 22: 429–593.
78. Wilson WH (1990) Competition and predation in marine soft-sediment communities. Annual Review in Ecology and Systematic 21: 221–241.
79. Hooper DU, Chapin FS, Ewel JJ, Hector A, Inchausti P, et al. (2005) Effects of biodiversity on ecosystem functioning: a consensus of current knowledge. Ecological Monographs 75: 3–35.
80. Sanders HL (1968) Marine benthic diversity: a comparative study. American Naturalist 102: 243–282.
81. Wilson JB (1982) Shelly faunas associated with temperate offshore tidal deposits. In: Stride AH, editor. Offshore tidal sands - processes and deposits. London: Chapman and Hall. pp.126–171.
82. Huston MA (1996) Biological diversity. Cambridge: Cambridge University Press. 681 p.

Relationships between Degree of Polymerization and Antioxidant Activities: A Study on Proanthocyanidins from the Leaves of a Medicinal Mangrove Plant *Ceriops tagal*

Hai-Chao Zhou[1,2,3]*, Nora Fung-yee Tam[2,3]*, Yi-Ming Lin[1], Zhen-Hua Ding[1], Wei-Ming Chai[1], Shu-Dong Wei[1]

1 Key Laboratory of the Ministry of Education for Coastal and Wetland Ecosystems, Xiamen University, Xiamen, China, **2** Department of Biology and Chemistry, City University of Hong Kong, Hong Kong SAR, China, **3** Futian-CityU Mangrove R&D Centre, City University of Hong Kong Shenzhen Research Institute, Shenzhen, China

Abstract

Tannins from the leaves of a medicinal mangrove plant, *Ceriops tagal*, were purified and fractionated on Sephadex LH-20 columns. ^{13}C nuclear magnetic resonance (^{13}C-NMR), reversed/normal high performance liquid chromatography electrospray ionization mass spectrometry (HPLC-ESI MS) and matrix-assisted laser desorption/ionization time-of-flight mass spectrometry (MALDT-TOF MS) analysis showed that the tannins were predominantly B-type procyanidins with minor A-type linkages, galloyl and glucosyl substitutions, and a degree of polymerization (DP) up to 33. Thirteen subfractions of the procyanidins were successfully obtained by a modified fractionation method, and their antioxidant activities were investigated using 2,2-diphenyl-1-picrylhydrazyl (DPPH) scavenging capacity and ferric reducing antioxidant power (FRAP) method. All these subfractions exhibited potent antioxidant activities, and eleven of them showed significantly different mean DP (mDP) ranging from 1.43 ± 0.04 to 31.77 ± 1.15. Regression analysis demonstrated that antioxidant activities were positively correlative with mDP when around mDP <10, while dropped and then remained at a level similar to mDP $=5$ with around 95 µg ml^{-1} for DPPH scavenging activity and 4 mmol AAE g^{-1} for FRAP value.

Editor: Silvia Mazzuca, Università della Calabria, Italy

Funding: This present work was supported by the National Natural Science Foundation of China (URL: http://isisn.nsfc.gov.cn/egrantweb/): no. 41306084 (Hai-Chao Zhou's funding), no. 41176090 (Zhen-Hua Ding's funding), and no. 31070522 (Yi-Ming Lin's funding). The funders had no role in study design, data collection and analysis, decision to publish, or preparation of the manuscript.

Competing Interests: The authors have declared that no competing interests exist.

* Email: zhouhc2013@gmail.com (HCZ); bhntam@cityu.edu.hk (NFYT)

Introduction

Ceriops tagal, a mangrove plant species in the Family of Rhizophoraceae, is one of the medicinal plants in East and Southeast Asia and is often used as a tradition herb to treat diseases such as hemorrhages and malignant ulcers [1,2]. The phytochemistry of *C. tagal* has been reported, however, the work mainly focused on the isolation and identification of terpenoid compounds in root, with little attention on leaf tannins [3,4]. The high level of tannins in the leaves of Rhizophoraceae is known to deter feeding by herbivores [5], but leaf tannins also show a diversity of other biological activities [6]. The unexplored tannins could be novel potential resources of bioactive compounds in mangrove plants. So far, the chemical properties of *C. tagal* tannins have not yet been determined and the structure-activity relationships of tannins are still not clear.

Tannins comprised as much as 20–40% dry weight in the leaf and bark of mangrove plants [7,8]. Compared with hydrolysable tannins, condensed tannins (proanthocyanidins) are more abundant. They are commonly found in mangrove plants [9,10,11,12,13,14] and are also the main component of the polyphenols in our diet [15]. Because of their antioxidant activities and other potentially health-promoting qualities, proanthocyanidins have attracted more and more research interests in recent years [16,17,18,19]. Proanthocyanidins are oligomers and polymers of flavan-3-ol that are bound together with B-type and A-type linkages [15]. The chemistry and biological features of proanthocyanidins largely depend on their structure, particularly the molecular weight that is also expressed as degree of polymerization (DP) [20,21]. Ariga et al. [22] found that the ability to scavenge free radicals was proportional to DP for simple flavonoid oligomers. Hagerman et al. [23] proved that tannins with highly polymerized and many hydroxyl groups are more potent antioxidants than the simple phenolics. Recently, some findings reported that the increasing DP may enhance the antioxidant power of condensed tannins [24,25]. However, those previous works either limitedly detected the simple oligomers or directly studied the bulk mixture of polymers. The relationships between DP and antioxidant activity of proanthocyanidins, therefore, remained largely unknown.

The present study, therefore, aims to (I) achieve a complete structural characterization of tannins from the leaf of a mangrove plant (*C. tagal*) by a combination of analytical techniques, including ^{13}C-NMR, MALDI-TOF MS, thiolysis degradation, and reversed/normal-phase HPLC-ESI MS; and (II) establish an efficient fractionation method to obtain a series of proanthocyanidins with different DP; and (III) explore the possible relationships between DP and antioxidant activity of condensed tannins.

Materials and Methods

2.1. Chemicals and Materials

Water used in this experiment was purified on a Millipore Milli-Q apparatus. HPLC grade dichloromethane, acetonitrile (CH_3CN), methanol, trifluoroacetic acid (TFA), acetic acid, and all analytical grade solvents (acetone, methanol, n-Butanol etc.) were obtained from Sinopharm (Sinopharm, Shanghai, China). Sephadex LH-20, Folin-Ciocalteu reagents, acetone-d_6, deuteroxide (D_2O), Amberlite IRP-64 cation-exchange resin, cesium chloride (CsCl), 2,5-dihydroxybenzoic acid (DHB), benzylmercaptan, 2,2-diphenyl-1-picrylhydrazyl (DPPH), 2,4,6-tripyridyl-S-triazine (TPTZ), ascorbic acid (AA), and all HPLC standards were purchased from Sigma (St. Louis, MO, USA). The mature leaves (the third pair with fully expanded and dark green) of a mangrove plant *C. tagal* (Rhizophoraceae *Ceriops*), were collected from a mangrove forest in the Dongzhai harbor (19°56′N, 110°34′E), Hainan, China. The leaves were immediately freeze-dried, ground, and stored at −20°C prior to analysis.

2.2. Extraction, Purification and Fractionation of Mangrove Tannins

The extraction procedure was conducted according to the method of Zhou et al. [25] The crude tannins extract (Fc) obtained was applied to a Sephadex LH-20 column (50×1.5 cm i.d.). The procedure of purification and fractionation of F_C was shown in Figure 1. Purified tannins (Fp) and nine subfractions (F1 to F9) were evaporated under vacuum to remove organic solvents followed by freeze-dried. The last two subfractions (F8 and F9) were re-loaded on a Sephadex LH-20 column (50×1.0 cm i.d.) to yield seven more subfractions as shown below.

2.3. Determination of Total Phenolics and Extractable Condensed Tannins

The procedure described by Zhou et al. [25] to estimate the total phenolics and extractable condensed tannins of the entire leaf of *C. tagal* was used, and the concentrations were determined by Folin-Ciocalteu [26] and butanol-HCl method [27], respectively. The content of the total phenolics (TP) and extractable condensed tannins (ECT) were calculated using the purified tannins (Fp) obtained from the purification of crude tannins extract (Fc) as the standard. Both of them were expressed as mg Fp equivalents per gram dry weight of leaves.

2.4. ^{13}C-NMR Analysis

The obtained purified tannins from *C. tagal* leaves were dissolved in acetone-d_6/D_2O. A Varian Metcury-600 spectrometer (Palo Alto, CA, USA) was employed in the 150 MHz mode [10].

2.5. Reversed-phase HPLC-ESI MS Analysis Followed by Thiolysis

The modified method described by Zhou et al. [9] was carried out to characterize Fp and obtain the mDP for each subfraction. Condensed tannins were thiolysis degraded by benzylmercaptan, and then the degradation products were analyzed on an Agilent 1200 system (Agilent, Palo Alto, CA, USA) interfaced to a QTRAP 3200 (Applied Biosystems, Foster, USA) with a 250 mm×4.6 mm i.d. 5.0 μm Hypersil ODS column (Elite, Dalian, China).

2.6. Normal-phase HPLC-ESI MS Analysis

Normal-phase HPLC-ESI-MS analysis was conducted according to the method of Hellstrom et al. [28] with minor modifications. The HPLC-MS equipment consisted of an Agilent

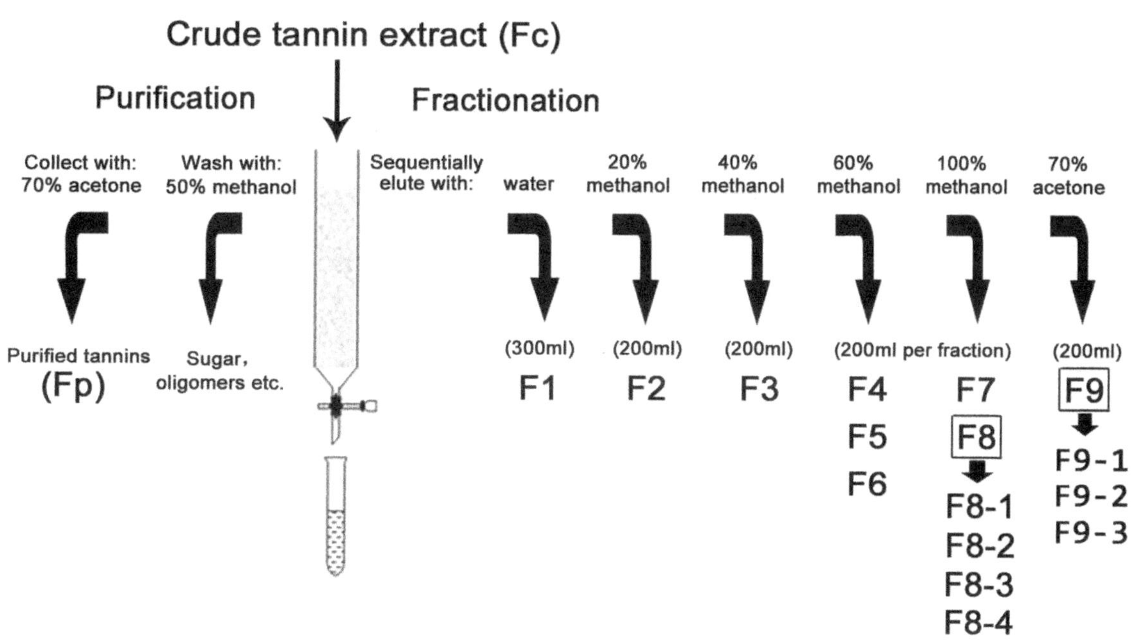

Figure 1. Sephadex LH-20 column chromatography for purification and fractionation of tannins from *C. tagal* leaves. Subfraction 8 (F8) and Subfraction 9 (F9) were further fractionated into F8-1 to F8-4 and F9-1 to F9-3, respectively.

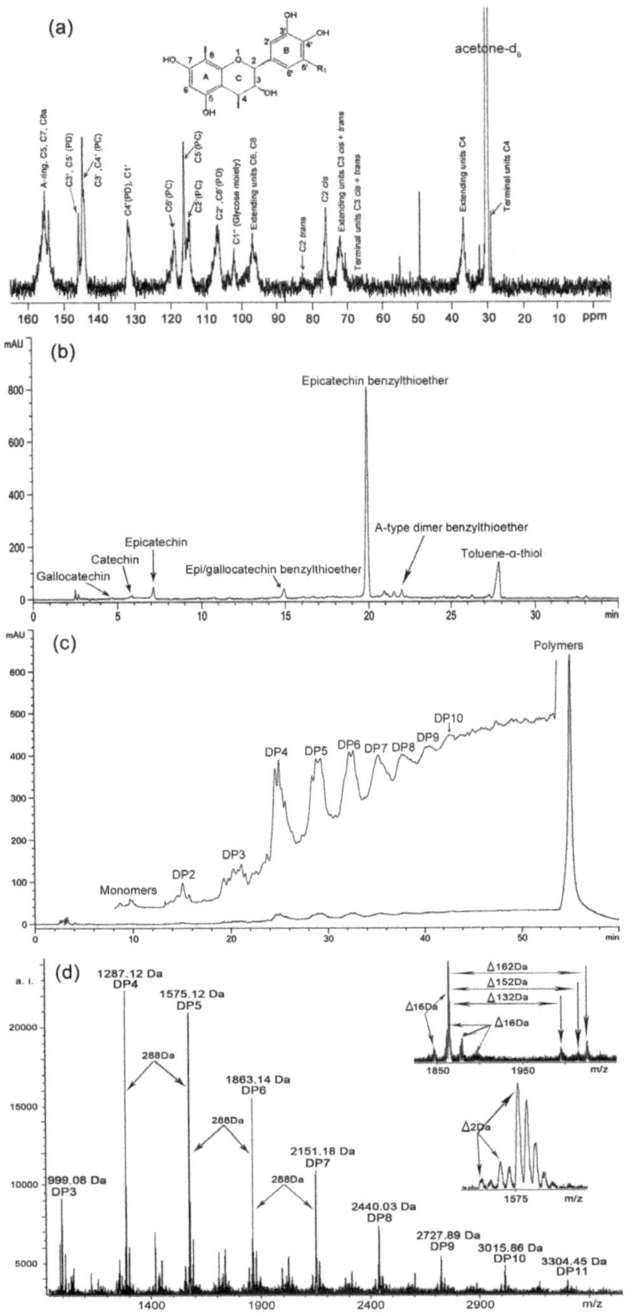

Figure 2. Structural characterization of purified tannins (Fp) from *C. tagal* leaves. Analyzed by (a) ^{13}C-NMR, (b) reversed-phase HPLC-ESI-MS, (c) normal-phase HPLC-ESI-MS and (d) MALDI-TOF MS.

1200 liquid chromatograph system as described above with a 250 mm×4.6 mm i.d. 5.0-μm Silica Luna column (Phenomenex, Darmstadt, Germany).

2.7. MALDI-TOF MS Analysis

The MALDI-TOF MS analyses were performed on a Bruker Reflex III instrument (Germany). The measurements were carried out using the conditions reported in the previous work [29]. Reflectron modes coupled with linear modes were further applied to show the different DP distribution of condensed tannins.

2.8. Determination of Antioxidant Activities

Antioxidant activities of each subfraction were evaluated by DPPH radical scavenging activity (DPPH method) [30] and ferric ion reducing antioxidant power (FRAP method) [31]. Their results were expressed in EC_{50} value (μg mL^{-1}) and mmol ascorbic acid equivalents per gram of each subfraction (mmol AAE g^{-1}), respectively. Procedures were modified according the method of Zhou et al. [32].

2.9. Statistical Analysis

Results of antioxidant activity were expressed as mean ± standard deviation of three independent determinations. A parametric one-way analysis of variance (ANOVA) was used to test any significant difference among subfractions at $P<0.05$, and all the data without transformation fulfilled the two assumptions, normal distribution and homogeneity in variance of the parametric test. All statistical analyses were performed with SPSS 17.0 for Windows.

2.10. Ethics Statement

Field permit issued by The Mangroves of Dongzhai Gang National Reserve in Dongzhai harbor, Hainan, China, allowed us to collect the mature leaves of *C. tagal* in this site. They also confirmed that our study did not involve any endangered or protected species.

Results and Discussion

3.1. Contents of Total Phenolics and Extractable Condensed Tannins

The total phenolics and extractable condensed tannins content of *C. tagal* leaves was 27.4±2.8% and 20.1±2.0%, respectively. Phenolic compounds (including proanthocyanidins) were considered to be the major contributor to the antioxidant activity of vegetables, fruits or medicinal plants [25,33,34]. The high level of total phenolics and extractable condensed tannins in *C. tagal* leaves strongly suggested that some of their pharmacological effects could be attributed to the presence of these valuable constituents.

3.2. Structural Characterization of Purified Tannins (Fp)

Fp from *C. tagal* leaves were initially analyzed by ^{13}C-NMR spectroscopy, and the signal assignment was performed according to the method of Czochanska et al. [35] The ^{13}C-NMR spectrum (Figure 2a) showed that the characteristic ^{13}C peaks are consistent with that of condensed tannins with dominant procyanidin units and a minor amount of prodelphinidins. The signals near 145 ppm, which arise from quaternary C3′ and C5′ in prodelphinidins (PD), and C3′ and C4′ in procyanidin (PC) units were used to estimate the PD:PC ratio [35]. In this experiment, direct integration of the signals at 146.2 ppm (PD units) and 145.2 ppm (PC units) gave a PD:PC ratio of about 8:92. The signals at 115.1 ppm (C2′), 116.1 ppm (C5′) and 119.0 ppm (C6′) further supported the presence of PC, so does signals at 107–108 ppm (C2′ and C6′) and 133.0 ppm (C4′) for PD. The structural diversity of the linkage (A type and B type) and the stereoisomer of catechin and epicatechin units were apparent from the spectrum. Specially, C5, C7, and C8a carbons of procyanidins appeared at 150 to 160 ppm [36]. The region between 70 and 90 ppm was sensitive to the stereochemistry of the C-ring. The ratio of the 2,3-*cis* to 2,3-*trans* isomers could be determined through the distinct differences in their respective C2 chemical shifts. C2 gave a resonance at 76.7 ppm for the *cis* form and

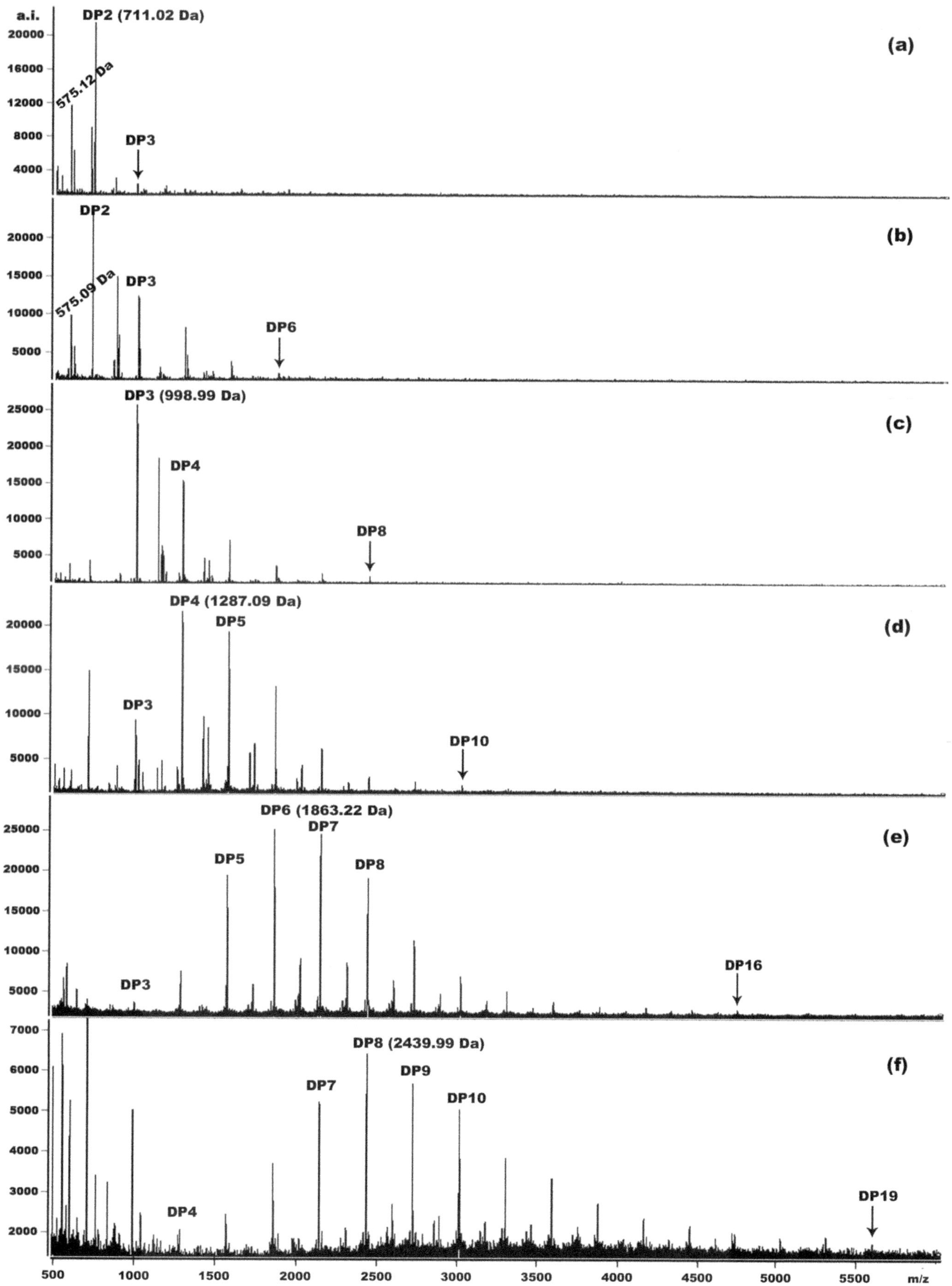
a.i.
DP2 (711.02 Da)
575.12 Da
DP3
(a)
DP2
575.09 Da
DP3
DP6
(b)
DP3 (998.99 Da)
DP4
DP8
(c)
DP4 (1287.09 Da)
DP5
DP3
DP10
(d)
DP6 (1863.22 Da)
DP7
DP5
DP8
DP3
DP16
(e)
DP8 (2439.99 Da)
DP9
DP7
DP10
DP4
DP19
(f)
m/z

Figure 3. Reflectron mode MALDI-TOF mass spectra for tannin subfraction F4 (a) to F9 (f). Degree of polymerization (DP) of the predominant polymers and largest polymer in each mass spectrum of subfraction are labeled.

83.6 ppm for the *trans* form. From the peak areas, it was estimated that the *cis* isomer is dominant. The C3 generally have their chemical shift 66–68 ppm in terminal units, and 72–74 in extending units [37]. Theoretically, the intensity of the C3 signal in terminal units relative to that of the signal in extension units could be used for elucidating the polymer chain length [35]. However, the application of this technique for the quantification of molecular weight suffered from inaccuracy due to the low signal-to-noise ratio [24]. Similarly, the spectra obtained in the present study also showed low signal-to-noise ratio (Figure 2a). The C4 atoms of the extension units showed a broad peak at around 37 ppm, while the terminal C4 exhibits 29 ppm [37], next to the solvent peak acetone-d_6. The signal at 103.1 ppm would attribute to C1″ of the glycose moiety connected to the C3 position, while the signals of C2″ (74.0 ppm), C3″ (76.3 ppm), C4″ (71.4 ppm) and C5″ (75.7 ppm) were overlapped with the chemical shifts 70–80 ppm referred above [38]. These results thus showed that the mangrove condensed tannins of *C. tagal* leaves were predominantly constituted of epicatechin, the main constitutive monomer with some glycosides.

Although the ^{13}C-NMR spectrum revealed complex structural characteristics of the condensed tannins of *C. tagal* leaves, this analytical technique hardly allows the determination of the polymer chain length and the chemical constitution of individual chains. Moreover, the sequential succession of monomer units in individual chains could not be elucidated [17,36,39]. To overcome these problems, further characterization was continued by means of HPLC-ESI-MS and MALDI-TOF MS.

The nature of flavan-3-ol units within proanthocyanidins was analyzed by acid catalysis in the presence of benzylmercaptan as nucleophile, which had been widely used [18,21,36]. In the thiolysis reaction, terminal units of condensed tannins are released as free flavanoids, while external units are distinguished as benzylthioether adducts. The extension and terminal units of proanthocyanidins could be distinguished by reversed-phase HPLC analysis (Figure 2b). The dominant peak observed was epicatechin benzylthioether along with five other smaller peaks for epicatechin, epi/gallocatechin benzylthioether, A-type dimer benzylthioether, catechin and gallocatechin. These results suggested that epicatechin was the dominant flavan-3-ol unit, which occurred as both terminal and extension units in proanthocyanidins of *C. tagal* leaves.

Normal-phase HPLC analysis was able to further reflects the heterogeneity of the proanthocyanidin mixture [28]. Although the peak band became broader in higher DP and was hard to be resolved, this normal-phase HPLC chromatogram of Fp (Figure 2c) indicated the presence of proanthocyanidins with DP from monomers to decamers (DP10). The results suggested the heterogeneity of proanthocyanidins in their constituent units and the linkages between them.

Figure 2d showed the MALDI-TOF mass spectra of Fp, recorded as Cs^+ adducts in the positive ion reflectron mode, and the enlarged spectrums demonstrated the good resolution. These results showed large structural heterogeneity of condensed tannins in the mangrove plant *C. tagal*. The mass increased from DP3 (999.08 Da) to DP11 (3304.45 Da) by 288 Da, which corresponded to one epi/catechin monomer unit. For each multiple, substructures with 16 Da, both higher and lower, could be found (Figure 2d). These masses had been identified as heteropolymers of repeating flavan-3-ol units, which demonstrated the coexistence of epi/gallocatechin and epi/afzelechin [39]. Additionally, there were three series peaks with distances of 162 Da, 152 Da, and 132 Da. The first series (162 Da) represented glucosylated heteropolyflavans containing the flavanone [39]. The second series (152 Da) was corresponding to the addition of one galloyl group at the heterocyclic C-ring [12,24]. And the 132 Da distance may have two different interpretations according to previous works. Reed et al. [39] interpreted it as the substitutions by pentoside (132 Da), while Xiang et al. [40] explained as quasimolecular ions $[M+2Cs^+-H]^+$ that generated by simultaneous attachment of two Cs^+ and loss of a proton. The distinct decrease of 2 Da illustrated the nature of interflavan bonds including A-type and B-type linkages. The data from MALDI-TOF MS well agreed with the results obtained by ^{13}C-NMR, reversed and normal-phase HPLC-ESI-MS analyses.

3.3. MALDI-TOF MS Analysis of Proanthocyanidin Subfractions

For the first step, ten subfractions (F1-F9) were obtained by fractionation on Sephadex LH-20 as described in methods section. After analysis by MALDI-TOF MS, fractions F1-F3 showed a cluster of unresolved peak lower than 500 Da (data not shown), which could be the sugars and other impurities. The spectra obtained by reflectron modes of the seven collected subfractions from fractions F4 (Figure 3a) to F9 (Figure 3f) clearly displayed the distinct predominant polymers and polymer range. The signals of lower oligomers were firstly detected in fraction F4 dominated by dimers (Figure 3a). Fractionation of the lower oligomers that could lead to the saturation of detector can significantly enhance the sensitivity of the detection of large polymers of proanthocyanidins under MALDI-TOF analysis [17]. Therefore, the last subfraction (F9) detected in the present study reached as high as DP19 polymer (Figure 3f), which demonstrated that the fractionation method could improve the detection of high DP polymers compared with the spectra of Fp (Figure 2d), even though it did not completely overcome the discrimination of high molecular weight polymers [17].

Since F8 and F9 still possessed of large DP distribution (with DP3-DP16 and DP4-DP19, respectively), a refined fractionation method was conducted to re-fractionate F8 (Figure 4) and F9 (Figure 5), which yielded other four and three subfractions, respectively, as well as with increasing trends for their predominant polymers and polymer range.

After refined fractionation for F8 and F9, as high as DP27 and DP33 polymer were detected in linear spectra of F8-4 (Figure 4) and F9-3 (Figure 5), respectively. It suggested that proanthocyanidins in *C. tagal* leaves can reach DP33, or even more. The present study clearly demonstrates a higher DP and clearer DP distribution could be obtained than many previous works reporting a high DP in various plant materials using MALDI-TOF MS [17]. In the present study, however, we showed a higher DP and clearer DP distribution than those described. Both the reflectron and linear modes of MALDI-TOF MS have been successfully applied to the analysis of proanthocyanidins. The high mass resolution power of reflectron modes allows distinguishing the mass of the isotopic peaks with enough accuracy to compare with the theoretical calculated mass, while the linear mode provides better information about the mass distribution, especially for the high DP of proanthocyanidins [17,29].

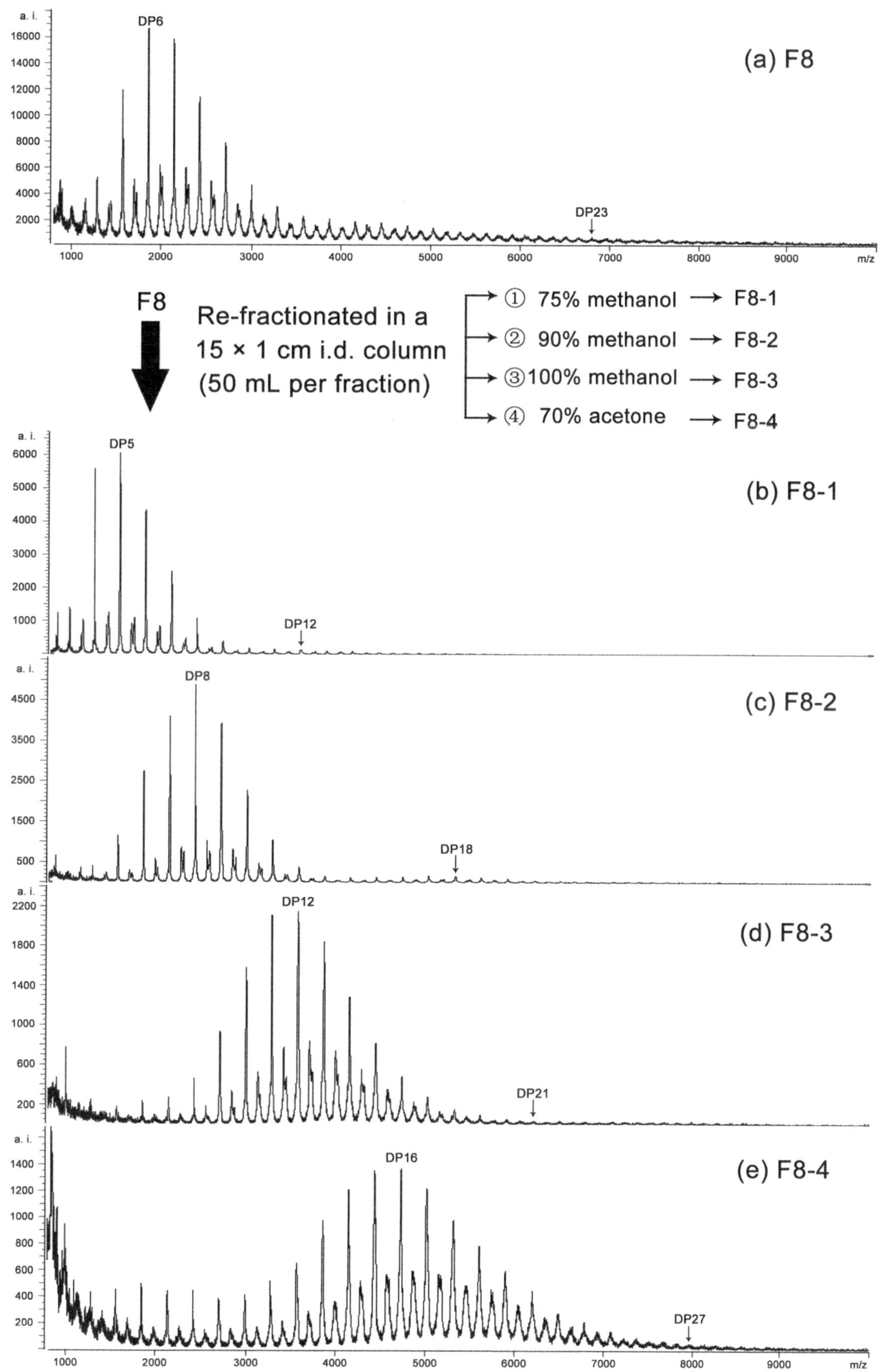

Figure 4. Linear mode MALDI-TOF mass spectra for tannin subfraction F8 (a). F8-1 (b) to F8-4 (e) were obtained by the further re-fractionation of F8. Degree of polymerization (DP) of the predominant polymers and largest polymer in each mass spectrum of subfraction are labeled.

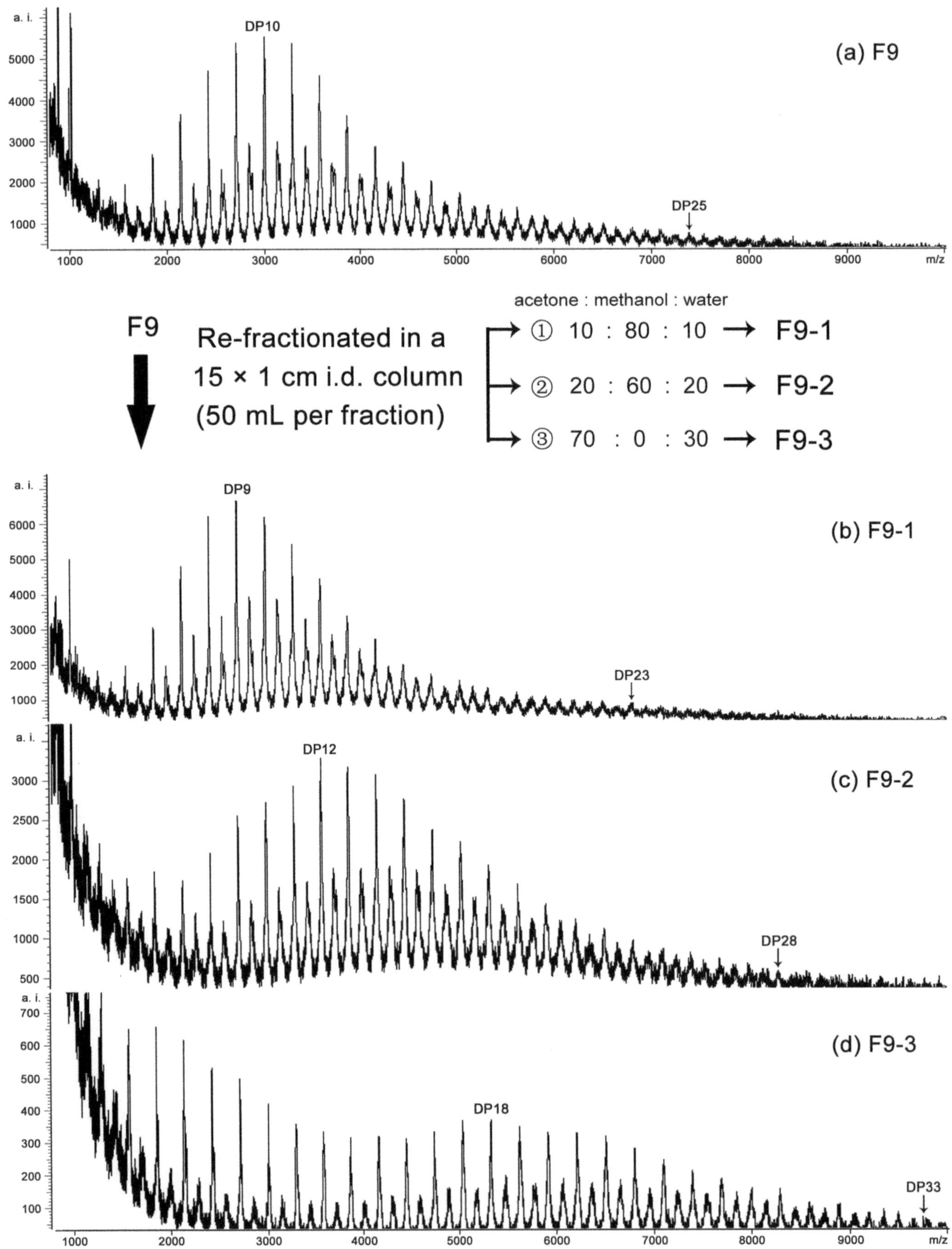

(a) F9
DP10
DP25
F9
Re-fractionated in a
15 × 1 cm i.d. column
(50 mL per fraction)
acetone : methanol : water
① 10 : 80 : 10 → F9-1
② 20 : 60 : 20 → F9-2
③ 70 : 0 : 30 → F9-3
(b) F9-1
DP9
DP23
(c) F9-2
DP12
DP28
(d) F9-3
DP18
DP33
a. i.
m/z

Figure 5. Linear mode MALDI-TOF mass spectra for tannin subfraction F9 (a). F9-1 (b) to F9-3 (d) were obtained by the further re-fractionation of F9. Degree of polymerization (DP) of the predominant polymers and largest polymer in each mass spectrum of subfraction are labeled.

3.4. Mean Degree of Polymerization (mDP) Analysis of Proanthocyanidin Subfractions

Although the above MALDI-TOF mass spectra graphically showed the differences among each proanthocyanidin subfraction, it only qualitatively illustrated the changes of DP. Therefore, mDP was calculated using thiolysis degradation coupled with reversed-phase HPLC-ESI MS. Table 1 listed the mDP of 13 proanthocyanidin subfractions from *C. tagal* leaves. The mDP increased with the eluted order on Sephadex LH-20: ranging from 1.43±0.04 (F4) to 16.14±0.58 (F9), from 7.26±0.31 (F8-1) to 16.14±0.58 (F8-4) and from 14.19±0.50 (F9-1) to 31.77±1.15 (F9-3), which agreed with the results of MALDI-TOF analysis.

These results confirmed that the fractionation method used in this study was effective to gradually separate the plant proanthocyanidins according to DP. The occurrence of proanthocyanidins in complex mixtures makes it difficult to firmly establish the effect of molecular size (or DP) on the biologically important properties of proanthocyanidins [41]. Therefore, the distinct subfractions obtained from this fractionation method were subjected to explore the relationship between DP and antioxidant activities for proanthocyanidins.

3.5. Antioxidant Activities and Their Relationships with mDP

DPPH method is a reliable and reproducible method to determine the radical scavenging activity of proanthocyanidins extracted from plant or fruits [24,42]. FRAP method is based on the redox reaction of ferric ion in the presence of a reducer. The reduction capacity of a compound may serve as a significant indicator of its potential antioxidant activity [43]. Table 1 summarized the antioxidant activities (DPPH and FRAP method) of proanthocyanidin subfractions from *C. tagal* leaves. The highest antioxidant activities were found in F8 with $EC_{50/DPPH}$ value 78.13±1.14 μg ml^{-1} and FRAP value 5.87±0.08 mmol AAE g^{-1}, while the lowest in F4 with $EC_{50/DPPH}$ value 133.45±2.24 μg ml^{-1} and FRAP value 2.58±0.04 mmol AAE g^{-1}. FRAP showed a similar trend to the results of DPPH assay. Although these two antioxidant assays with different mechanisms were used in this study, other antioxidant assays, such as peroxyl radical scavenging capacity or even assays for peroxidative/prooxidative enzymes should be included in further research, such as peroxyl radical scavenging capacity or even assays for peroxidative/prooxidative enzymes. Compared with the results from other plant materials, such as *Delonix regia* [18], *Litchi chinensis* [25] and *Garcinia mangostana* [32], proanthocyanidins from *C. tagal* leaves also exhibited substantial radical scavenging activity and reduction capacity.

Regression analysis was performed to establish the relationship between antioxidant activities and mDP (Figure 6). It demonstrated that antioxidant activities was positively correlative with mDP when mDP <10, while dropped and then remained at a level with around 95 μg ml^{-1} for $EC_{50/DPPH}$ value and 4 mmol AAE g^{-1} for FRAP value (similar to the antioxidant level of mDP = 5). Although some researchers attempted to study this relationship of proanthocyanidins, the work was limited to low oligomers and low DP. For instance, Gaulejac et al. [44] found an increase in the activity of procyanidins, but their work only focused on the oligomers from 1 to 4 units. Es-Safi et al. [24] simply compared the commercial standard substances (catechin and B_3 procyanidin dimer) with pear juice polymeric proanthocyanidins. Jerez et al. [42] showed an increase of the antiradical activity of *Pinus pinaster* and *Pinus radiata* procyanidins up to 6–7 mDP (the mDP obtained by thiolysis with cysteamine). Recently, Zhou et al. [32]

Table 1. Mean degree of polymerization (mDP) and antioxidant activities of each subfraction obtained by fractionation of proanthocyanidins from *C. tagal* leaves.

Fractions	mDP	$EC_{50/DPPH}$ value[I] (μg ml^{-1})	FRAP value[II] (mmol AAE g^{-1})
F4	1.43±0.04 k	133.45±2.24 a	2.58±0.04 h
F5	2.49±0.14 j	114.74±2.24 b	3.60±0.05 g
F6	3.52±0.34 i	103.56±0.79 c	4.05±0.13 ef
F7	7.78±0.29 h	83.60±0.62 g	5.46±0.05 b
F8	10.52±0.62 g	78.13±1.14 h	5.87±0.08 a
F8-1	7.26±0.31 h	84.11±1.41 g	5.19±0.05 c
F8-2	11.72±0.56 f	97.04±1.04 d	3.83±0.07 fg
F8-3	16.84±0.82 d	86.40±0.24 fg	4.28±0.11 e
F8-4	24.62±1.15 b	98.49±1.48 d	3.80±0.01 fg
F9	16.14±0.58 d	87.14±1.68 f	4.52±0.23 d
F9-1	14.19±0.50 e	96.49±1.55 d	3.80±0.16 fg
F9-2	22.01±0.81 c	93.17±2.01 e	4.05±0.08 ef
F9-3	31.77±1.15 a	105.37±1.22 c	3.66±0.09 g

[I]The $EC_{50/DPPH}$ value (μg ml^{-1}) is the concentration of each subfraction at which the scavenging activity is 50%.Values with different letters in the same column are significantly different at $P<0.05$ level.

[II]The FRAP value (mmol AAE g^{-1}) is expressed in mmol ascorbic acid equivalents per gram of each subfraction. Values with different letters in the same column are significantly different at $P<0.05$ level.

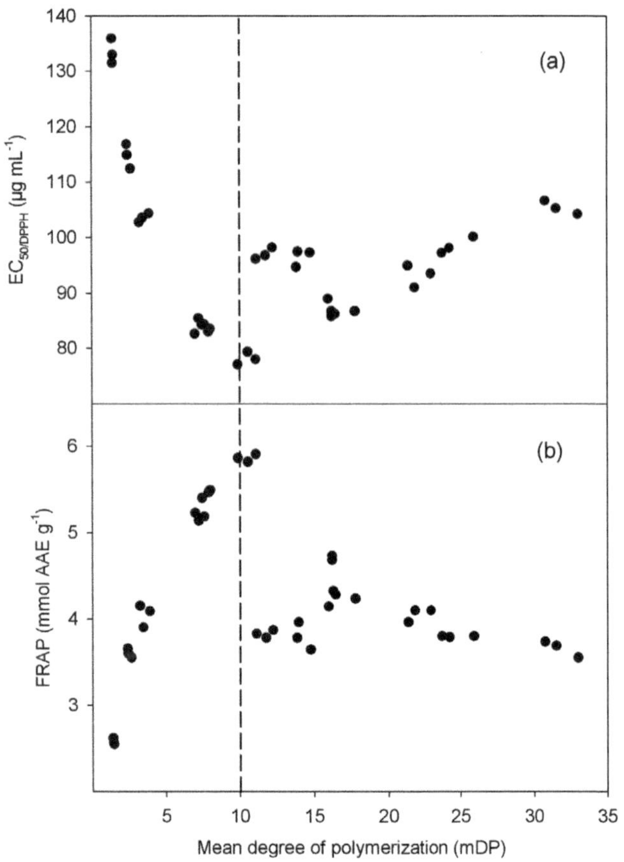

Figure 6. Regression analysis for mean degree of polymerization (mDP) and antioxidant activity. (a) $EC_{50/DPPH}$ value and (b) FRAP value correlated with mDP for each proanthocyanidin subfraction from *C. tagal* leaves.

reported an increasing antioxidant activity of mangosteen tannins with mDP between 2.71 and 9.27, but decreasing in mDP = 16.80, and proposed that 9–10 mDP could be considered as a dividing and critical point for predicting the structure-activity of mangosteen condensed tannins, which was confirmed with the present results. However, all those previous works could not provide more information for the higher DP. In this study, a decrease and then remain at a lower antioxidant level was showed when mDP >10. However, the antioxidant activity based on the DPPH test in the present study could be influenced by the amount of total polyphenols. Zhou et al. [32] also showed linear relationships between DPPH and total polyphenol content. Similarly, antioxidant tests were positively correlated with total polyphenols [32,45,46,47]. So only the mDP data may not be sufficient to explain the differences in EC_{50} and it is better to explain the differences in EC_{50} with reference to total phenolics and ECT in each fraction. Further investigations on the relationships between antioxidant activity and polyphenols in mangrove leaves and their mechanisms are required. Further investigations on the relationships between antioxidant activity and polyphenols in mangrove leaves and their mechanisms are required.

Mangrove tannins from *C. tagal* leaves were successfully characterized by ^{13}C-NMR, MALDI-TOF MS, HPLC-ESI MS and column chromatographic fractionation. They had substantial DPPH free radical scavenging ability and ferric reducing antioxidant power, which could be used as a new source of antioxidants. The major challenge in the research on condensed tannins is probably the difficulty in obtaining them in an individual molecular form. The complete purification of a procyanidin with a DP above five is almost impossible. Therefore, for studying their structures and properties, more or less mixtures polymerized are often employed [48]. More, the synergistic effects of active mixtures make plant extracts and fractions more interesting than the pure compounds for functional food applications [42]. In the present study, an effective fractionation method was established to elucidate more about the relationships between DP and antioxidant activities. The established relationships could be used as a theoretical method for predicting the structure-activity of proanthocyanidins.

Author Contributions

Conceived and designed the experiments: HCZ NFYT. Performed the experiments: HCZ WMC SDW. Analyzed the data: HCZ YML. Contributed reagents/materials/analysis tools: HCZ YML ZHD. Wrote the paper: HCZ NFYT.

References

1. Lin P (2006) Marine higher plant ecology. Science Press, Beijing: 76–80.
2. Rastogi RP, Mehrotra BN (1991) Compendium of Indian medicinal plants. Publications & Information Directorate, New Delhi 1.
3. Chen JD, Qiu Y, Yang ZW, Lin P, Lin YM (2008) Dimeric diterpenes from the roots of the mangrove plant *Ceriops tagal*. Helvetica Chimica Acta 91: 2292–2298.
4. Chacha M, Mapitse R, Afolayan AJ, Majinda RRT (2008) Antibacterial diterpenes from the roots of *Ceriops tagal*. Natural Product Communications 3: 17–20.
5. Feller IC, Whigham DF, O'Neill JP, McKee KL (1999) Effects of nutrient enrichment on within-stand cycling in a mangrove forest. Ecology 80: 2193–2205.
6. Mainoya J, Mesaki S, Banyikwa F (1986) The distribution and socio-economic aspects of mangrove forests in Tanzania. Man in the Mangroves: the Socio-Economic Situation of Human Settlements in Mangrove Forests: 87–95.
7. Kraus TEC, Dahlgren RA, Zasoski RJ (2003) Tannins in nutrient dynamics of forest ecosystems - a review. Plant and Soil 256: 41–66.
8. Lin YM, Liu JW, Xiang P, Lin P, Ye GF, et al. (2006) Tannin dynamics of propagules and leaves of *Kandelia candel* and *Bruguiera gymnorrhiza* in the Jiulong River Estuary, Fujian, China. Biogeochemistry 78: 343–359.
9. Zhou HC, Tam NYF, Lin YM, Wei SD, Li YY (2012) Changes of condensed tannins during decomposition of leaves of *Kandelia obovata* in a subtropical mangrove swamp in China. Soil Biology and Biochemistry 44: 113–121.
10. Zhang LL, Lin YM, Zhou HC, Wei SD, Chen JH (2010) Condensed tannins from mangrove species *Kandelia candel* and *Rhizophora mangle* and their antioxidant activity. Molecules 15: 420–431.
11. Maie N, Pisani O, Jaffe R (2008) Mangrove tannins in aquatic ecosystems: Their fate and possible influence on dissolved organic carbon and nitrogen cycling. Limnology and Oceanography 53: 160–171.
12. Oo CW, Pizzi A, Pasch H, Kassim MJ (2008) Study on the structure of mangrove polyflavonoid tannins with MALDI-TOF mass spectrometry. Journal of Applied Polymer Science 109: 963–967.
13. Hernes PJ, Benner R, Cowie GL, Goni MA, Bergamaschi BA, et al. (2001) Tannin diagenesis in mangrove leaves from a tropical estuary: A novel molecular approach. Geochimica Et Cosmochimica Acta 65: 3109–3122.
14. Wang Y, Zhu H, Tam NFY (2014) Polyphenols, tannins and antioxidant activities of eight true mangrove plant species in South China. Plant and Soil 374: 549–563.
15. Khanbabaee K, van Ree T (2001) Tannins: Classification and definition. Natural Product Reports 18: 641–649.
16. Alasalvar C, Karamac M, Kosinska A, Rybarczyk A, Shahidi F, et al. (2009) Antioxidant Activity of Hazelnut Skin Phenolics. Journal of Agricultural and Food Chemistry 57: 4645–4650.
17. Monagas M, Quintanilla-Lopez JE, Gomez-Cordoves C, Bartolome B, Lebron-Aguilar R (2010) MALDI-TOF MS analysis of plant proanthocyanidins. Journal of Pharmaceutical and Biomedical Analysis 51: 358–372.
18. Chai WM, Shi Y, Feng HL, Qiu L, Zhou HC, et al. (2012) NMR, HPLC-ESI-MS, and MALDI-TOF MS analysis of condensed tannins from *Delonix regia* (Bojer ex Hook.) Raf. and their bioactivities. Journal of Agricultural and Food Chemistry 60: 5013–5022.

19. Chen XX, Shi Y, Chai WM, Feng HL, Zhuang JX, et al. (2014) Condensed tannins from *Ficus virens* as tyrosinase inhibitors: structure, inhibitory activity and molecular mechanism. Plos One 9: e91809.
20. Gu LW, Kelm M, Hammerstone JF, Beecher G, Cunningham D, et al. (2002) Fractionation of polymeric procyanidins from lowbush blueberry and quantification of procyanidins in selected foods with an optimized normal-phase HPLC-MS fluorescent detection method. Journal of Agricultural and Food Chemistry 50: 4852–4860.
21. Li CM, Leverence R, Trombley JD, Xu SF, Yang J, et al. (2010) High molecular weight persimmon (*Diospyros kaki* L.) proanthocyanidin: a highly galloylated, A-linked tannin with an unusual flavonol terminal unit, myricetin. Journal of Agricultural and Food Chemistry 58: 9033–9042.
22. Ariga T, HAMANO M (1990) Radical scavenging action and its mode in procyanidins B-1 and B-3 from azuki beans to peroxyl radicals. Agricultural and Biological Chemistry 54: 2499–2504.
23. Hagerman AE, Riedl KM, Jones GA, Sovik KN, Ritchard NT, et al. (1998) High molecular weight plant polyphenolics (tannins) as biological antioxidants. Journal of Agricultural and Food Chemistry 46: 1887–1892.
24. Es-Safi NE, Guyot S, Ducrot PH (2006) NMR, ESI/MS, and MALDI-TOF/MS analysis of pear juice polymeric proanthocyanidins with potent free radical scavenging activity. Journal of Agricultural and Food Chemistry 54: 6969–6977.
25. Zhou HC, Lin YM, Li YY, Li M, Wei SD, et al. (2011) Antioxidant properties of polymeric proanthocyanidins from fruit stones and pericarps of Litchi chinensis Sonn. Food Research International 44: 613–620.
26. Makkar HPS, Blümmel M, Borowy NK, Becker K (1993) Gravimetric determination of tannins and their correlations with chemical and protein precipitation methods. Journal of Agricultural and Food Chemistry 61: 161–165.
27. Terrill TH, Rowan AM, Douglas GD, Barry TN (1992) Determination of extractable and bound condensed tannin concentrations in forage plants, protein concentrate meals and cereal grains. Journal of the Science of Food and Agriculture 58: 321–329.
28. Hellstrom J, Sinkkonen J, Karonen M, Mattila P (2007) Isolation and structure elucidation of procyanidin oligomers from saskatoon berries (*Amelanchier alnifolia*). Journal of Agricultural and Food Chemistry 55: 157–164.
29. Zhou HC, Lin YM, Chai WM, Wei SD, Liao MM (2011) Characterization of condensed tannins from litchi seed by reflectron modes and linear modes of MALDI-TOF MS. Acta Chimica Sinica 69: 2981–2986.
30. Brand-Williams W, Cuvelier ME, Berset C (1995) Use of a free radical method to evaluate antioxidant activity. Lebensmittel Wissenschaft and Technologie 28: 25–30.
31. Benzie IFF, Strain JJ (1996) The ferric reducing ability of plasma as a measure of "antioxidant power": the FRAP assay. Analytical Biochemistry 239: 70–76.
32. Zhou HC, Lin YM, Wei SD, Tam NFY (2011) Structural diversity and antioxidant activity of condensed tannins fractionated from mangosteen pericarp. Food Chemistry 129: 1710–1720.
33. Cai YZ, Sun M, Xing J, Luo Q, Corke H (2006) Structure-radical scavenging activity relationships of phenolic compounds from traditional Chinese medicinal plants. Life Sciences 78: 2872–2888.
34. Chirinos R, Rogez H, Campos D, Pedreschi R, Larondelle Y (2007) Optimization of extraction conditions of antioxidant phenolic compounds from mashua (*Tropaeolum tuberosum* Ruiz & Pavo) tubers. Separation and Purification Technology 55: 217–225.
35. Czochanska BZ, Foo YL, Newman RH, Porter LJ (1980) Polymeric proanthocyanidin. stereochemistry, structural units, and molecular weight. Journal of the Chemical Society, Perkin Transactions 1: 2278–2286.
36. Fu C, Loo AEK, Chia FPP, Huang D (2007) Oligomeric proanthocyanidins from mangosteen pericarps. Journal of Agricultural and Food Chemistry 55: 7689–7694.
37. Kraus TEC, Yu Z, Preston CM, Dahlgren RA, Zasoski RJ (2003) Linking chemical reactivity and protein precipitation to structural characteristics of foliar tannins. Journal of Chemical Ecology 29: 703–730.
38. Castillo-Munoz N, Gomez-Alonso S, Garcia-Romero E, Gomez MV, Velders AH, et al. (2009) Flavonol 3-O-glycosides series of *Vitis vinifera* cv. *Petit Verdot* red wine grapes. Journal of Agricultural and Food Chemistry 57: 209–219.
39. Reed JD, Krueger CG, Vestling MM (2005) MALDI-TOF mass spectrometry of oligomeric food polyphenols. Philadelphia, PA. Pergamon-Elsevier Science Ltd: 2248–2263.
40. Xiang P, Lin YM, Lin P, Xiang C, Yang ZW, et al. (2007) Effect of cationization reagents on the matrix-assisted laser desorption/ionization time-of-flight mass spectrum of Chinese gallotannins. Journal of Applied Polymer Science 105: 859–864.
41. Taylor AW, Barofsky E, Kennedy JA, Deinzer ML (2003) Hop (*Humulus lupulus* L.) proanthocyanidins characterized by mass spectrometry, acid catalysis, and gel permeation chromatography. Journal of Agricultural and Food Chemistry 51: 4101–4110.
42. Jerez M, Tourino S, Sineiro J, Torres JL, Nunez MJ (2007) Procyanidins from pine bark: Relationships between structure, composition and antiradical activity. Food Chemistry 104: 518–527.
43. Meri S, Kanner J, Akiri B, Hadas SP (1995) Determination and involvement of aqueous reducing compounds in oxidative defense systems of various senescing leaves. Journal of Agricultural and Food Chemistry 43: 1813–1815.
44. de Gaulejac N, Vivas N, de Freitas V, Bourgeois G (1999) The influence of various phenolic compounds on scavenging activity assessed by an enzymatic method. Journal of the Science of Food and Agriculture 79: 1081–1090.
45. Cai YZ, Luo Q, Sun M, Corke H (2004) Antioxidant activity and phenolic compounds of 112 traditional Chinese medicinal plants associated with anticancer. Life Sciences 74: 2157–2184.
46. Silva E, Souza J, Rogez H, Rees J, Larondelle Y (2007) Antioxidant activities and polyphenolic contents of fifteen selected plant species from the Amazonian region. Food Chemistry 101: 1012–1018.
47. Lizcano LJ, Bakkali F, Ruiz-Larrea MB, Ruiz-Sanz JI (2010) Antioxidant activity and polyphenol content of aqueous extracts from Colombian Amazonian plants with medicinal use. Food Chemistry 119: 1566–1570.
48. Guyot S, Marnet N, Drilleau JF (2001) Thiolysis-HPLC characterization of apple procyanidins covering a large range of polymerization states. Journal of Agricultural and Food Chemistry 49: 14–20.

10

Long-Term Fluctuations in Circalunar Beach Aggregations of the Box Jellyfish *Alatina moseri* in Hawaii, with Links to Environmental Variability

Luciano M. Chiaverano[1]*, Brenden S. Holland[2], Gerald L. Crow[3], Landy Blair[4], Angel A. Yanagihara[5]

1 Kewalo Marine Laboratory, Pacific Biosciences Research Center, University of Hawaii at Mānoa, Honolulu, Hawaii, United States of America, **2** Center for Conservation Research & Training, Pacific Biosciences Research Center, University of Hawaii at Mānoa, Honolulu, Hawaii, United States of America, **3** Waikiki Aquarium, University of Hawaii at Mānoa, Honolulu, Hawaii, United States of America, **4** Ocean Safety and Lifeguard Services, City and County of Honolulu, Honolulu, Hawaii, United States of America, **5** Bekesy Laboratory, Pacific Biosciences Research Center, and Department of Tropical Medicine, John A. Burns School of Medicine, University of Hawaii at Mānoa, Honolulu, Hawaii, United States of America

Abstract

The box jellyfish *Alatina moseri* forms monthly aggregations at Waikiki Beach 8–12 days after each full moon, posing a recurrent hazard to swimmers due to painful stings. We present an analysis of long-term (14 years: Jan 1998– Dec 2011) changes in box jellyfish abundance at Waikiki Beach. We tested the relationship of beach counts to climate and biogeochemical variables over time in the North Pacific Sub-tropical Gyre (NPSG). Generalized Additive Models (*GAM*), Change-Point Analysis (*CPA*), and General Regression Models (*GRM*) were used to characterize patterns in box jellyfish arrival at Waikiki Beach 8–12 days following 173 consecutive full moons. Variation in box jellyfish abundance lacked seasonality, but exhibited dramatic differences among months and among years, and followed an oscillating pattern with significant periods of increase (1998–2001; 2006–2011) and decrease (2001–2006). Of three climatic and 12 biogeochemical variables examined, box jellyfish showed a strong, positive relationship with primary production, >2 mm zooplankton biomass, and the North Pacific Gyre Oscillation (NPGO) index. It is clear that that the moon cycle plays a key role in synchronizing timing of the arrival of *Alatina moseri* medusae to shore. We propose that bottom-up processes, likely initiated by inter-annual regional climatic fluctuations influence primary production, secondary production, and ultimately regulate food availability, and are therefore important in controlling the inter-annual changes in box jellyfish abundance observed at Waikiki Beach.

Editor: Graeme Clive Hays, University of Wales Swansea, United Kingdom

Funding: Field components of this project were partially supported by grants from the National Institutes of Health (U54 NS039406, P20 RR016453, G12 RR003061, P20 RR016453, R21 DA024444) and the Victoria S and Bradley L Geist Foundation of Hawaii Community Foundation (HCF 958935, 991879, 20001741, 20011908, 20061497, 20071368 and 47031) (AY). BSH and LMC were supported by the Oahu Army Natural Resources Program during preparation of this manuscript. The funders had no role in study design, data collection and analysis, decision to publish, or preparation of the manuscript.

Competing Interests: The authors have declared that no competing interests exist.

* E-mail: lchiave@hawaii.edu

Introduction

Jellyfish (cubozoans, hydrozoans and scyphozoans) are conspicuous, ecologically important constituents of coastal and oceanic systems. In the last three decades, jellyfish have received growing attention due to fluctuation in abundance often resulting in population explosions (e.g., blooms) in marine ecosystems worldwide, and frequently interfering directly with human activities [1]. When abundant, jellyfish cause widespread problems by clogging fishing nets [2,3], causing fish mortality in aquaculture pens [4,5], clogging intake screens in power generation and desalination plants [6], and impact tourism by stinging swimmers [7].

Predicting changes in jellyfish aggregations over time has been a difficult task due to a number of factors, including difficulties associated with sampling [8], scarcity of historical records and long-term time series datasets [1,9], and the unusual characteristics of their life cycle [10,11]. Many hydrozoans and scyphozoans, and all cubozoans, have complex life cycles including a benthic sessile polyp phase and a planktonic medusa phase [12]. Under favorable conditions, polyps can asexually produce large numbers of new polyps and ephyrae (i.e., immature medusae) [13], which can quickly grow into adult medusae [14]. Under adverse conditions, polyps can form dormant cysts [15] and medusae can also stop somatic and reproductive growth [16,17] until favorable conditions return [18,19]. This remarkable plasticity makes jellyfish suited to highly variable environments, and can result in large temporal and spatial fluctuations in abundance at various time scales [20].

Although many jellyfish populations appear to respond to anthropogenic processes, such as coastal eutrophication, overfishing, translocation of species, benthic-trawling and increased substrate availability for polyp settlement [1,19,21,22,23,24], there is increasing evidence that jellyfish populations are affected by large-scale climate variation and regional environmental conditions associated with climate fluctuations [25,26,27]. In the Irish Sea, 68% of jellyfish abundance was explained by variation in the North Atlantic Oscillation, sea surface temperature, zooplankton biomass, and precipitation [28]. Within the northern California current the highest catches of medusae correlated with cool temperatures during spring-summer and low winter-summer runoff of the Columbia River, and negative anomalies of the Pacific Decadal Oscillation (PDO) [29].

Understanding the relationships among large-scale climate forces and associated regional environmental variables with fluctuations in jellyfish populations over time is crucial to explaining and predicting trends in jellyfish abundance on a global scale [9,30]. Long-term studies on climate fluctuations and jellyfish have increased in the last decade. However, most of these studies have taken place in highly productive, temperate, coastal environments, and only a few attempts have been made to assess trends in jellyfish abundances in oligotrophic, open-ocean gyres [31,32]. In addition, long-term studies of jellyfish abundance and climate exist for only a few species of scyphozoans, hydrozoans, and ctenophores [1], and no studies exist for cubozoan species. This is somewhat surprising as box jellyfish are among the most venomous animals in the world [33]. Quite often, their habitat overlaps with areas of human recreation resulting in dangerous encounters due to painful, even lethal stings [34,35], causing beach closures at various global localities [33].

We have assessed long-term trends in abundance of box jellyfish and investigated the role of environmental conditions in an oligotrophic environment. The North Pacific Sub-tropical Gyre (NPSG) is the earth's largest contiguous oligotrophic biome, extending from 15°N to 35°N and from 135°E to 135°W, encompassing the entire Hawaiian Archipelago [36]. *Alatina moseri* Gershwin, 2005 (previously *Carybdea alata* Reynaud, 1830) is a tropical box jellyfish (Cubozoa) with a complex metagenic life cycle involving a benthic polyp and pelagic medusae [37]. On the Hawaiian Island of Oahu, reproductive cubomedusae have been observed near shore for a 2–4 day period 8–12 days after each full moon (i.e., during the waning crescent phase) along Waikiki beach since 1994 [38], and sporadically along beaches of other leeward Oahu bays. Box jellyfish aggregations often cause mass stinging events and beach closures [39]. Occasional circalunar sightings of box jellyfish on the south shore of Oahu (Waikiki Beach and surrounding areas) were first noted the late 1980s [38], and the timing of this monthly phenomenon has remained consistent. However, questions regarding whether the number of box jellyfish arriving at the beach has changed over time, or if fluctuations are affected by environmental conditions, have not been previously addressed. We have conducted a 14-year daily beach census of *Alatina moseri* medusae occurrence in leeward Oahu focused on Waikiki in order to: 1) assess seasonal patterns in abundance, 2), determine whether abundance of box jellyfish arriving at the beach has changed over time, and 3) explore potential links in variance of box jellyfish abundance at the beach to fluctuations in environmental conditions within the NPSG. This approach will provide useful information regarding the ecology of the Cubozoa, insights into jellyfish responses to environmental conditions in open-ocean, subtropical gyres, and will ultimately enhance our ability to predict the extent of onshore influxes of venomous jellyfish species in areas where they impose a health hazard.

Methods

Box jellyfish abundance and bell height at Waikiki Beach

Box jellyfish beach surveys were carried out monthly at Waikiki Beach, Oahu, Hawaii (21° 16.2′ N, 157° 49.4′ W; Figure 1) by University of Hawaii, Ocean Safety and Lifeguard Services (OSLS) personnel, and volunteers from January 1998 through December 2011. Waikiki Beach, a ~3 km beach on the leeward end of Mamala Bay along the south shore of Oahu (Figure 1), was chosen due to the consistent majority prevalence of box jellyfish arrivals in this area, and its popularity as a tourist destination; it is also the most visited beach in the state of Hawaii. Waikiki Beach was patrolled and checked for the presence of box jellyfish on a daily basis. When *Alatina moseri* medusae were present on or near the beach (typically 8–12 days after each full moon), surveys were initiated. Box jellyfish arrive at Waikiki Beach exclusively during the early morning ebbing tide. Thus, surveys were typically performed from 0200 to 0700 in order to ensure that observers were present for the duration of the influx. Surveys consisted of walking and wading into the shallow water, along the designated portion of shoreline (see below), in one direction. When one end of the section was reached, the search proceeded in the opposite direction. This protocol was repeated until all jellyfish that arrived during a given influx were collected and counted. Individuals were recently beached or swimming in the shallow waters at the time of collection, and visualized with head-lamps and/or flashlights, then collected by hand. Since arrival of medusae to Waikiki Beach can occur on multiple consecutive nights, searches were carried out daily until no medusae were found. Abundance of box jellyfish at the beach was determined by standardizing: 1) *Area searched*: all box jellyfish included in this study were collected from a 400 m portion of Waikiki, up to 1.5 m deep (Figure 1), via 14 years of monthly surveys. The study area was selected because of regular arrival of box jellyfish each month (hereafter referred to as Waikiki Beach). The geographic distribution of box jellyfish arrival on Oahu is not uniform around the island, such that jellyfish counts from Waikiki Beach cannot be extrapolated to other Hawaiian beaches. During the study period, OSLS personnel monitored box jellyfish influxes in other beaches along leeward Oahu, such as Hanauma Bay, Pokai Bay, and Yokohama Bay (Figure 1), and only Mamala Bay (Waikiki Beach) showed consistent monthly influx events (AAY, unpubl. data). 2) *Number of participants*: All box jellyfish medusae were collected within the designated area by 2–4 people. Bell height was recorded for all individuals collected monthly during 2001. As *Alatina moseri* medusae arriving at the beach are relatively large adult individuals (mean bell height = 64 mm, *SD*: 9; Table S1 & Figure S1 in File S1), they can be easily spotted and collected in the relatively small study area. Adult specimens of *A. moseri* can be easily distinguished from adults of the other two cubozoan species occurring in Hawaii, *Carybdea sivickisi* [40] and *Carybdea arborifera* [41]. Although not all individuals counted during this 14 year study were measured, all were examined and confirmed to be adult *A. moseri*. This research did not require collection permits, since the box jellyfish *A. moseri* is neither an endangered nor a protected species.

Climate data

Climatic phenomena within the subtropical North Pacific were described using bi-monthly indices of the Multivariate ENSO Index (MEI), and monthly indices of the Pacific Decadal Oscillation (PDO) and the North Pacific Gyre Oscillation (NPGO), for 1998–2011. The MEI is used to evaluate the variability in ENSO forcing, and it is obtained as the first unrotated principal component of six observed variables (sea-level pressure, zonal and meridional components of surface winds, sea surface temperature, surface air temperature, and total cloud cover) confined to the tropical Pacific [42]. This index is available online from NOAA Climate Diagnostic Center at http://www.cdc.noaa.gov/people/klaus.wolter/MEI/table.html. MEI values refer to El Niño and La Niña conditions, respectively. The PDO forcing emerges as the first mode of sea surface temperature (SST) and sea surface height (SSH) anomalies in the North Pacific Ocean poleward of 20° N [43], and is strongly correlated with variability in atmospheric circulation around the Aleutian low pressure system. Positive and negative PDO values refer to warm and cool

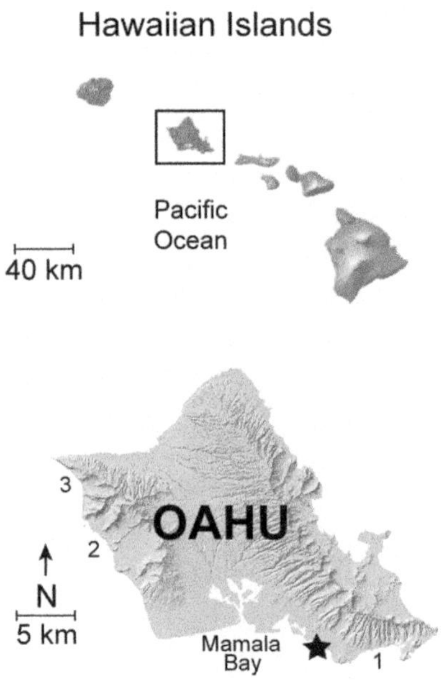

Figure 1. Map of the island of Oahu showing study area along Waikiki Beach. Box jellyfish, *Alatina moseri*, were collected monthly from 1998–2011 at Waikiki Beach, which is represented by the star symbol. Numbers indicate beach areas along leeward Oahu where *A. moseri* medusae arrive sporadically: 1) Hanauma Bay, 2) Pokai Bay, 3) Yokohama Bay.

phases, respectively. The PDO indexes were obtained from the Joint Institute for the Study of the Atmosphere and Ocean website at http://jisao.washington.edu/pdo/PDO.latest. The NPGO index represents the second principal component of sea surface height variability in the Northeast Pacific and, as such, is strongly correlated with fluctuations in the intensity of the geostrophic circulation in this region and its signal is strong south of 38° N [44]. The NPGO indices are available online from Dr. Emanuele Di Lorenzo's website at http://www.o3d.org/npgo/npgo.php. Positive and negative index values refer to more and less intense gyre circulations, respectively.

Biogeochemical data

Thirteen biogeochemical variables collected at station ALOHA (A Long-term Oligotrophic Habitat Assessment), 22°45′ N, 158°W, were downloaded from the HOT Data Organization and Graphical System (HOT-DOGS©) website at http://hahana.soest.hawaii.edu/hot/hot-dogs/interface.html. The location of station ALOHA was carefully chosen to be representative of biogeochemical processes characteristic of the North Pacific Subtropical Gyre (NPSG) [45], which encompasses all main Hawaiian Islands. Variables included: sea surface temperature (SST, °C), surface salinity (S, ppt), particulate nitrogen (PN mmol m^{-2}), and carbon (PC, mmol m^{-2}) concentrations, chlorophyll a concentration (CHa, mg m^{-2}), *Prochloroccocus* (PL, $\# \times 10^{11}$ m^{-2}), eukaryotes (EU, $\# \times 10^{11}$ m^{-2}) and heterotrophic bacteria (HB, $\# \times 10^{11}$ m^{-2}) abundance, primary production (PRI, mgC m^{-2} day^{-1}) and zooplankton biomass (ZO, g DW m^{-2}), which included "day" and "night" zooplankton biomass. Each group was separated into two range sizes: <2 mm and >2 mm (for zooplankton collection and processing methods, see [46]). We integrated values (100 m to the surface, except zooplankton biomass, which was integrated from 160 m to the surface [46]) for all variables except SST and S, for which we obtained surface (0–5 m) monthly means. Variables downloaded spanned the study period from January 1998 to December 2010. Information and description of protocols used for sampling, experimentation and analysis are available from the HOT-DOGS© website at http://hahana.soest.hawaii.edu/hot/hot-dogs/interface.html.

Local weather data

Weather conditions for the Waikiki Beach area were examined using seven local weather parameters, acquired from the National Climatic Data Center (NCDC) at http://www.ncdc.noaa.gov/ including: daily air temperature (°C), rainfall (mm), cloud cover (%), humidity (%), atmospheric pressure (mm Hg), wind speed (km/hr), and wind direction (degrees). Daily values for the period from January 1998 to December 2011 were used to calculate monthly and annual means. For monthly and annual wind direction, we used the method described by Gilhousen (1987) [47] to obtain daily "x" and "y" vector components of wind direction. Average x and y values were then calculated per month and per year, and the resultant wind direction (i.e., vector) was derived from arc-tangent (y/x). Air temperature and rainfall data were obtained from the weather station Waikiki 717.2 (USC00519397: 21° 16′ 17.8" N, 157° 48′ 57.6" W), the remaining parameters were recorded at the weather station at Honolulu International Airport (USW00022521: 21° 19′ 26 N, 157° 55′ 46" W).

Statistical analysis

Box jellyfish seasonality, medusa size, and long-term trends of abundance. Prior to analysis, beach counts were $\log_{10}$-transformed to achieve normal distribution (*Kolmogorov-Smirnov* test, $d = 0.74$, $p = 0.25$). In order to evaluate seasonality of box jellyfish abundance, monthly data from all years were combined into 12 months, and a one-way ANOVA was used to test for significant differences among months. An additional one-way ANOVA was applied to assess seasonality in the bell height of *Alatina moseri* medusae collected throughout 2001. For trend analysis, we used monthly data and mean annual abundances (log-transformed). Potential long-term trend changes in box jellyfish abundance (i.e., net increase or decrease) were assessed using *General Linear Regression Models* (*GRM*). *Generalized Additive Models* (*GAM*) were used to assess potential temporal trends (monthly and annual scales) in the dataset. The *GAM* is a nonlinear regression technique that uses non-parametric smoothing functions (i.e., cubic splines) to model the relationships among response and forcing variables [48,49,50]. An appealing feature of *GAMs* is that they do not assume a functional form for the fitted curve *a priori*. Instead, each predictor variable in the model is separated into sections delimited by "knots", and polynomial functions are fit to each section separately, revealing underlying trends in the observed data [48]. Once periods of decrease and increase in jellyfish abundance were determined by *GAM*, *GRM* models were used to test whether the linear slope deviates significantly from zero within each period. Trend analysis was carried out using the statistical software package Statistica© v7 with the associated distribution set to "normal" and the default smoothing function (i.e., cubic spline). Non-linear r – values were tested for significance at $\alpha = 0.05$.

To further evaluate whether significant shifts in monthly abundances (log-transformed) occurred within the periods determined by the *GAM*, we performed a change-point analysis (*CPA*) using the software Change-Point Analyzer © v2.3 [51]. The *CPA* utilizes a combination of time varying cumulative sum charts (*CUSUM*) and bootstrapping to detect significant shifts in the mean of time-ordered datasets, providing number of shifts, direction of each change, and the time of each shift with respective confident levels and confident intervals. This procedure tests the null hypothesis of "no change" across the dataset. Number of bootstrap samples was set to 2,000, without replacement. The confidence level for identifying candidate changes was set at 90%; however, a regime shift was considered to be significant only when confidence levels were higher than 95%, and confidence intervals around a significant change were set at 99%.

Climate indices, biogeochemical variables, and weather parameters versus box jellyfish abundance. Since non-stationary effects are the rule rather than the exception in ecology, box jellyfish abundance, climate, biogeochemical variables, and weather parameters were detrended (except wind direction) by linearly regressing each variable (log_{10}-transformed) against time (monthly scale) to remove temporal trends, and the residuals were saved for subsequent analyses [52]. No significant autocorrelations were detected in the dataset (*Chelton Method*; [53]). Mean annual residual values were then obtained. Potential associations among detrended variables (monthly and annual scales) were evaluated using *GAM* with a univariate approach, with jellyfish abundance as the response variable. When a relationship was detected as non-significant by the *GAM* (non-linear $p>0.05$), a *GRM* was used. The *GAM* and *GRM* analyses were performed using Statistica v7. In order to test for delayed response, or lag of jellyfish to environmental variables, we performed a *Cross-correlation* analysis considering ±1-year (annual scale). All cross-correlation analyses were carried out in Minitab v13. In order to test the potential association between box jellyfish counts and wind direction (monthly and annual means), *Circular-Linear* correlation analyses were performed using the circular statistics software Oriana© v.4.

Results

Seasonality, medusa size, and long-term trends of box jellyfish abundance at Waikiki Beach

Abundance of *Alatina moseri* medusae at Waikiki Beach did not vary significantly among months, indicating no seasonality ($F_{11, 156}=0.31, p=0.98$). During this study, *A. moseri* medusae arrived at Waikiki Beach 8–12 days after each full moon (i.e., during waning crescent phase of the moon). Arrivals of box jellyfish to the beach happened on 1–4 consecutive nights, with the exception of February 2002 when medusae arrived on 9 consecutive nights. Circalunar aggregations frequently consisted of 100–1000 box jellyfish (65% of the counts), and arrivals of less than 100 individuals, or more than 1,000, occurred 28% and 7% of the time, respectively. A total of 66,605 medusae were counted between January 1998 and December 2011, with an average of 396.5 individuals (range: 5–2,365; median: 271.5) arriving 8–12 days after each full moon and 4,757 per year (range: 2,155–8,696; median: 4,773) (Table S2 in File S1).

The bell height of individuals ($n=4228$) measured during beach surveys in 2001 varied significantly among months ($F_{1, 11}=70$, $p<0.0001$); however differences in mean bell height throughout the year did not exceed 1.2 cm. The significant differences detected by the ANOVA were due to the consistently small variation in medusae size observed within months (Table S1 & Figure S1 in File S1).

The *GRM* did not reveal a long-term trend in box jellyfish abundance at the beach (i.e., a net increase or decline over time); the linear slope of box jellyfish abundance (log-transformed) did not deviate significantly from zero during the period of study (1998–2011: Slope: 0.0012±0.002; $p=0.51$; $n=168$). However, the *GAM* indicated that box jellyfish abundance co-varied with time in a cyclical, non-linear fashion, explaining 12% of the variation in monthly abundance (non-linear $p=0.003$), and 53% of the variation in mean annual abundance (non-linear $p=0.01$) during the 14-year study, respectively (Figure 2A & 2B). Thus, based on periods determined by the *GAM*, box jellyfish aggregations: 1) significantly increased from 1998 to 2001 (*GRM*, $r=0.46$, $r^2=0.26$; $p<0.01$), 2) significantly decreased from 2001 to 2006 (*GRM*, $r=-0.45$, $r^2=-0.21$, $p<0.001$), and 3) then significantly increased until the end of data collection in 2011 (*GRM*, $r=0.29$, $r^2=0.18$; $p<0.001$). Thus, the 14-year dataset was characterized by four intermittent periods of increase and decrease in box jellyfish abundance, each period lasting approximately four years (Figure 2A).

Change-Point Analysis (*CPA*) detected three significant shifts in average abundance throughout the 14-year period, each one within periods of increase or decrease previously detected by the *GAM*. The first shift was detected in 2000 (October ±6 months, 100% confidence level), with average abundance increasing from 273 to 534. The second shift was detected in 2004 (August ±6 months, 100% confidence level), with average abundance decreasing from 534 to 227, and the third one was detected in

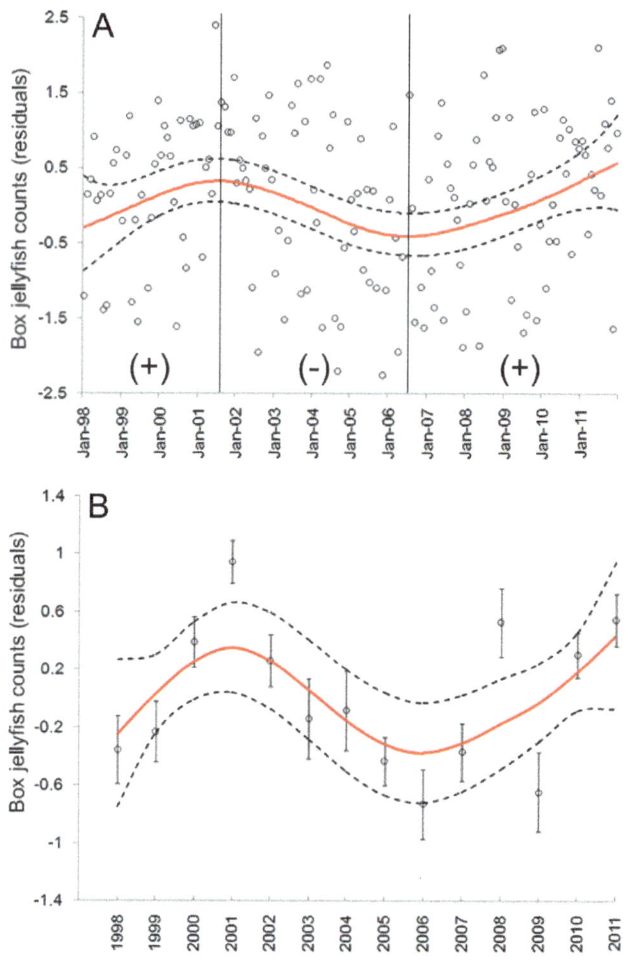

Figure 2. Long-term changes in abundance of box jellyfish at Waikiki Beach. Trends of box jellyfish abundance (solid line) over time was determined by the Generalized Additive Model (*GAM*) using (A) monthly and (B) annual mean counts. Data were log-transformed and detrended before analysis. Dashed lines indicate 95% confidence limits. Vertical solid lines indicate periods of significant (*Linear regression*, $p<0.05$) increase (+) or decrease (−) in box jellyfish counts. Bars show standard errors.

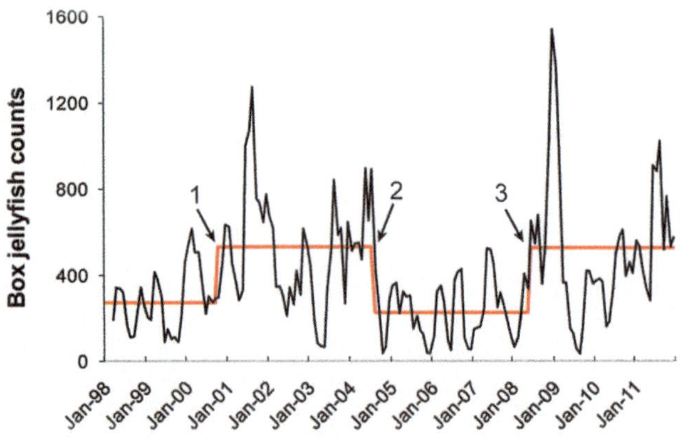

Figure 3. Monthly time-series of box jellyfish counts at Waikiki Beach. A three-month moving average smoothing approach was used to enhance visualization of the trend, depicted by the solid back line. The solid red line indicates mean box jellyfish beach counts during periods of no significant change determined by the Change Point Analysis (CPA). Three significant regime shifts in jellyfish numbers occurred during the 14-year period of study, corresponding to the periods previously identified by the *GAM* analysis: 1) 2000, 2) 2004, and 3) 2008. Dataset had no zeroes.

2008 (June ±8 months, 98% confidence level), with increased average beach abundance from 227 to 526 (Figure 3).

Relationships between box jellyfish counts and climate indices, biogeochemical variables, and weather parameters

Of a total of three climate indices, 13 biogeochemical variables, and seven weather parameters analyzed with a univariate approach using the *GAM* and *GRM*, one climatic and two biogeochemical variables showed a significant relationship with beach counts (Figure 4). No significant relationships were detected between beach counts and any weather parameters examined at both monthly and annual scales (Table S3 in File S1). In addition, the *Circular-Linear* correlation analyses did not detect a significant relationship between jellyfish counts and wind direction at any time scale (Table S3 in File S1). On a monthly scale, the *GRM* indicated a weak positive, significant linear relationship between box jellyfish abundance and both the NPGO (*GRM*, $r^2 = 0.27$, $p<0.001$; Figure 4A) and >2 mm night zooplankton biomass (*GRM*, $r^2 = 0.19$, $p = 0.036$; Figure 4E). However, on an annual scale, the *GRM* revealed strong, positive linear relationships of beach counts with the NPGO (*GRM*: $F_{1,\ 12} = 13$, $r = 0.72$, $r^2 = 0.52$; $p = 0.003$; Figures 4B & 5A), primary production (*GRM*: $F_{1,\ 11} = 13.5$, $r = 0.74$, $r^2 = 0.54$; $p = 0.004$; Figures 4D & 5B), and >2 mm night zooplankton biomass (*GRM*: $F_{1,11} = r = 0.6$; $r^2 = 0.36$; $p = 0.03$; Figures 4F & 5C). Significant inter-relationships (on an annual scale) were detected among variables that correlated significantly with jellyfish abundance at Waikiki Beach. Primary production (as covariate) showed a strong positive linear relationship with >2 mm night zooplankton biomass (*GRM*, $r = 0.62$; $r^2 = 0.39$; $p = 0.01$; Figure 6A). Interestingly, the *GAM* detected a strong non-linear relationship (i.e., "u" shape) between the NPGO (covariate) and primary production (*GAM*, $r^2 = 0.75$, non-linear $p = 0.001$; Figure 6B). Further analysis indicated that the non-linear relationship between primary production and NPGO was caused by an unusual pairwise observation for 2005. Disregarding this outlier, there was a strong, positive linear relationship between NPGO and primary production (*GRM*: $r = 0.86$; $r^2 = 0.7$; $p = 0.0004$).

Discussion

This study represents the first long-term assessment of trends in abundance of cubozoan medusae, and the first to evaluate potential links between box jellyfish abundance and environmental conditions in a highly oligotrophic environment. Although *Alatina moseri* medusae arrived consistently at Waikiki Beach 8–12 days after full moon throughout the 14-year period, the number of medusae at the beach showed no seasonality and size (bell height) was relatively uniform throughout 2001 (Table S1 & Figure S1 in File S1). Given the lack of seasonality in the NPSG, where sea surface temperature (SST) varies only 3°C throughout the year (HOT data), this may not be surprising. In temperate regions, occurrence of medusae is usually seasonal as onset of strobilation is associated with seasonal variation in SST; however, in tropical and sub-tropical areas, jellyfish polyps typically exhibit continuous or semi-continuous strobilation thus perennial presence of medusae [54]. Continuous or semi-continuous production of medusae from polyps likely explains the lack of seasonality observed in the size of beach aggregations of box jellyfish at Waikiki. In Australia's Great Barrier Reef region, medusae of *Alatina moseri* (= *A. mordens* [41]) also occur year-round, and laboratory observations suggest that polyps are capable of continuous asexual reproduction, and metamorphosis into young medusae (ephyrae) is independent of water temperature variation [55]. In Puerto Rico (Caribbean Sea), polyps of *A. moseri* (= *Carybdea alata*) are capable of year-round asexual reproduction, and year-round occurrence of medusae of this species has been documented in this area [37]. In addition, polyps of scyphozoan jellyfish, such as *Aurelia* spp. [56] and *Mastigias* spp. [57] from tropical and sub-tropical regions, release ephyrae throughout the year in the absence of significant SST variation. Our data support earlier observations that medusae of *A. moseri* arrive to Waikiki Beach 8–12 days after each full moon [38]. Field observations indicate that box jellyfish arrive at Waikiki Beach with fully mature and active gonads, and medusae become senescent a few hours post-reproduction (sperm are broadcast, strands of fertilized eggs are expelled; AAY, unpubl. data), suggesting medusae of this species likely reproduce once in their lifetime (~1 year, [37]). Interestingly, polyps of *A. moseri* go through a metamorphosis (at ~75 days of age, raised at 26–29°C in the laboratory) in which the whole polyp becomes a single young medusa, and no remnant is left behind [37]. Therefore, continuous or semi-continuous metamorphosis from polyps to medusae may be required to maintain the adult medusae cohort over time, while monthly sexual reproductive events may be necessary to sustain an ephemeral polyp population.

Box jellyfish abundance at Waikiki Beach showed neither net increase nor net decrease throughout the 14-year period of study; instead, it showed dramatic monthly and annual variability in a cyclical, non-linear pattern with intermittent periods (pulses) of increase (positive phase) and decrease (negative phase) (Figures 2 & 3). These results are consistent with recent long-term studies suggesting that jellyfish populations, on a global scale, oscillate over different time scales with no apparent long-term trend [20,30]. The arrival of box jellyfish medusae at Waikiki Beach was recurrent with consistent, predictable timing, however, monthly aggregation size varied substantially. Thus the model (*GAM*) showed a poor fit (12%) with monthly beach counts from Jan 1998 to Dec 2011. Month to month comparison of jellyfish beach counts are notorious for their high temporal variability [58,59], and a number of physical factors, including waves, currents,

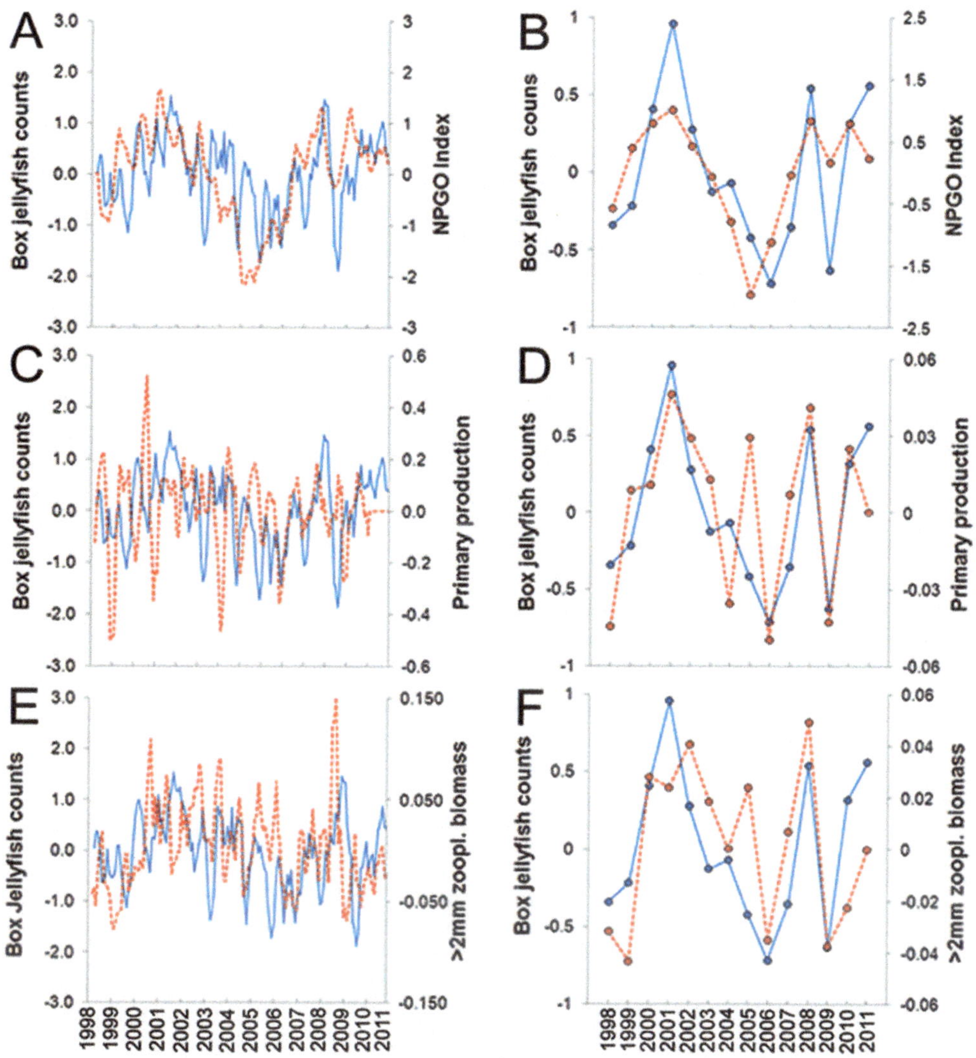

Figure 4. Time-series of climatic and biogeochemical variables with box jellyfish beach counts at Waikiki. Log-transformed, detrended variables are shown, Y-axes are residual values. Graphs A, C, and E represent monthly time-series (three-month average smoothing), while graphs B, D, and F represent annual means. Solid blue lines represent box jellyfish numbers, dashed red lines represent climate and biogeochemical variables, including NPGO (A, B), primary production (C, D), and zooplankton biomass (E, F). Note different scale in the Y-axis of variables between monthly and annual analyses. Error bars were not included in order to enhance visualization. Biogeochemical variables were obtained at Station ALOHA, from the Hawaiian Oceanographic Time-series program (HOT) at http://hahana.soest.hawaii.edu/hot/.

topography, individual behavior and life history, could all play a role in the monthly variation in numbers of medusae at the beach [58,59,60]. Predictability of our model, however, increased (53%) when applied to annual mean beach counts, which are typically considered a more reliable proxy for population abundance for beach counts [61,62].

Our understanding of inter-annual changes in abundance of box jellyfish at Waikiki Beach was enhanced by comparison with long-term local and regional environmental heterogeneity: counts at Waikiki Beach did not correlate significantly with any of the local weather parameters studied (Table S3 in File S1), whereas years of high jellyfish beach counts coincided with years of elevated NPGO (North Pacific Gyre Oscillation) index, primary production, and macro-zooplankton abundance. Thus, fluctuations in abundance of *Alatina moseri* medusae at Waikiki Beach appear to follow decadal climatic oscillations occurring within the sub-tropical North Pacific, independent of local weather conditions (wind speed, wind direction, air temperature, rainfall, cloud cover, humidity, and atmospheric pressure; Table S3 in File S1). Beach counts showed a strong, positive correlation with the North Pacific Gyre Oscillation (NPGO), while no significant relationship was detected with any other large-scale climate force, such as the Pacific decadal Oscillation (PDO) and the Multivariate ENSO Index (MEI). Unlike the PDO and MEI [42,43], the NPGO is associated with variation in wind-driven circulation including upwelling and horizontal advection at both regional and basin scales, and physical processes controlling salinity and nutrient concentrations [44], which in turn drive changes in phytoplankton levels, leading to variability at higher trophic levels. Thus the NPGO is an indicator of fluctuations in the mechanisms controlling planktonic ecosystem dynamics [44]. The NPGO is considered the primary climate index linked to decadal biological variability in the NPSG. Recent studies [63,64] using biogeochemical ocean circulation models, showed NPGO to be the only climate index significantly correlated with primary production at Station ALOHA, in agreement with results of this study. In fact,

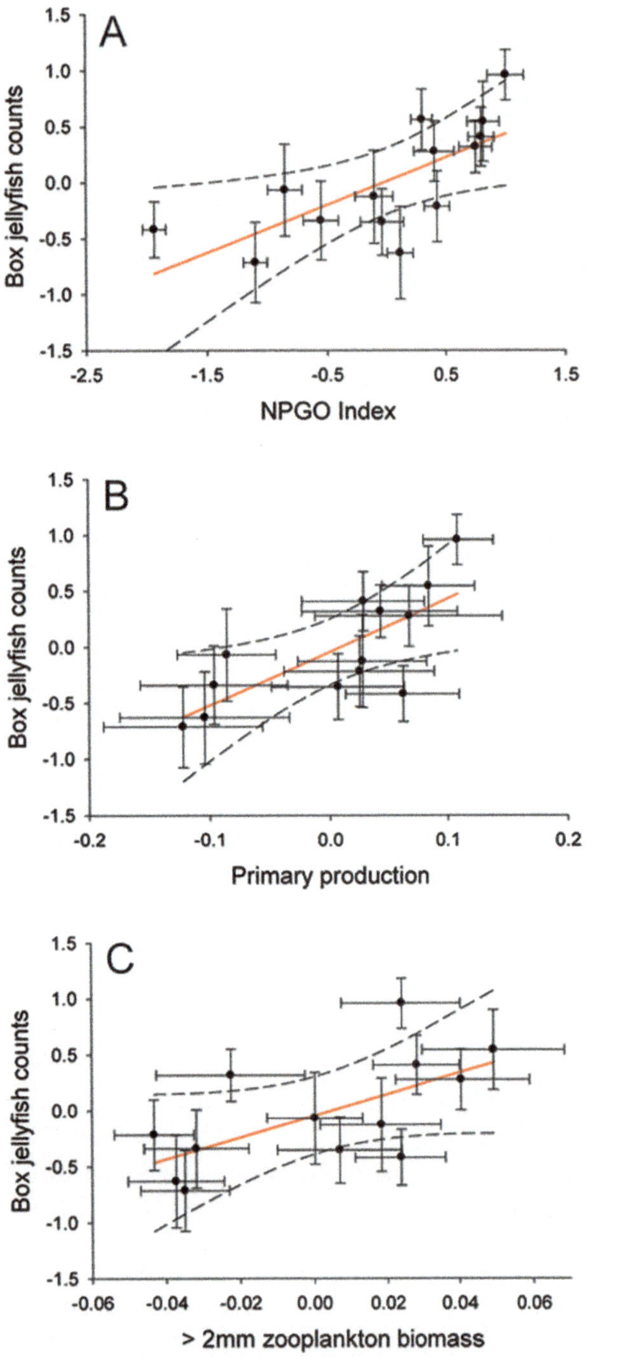

Figure 5. Significant relationships among box jellyfish abundance at Waikiki Beach and environmental variables. A: North Pacific Gyre Oscillation (NPGO), B: Primary production, C: zooplankton biomass within the sub-tropical North Pacific. Significant linear relationships were detected using General Regression Models (*GRM*). Biogeochemical variables were obtained at Station ALOHA, from the Hawaiian Oceanographic Time-series program (HOT) at http://hahana.soest.hawaii.edu/hot/. Solid red line represents the model fit, and dashed lines represent 99% confidence levels. Bars represent standard errors.

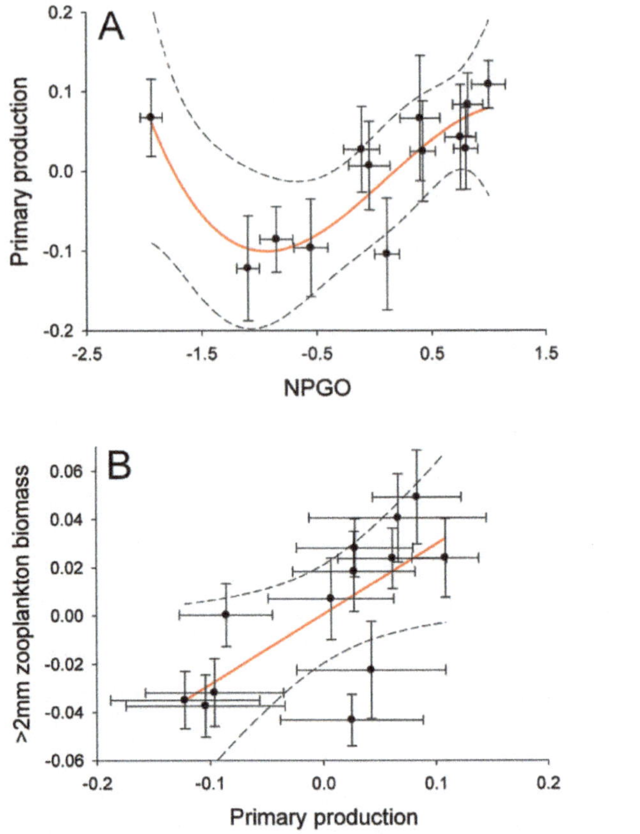

Figure 6. Significant relationships among environmental variables that correlated significantly with box jellyfish abundance at Waikiki Beach. A) Non-linear relationship between primary production and NPGO detected by using Generalized Additive Models (*GAM*). B) Linear relationship between zooplankton biomass and primary production detected by using General Regression Models (*GRM*). Biogeochemical variables were obtained at Station ALOHA, from the Hawaiian Oceanographic Time-series program (HOT) at http://hahana.soest.hawaii.edu/hot/. Solid red line represents the model fit, and dashed lines represent 99% confidence levels. Bars represent standard errors.

neither PDO nor MEI was significantly correlated with beach counts or with any biogeochemical variable that showed a significant relationship with box jellyfish abundance. Although the actual mechanism has not been determined, Dave and Lozier (2010) [64] suggest that the correlation between the NPGO and primary productivity in the NPSG may be driven by changes in diazotrophic activity (N_2-fixation). Approximately 80% of all nitrogen fixation at Station ALOHA occurs within 60 m of the surface, and evidence suggests that diazotrophs account for up to half of vertically integrated new production in the region [65,66,67]. An increase in N-fixation activity translates into introduction of new nitrogen in the system, which can in turn cause an increase in zooplankton abundance via direct trophic transfer [68,69].

The strong positive relationship between annual primary production and zooplankton biomass (secondary production) observed in this study may reflect trophic interactions previously observed at Station ALOHA [69], where copepods comprise the dominant meso-zooplankton taxa (~175 species) as well as the primary consumers. Such periods of high primary production can support elevated secondary productivity [69]. In this study, we detected a strong positive correlation between box jellyfish counts and zooplankton, suggesting that abundance of the box jellyfish *Alatina moseri* at Waikiki Beach could be influenced by regional prey availability. Cubomedusae are predators that feed on a variety meso- and macro-zooplankton taxa [70,71,72,73]. Although detailed prey preference data for *A. moseri* is not currently available, preliminary molecular analysis of gut contents has confirmed the presence of oceanic meso- and macro-zooplankton taxa (Holland et al., in prep). Jellyfish are known to increase growth rates and reach larger adult sizes in response to increased food availability [74], and since body size positively correlates with fecundity in jellyfish, more eggs (and more larvae) are produced when food is readily available [75,76], potentially increasing larval recruitment [54]. At the same time, increased food availability in the water column can also affect the amount of food reaching the benthos [77,78,79], and jellyfish polyps are known to increase budding rates (i.e., asexual production of new polyps) in response to an increase in food supply [80,81,82,83], potentially translating into higher numbers of medusae released via strobilation during periods with high zooplankton biomass compared to years of low zooplankton biomass. In polyps of the box jellyfish *A. moseri* (= *A. mordens*, [41]) increased feeding frequency resulted in increased budding rates [55], with similar results obtained for a related species [84]. As previously mentioned, abundance of *A. moseri* medusae is not seasonal in Hawaii (US, Waikiki Beach), the Caribbean (Puerto Rico) [37], and the Great Barrier Reef (Australia) [55], probably due to continuous production of medusae via polyp metamorphosis, which may account for the lack of a significant lag between jellyfish abundance of jellies arriving at the beach and zooplankton biomass observed in this study.

In spite of the distance between collection localities for jellyfish (Waikiki) and biogeochemical variables (Station ALOHA: 22.45°N, 158°W, ~200 km north of Oahu), correlations are of value for a number of reasons. Station ALOHA was selected because it is representative of oceanographic processes occurring throughout the NPSG [45]. In fact, *Alatina moseri* is one of the few pelagic cubozoan medusae [41], and is seen near shore for only 2–4 days, 8–12 days after each full moon. During non-aggregation periods, medusae have been collected from near the surface offshore, in waters several hundred meters deep [85], and on the day of arrival at the beach, *A. moseri* medusae have been observed ~2 km south of Waikiki Beach actively swimming straight towards the shore against or across currents (AAY, unpubl. data), evidencing the swimming capabilities of large cubozoans [86]. In addition, the Hawaiian Islands were formed via volcanic hotspot located near the center of the NPSG, where oceanic conditions occur close to shore, and depths >400 m are reached within 5 km of the coastline. Thus, regional offshore climatic and biogeochemical processes recorded at Station ALOHA likely have relevance to inter-annual fluctuations in numbers of cubozoan medusae arriving at Waikiki Beach. Given the fact that only mature *A. moseri* medusae arrive at Waikiki Beach from offshore 8–12 days after each full moon, annual fluctuation in beach counts could reflect inter-annual changes in adult off-shore population size. However, the possibility that beach counts do not reflect offshore abundance cannot be ruled out. In the Celtic Sea, for example, medusae of *Chrysaora hysoscella* were shown to be abundant offshore but were rarely observed at the shoreline [58]. However, our observations of *A. moseri* actively swimming towards the beach during the ebbing tide, against as well as across surface currents and tidal flow, contrasts with passive stranding events observed for instance, in scyphozoan medusae [58,59]. Lunar cycles have previously been shown to affect swimming behavior of jellyfish [33] and reproductive cycles of marine Cnidaria [87]. We suggest that the moon cycle may play a role in synchrony of gonad development and timing of *Alatina moseri* arrival at the shore, while environmental variables, such as regional food availability may play a role in fluctuations in Waikiki beach counts.

Long-term, optimally multi-decadal studies are necessary to understand complex ecological factors driving population fluxes in pelagic marine taxa. Box jellyfish aggregations are a major, and poorly understood public safety hazard in populated coastal areas, and present persistent challenges to resource managers and lifeguard services. Only through an enhanced understanding of the relevant ecological drivers will we be able to generate predictive models to minimize public health impacts. Perhaps the most important finding here is that annual Waikiki beach counts appear to be influenced by fluctuation in environmental conditions within the NPSG, including climate (NPGO) and certain biological variables, independent of local weather conditions. These findings suggest that food availability within the NPSG may be an important predictor of inter-annual fluctuations in abundance of *A. moseri* medusae during beach aggregations in Hawaii, likely via bottom-up control. Additional studies will be required to test whether observed fluctuations in beach counts correspond to abundance offshore for Waikiki box jellyfish. Our work will continue to focus on the oscillating aggregation patterns revealed in this study, and on elucidation of the environmental mechanisms governing fluctuations of this important and poorly understood species.

Supporting Information

File S1 Monthly average bell height of Alatina moseri medusae collected at Waikiki Beach throughout 2001 (Table S1), bell height frequency distribution of Alatina moseri medusae collected at Waikiki Beach throughout 2001 (Figure S1), monthly counts (raw data) of Alatina moseri medusae recorded at Waikiki Beach from January 1998 to December 2011 (Table S2), and non-significant correlations between numbers of Alatina moseri medusae at Waikiki Beach and climate indices, biogeochemical variables, and weather parameters (Table S3).

Acknowledgments

This paper is dedicated to the memory of Charles Edmondson, renowned Hawaii invertebrate specialist. We recognize contributions of Kim Bridges and Mahealani Kaneshiro-Pineiro to an early draft of this manuscript, and M. Kaneshiro-Pineiro's presentation at the 3rd International Jellyfish Blooms Symposium in 2010. We are grateful to John Chung, Karen Quinn, Darin Hayakawa, Louise Oliver, Kikiana Hurwitz, Amanda Lee, Cole Nishikawa, Paige Baum, Brian Hew, Karen Tatum and Teresa Carrette for field assistance. The support of Rob Miller, Ralph Goto and OSLS is greatly appreciated. We gratefully acknowledge the valuable and ongoing work of the Hawaiian Ocean Time-series (HOT) program. Thanks also to Lu Eldredge, Daniela del Valle, Shinji Sugiura, and Kelly Robinson for helpful discussions, and Jennifer Crites, Robert Kinzie III, Zachary Williams, Jamie Seymour and Andrew Rossiter for reviewing drafts of this manuscript.

Author Contributions

Conceived and designed the experiments: AAY GLC. Performed the experiments: GLC AAY LB. Analyzed the data: LMC. Wrote the paper: LMC BSH.

References

1. Purcell JE (2012) Jellyfish and ctenophore blooms coincide with human proliferations and environmental perturbations. Annual Review of Marine Science 4: 209–235.
2. Nagata RM, Haddad MA, Nogueira M Jr. (2009) The nuisance of medusae (Cnidaria, Medusozoa) to shrimp trawls in central part of southern Brazilian Bight, from the perspective of artisanal fishermen. Pan-American Journal of Aquatic Science 4: 312–325.
3. Dong Z, Liu D, Keesing JK (2010) Jellyfish blooms in China: dominant species, causes and consequences. Marine Pollution Bulletin 60: 954–963.
4. Doyle TK, De Haas H, Cotton D, Dorschel B, Cummins V, et al (2008) Widespread occurrence of the jellyfish *Pelagia noctiluca* in Irish coastal and shelf waters. Journal of Plankton Research 30: 963–968.
5. Delannoy CMJ, Houghton JDR, Fleming NEC, Ferguson HW (2011) Mauve stingers (*Pelagia noctiluca*) as carriers of the bacterial fish pathogen *Tenacibaculum maritimum*. Aquaculture 311: 255–257.
6. Daryanabard R, Dawson MN (2008) Jellyfish blooms: *Crambionella orsini* (Scyphozoa: Rhizostomeae) in the Gulf of Oman, Iran, 2002–2003. Journal of the Marine Biological Association of United Kingdom 88: 477–483.
7. Fenner PJ, Lippmann J, Gershwin L-A (2010) Fatal and nonfatal severe jellyfish stings in Thai waters. Journal of Travel Medicine 17: 133–138.
8. Pierce J (2009) Prediction, location, collection and transport of jellyfish (Cnidaria) and their polyps. Zoo Biology 28: 163–176.
9. Brotz L, Cheung WWL, Kleisner K, Pakhomov E, Pauly D (2012) Increasing jellyfish populations: trends in large marine ecosystems. Hydrobiologia 690: 3–20.
10. Dawson MN, Hamner WM (2009) A character-based analysis of the evolution of jellyfish blooms: adaptation and exaptation. Hydrobiologia 616: 193–215.
11. Hamner WM Dawson MN (2009) A review and synthesis on the systematics and evolution of jellyfish blooms: advantageous aggregations and adaptive assemblages. Hydrobiologia 616: 161–191.
12. Collins AG (2002) Phylogeny of Medusozoa and the evolution of cnidarians life cycles. Journal of Evolutionary Biology 15: 418–432.
13. Arai MN (1997) A Functional Biology of Scyphozoa. Chapman and Hall, London: 300.
14. Palomares MLD, Pauly D (2009) The growth of jellyfishes. Hydrobiologia 616: 11–21.
15. Arai MN (2009) The potential importance of podocysts to the formation of scyphozoan blooms: a review. Hydrobiologia. 616: 241–246.
16. Hamner WM, Jensen RM (1974) Growth, degrowth, and irreversible cell differentiation in *Aurelia aurita*. America Zoologist 14: 833–849.
17. Chiaverano LM, Mianzan HW, Ramirez F (2004) Gonad development and somatic growth patterns of *Olindias sambaquiensis* (Limnomedusae, Olindiidae). Hydrobiologia 530–31: 373: 381.
18. Boero F, Bouillon J, Gravili C, Miglietta MP, Parsons T, et al (2008) Gelatinous plankton: irregularities rule the world (sometimes) Marine Ecolgy Progress Series 356: 299–310.
19. Richardson AJ, Bakun A, Hays GC, Gibbons MJ (2009) The jellyfish joyride: causes, consequences and management responses to a more gelatinous future. Trends in Ecolgy and Evolution 24: 312–322.
20. Condon RB, Graham WM, Duarte CM, Pitt KA, Lucas CH, et al (2012) Questioning the rise of gelatinous zooplankton in the world's oceans. Bioscience 62: 160–169.
21. Mills C (2001) Jellyfish blooms: are populations increasing globally in response to changing ocean conditions? Hydrobiologia 451: 55–68.
22. Graham WM, Martin DL, Felder DL, Asper VL, Perry HM (2003) Ecological and economic implications of a tropical jellyfish invader in the Gulf of Mexico. Biological Invasions 5: 53–69.
23. Purcell JE, Uye S, Lo W (2007) Anthropogenic causes of jellyfish blooms and their direct consequences for humans: a review. Marine Ecology Progress Series 350: 153–174.
24. Pauly D, Alder J, Booth S, Cheung WWL, Christensen V, et al (2009) Fisheries in Large Marine Ecosystems: Descriptions and Diagnoses, 23–40 in *The UNEP Large Marine Ecosystem Report: a Perspective on Changing Conditions in LMEs of the World's Regional Seas*, Sherman K and Hempel G (ed). UN Environment Programme Regional Seas Report and Studies No. 182, Nairobi, Kenya.
25. Molinero JC, Buecher E, Lučić D, Malej A, Miloslavić M (2008) Climate and Mediterranean jellyfish: assessing the effect of temperature regimes on jellyfish outbreak dynamics. Annales Series Historia Naturalis 19: 11–18.
26. Lynam CP, Attrill MJ, Skogen MD (2010) Climatic and oceanic influences on the abundance of gelatinous zooplankton in the North Sea. Journal of the Marine Biological Association of United Kingdom 90: 1153–1159.
27. Robinson KL, Graham WM (2013) Long-term change in the abundances of northern Gulf of Mexico scyphomedusae *Chrysaora* sp. and *Aurelia* spp. with links to climate variability. Limnology and Oceanography 58: 235–253.
28. Lynam CP, Lilley MKS, Bastian T, Doyle TK, Beggs SE, et al (2011) Have jellyfish in the Irish Sea benefited from climate change and overfishing? Global Change Biology 17: 767–82.
29. Suchman CL, Brodeur RD, Emmett RL, Daly EA (2012) Large medusae in surface waters of the Northern California Current: variability in relation to environmental conditions. Hydrobiologia 690: 113–125.
30. Condon RH, Duarte CM, Pitt KA, Robinson KL, Lucas CH, et al (2013) Recurrent jellyfish blooms are a consequence of global oscillations. Proceedings of the National Academy of Science 110: 1000–1005.
31. Ferraris M, Berline L, Lombard F, Guidi L, Elineau A, et al (2012) Distribution of *Pelagia noctiluca* (Cnidaria, Scyphozoa) in the Ligurian Sea (NW Mediterranean Sea). Journal of Plankton Research 34: 874–885.
32. Lilley MKS, Beggs SE, Doyle TK, Hobson VJ, Stromberg KHP, et al (2011) Global patterns of epipelagic gelatinous zooplankton biomass. Marine Biology 158: 2429–2436.
33. Fenner PJ, Williamson JA (1996) Worldwide deaths and severe envenomation from jellyfish stings. Medical Journal of Australia 165: 658–661.
34. Fenner PJ, Hadok JC (2002) Fatal envenomation by jellyfish causing Irukandji syndrome. Medical Journal of Australia 177: 362–363.
35. Huynh TT, Pereira P, Mulcahy R, Cullen P, Seymour J, et al (2003) Severity of Irukandji syndrome and nematocyst identification from skin scrapings Medical Journal of Australia 178: 38–41.
36. Karl DM (1999) A sea of change: Biogeochemical variability in the North Pacific Subtropical Gyre. Ecosystems 2: 181–214.
37. Arneson AC, Cutress CE (1976) Life history of *Carybdea alata* Reynaud, 1830 (Cubomedusae). In: Mackie GO (ed) Coelentrate Ecology and Behavior. Plenum Press, New York.
38. Thomas CS, Scott SA, Galanis DJ, Goto RS (2001) Box jellyfish (*Carybdea alata*) in Waikiki: Their influx cycle plus the analgesic effect of hot and cold packs on their stings to swimmers at the beach: A randomized, placebo-controlled, clinical trial. Hawaii Medical Journal 60: 100–107.
39. Yoshimoto CM, Yanagihara AA (2002) Cnidarian (coelenterate) envenomations in Hawaii improve following heat application. Transactions of the Royal Society of Tropical Medicine and Hygiene 96: 300–303.
40. Matsumoto GI, Crow GL, Cornelius PFS, Carlson BA (2002) Discovery of the Cubomedusa *Carybdea sivickisi* (Cubozoa:Carybdeidae) in the Hawaiian Islands. Bishop Museum Occassional Papers 69: 44–46.
41. Bentlage B, Cartwright P, Yanagihara AA, Lewis C, Richards GS, et al (2010) Evolution of box jellyfish (Cnidaria: Cubozoa), a group of highly toxic invertebrates. Proceedings of the Royal Society of London 277: 493–501.
42. Wolter K, Timlin MS (1998) Measuring the strength of ENSO: How does 1997/98 rank? Weather 53: 315–324.
43. Mantua N, Hare S, Zhang Y, Wallace J, Francis R (1997) A Pacific interdecadal climate oscillation with impacts on salmon production, Bulletin of the Americal Meteorolgy Society 78: 1069–1079.
44. Di Lorenzo E, Schneider N, Cobb KM, Chhak K, Franks PJS, et al (2008) North Pacific Gyre Oscillation links ocean climate and ecosystem change. Geophysical Research Letters 35 L08607, doi:10.1029/2007GL032838.
45. Karl D, Lukas R (1996) The Hawaii Ocean Time-series (HOT) program: Background, rationale and field implementation. Deep-Sea Research II 43: 129–156.
46. Sheridan C, Landry M (2004) A 9-year increasing trend in mesozooplankton biomass at the Hawaii Ocean Time-series Station ALOHA. ICES Journal of Marine Science 61: 57–463.
47. Gilhousen DB (1987) A field evaluation of NDBC moored buoy winds. Journal of Atmospheric and Oceanic Technology 4: 94–104.
48. Hastie TJ, Tibshirani RJ (1990) Generalized Additive Models, Chapman and Hall, New York.
49. Wood SN (2004) Stable and efficient multiple smoothing parameter estimation for generalized additive models. Journal of America Statistics Association 99: 637–686.
50. Wood SN (2006) Generalized additive models: an introduction with *R*: Chapman and Hall, London.
51. Taylor WA (2003) Change-Point Analyzer, Version 2.3, Taylor Enterprises, Inc., Libertyville, IL (www.variation.com*cpa*).
52. Molinero JC, Ibanez F, Nival P, Buecher E (2005) North Atlantic climate and northwestern Mediterranean plankton variability. Limnology and Oceanography 50: 1213–1220.
53. Pyper BJ, Peterman RM (1998) Comparison of methods to account for autocorrelation in correlation analyses of fish data. Canadian Journal of Fisheries and Aquatic Sciences 55: 2127–2140.
54. Lucas CH (2001) Reproduction and life history strategies of the common jellyfish, *Aurelia aurita*, in relation to its ambient environment. Hydrobiologia 451: 229–246.
55. Courtney R (2012) Seasonality in the tropical *Alatina mordens*. Master thesis 74.
56. Schroth W, Jarms G, Streit B, Schierwater B (2002) Speciation and phylogeography in the cosmopolitan marine moon jelly, *Aurelia* sp. BMC Evolutionary Biology 2: 1–10.
57. Hamner WM, Gilmer RW, Hamner PP (1982) The physical, chemical, and biological characteristics of a stratified, saline, sulfide lake in Palau. Limnology and Oceanography 27: 896–909.
58. Houghton JDR, Doyle TK, Davenport J, Lilley MK, Wilson RP, et al (2007) Stranding events provide indirect insights into the seasonality and persistence of jellyfish medusae (Cnidaria: Scyphozoa). Hydrobiologia 589: 1–13.
59. Doyle TK, Houghton JD, Buckley SM, Hays GC, Davenport J (2007) The broad-scale distribution of five jellyfish species across a temperate coastal environment. Hydrobiologia 579: 29–39.

60. Graham WM, Pages F, Hamner WM (2001) A physical context for gelatinous zooplankton aggregations: a review. Hydrobiologia 451: 199–212.
61. Baker JD, Howell EA, Polovina JJ (2012) Relative influence of climate variability and direct anthropogenic impact on a Sub-tropical Pacific top predator, the Hawaiian monk seal. Marine Ecology Progress Series 469: 175–189.
62. Eberhardt LL, Garrott RA, Becker BL (1999) Using trend indices for endangered species. Marine Mammals Science 15: 766–785.
63. Saba VS, Friedrichs MAM, Carr M-E, Antoine D, Armstrong RA, (2010) The challenges of modeling depth-integrated marine primary productivity over multiple decades: A case study at BATS and HOT. Global Biogeochemical Cycles 24: GB3020.
64. Dave AC, Lozier SM (2010) Local stratification control of marine productivity in the subtropical North Pacific. Journal of Geophysical Research 115: C12032.
65. Karl DM, Letelier R, Tupas L, Dore J, Christian J, et al (1997) The role of nitrogen fixation in biogeochemical cycling in the subtropical North Pacific Ocean. Nature 388: 533–538.
66. Dore JE, Brum JR, Tupas LM, Karl DM (2002) Seasonal and interannual variability in sources of nitrogen supporting export in the oligotrophic subtropical North Pacific Ocean. Limnology and Oceanography 47: 1595–1607.
67. Dore JE, Letelier RM, Church MJ, Lukas R, Karl DM (2008) Summer phytoplankton blooms in the oligotrophic North Pacific Subtropical Gyre: Historical perspective and recent observations, Progress in Oceanography 76: 2–38.
68. Valiela I (1995) Marine Ecological Processes, 2nd edn. New York: Springer-Verlag, 686.
69. Hannides CCS, Popp BN, Landry MR, Graham BS (2009) Quantification of zooplankton trophic position in the North Pacific Subtropical Gyre using stable nitrogen isotopes. Limnology Oceanography 54: 50–61.
70. Stewart S (1996) Field behavior of *Tripedalia cystophora* (Class Cubozoa). Marine and Freshwater Behavior and Physiology 27: 175–188.
71. Carrette T, Alderslade P, Seymour J (2002) Nematocyst and prey in two Australian cubomedusans, *Chironex fleckeri* and *Chiropsalmus* sp. Toxicon 40: 1547–1551.
72. Underwood AH, Seymour JE (2007) Venom ontogeny, diet and morphology in *Carukia barnesi*, a species of Australian box jellyfish that causes Irukandji syndrome. Toxicon 49: 1073–1082.
73. Nogueira M, Haddad A (2008). The diet of cubomedusae (Cnidaria, Cubozoa) in Southern Brazil. Brazilian Journal of Oceanography 56: 157–164.
74. Ishii H, Båmstedt U (1998) Food regulation of growth and maturation in a natural population of *Aurelia aurita* (L.). Journal of Plankton Research 20: 805–816.
75. Lucas CH (1996) Population dynamics of *Aurelia aurita* (Scyphozoa) from an isolated brackish lake, with particular reference to sexual reproduction. Journal of Plankton Research 18: 987–1007.
76. Lucas CH, Lawes S (1998) Sexual reproduction of the scyphomedusa *Aurelia aurita* in relation to temperature and variable food supply. Marine Biology 131: 629–638.
77. Boero F, Belmonte G, Fanelli G, Piraino S, Rubino F (1996) The continuity of living matter and the discontinuities of its constituents: do plankton and benthos really exist? Trends in Ecology and Evolution 11: 177–180.
78. Kahn AS, Ruhl HE, Smith KL (2012) Temporal changes in deep-sea sponge populations are correlated to changes in surface climate and food supply. Deep Sea Research I: Oceanographic Research Papers 70: 36–41.
79. Perea-Blázquez A, Davy SK, Bell JJ (2012) Estimates of particulate organic carbon flowing from the pelagic environment to the benthos through sponge assemblages. PLoS ONE 7 e29569. doi:10.1371/journal.pone.0029569.
80. Ma X, Purcell JE (2005) Temperature, salinity, and prey effects on polyp versus medusa bud production by the invasive hydrozoan *Moerisia lyonsi*. Marine Biology 147: 225–234.
81. Hoover RA, Purcell JE (2009) Substrate preferences of scyphozoan *Aurelia labiata* polyps among common dock-building materials. Hydrobiologia 616: 259–67.
82. Han CH, Uye I (2010) Combined effects of food supply and temperature on asexual reproduction and somatic growth of polyps of the common jellyfish *Aurelia aurita*. Plankton and Benthos Research 5: 98–105.
83. Webster CN, Lucas CH (2012) The effects of food and temperature on settlement of *Aurelia aurita* planula larvae and subsequent somatic growth. Journal of Experimental Marine Biology and Ecology 436–437: 50–55.
84. Fischer AB, Hofmann DK (2004) Budding, bud morphogenesis, and regeneration in *Carybdea marsupialis* Linnaeus, 1758 (Cnidaria: Cubozoa). Hydrobiologia 530–531: 331–337.
85. Garm A, Bielecki J, Petie R, Nilsson D-E (2012). Opposite patterns of diurnal activity in the box jellyfish *Tripedalia cystophora* and *Copula sivickisi*. Biological Bulletin 222: 35–45.
86. Gordon MR, Seymour JE (2009) Quantifying movement of the tropical Australian cubozoan *Chironex fleckeri* using acoustic telemetry. In Jellyfish Blooms: Causes, Consequences, and Recent Advances, Springer Netherlands (87–97).
87. Babcock RC, Bull GD, Harrison PL, Heyward AJ, Oliver JK, et al (1986) Synchronous spawnings of 105 scleractinian coral species on the Great Barrier Reef. Marine Biology 90: 379–394.

11

Cell Turnover and Detritus Production in Marine Sponges from Tropical and Temperate Benthic Ecosystems

Brittany E. Alexander[1,2]*, Kevin Liebrand[1], Ronald Osinga[2], Harm G. van der Geest[1], Wim Admiraal[1], Jack P. M. Cleutjens[3], Bert Schutte[4], Fons Verheyen[5], Marta Ribes[6], Emiel van Loon[7], Jasper M. de Goeij[1,2]

1 Department of Aquatic Ecology and Ecotoxicology, Institute for Biodiversity and Ecosystem Dynamics, University of Amsterdam, Amsterdam, The Netherlands, **2** Porifarma B.V. Poelbos 3, Ede, The Netherlands, **3** Department of Pathology, Cardiovascular Research Institute Maastricht, Maastricht University, Maastricht, The Netherlands, **4** Department of Molecular Cell Biology, Research Institute Growth and Development, Maastricht University, Maastricht, The Netherlands, **5** Electron Microscopy Unit, CRISP, Maastricht, The Netherlands, **6** Institut de Ciències del Mar-Consejo Superior de Investigaciones Científicas (ICM-CSIC), Barcelona, Spain, **7** Department of Computational Geo-Ecology, Institute for Biodiversity and Ecosystem Dynamics, University of Amsterdam, Amsterdam, The Netherlands

Abstract

This study describes *in vivo* cell turnover (the balance between cell proliferation and cell loss) in eight marine sponge species from tropical coral reef, mangrove and temperate Mediterranean reef ecosystems. Cell proliferation was determined through the incorporation of 5-bromo-2′-deoxyuridine (BrdU) and measuring the percentage of BrdU-positive cells after 6 h of continuous labeling (10 h for *Chondrosia reniformis*). Apoptosis was identified using an antibody against active caspase-3. Cell loss through shedding was studied quantitatively by collecting and weighing sponge-expelled detritus and qualitatively by light microscopy of sponge tissue and detritus. All species investigated displayed substantial cell proliferation, predominantly in the choanoderm, but also in the mesohyl. The majority of coral reef species (five) showed between 16.1±15.9% and 19.0±2.0% choanocyte proliferation (mean±SD) after 6 h and the Mediterranean species, *C. reniformis*, showed 16.6±3.2% after 10 h BrdU-labeling. *Monanchora arbuscula* showed lower choanocyte proliferation (8.1±3.7%), whereas the mangrove species *Mycale microsigmatosa* showed relatively higher levels of choanocyte proliferation (70.5±6.6%). Choanocyte proliferation in *Haliclona vansoesti* was variable (2.8–73.1%). Apoptosis was negligible and not the primary mechanism of cell loss involved in cell turnover. All species investigated produced significant amounts of detritus (2.5–18% detritus bodyweight^{-1}·d^{-1}) and cell shedding was observed in seven out of eight species. The amount of shed cells observed in histological sections may be related to differences in residence time of detritus within canals. Detritus production could not be directly linked to cell shedding due to the degraded nature of expelled cellular debris. We have demonstrated that under steady-state conditions, cell turnover through cell proliferation and cell shedding are common processes to maintain tissue homeostasis in a variety of sponge species from different ecosystems. Cell turnover is hypothesized to be the main underlying mechanism producing sponge-derived detritus, a major trophic resource transferred through sponges in benthic ecosystems, such as coral reefs.

Editor: Nicole Webster, Australian Institute of Marine Science, Australia

Funding: Porifarma B.V. received funding from the European Union Seventh Framework Programme (FP7/2007–2013) under grant agreement no. KBBE-2010-266033 to undertake the research leading to these results. Funding was also received from The Innovational Research Incentives Scheme of the Netherlands Organization for Scientific Research (NWO-VENI; 863.10.009; personal grant to JMdG). Mediterranean sampling was partially funded by the grant CGL2010-18466 from the Spanish Government to MR. BEA, RO, and JMdG are affiliated with Porifarma B.V. Porifarma B.V. provided support in the form of salaries for authors BEA, RO, and JMdG, but no employees of Porifarma B.V. other than BEA, RO, and JMdG had any additional role in the study design, data collection and analysis, decision to publish, or preparation of the manuscript. The specific roles of these authors are articulated in the 'author contributions' section.

Competing Interests: BEA, RO, and JMdG are affiliated with Porifarma B.V. Porifarma B.V. received funding from the European Union Seventh Framework Programme (FP7/2007–2013) under grant agreement no KBBE-2010-266033 to undertake the research leading to these results. BEA is employed by Porifarma B.V. under grant agreement no KBBE-2010-266033 from the European Union Seventh Framework Programme (FP7/2007–2013). There are no patents, products in development or marketed products to declare.

* Email: B.E.Alexander@uva.nl or britt.e.alexander@gmail.com

Introduction

Sponges are key components of aquatic ecosystems. On coral reefs, a large proportion of the available suspended [1] and dissolved [2] organic energy and nutrients are retained by sponges and subsequently transferred to higher trophic levels through the so-called 'sponge loop' [3]. When including dissolved organic matter (DOM) in the energy budgets of sponges, the majority (81–95%) of the daily diet of investigated sponges on coral reefs consists of DOM [2], [4], [5]. The conversion of DOM into particulate organic matter (POM, also referred to as detritus) through rapid proliferation and shedding of sponge cells is proposed to be the main underlying mechanism involved in the transfer of DOM to higher trophic levels [3]. This proposition is based on the study of cell turnover in a single species of tropical coral reef sponge, *Halisarca caerulea*. The filter cells (choanocytes) of *H. caerulea* proliferate rapidly, with a cell cycle duration of only 5.4 h, one of the fastest described to date in any multi-cellular animal *in vivo* [6]. Under steady-state, i.e. non-growing conditions, rapid cell proliferation is balanced with massive amounts of cell shedding in order to maintain tissue homeostasis in the choanocyte compartment. The 'old' choanocytes are thought to be expelled as detritus into the water column [6]. Tissue turnover is evident in *H. caerulea* and in three additional coral reef sponge species; *Chondrilla caribensis, Scopalina ruetzleri* and *Haliclona implexiformis* (now re-identified as *Haliclona vansoesti*), which have been shown to convert assimilated ^{13}C- and ^{15}N-enriched DOM into ^{13}C- and ^{15}N-enriched detritus [3]. However, cell proliferation and shedding have not yet been investigated in species other than *H. caerulea*.

Sponges are widely recognized as one of the earliest metazoan life forms, e.g. [7], [8]. However, little is known about tissue homeostasis in sponges, which is one of the basal prerequisites of multi-cellularity. Tissue homeostasis is the regulation of cell proliferation and cell loss, also known as cell turnover, which acts to maintain a healthy population of cells. Cell turnover occurs in many fully developed metazoan tissues and keeps tissues in a constant state of cellular flux. This is mediated through the elimination of differentiated cells that are replaced by the proliferation of adult stem cells [9], [10]. Sponges display slow growth rates both *in situ* and *ex situ* [11], however, they have potential for rapid proliferation as demonstrated by their ability to grow new tissue at up to 2,900 times their normal growth rate during tissue regeneration in response to wound infliction [12]. Division of choanocytes by mitosis has previously been reported in a few sponge species, including the marine sponge *Hymeniacidon sinapium* [13] and the freshwater sponge *Ephydatia fluviatilis* [14].

Apoptosis (programmed cell death) and cell shedding are widely regarded as common mechanisms of cell loss involved in the homeostasis of metazoan tissues [10], [15]. Apoptotic pathways have been identified in sponges, e.g. cDNA's encoding predicted caspase-3-related proteins have been isolated from specimens of *Geodia cydonium* and *Suberites domuncula* [16] and apoptosis involving caspase-3 has been measured in sponge cell assays [17]. Additionally, anti-active caspase-3 antibodies have been used to detect caspase-3 activity in sponge tissue sections [6], [18]. However, only low amounts of apoptosis have been found in *H. caerulea*, located predominantly in the mesohyl compartment and not in the proliferative choanoderm [6]. Apoptosis is therefore not considered to be the principle mechanism of cell loss involved in cell turnover under steady-state conditions for this species. The role of apoptosis in cell turnover is unknown for other sponge species.

Cell shedding, as described in *H. caerulea* [6], is a common mechanism of cell loss for the maintenance of tissue homeostasis in highly proliferative tissues, such as the epithelium of the mammalian gastrointestinal tract, e.g. [19–21]. Evidence for cell shedding in other sponge species is still lacking, but the link between sponge biomass turnover and the production of sponge-derived detritus is becoming more evident. Direct evidence shows the production of significant amounts of detritus as a result of dissolved organic matter assimilation in four species of tropical coral reef sponge [3]. Detritus production has also been observed in a variety of additional tropical marine sponges [4], [22] and in sponges from other marine ecosystems, such as deep-sea cold water [23], Mediterranean [24] and freshwater ecosystems [25]. However, this detritus production has generally been linked to feeding waste, and cellular debris derived from cell turnover has not been taken into account.

Sponges have abundant and diverse communities of associated microbes in their tissues, e.g. [26], [27], and species are generally categorized as either low microbial abundance (LMA) sponges, which have microbial abundances similar to seawater, or high microbial abundance (HMA) sponges, with microbial concentrations 2–4 orders of magnitude higher than seawater [28–30]. According to genomic analysis, sponge-associated microbes should be capable of a broad range of metabolic transformations [27]. The question is whether these microbes in turn affect tissue homeostasis of the sponge holobiont.

We hypothesize that cell turnover through cell proliferation (predominantly in choanocytes) and cell shedding is a general mechanism for maintaining tissue homeostasis under steady-state conditions in a wide variety of sponge species from various benthic marine ecosystems. We therefore investigated *in vivo* cell proliferation, cell loss (through cell shedding and apoptosis) and detritus production in eight sponge species, with different abundances of associated microbes, from tropical coral reef, mangrove, and temperate Mediterranean reef ecosystems. Cell proliferation in sponge tissue was investigated by *in vivo* labeling with the thymidine-analogue 5-bromo-2′-deoxyuridine (BrdU) and subsequent immunohistochemical staining of tissue sections. Cell loss through apoptosis was investigated by immunohistochemistry using an antibody against active caspase-3. Cell loss through shedding was assessed qualitatively in histological sections and the dry weight of detritus produced daily by sponges was determined.

Materials and Methods

Ethics statement

Research on Curaçao was performed under the research permit (#2012/48584) issued by the Curaçaoan Ministry of Health, Environment and Nature (GMN) to the CARMABI foundation.

Sponge species and collection

We studied eight demosponge (Porifera: Demospongiae) species; six tropical coral reef species (Halisarca caerulea, Chondrilla caribensis, Scopalina ruetzleri, Clathria sp., Haliclona vansoesti and Monanchora arbuscula), one mangrove species (Mycale microsigmatosa) and one temperate Mediterranean reef species (Chondrosia reniformis). Tropical reef and mangrove species were collected by SCUBA diving or snorkeling on the reefs of the Caribbean island of Curaçao (12°12′N, 68°56′W), between February and April 2011 and 2013. The Mediterranean reef species was collected at the Medes Islands, Catalunya, Spain (42°05′N, 3°23′W) between August and September 2011. Sponges were chiseled from the (coral) rock or mangrove root and collected attached to their substrate, which was cleared of other organisms. All sponges were trimmed to a size of approximately 25 cm^2 with no available substrate for growth in order to induce steady-state conditions. Specimens were kept in 100 L running seawater aquaria with a flow rate of 3 L min^{-1} (exchange rate of 33 min) at ambient temperature (26–27°C for tropical aquaria and 18–20°C for temperate aquaria). Sponges were allowed to acclimatize for a minimum of one week prior to incubation experiments. Any changes in the shape and size of sponges were noted during the experimental period of up to 4 weeks in order to ensure steady-state conditions.

BrdU-labeling, fixation and embedding

Individual sponges (n = 3 per species, *H. vansoesti* n = 8) were enclosed in incubation chambers (3 L) with magnetic stirring devices [3,6]. Incubation chambers were kept in the aquaria to maintain ambient seawater temperature. In order to measure cell proliferation, 5-bromo-2′-deoxyuridine (BrdU, Sigma) was added to incubation chambers containing the sponges. Sponges were incubated in seawater containing 50 μmol L^{-1} BrdU (optimal BrdU concentration was determined during preliminary tests) for 6 h (continuous labeling) in order to estimate and compare the relative amount of cell proliferation between numerous species and to test if the choanoderm is the predominant location of proliferating cells. According to the one population model [31], BrdU-labeling over time results in a linear increase in the number of BrdU-positive cells until the growth fraction is reached, after which there is no further increase in the number of BrdU-positive cells, (see also [6]). The length of the cell cycle is determined by the length of the linear increase and the growth fraction indicates the proportion of cells actively involved in proliferation. After 6 h of BrdU-labeling, choanocytes have either reached their growth fraction, or not, in which case the number of BrdU-positive choanocytes is still increasing linearly with time. If after 6 h of BrdU-labeling choanocytes have not yet reached their growth fraction, the number of cells labeled gives an indication of the rate of choanocyte proliferation, i.e. the number of choanocytes replaced every 6 h. Based on the 5.4 h cell cycle in *H. caerulea* [6], we assume that the cell cycle duration of choanocytes in other sponge species under steady-state conditions is unlikely to be much shorter than 6 h. If the cell cycle is shorter than 6 h, proliferation rates given here are slightly underestimated. For specimens of *C. reniformis* we were only able to access data from 10 h BrdU-incubations. Additional specimens of *H. caerulea* were incubated with BrdU for 10 h to determine if there are differences in the percentage of BrdU-positive cells between 6 h and 10 h of labeling in this species, i.e. to determine whether the growth fraction for choanocytes has been reached.

Immediately after the incubations, sponge tissue samples (~0.5 cm^2) were fixed in 4% paraformaldehyde in phosphate-buffered saline (PFA/PBS; 4 h at 4°C), rinsed in PBS, dehydrated through a graded series of ethanol and stored in 70% ethanol at 4°C until further processing. For histological investigations of *in situ* shedding, sponge tissue samples (~1 cm^2) were collected from all species (n = 3 per species) and placed in 15 mL syringes *in situ* (underwater) with the plunger removed. Seawater was pushed out using the plunger and a second syringe was used to transfer 4% PFA/PBS to the syringe containing sponge tissue, also performed *in situ* (underwater). Samples were transported to the lab within 2 h and post-fixed in 4% PFA/PBS (4 h at 4°C), rinsed in PBS, dehydrated through a graded series of ethanol and stored in 70% ethanol at 4°C until further processing. All tissue samples were embedded in butyl-methyl-methacrylate (BMM).

BrdU-immunohistochemistry

Semi-thin sections (2 μm) of the embedded sponge tissue were cut on an ultramicrotome (Leica EM UC6) using a diamond knife (dEYEmond Histo 5 mm, Isselburg, Germany) and collected on glass slides (StarFrost, Knittelglass). BMM was removed in acetone and endogenous peroxidase activity was blocked by incubating slides in methanol containing 0.3% H_2O_2 (20 min). Slides were washed in tris-buffered saline (TBS) and incubated in citric acid (0.2% pH 6.0, 30 min at 85°C). After subsequent washing in TBS, DNA was denatured in HCl (2 mol L^{-1}, 30 min at 37°C), pH-neutralized in sodium borate buffer (pH 8.5), and washed with TBS. Slides were incubated with mouse anti-BrdU monoclonal antibody (Nordic-MUbio MUB0200S, 1:50 in TBS with 1% BSA, 0.1% Tween 20, 60 min) then washed in TBS. Primary antibody was detected using an avidin-biotin enzyme complex (Vectastain Elite ABC Kit, Vector laboratories). Slides were incubated with biotinylated rabbit anti-mouse antibody (in TBS with 1% BSA, 0.1% Tween 20, 30 min), washed in TBS and then incubated in avidin-biotin-peroxidase complex (in TBS with 1% BSA, 0.1% Tween 20, 30 min). Peroxidase activity was visualized with DAB (DAKO; 5–10 min; positive cells have brown-stained nuclei). Sections were washed in distilled water then counterstained in haematoxylin to visualize BrdU-negative nuclei (stained dark blue), dehydrated through a graded series of ethanol and mounted in Entellan (Merck). BrdU-labelled mouse intestinal tissue was used as a positive control and immunohistochemistry without primary antibody (on both mouse and sponge tissue) served as a negative control.

Analysis of cell proliferation

All slides were examined under a light microscope (Olympus BH-2) and photographs were taken using an Olympus DP70 camera. To quantify the number of proliferative cells per species, two tissue samples were taken per specimen (middle section with and without osculum). From each tissue sample three areas of the sponge were sectioned, each approximately 100 μm apart. The percentage of BrdU-positive cells in the choanoderm (choanocytes) and in the mesohyl (all cells located between the pinacoderm and the choanoderm) was calculated per section. At least 250 choanocytes and 250 mesohyl cells were counted from each section making a total of at least 1500 (2 tissue areas x 3 sections x 250 cells) of each cell type counted per sponge.

Cell shedding, detritus collection and quantification

Cell shedding was assessed qualitatively using light microscopy to observe the presence of shed cells in histological sections of sponge tissue sampled *in situ*. To quantify the production of detritus by sponges, five species were selected; *H. caerulea* (n = 7), *C. caribensis* (n = 5), *S. ruetzleri* (n = 8), *H. vansoesti* (n = 5), and *M. arbuscula* (n = 2). Sponges were placed in individual open-topped plastic pots (9.4 cm x 4.2 cm) within the aquaria and detritus was collected every 24 h for up to six consecutive days from the bottom surface of the pots and from the surface of the sponge using a glass Pasteur pipette. Detritus was centrifuged (340 x g, 5 min) and the supernatant discarded. Samples were freeze-dried overnight. The sponges were removed from their substrates and oven-dried for 24 h at 60°C. Dry-weights (DW) of both detritus and sponge tissue were determined. Pots without sponge (n = 3) served as controls from which detritus was collected for six consecutive days. The average percentage body weight (DW detritus DW sponge $^{-1}$) produced as detritus over 24 h was calculated for each of the five species tested. Additionally, detritus samples from *H. caerulea* (n = 5) were fixed in 4% PFA for 4 h for light microscopy analysis. Samples were rinsed in PBS, dehydrated through a graded series of ethanol and stored in 70% ethanol until further processing. Samples were embedded in gelatin prior to paraffin embedding and sectioning (5 μm). Sections were stained with haematoxylin and viewed using a light microscope in order to identify cellular debris within the detritus.

Active caspase-3 immunohistochemistry and analysis of tissue sections

After removing BMM and blocking endogenous peroxidase activity, slides were washed in TBS and incubated in citric acid (see BrdU-immunohistochemistry). Slides were washed in TBS and blocked with goat serum (2% in TBS, 0.1% Tween 20, 30 min). After washing in TBS, slides were incubated with rabbit anti-active caspase-3 monoclonal antibody (BD Pharmingen, 1:200 in TBS with 1% BSA, 0.1% Tween 20, 60 min). Slides were washed in TBS, then incubated with biotinylated goat anti-rabbit antibody (Vectastain Elite ABC Kit, Vector Laboratories; in TBS with 1% BSA, 0.1% Tween 20, 30 min), washed in TBS and then incubated in avidin-biotin-peroxidase complex. Peroxidase activity was visualized with DAB (active caspase-3 positive cells contain brown stained cytoplasm). Slides were counterstained in haematoxylin, dehydrated through a graded series of ethanol and mounted in Entellan. Human tonsil tissue served as a positive control and immunohistochemistry without primary antibody (on both human and sponge tissue) served as a negative control. The percentages of active caspase-3 positive cells were counted in at least 3 tissue sections per species, each from a different sponge individual. At least 250 mesohyl cells were counted per section.

Sponge-associated microbial abundance classification using transmission electron microscopy (TEM)

Samples were fixed in 2.5% glutaraldehyde, 0.1 mol L^{-1} cacodylic acid in 0.2 µm-filtered seawater for 1 h at room temperature. Samples were post-fixed in 1% osmium tetroxide (OsO_4) in filtered seawater for 1 h, dehydrated in a graded series of ethanol and stored in 70% ethanol until further processing. Tissue was embedded in epon-araldite. TEM images were used to categorize sponges as high microbial abundance (HMA) or low microbial abundance (LMA) as done by Hentschel and colleagues [30].

Statistical analysis

In order to test if the choanoderm is the predominant location of proliferating cells compared to the mesohyl, the difference between the percentages of proliferating choanocytes versus mesohyl cells was investigated. This was done as a function of different variables: species identity, observed cell shedding (high or low) and tissue area (middle with or without osculum). For this purpose a linear mixed model was used [32], using the individual organism as a random effect. To determine the effects of these models, the p-values are reported, based on an F-test with a Kenward-Roger df approximation [33], which compares the full model with a model involving only a random effect. In addition, the 95%-confidence intervals (95%-CI) of the model coefficients are reported for the significant models. Additionally, a linear mixed model was used to test the difference in the percentage of BrdU-positive choanocytes between 6 h and 10 h of continuous labeling in *H. caerulea*. A linear mixed model was also used to investigate the difference in detritus production between test pots with sponge and control pots without sponge, using the individual organism as a random effect. The relationship between the percentage detritus biomass^{-1} d^{-1} and observed cell shedding (high or low) was investigated using a linear model. All calculations were conducted in R, using the lme4 package (R Development Core Team 2013).

Results

Cell proliferation

BrdU-positive choanocytes and mesohyl cells were observed in all species (Fig. 1, Fig. 2B, Fig. S1B, Table 1). The mouse intestine positive control tissue (Fig. S2A) and sponge tissue sections clearly showed BrdU-positive nuclei (brown-stained) and BrdU-negative nuclei (blue-stained), whereas only small amounts of non-specific BrdU-labeling were present in the cytoplasm or extracellularly (Fig. 2B, Fig. S1B). Negative controls (no primary antibody) showed no BrdU-positive cells (Fig. S2B). In general, choanocytes were more proliferative than mesohyl cells, and the effect varied with species (linear mixed model, $p<0.001$). There was a significant difference in the percentage of proliferating choanocytes versus mesohyl cells for six out of the eight species studied; *H. caerulea*, *C. caribensis*, *C. reniformis*, *S. ruetzleri*, *H. vansoesti* and *M. microsigmatosa* (Fig. 1, linear mixed model; mean differences and 95%-CI in Table 2). *Clathria* sp. and *M. arbuscula* showed no significant difference in the percentage of proliferating choanocytes versus mesohyl cells (Fig. 1, linear mixed model; mean differences and 95%-CI in Table 2).

Five out of seven sponge species had similar levels of choanocyte proliferation (mean±SD throughout) after labeling with BrdU for 6 h, ranging between 16.1% and 19.0%; *H. caerulea* (17.6±3.3%), *C. caribensis* (18.0±5.4%), *S. ruetzleri* (19.0±2.0%), and *Clathria* sp. (17.3±2.2%) and *H. vansoesti* (16.1±15.9%) (Table 1, Fig. 1, Fig. 2B, Fig. S1B). The tropical coral reef sponge, *M. arbuscula*, had lower levels of choanocyte proliferation compared to other species, with a mean of 8.1±3.7% (Table 1, Fig. 1, Fig. S1B). The highest percentage of BrdU-positive choanocytes (70.5±6.6%) was found in the mangrove species *M. microsigmatosa* (Table 1, Fig. 1, Fig. 2B). The tropical coral reef species *H. vansoesti* displayed a high variation in the amount of choanocyte proliferation ranging from as low as 2.8% to as high as 73.1% after 6 h of continuous labeling (Table 1, Fig. 1). This variability was found between different individuals and within tissue samples from the same individual. The Mediterranean species, *C. reniformis*, had 16.6±3.2% choanocyte proliferation after a 10 h BrdU-labeling (Table 1, Fig. 1, Fig. 2B). No difference was found in choanocyte proliferation between 6 h (17.6±3.3%) and 10 h (16.9±3.9%) of BrdU-labeling for *H. caerulea* (linear mixed model, $p = 0.697$). For all species, there was no significant difference in cell proliferation between tissue areas with and without an osculum (linear mixed model, $p = 0.586$). No notable change in shape or biomass increase was observed in sponges over time. No apparent relation was found between cell proliferation and microbial abundance classification of sponge species. Species with similar levels of choanocyte proliferation after 6 h of BrdU-labeling (between 16.1% and 19.0%), were either classified as HMA (*H. caerulea* and *C. caribensis*) or LMA (*S. ruetzleri* and *Clathria* sp.) (Table 1).

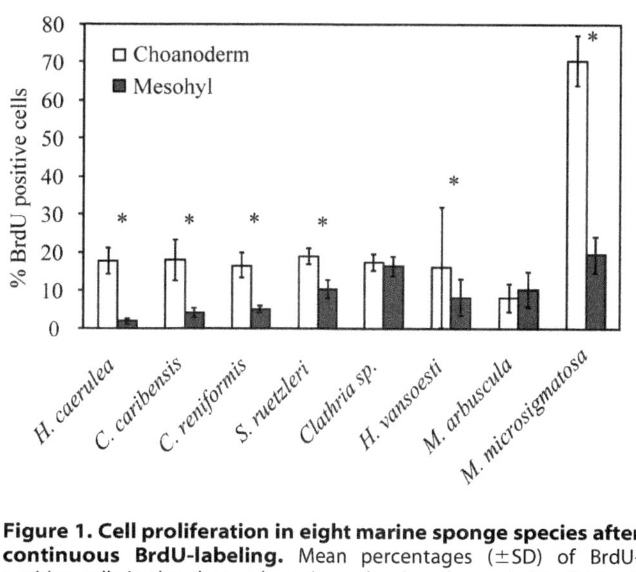

Figure 1. Cell proliferation in eight marine sponge species after continuous BrdU-labeling. Mean percentages (±SD) of BrdU-positive cells in the choanoderm (open bars) and mesohyl (solid bars) are shown. Significant differences between cell proliferation in the choanoderm versus mesohyl are indicated for each species (* $p<0.001$; 95%-CI are given in Table 2). All species were labeled with BrdU for 6 h except for *C. reniformis* which was labeled for 10 h.

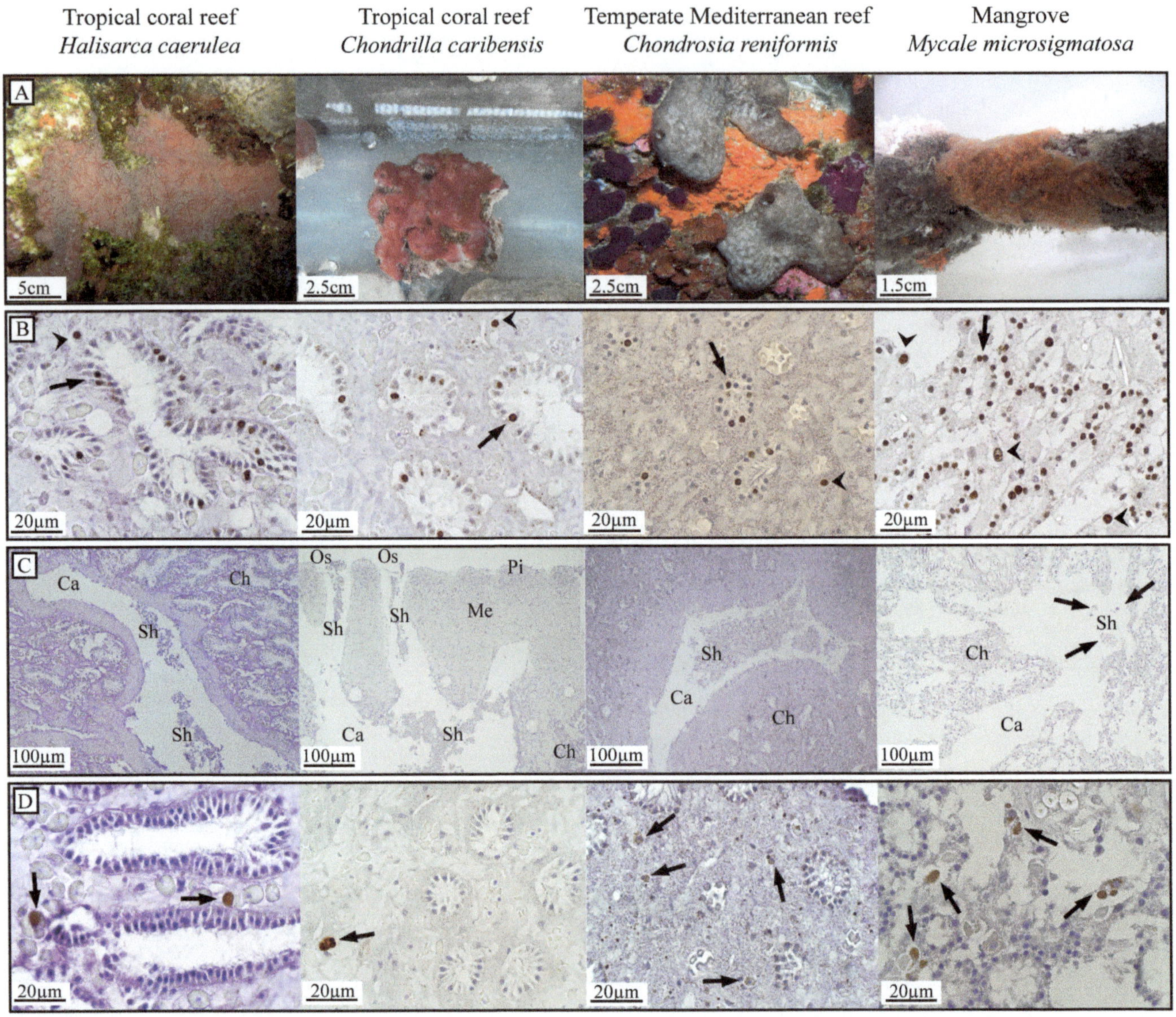

Figure 2. Cell proliferation and cell loss in a selection of four sponge species. Sponge species from three benthic ecosystems; tropical coral reef, temperate Mediterranean reef, and mangroves. (A) *In situ* (*H. caerulea, C. reniformis and M. microsigmatosa*) and *ex situ* (*C. caribensis*) photographs of test species. (B) BrdU-positive choanocytes (arrows) and mesohyl cells (arrowheads) of sponges BrdU-labeled for 6 h (10 h for *C. reniformis*) *in vivo* as a measure for proliferation. Areas of non-specific BrdU-labeling are occasionally seen in the cytoplasm of cells or extracellularly. (C) High amounts of cell shedding (Sh) in the lumen of excurrent canals (Ca) in specimens of *H. caerulea, C. caribensis* and *C. reniformis* sampled *in situ*. Choanocyte chambers (Ch), oscula (Os), the mesohyl (Me) and pinacoderm (Pi) are shown. Minor amounts of cell shedding (arrows) in the tropical mangrove sponge *M. microsigmatosa* sampled *in situ*. (D) Active caspase-3 activity of *in vivo* tissue was found in cells located in the mesohyl (arrows) resembling spherulous cells and, occasionally, archeocytes.

Cell shedding

Cell shedding was observed in tissue sections of all sponge species investigated, except *H. vansoesti*, for which cell shedding could not be determined (Table 1, Fig. 2C, Fig. S1C). Cellular debris was located inside the lumen of excurrent canals leading to the oscula (outflow openings) (Fig. 2C). BrdU-positive shed choanocytes (identified based on their size and shape) were found in all species in which cell shedding was observed, indicating the choanocyte origin of the shed cellular debris and fast cell turnover of choanocytes (within 6–10 h). In addition to choanocytes, cells resembling spherulous cells were also shed into the lumen of excurrent canals (Fig. 3A, Fig. 3B, Fig. 3C). Closer to the outflow opening, shed cells transform into mucal sheets of cellular debris (Fig. 3D). Light microscopy analysis of detritus samples indicate that sponge-derived cellular debris degrades further once shed into the ambient water (Fig. 3E).

In histological sections of each species, observed shedding was categorized as either 'high' or 'low' (Table 1). For example, high amounts of cell shedding were observed in *H. caerulea, C. caribensis* and *C. reniformis* (Fig. 2C), whereas only low amounts of cell shedding were observed in *M. microsigmatosa* (Fig. 2C), *S. ruetzleri*, *Clathria* sp. and *M. arbuscula* (Fig. S1C). No relation was found between the categorization of observed *in situ* cell shedding as high or low and the percentage of choanocyte proliferation measured *in vivo* (linear mixed model, $p = 0.295$). However, high amounts of cell shedding were only observed in

Table 1. Cell proliferation (choanocytes and mesohyl cells), cell loss (shedding and detritus production), and microbial abundance categorization (HMA/LMA) in eight demosponge species (Porifera: Demospongia) from three benthic ecosystems (tropical coral reef, temperate Mediterranean reef and mangrove).

Species	*Halisarca caerulea*	*Chondrilla caribensis*	*Chondrosia reniformis*	*Scopalina ruetzleri*	*Clathria* sp.	*Haliclona vansoesti*	*Monanchora arbuscula*	*Mycale microsigmatosa*
Habitat	Tropical coral reef	Tropical coral reef	Temperate Mediterranean reef	Tropical coral reef	Tropical coral reef	Tropical coral reef	Tropical coral reef	Mangrove
% BrdU-positive choanocytes	17.6±3.3	18.0±5.4	16.6±3.2*	19.0±2.0	17.3±2.2	16.1±15.9	8.1±3.7	70.5±6.6
% BrdU-positivemesohyl cells	2.0±0.8	4.2±1.3	5.0±0.9*	10.4±2.4	16.4±2.5	8.3±4.8	10.3±4.7	19.5±4.9
Observed shedding	high	high	high	low	low	NA	low	low
Microbial abundance classification	L/HMA [56]	HMA	HMA [54]	LMA	LMA	LMA	LMA	LMA
Detritus production								
mg DW d^{-1}	16.3±11.4	12.8±9.5	NA	15.0±8.2	NA	14.3±4.5	18.0±5.8	NA
% bodymass^{-1} d^{-1}	15.1±6.7	2.5±2.2	NA	5.2±3.8	NA	5.6±3.4	18.0±5.7	NA

mean±SD are shown; NA: not applicable; * indicates 10 h BrdU labeling.

species identified as HMA, whereas low amounts of shedding were only observed in species identified as LMA (Table 1). TEM images confirmed *C. caribensis* to be a HMA species (Fig. 4A) and *S. ruetzleri, H. vansoesti, M. arbuscula* and *M. microsigmatosa* (Fig. 4B) to be LMA species.

Detritus production

Daily detritus production (mean±SD throughout) in pots containing sponges (all species combined; 12.3±10.5 mg DW) was considerably higher than controls without sponge (2.0±0.7 mg DW) and the difference was highly significant (linear mixed model, $p<0.001$, Fig. 5A). After adjusting for controls, individual sponge species produced between 12.8±9.5 mg and 18.0±5.8 mg detritus per day (Table 1, Fig. 5A) amounting to an average daily detritus production per sponge biomass of 9.3±6.8% (range 2.5–18% detritus biomass^{-1} d^{-1}, Table 1, Fig. 5B). There was no significant relationship between percentage detritus biomass^{-1} d^{-1} and observed cell shedding (categorized as high or low) per species ($p = 0.337$).

Apoptosis

The human tonsil positive control tissue showed active caspase-3 positive cells with cytoplasmic staining (Fig. S2C). No active caspase-3 staining was found in negative controls (no primary antibody) (Fig. S2D). The number of active caspase-3-positive cells acted as a measure for apoptosis. Active caspase-3-positive cells were predominantly found in the mesohyl and accounted for <1% of total cells (Fig. 2D, Fig. S1D), and resembled spherulous cells and archeocytes based on their morphology, size and location in the tissue. Active caspase-3-positive choanocytes were rarely found. Specific staining was present in the cytoplasm of cells and often in inclusions of presumed spherulous cells with intact nuclei. No active caspase-3 positive cells were found in the tissue of *H. vansoesti*.

Discussion

This study demonstrates that, similar to *H. caerulea* [6], tissue homeostasis in sponges under steady-state (non-growing) conditions occurs through cell turnover via high levels of cell proliferation (dominated by choanocytes) and cell shedding. This occurs in a variety of sponge species from different tropical and temperate benthic ecosystems. Understanding cellular processes of tissue homeostasis in sponges allows insight into the physiological state of sponges and mechanisms of sponge-mediated energy and nutrient transfer within benthic ecosystems.

Cell proliferation

C. caribensis, S. ruetzleri, Clathria sp., and *H. vansoesti* all show similar percentages of choanocyte proliferation after 6 h of BrdU-labeling (between 16.1% and 19.0%) to that observed in *H. caerulea* (17.6%), although the variation in *H. vansoesti* was high. The two remaining species showed either consistently lower (*M. arbuscula*; 8.1%) or higher (*M. microsigmatosa*; 70.5%) percentages of choanocyte proliferation after 6 h. We were only able to measure cell proliferation after 10 h of BrdU-labeling in the Mediterranean species *C. reniformis*, and therefore it is difficult to compare the estimated rate of choanocyte proliferation (16.6% in 10 h) to the tropical species. The amount of choanocyte proliferation found in all species and the identification of the choanoderm as the predominant location of proliferating cells in the majority of species demonstrates that a high level of choanocyte proliferation is common in sponges and spans diverse ecosystems and sponge taxa. Estimates of choanocyte proliferation

Table 2. Mean differences in the percentage of cell proliferation between the choanoderm and mesohyl (see Fig. 1 for percentages of cell proliferation in choanoderm and mesohyl).

	% Proliferation difference	
Species	**mean**	**95%-CI**
H. caerulea	15.7	11.98–19.43*
C. caribensis	13.8	10.03–17.48*
C. reniformis	10.5	7.45–15.61*
S. ruetzleri	8.6	4.84–12.29*
Clathria sp.	1.0	−2.87–4.79
H. vansoesti	7.9	6.84–5.62*
M. arbuscula	−1.3	−5.04–2.41
M. microsigmatosa	51.0	47.30–54.75*

The 95%-CI are estimated using a linear mixed model based on an F-test with Kenward-Roger approximation.
* = Significant difference in percentage proliferation between choanocytes and mesohyl cells.

rates given here are considered high since, under steady-state conditions, the time to replace the number of choanocytes equivalent to the entire choanocyte compartment ranges from approximately 8.5 h in *M. microsigmatosa* to 74 h in *M. arbuscula*. The slowest rate (74 h or 3 days) is comparable to known high turnover rates in cell populations of other metazoans, such as the endodermal gland cells of *Hydra attenuata* (3.5 days) [34] and the epithelium of the mammalian gastrointestinal tract (3–5 days) [35–37]. Proliferation rates presented here may be underestimated if the sponges investigated would have cell cycle durations considerably shorter than 6 h. To determine the cell cycle duration and growth fraction of choanocytes in each species, cell cycle analyses should be carried out by labeling sponges with BrdU at multiple time points [6], [31].

In this study, we found a considerably lower percentage (17.6±3.3%) of choanocyte proliferation in *H. caerulea* after 6 h of continuous BrdU-labeling compared with the percentage of proliferation (46.6±2.6%) previously found *in vivo* and *in situ* for the same species [6]. Suboptimal nutritional conditions in the aquaria compared to the previous study are arguably the cause of this difference. Food intake is one of the most important stimuli influencing homeostatic cell proliferation in uptake systems such as

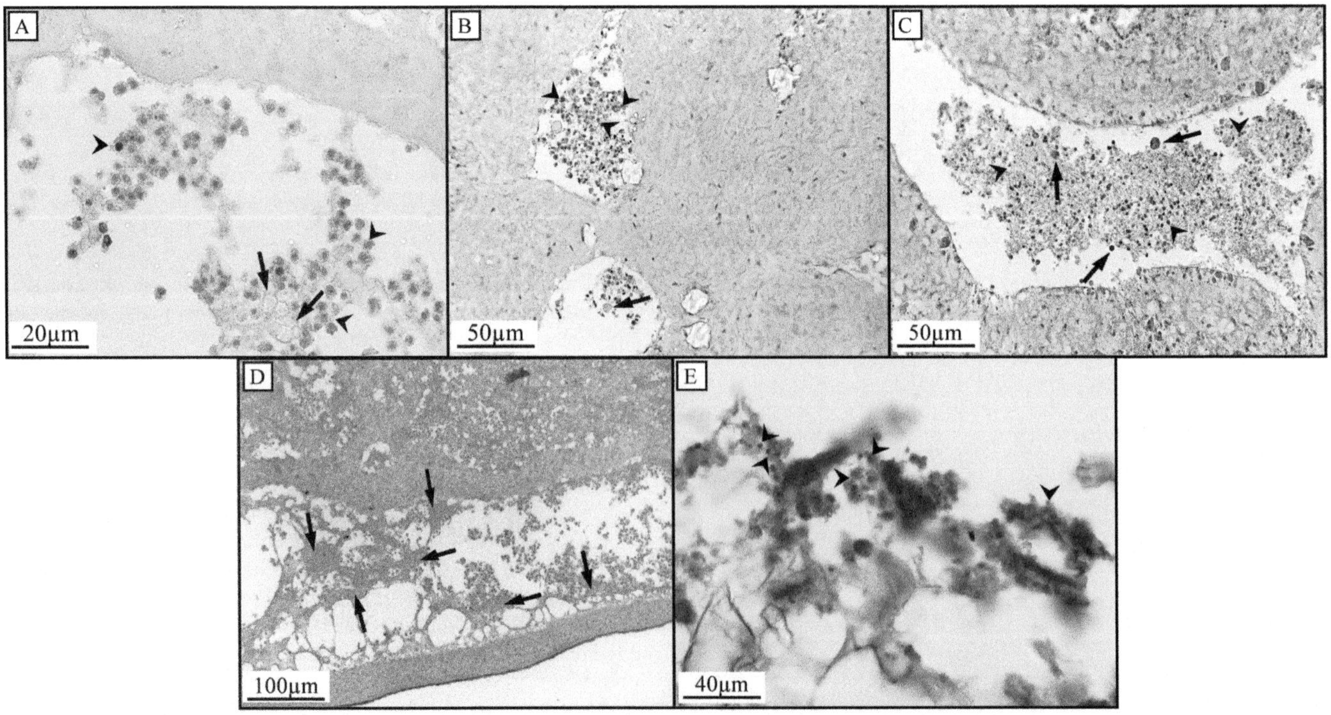

Figure 3. Histological investigations of cell shedding and identification of cellular debris in sponge-derived detritus. Arrowheads indicate shed choanocytes and arrows indicate shed spherulous cells in *H. caerulea* (A), *C. caribensis* (B) and *C. reniformis* (C). Shed cells in *H. caerulea* are present as mucal sheets (arrows) of cellular debris when close to the outflow openings (D). Light microscopy of detritus samples shows the presence of degraded cellular material, indicated by arrowheads (E).

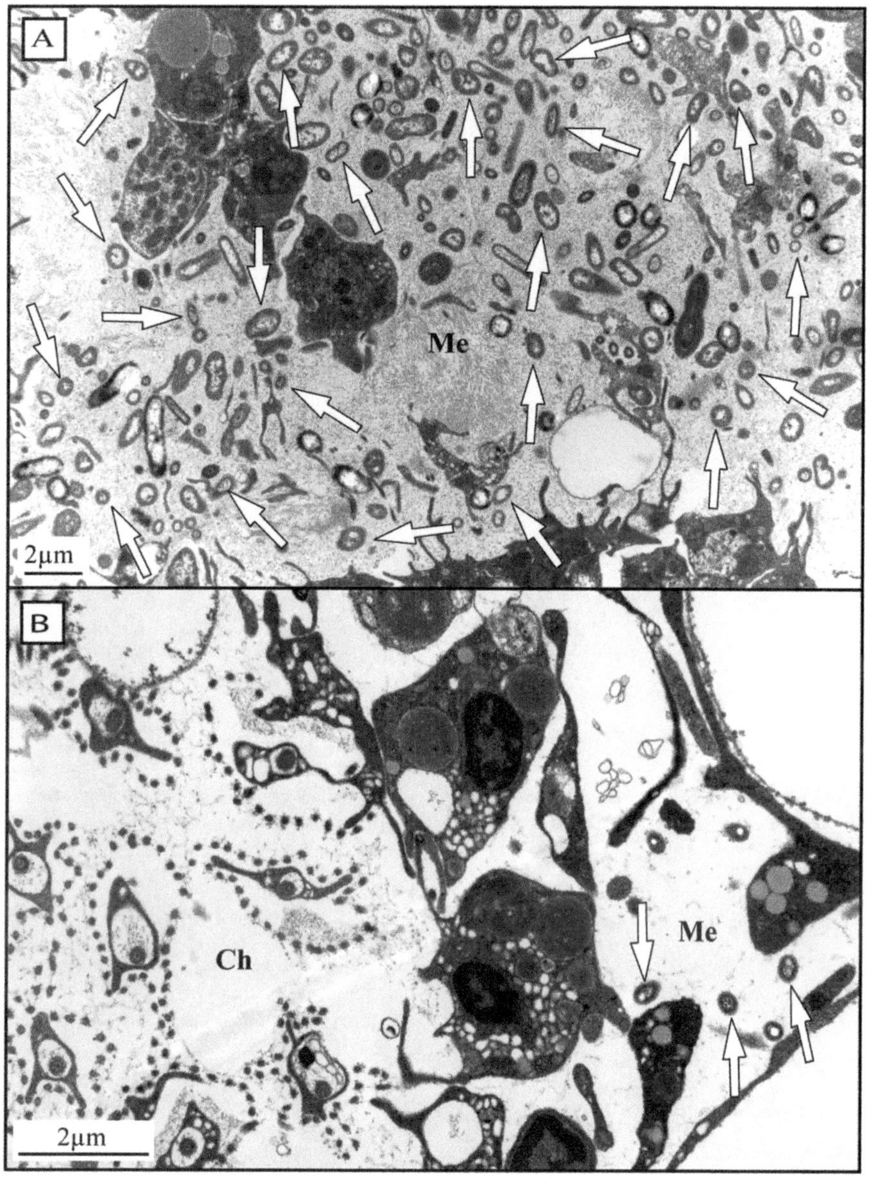

Figure 4. TEM images indicating microbial abundances in two sponge species. Arrows indicate selected bacterial symbionts. (A) Many associated microorganisms are present in the mesohyl (Me) of the high microbial abundance (HMA) species *C. caribensis.* (B) Choanoderm (Ch) and mesohyl (Me) of the low microbial abundance (LMA) species *M. microsigmatosa.* Only a few associated microorganisms are present.

the mammalian gastrointestinal tract [38], [39]. Proliferation rates of mouse and rat intestinal epithelial cells decrease during starvation and subsequently increase after re-feeding [39], [40]. Cell renewal is an energetically-costly process and a reduction in proliferation is an adaptive mechanism enabling an organism to survive during periods of reduced food intake. During fieldwork campaigns for this study, bacterial concentrations (as a proxy for food concentration) in the tropical aquaria had decreased to only one-third of ambient reef water concentrations due to long-term bio-fouling of sponges and other filter feeders in the water inlet pipes. Reduced cell proliferation following starvation has been found to be due to a prolonged cell cycle, e.g. [40], or a decrease in the number of actively proliferating cells, e.g. [41]. No difference was found between the percentage of proliferative choanocytes after 6 h and 10 h of BrdU-labeling in *H. caerulea*, suggesting that the choanocyte cell population had reached its growth fraction. Therefore, we hypothesize that for *H. caerulea*, the growth fraction becomes reduced with a decline in food availability. The relationship between nutrient concentration and cell proliferation should be investigated in future work to confirm this hypothesis. We conclude that the percentages of cell proliferation and subsequent amount of cell loss presented here may be underestimated compared to *in situ* physiological conditions, with one exception: the mangrove species, *M. microsigmatosa*, had extremely high proliferative activity in both choanocytes (70.5±6.6%) and mesohyl cells (19.5±4.9%) compared to other species. This may have been caused by additional nutrients obtained directly from the root substrate. Mangrove sponges have a facultative mutualistic relationship with the roots they overgrow, directly using carbon derived from roots as an energy source [42], [43]. When experimentally attached to mangrove prop roots, growth rates of *Tedania ignis* and *Haliclona implexiformis*

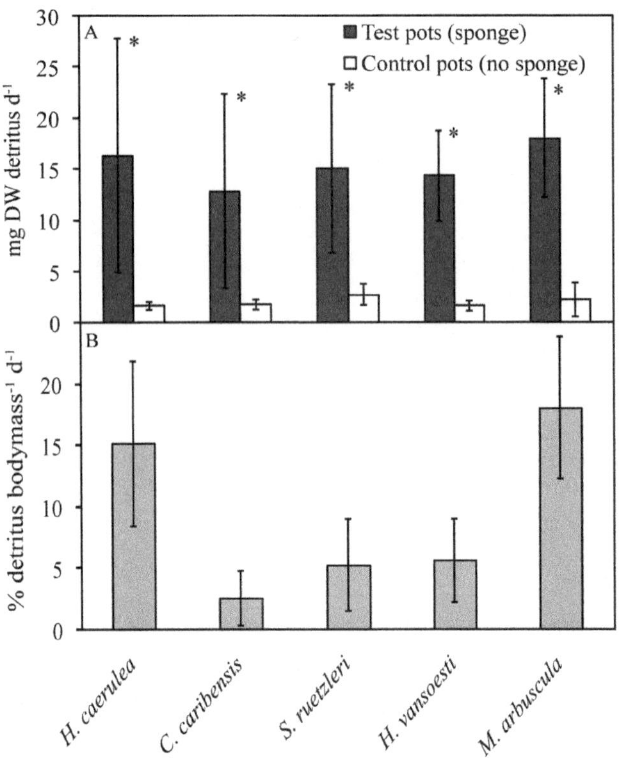

Figure 5. Daily detritus production of five tropical coral reef sponge species. (A) Daily detritus production (mg) and controls (mean±SD). Significant differences were found between detritus collected from pots containing sponge and control pots without sponge (* $p<0.001$). (B) Percentage bodyweight produced in detritus per day in five tropical coral reef species.

increased 1.4–10 times compared to sponges grown on PVC tubes due to nutrient exchange between the roots and the sponges [43]. A direct supply of nutrients from the root substrate may have caused *M. microsigmatosa* to be less affected by sub-optimal aquarium conditions during the experimental period compared to the reef species, and thus facilitated higher levels of cell proliferation. It is unknown whether an extremely high level of cell proliferation is a characteristic intrinsically specific to this species or a characteristic of all sponges inhabiting mangroves.

We assumed that sponges were in steady-state conditions based on observations that they only changed marginally in shape or size during the experimental period (1–4 weeks). If proliferation would have been allocated to growth (biomass increase), the levels of cell proliferation measured would indicate biomass doubling rates in terms of days. However, we cannot fully exclude that some sponges, or parts of the sponge, may have been in an alternative physiological state such as tissue regression [44], tissue regeneration, e.g. [12], [45], or somatic growth, e.g. [46], [47]. Tissue regeneration and somatic growth may influence homeostatic cell turnover due to competing demands on limited energy resources [48]. An alternative physiological state occurring in some parts of a sponge may explain the high variation (2.8–73.1%) in choanocyte proliferation in *H. vansoesti* and may also be the cause of the low choanocyte proliferation observed for *M. arbuscula*. Choanocytes were not the predominant population of proliferating cells in *M. arbuscula* and there was no difference in proliferation between choanocytes and mesohyl cells. To gain insight into the relationship between the physiological state of a sponge and cell proliferation, the cell kinetics of both choanocytes and other cell types, such as mesohyl cells, should be further examined.

Proliferative cells located in the mesohyl presumably consist primarily of archeocytes based on their size, shape, location and mitotic activity [46]. Choanocytes and archeocytes are both known to be the predominant populations of proliferating cells in sponges [6], [49], [50]. The division of archeocytes may be related to self-renewal of the choanocyte population to replace cells lost due to shedding during steady-state tissue homeostasis. Choanocyte self-renewal may either be the result of the division of archeocytes committed to the choanocyte lineage or choanocyte division [51]. Proliferation in archeocytes may also correspond to somatic growth [46] or regenerative growth [48], both of which require the division of archeocytes for cell differentiation and for self-renewal of the archeocyte cell population [50].

Cell loss: shedding

Sponges require a mechanism of cell loss to counterbalance the high levels of cell proliferation under non-growing conditions. The presence of a large number of choanocytes (including BrdU-positive choanocytes) in the canals leading to the outflow openings of the sponges provides evidence that cell shedding is the principle mechanism of cell loss to maintain homeostasis in the choanocyte compartment. Aside from choanocytes, we found that cells from the mesohyl resembling spherulous cells were also shed. This corroborates with previous studies on *H. caerulea* [6], [52] and TEM studies describing shed cells (including various mesohyl cell types) in the excurrent canals of other sponge species (Maldonado, unpublished data). Cell shedding could not be analyzed in *H. vansoesti* due to its loose tissue morphology, making it difficult to distinguish between cells within the mesohyl and shed cells in histological sections.

The amount of shed cells observed in histological sections is not always as much as expected based on cell proliferation rates. For example, *S. ruetzleri*, *Clathria* sp. and *M. microsigmatosa* displayed substantial proliferation in choanocytes but only small amounts of cell shedding were observed in their tissue sections compared to *H. caerulea*, *C. caribensis* and *C. reniformis*. Interestingly, the amount of observed shedding in histological sections of sponges seems to be related to their microbial abundances and may be influenced by the residence time of cellular debris within the canal system. Water pumping rates and tissue densities of sponges have been found to be related to the abundance of sponge-associated microbes. High microbial abundance (HMA) species have more dense collagenous tissues and slower pumping rates than their low microbial abundance (LMA) counterparts, which generally have loose tissue morphologies and high pumping rates [28], [53]. All species in which low amounts of cell shedding were observed have been identified as LMA sponges and species in which high amounts of cell shedding were observed are HMA sponges. Slow pumping rates may cause longer residence times of cellular debris in the canals, which subsequently may be more easily visualized in tissue sections. Higher pumping rates in the well-irrigated LMA species may cause the shed material to be quickly removed from the sponge tissue with the water flow. Oscular outflow measurements are required to confirm this hypothesis.

Cell loss: detritus production

We attempted to directly link the process of cell shedding and the production of detritus in sponges. Isotope tracer studies for *H. caerulea*, *C. caribensis*, *S. ruetzleri* and *H. vansoesti* have shown that these species convert assimilated $DO^{13}C$ and $DO^{15}N$ into detrital $PO^{13}C$ (11–24% in 3 h) and $PO^{15}N$ (18–36% in 3 h) respectively [3]. Although all sponge species investigated here produced significant amounts of detritus, there was high variability in the amount of detritus produced, and no apparent relation between detritus production and the amount of cell shedding observed in tissue sections per species. This may be due to variation in the composition of detritus between species and the methodology used. It was impossible in practical terms to be certain that all sponge-produced detritus was captured and collected in the experimental pots. Lighter detrital particles may have been lost in the flow-through aquarium set-up. Detritus has been described as being derived from feces, mucus and other indigestible waste (concentrated filtered non-digested particles) [4], [22–25], and therefore detrital composition is likely to be species dependent. Furthermore, due to the degraded nature of the detritus (only cell remnants could be identified), it was impossible to quantify the amount and type of cells within the detritus. Therefore, at present we cannot predict to what extent shed choanocytes, along with shed mesohyl cells, contribute to detritus production.

Cell loss: apoptosis

Apoptosis is clearly not the principal mechanism by which cells are lost in order to maintain tissue homeostasis in the choanocyte compartment of sponges under steady-state conditions. In all species, apoptotic activity was largely absent in the proliferative choanocyte compartment and only found in a few cells, morphologically resembling spherulous cells or occasionally resembling archeocytes, as previously found for *H. caerulea* [6]. Many of the active caspase-3 positive cells had intact nuclei. These cells may have phagocytized apoptotic bodies or may be in the early stages of apoptosis in which nuclear fragmentation has not yet occurred. Spherulous cells are expelled into excurrent canals [6], [52] and it has been suggested that apoptosis acts as part of the waste control system in sponges rather than acting to maintain tissue homeostasis [6]. Although the positive and negative controls provide evidence for specific active caspase-3 staining, a western blot is required to validate the specificity of the anti-active caspase-3 antibody in sponge tissue.

Cell turnover and the sponge loop

Recently, one of the major energy and nutrient cycling pathways occurring on coral reefs, the so-called 'sponge loop', has been found to depend on three prerequisites: (1) the uptake of DOM by sponges, (2) the conversion of DOM into detritus and (3) the uptake of sponge-derived detritus by detritivores [3]. Cell turnover (cell proliferation and cell shedding leading to detritus production) is proposed to be the main underlying mechanism involved in the transfer of DOM to higher trophic levels by sponges [3]. In this study, we found that all four sponge species investigated by De Goeij and colleagues [3] exhibit high levels of cell proliferation along with detritus production. Additionally, cell turnover, predominantly through choanocyte proliferation and cell shedding, has been found in sponges from other sponge-dominated benthic ecosystems, namely mangroves and temperate Mediterranean reefs. The ability of the Mediterranean sponge *C. reniformis* [54] and the mangrove sponge *Niphates erecta* [55] to take up DOM, another prerequisite of a sponge loop pathway, suggests that resource cycling within these ecosystems may be controlled by a cell turnover-driven sponge loop pathway. Cellular processes of tissue homeostasis provide insight into how sponges may convert assimilated (together with concentrated indigested) organic matter into detritus, which serves as a food source for higher trophic levels.

Supporting Information

Figure S1 Cell proliferation and cell loss in four species of tropical coral reef sponge. (A) *In situ* (*S. ruetzleri, H. vansoesti*) and *ex situ* (*Clathria* sp., *M. arbuscula*) photographs of test species. (B) BrdU-positive choanocytes (arrows) and mesohyl cells (arrowheads) of sponges BrdU-labeled for 6 h *in vivo* as measure of cell proliferation. Areas of non-specific BrdU-labeling can occasionally be seen in the cytoplasm of cells or extracellularly. (C) Minor amounts of cell shedding (Sh, arrows) shown for *S. ruetzleri, Clathria* sp. and *M. arbuscula*. No shedding could be identified in histological sections of *H. vansoesti*. Choanocyte chambers (Ch), the mesohyl (Me) and excurrent canals (Ca) are shown. (D) Active caspase-3 activity of *in vivo* tissue was confined to cells located in the mesohyl (arrows). No active caspase-3 positive cells were found in the tissue of *H. vansoesti*.

Figure S2 Positive and negative controls for BrdU and active caspase-3 immunohistochemistry. (A) BrdU-labeled mouse intestine positive control tissue. Arrows indicate BrdU-positive intestinal epithelial cells (brown-stained). (B) BrdU-labeled mouse intestine negative control tissue (no primary anti-BrdU antibody) showed no BrdU-positive cells. (C) Active caspase-3 positive cells (brown-stained, indicated by arrows) in human tonsil positive control tissue. (D) Human tonsil negative control (no primary anti caspase-3 antibody) showed no active caspase-3 positive cells.

Acknowledgments

We thank Paula Lopez, Fabiènne Doveren, and Ari Muskat for their assistance in the field and lab; Henk van Veen, Wikky Tigchelaar and the staff of the electron microscopy lab at the Academic Medical Center (AMC) Amsterdam for their support in histology; Henk Kieft for help with BMM embedding; Joana Xavier, Francisco Pires and Rob van Soest for the taxonomic identification of sponge species; the two anonymous reviewers for their valuable comments and Katie Breach for assistance with proof reading.

Author Contributions

Conceived and designed the experiments: BEA RO HGvdG WA JPMC BS JMdG. Performed the experiments: BEA KL MR JMdG. Analyzed the data: EvL BEA FV. Contributed reagents/materials/analysis tools: RO JPMC BS FV MR JMdG. Wrote the paper: BEA RO HGvdG WA JPMC BS MR FV EvL JMdG.

References

1. Richter C, Wunsch M, Rasheed M, Kötter I, Badran M (2001) Endoscopic exploration of Red Sea coral reefs reveals dense populations of cavity-dwelling sponges. Nature 413: 726–730.
2. De Goeij JM, van den Berg H, van Oostveen MM, Epping EHG, Van Duyl FC (2008) Major bulk dissolved organic carbon (DOC) removal by encrusting coral reef cavity sponges. Mar Ecol Prog Ser 357: 139–151.
3. De Goeij JM, van Oevelen D, Vermeij MJA, Osinga R, Middelburg JJ, et al. (2013) Surviving in a marine desert: The sponge loop retains resources within coral reefs. Science 342: 108–110.
4. Yahel G, Sharp J, Marie D, Hase C, Genin A (2003) *In situ* feeding and element removal in the symbiont-bearing sponge *Theonella swinhoei*: Bulk DOC is the major source for carbon. Limnol Oceanogr 48: 141–149.

5. Mueller B, de Goeij JM, Vermeij MJA, Mulders Y, van der Ent E, et al. (2014) Natural diet of coral-excavating sponges consists mainly of dissolved organic carbon (DOC). PLoS One 9 (2): e90152.
6. De Goeij JM, De Kluijver A, Van Duyl FC, Vacelet J, Wijffels RH, et al. (2009) Cell kinetics of the marine sponge *Halisarca caerulea* reveal rapid cell turnover and shedding. J Exp Biol 212: 3892–3900.
7. Müller WEG (1998) Origin of metazoa: Sponges as living fossils. Naturwissenschaften 85: 11–25.
8. Philippe H, Derelle R, Lopez P, Pick K, Borchiellini C, et al. (2009) Phylogenomics revives traditional views on deep animal relationships. Curr Biol 19: 706–712.
9. Biteau B, Hochmuth E, Jasper H (2011) Maintaining tissue homeostasis: dynamic control of somatic stem cell activity. Cell Stem Cell 9: 402–411.
10. Pellettieri J, Alvarado AS (2007) Cell turnover and adult tissue homeostasis: From humans to planarians. Annu Rev Genet 41: 83–105.
11. Schippers KJ, Sipkema D, Osinga R, Smidt H, Pomponi SA, et al. (2012) Cultivation of sponges, sponge cells and symbionts: Achievements and future prospects. Adv Mar Biol 62: 273–337.
12. Ayling AL (1983) Growth and regeneration rates in thinly encrusting demospongiae from temperate waters. Biol Bull 165: 343–352.
13. Shore RE (1971) Growth and renewal studies of the choanocyte population in *Hymeniacidon sinapium* (porifera: Demospongiae) using colcemid and 3-H thymidine. J Exp Zool 177: 359–363.
14. Tanaka K, Watanabe Y (1984) Choanocyte differentiation and morphogenesis of choanocyte chambers in the freshwater sponge, *Ephydatia fluviatilis*, after reversal of developmental arrest caused by hydroxyurea. Zool Sci: 561–570.
15. Denning DP, Hatch V, Horvitz R (2012) Programmed elimination of cells by caspase-independent cell extrusion in *C. elegans*. Nature 488: 226–230.
16. Wiens M, Müller WEG (2006) Cell death in Porifera: molecular players in the game of apoptotic cell death in living fossils. Can J Zool 84: 307–321.
17. Schippers KJ, Martens DE, Pomponi SA, Wijffels RH (2011) Cell cycle analysis of primary sponge cell cultures. In Vitro Cell Dev Biol -Anim 47: 302–311.
18. Tepsuporn S, Kaltenbach J, Kuhns W, Burger M, Fernandez-Busquets X (2003) Apoptosis in *Microciona prolifera* allografts. Biol Bull 205: 199–201.
19. Potten CS, Loeffler M (1990) Stem cells: attributes, cycles, spirals, pitfalls and uncertainties. Lessons for and from the crypt. Development 110: 1001–1020.
20. Bullen TF, Forrest S, Campbell F, Dodson AR, Hershman J, et al. (2006) Characterization of epithelial cell shedding from human small intestine. Lab Invest 86: 1052–1063.
21. Günther C, Neumann H, Neurath MF, Becker C (2013) Apoptosis, necrosis and necroptosis: Cell death regulation in the intestinal epithelium. Gut 62: 1062–1071.
22. Reiswig HM (1971) Particle feeding in natural populations of three marine demosponges. Biol Bull 141: 568–591.
23. Witte U, Brattegard T, Graf G, Springer B (1997) Particle capture and deposition by deep sea sponges from the norwegian-greenland sea. Mar Ecol Prog Ser 154: 241–252.
24. Ribes M, Coma R, Gili J (1999) Natural diet and grazing rate of the temperate sponge *Dysidea avara* (demospongiae, dendroceratida) throughout an annual cycle. Mar Ecol Prog Ser 176: 179–190.
25. Weissenfels N (1976) Bau und funktion des Süßwasserschwamms *Ephydatia fluviatilis* L. (porifera). III. nahrungsaufnahme, verdauung und defäkation. Zoomorphology: 73–88.
26. Taylor MW, Radax R, Steger D, Wagner M (2007) Sponge-associated microorganisms: Evolution, ecology, and biotechnological potential. Microbiol Mol Bio Rev 71: 295–347.
27. Hentschel U, Piel J, Degnan SM, Taylor MW (2012) Genomic insights into the marine sponge microbiome. Nat Rev Microbiol 10: 641–U75.
28. Vacelet J, Donadey C (1977) Electron microscopy study of the association between some sponges and bacteria. J Exp Mar Biol Ecol 30: 301–314.
29. Reiswig HM (1981) Partial carbon and energy budgets of the bacteriosponge Verongia fistularis (Porifera: Demospongiae) in Barbados. Mar Ecol 2: 273–293.
30. Hentschel U, Fieseler L, Wehrl M, Gernert C, Steinert M, et al. (2003) Microbial diversity of marine sponges. Prog Mol Subcell Biol 37: 59–88.
31. Nowakowski R, Lewin S, Miller M (1989) Bromodeoxyuridine immunohistochemical determination of the lengths of the cell-cycle and the DNA-synthetic phase for an anatomically defined population. J Neurocytol 18: 311–318.
32. Lee Y, Nelder JA, Pawitan Y (2006) Generalized linear models with random effects: Unified analysis via h-likelihood. London: Chapman & Hall/CRC. 396 p.
33. Kenward MG, Roger JH (2009) An improved approximation to the precision of fixed effects from restricted maximum likelihood. Comput Stat Data Anal 53: 2583–2595.
34. Schmidt T, David CN (1986) Gland cells in *Hydra*: cell cycle kinetics and development. J Cell Sci 85: 197–215.
35. Creamer R, Shorter RG, Bamforth J (1961) The turnover and shedding of epithelial cells Part I The turnover in the gastro-intestinal tract. Gut 2: 110–116.
36. Moore KA, Lemischka IR (2006) Stem cells and their niches. Science 311: 1880–1885.
37. van der Flier LG, Clevers H (2009) Stem cells, self-renewal, and differentiation in the intestinal epithelium. Annu Rev Physiol 71: 241–260.
38. Goodlad RA, Plumb JA, Wright NA (1987) The relationship between intestinal crypt cell proliferation and water absorption measured in vitro in the rat. Clin Sci 72: 297–304.
39. Chaudhary M, Mandir N, FitzGerald A, Howard J, Lord G, et al. (2000) Starvation, leptin and epithelial cell proliferation in the gastrointestinal tract of the mouse. Digestion 61: 223–229.
40. Aldewachi H, Wright N, Appleton D, Watson A (1975) Effect of starvation and refeeding on cell-population kinetics in rat small bowel mucosa. J Anat 119: 105–121.
41. Wong WM, Wright NA (1999) Cell proliferation in gastrointestinal mucosa. J Clin Pathol 52: 321–333.
42. Hunting ER, de Goeij JM, Asselman M, van Soest RWM, van der Geest HG (2010) Degradation of mangrove-derived organic matter in mangrove associated sponges. Bull Mar Sci 86: 871–877.
43. Ellison A, Farnsworth E, Twilley R (1996) Facultative mutualism between red mangroves and root-fouling sponges in Belizean mangal. Ecology 77: 2431–2444.
44. Luter HM, Whalan S, Webster NS (2012) The marine sponge *Ianthella basta* can recover from stress-induced tissue regression. Hydrobiologia 687: 227–235.
45. Koopmans M, van Rijswijk P, Martens D, Egorova-Zachernyuk TA, Middleburg JJ, et al. (2011) Carbon conversion and metabolic rate in two marine sponges. Mar Biol 158: 9–20.
46. Simpson TL (1984) The cell biology of sponges. New York: Springer-Verlag.
47. Sipkema D, Yosef NAM, Adamczewski M, Osinga R, Mendola D, et al. (2005) Hypothesized kinetic models for describing the growth of globular and encrusting demosponges. Mar Biotech 8: 40–51.
48. Henry L, Hart M (2005) Regeneration from injury and resource allocation in sponges and corals - A review. Int Rev Hydrobiol 90: 125–158.
49. Sun L, Song Y, Qu Y, Yu X, Zhang W (2007) Purification and in vitro cultivation of archaeocytes (stem cells) of the marine sponge *Hymeniacidon perleve* (demospongiae). Cell Tissue Res 328: 223–237.
50. Funayama N (2013) The stem cell system in demosponges: Suggested involvement of two types of cells: Archeocytes (active stem cells) and choanocytes (food-entrapping flagellated cells). Dev Genes Evol 223: 23–38.
51. Leys SP, Hill A (2012) The physiology and molecular biology of sponge tissues. In: Becerro MA, Uriz MJ, Maldonado M, Turon X, editors. Advances in sponge science: physiology, chemical and microbial diversity. Academic Press. pp.1–56.
52. Vacelet J, Donadey C (1987) A new species of Halisarca (Porifera, Demospongiae) from the Caribbean, with remarks on the cytology and affinities of the genus. In: Jones WC, editor. European contribution to the taxonomy of sponges. Sherkin Island Marine Station Publications.pp. 5–12.
53. Weisz JB, Lindquist N, Martens CS (2008) Do associated microbial abundances impact marine demosponge pumping rates and tissue densities? Oecologia 155: 367–376.
54. Ribes M, Jiménez E, Yahel G, López-Sendino P, Diez B, et al. (2012) Functional convergence of microbes associated with temperate marine sponges. Environ Microbiol 14: 1224–1239.
55. Granek EF, Compton JE, Phillips DL (2009) Mangrove-exported nutrient incorporation by sessile coral reef invertebrates. Ecosystems 12: 462–472.Slattery M, Gochfeld DJ, Easson CG, O'Donahue LRK (2013) Facilitation of coral reef biodiversity and health by cave sponge communities. Mar Ecol Prog Ser 476: 71–86.
56. Maldonado M, Ribes M, van Duyl FC (2012) Nutrient fluxes through sponges: biology, budgets, and ecological implications. In: Becerro MA, Uriz MJ, Maldonado M, Turon X, editors. Advances in sponge science: physiology, chemical and microbial diversity. Academic Press. pp.113–82.

12

Environmental Quality of Italian Marine Water by Means of Marine Strategy Framework Directive (MSFD) Descriptor 9

Chiara Maggi*, Serena Lomiri, Bianca Di Lorenzo, Marco d'Antona, Maria Teresa Berducci

ISPRA, Institute for Environmental Protection and Research, Roma, Italy

Abstract

ISPRA, on behalf of the Italian Ministry of Environment, carried out the initial assessment of environmental quality status of the 3 Italian subregions (Mediterranean Sea Region) on Descriptor 9. The approach adopted to define the GES started to verify that contaminants in fish and other seafood for human consumption did not exceed levels established by Community legislation (Reg. 1881/2006 and further updates). As the Marine Strategy Framework Directive (MSFD) requires to use health tools to assess the environment, Italy decided to adopt a statistical range of acceptance of thresholds identified by national (D.Lgs. 152/2006 concerning water quality required for mussel farms) and international legislation (Reg. 1881/2006 and further updates), which allowed to use the health results and to employ them for the assessment of environmental quality. Italy proposed that Good Environmental Status (GES) is achieved when concentrations are lower than statistical range of acceptance, estimated on samples of fish and fishery products coming from only national waters. GIS-based approach a to perform different integration levels for station, cell's grid and years, was used; the elaborations allowed to judge the environmental quality good.

Editor: Silvia Mazzuca, Università della Calabria, Italy

Funding: This work has been carried out by the Institute for Environmental Protection and Research (ISPRA) on behalf of the Italian Ministry of Environment (http://www.strategiamarina.isprambiente.it/). The funders had no role in study design, data collection and analysis, decision to publish, or preparation of the manuscript.

Competing Interests: The authors have declared that no competing interests exist.

* Email: chiara.maggi@isprambiente.it

Introduction

Marine waters have traditionally been used by population for different activities (e.g. fishing, aquaculture, shipping, tourism, discharges from agriculture and urban areas). Currently, new activities are being developed or increasing (e.g. renewable energies, extraction of minerals, etc.), and thus competing for space with traditional uses, causing many spatial conflicts and increasing human impacts on marine ecosystems [1,2,3]. On the other hand, numerous anthropogenic sources such as land-based industrial activity, urban discharge, pesticide use, nuclear accidents and discharges, aquaculture, heavy shipping lines might be responsible for elevated levels of contaminants in sediment and biota, including species for human consumption.

To face this status the European Parliament and the Council of the European Union on 17 June 2008 have issued the Framework Directive 2008/56/EC (MSFD, [4]) on the Strategy for the Marine Environment, subsequently implemented in Italy with the Legislative Decree n. 190 of 13 October 2010 [5].

The aim of the Marine Strategy is to protect more effectively the marine environment across Europe. The MSFD requires that Member States take measures to achieve or maintain Good Environmental Status (GES) by 2020. The goal of the MSFD is in line with the objectives of the Water Framework Directive (WFD–2000/60/EC Directive [6]), achieving good ecological status by 2015, for surface water including marine-coastal waters and in case of chemical status, also territorial waters, and providing for the first review of the River Basin Management Plans in 2020 [7,8]. MSFD Annex I includes a set of 11 descriptors, on the basis of which GES should be determined. Some descriptors deal with general status of the environment, others with specific habitat integrity or specific pressures. The Commission Decision of 1 September 2010 ([9] COM DEC 2010/477/EU) listed 29 criteria and 56 indicators, based on the scientific and technical assessment prepared by the Task Groups set up by the Joint Research Centre and the International Council on the Exploration of the Seas, to assess GES for each descriptor of the MSFD Annex I. These criteria and indicators include a factors' combination in relation to state, impacts and pressures.

Member States had to make an initial assessment of own marine waters, taking account of existing data, where available, specifying the essential features and characteristics and current environmental status of the waters. The initial assessment was provided by 2012, together with the definition of targets and monitoring indicators. For this evaluation, it was necessary to point out predominant pressures and impacts existing in each region or subregion (MSFD art. 8, [4]). By 2015, Member States will have to develop coherent and coordinated programmes of measures. To ensure clean, healthy and productive marine waters, Member

States strategies will have to be coordinated, consistent and properly integrated with regulations provided by existing Community legislation (such as transport, fisheries, tourism, infrastructure, research) and international agreements.

This work describes the initial assessment by means of criteria and indicators defined on Descriptor 9 ("*Contaminants in fish and other seafood for human consumption do not exceed levels established by Community legislation or other relevant standards*"), carried out by the Institute for Environmental Protection and Research (ISPRA) on behalf of the Italian Ministry of Environment.

Focus on Descriptor 9: Contaminants in fish and other seafood for human consumption

MSFD Descriptor 9 requires to use health tools to assess the marine environment; it starts from ensuring that contaminants in fish and other seafood for human consumption, detected in the sea, do not exceed levels established by Community legislation or other relevant standards in order to protect the public health and, as indirect effect, to avoid negative influence on the sustainable use of marine resources. Thus, Good Environmental Status should be achieved when contaminants are below the levels fixed for human consumption, although the absence of human health effects may however involve environmental pollution effects. Descriptor 9 necessarily requires a balance between health information and environmental assessment; so, it becomes necessary to identify possible relations between contaminant levels in seafood tissues and the status of marine environment. In this way, starting from health information, it should be possible to reconstruct a general description of the environmental status of the area where the analyzed organisms live.

Decision 2010/477/EU [9] defines the need to monitor the presence of contaminants and hazardous substances in edible tissues of fisheries species fit for human consumption, by the use of two indicators:

- 9.1.1: actual levels of contaminants that have been detected and number of contaminants which have exceeded maximum regulatory levels;
- 9.1.2: frequency of regulatory levels being exceeded.

So, the goal of the Descriptor 9 is to check the public health. Nevertheless the aim of the Marine Strategy Directive is providing to the achievement of the Good Environmental Status, taking into account all descriptors together.

Indeed, also regarding data retrieval, we observed that existing national health monitoring programs often do not cover the information needs for compliance to GES, because monitoring programs on fish and seafood for public health reasons focus on estimating consumer exposure rather than assessing environmental status. Sampling procedures include all sizes of fish sold for human consumption rather than focusing on a standardized sample (offering greater possibilities in comparing degrees of contamination in the marine environment), without considering biological factors [10] that can influence concentrations of contaminants in fish, such as seasonal variation, age and sex [11]. However, the standardized sampling shuld produced a loss of representativeness compared to the great variability of samples available for human consumption.

Anyway, GES for Descriptor 9 should be judged in view of the monitoring of Descriptor 8 ("*Concentrations of contaminants are at levels not giving rise to pollution effects*"), also dealing with contaminants in marine environment. Descriptors 8 and 9 both measure contaminants in the marine environment, but focus on different state indicators. The importance of pay serious attention to the public health comes from the increasing recognition that the body burdens of environmental contaminants in the European human population is a growing cause for concern and may be associated with alterations in the frequency of occurrence of a range of diseases including diabetes, heart attacks, strokes, neurological disorders and cancers [12,13]. Otherwise, the global per-capita demand for seafood has reached an all-time high, and is likely to continue to increase [14].

It was noted that almost all the contaminants thought to be of concern for marine vertebrates are also of concern for human health, due to the transfer along the food web. Thus, the relationship between the degree of chemical contamination and health of marine vertebrates is of great relevance to contaminant-related human diseases [15]. This is likely to be an even greater area of interest in the coming years as the human demographic ageing proceeds across Europe, allowing longer-term accumulation of contaminants in human tissues.

Principles for assessing Italian marine waters

Concerning to Mediterranean area, there were some studies, most of them conducted in non-Italian waters, stale and with a very local profile (i.e. [16,17,18,19,20,21]). Among them, European Environmental Agency published data (1985–2002) relating to concentrations of selected metals and organic contaminants in Mediterranean Sea mussels, highlighting a moderate variability and values non-exceeding regulatory levels. That being so, a more recent data collection was needed.

Italy used both results from health and environmental monitoring, so as to allow a proper evaluation and to cover the information needs for monitoring compliance to GES under Descriptor 9.

Human health programs monitored fish from different sizes and ages; so, higher levels can not automatically be interpreted as a negative status or evolution of the marine environment. In fact contamination appears to change according to the trophic level considered, with estimates higher in larger species, i.e. pelagic fish, containing more fat and occupying an higher food web position [22].

That being so, the "one out, all out" (OOAO) approach, based on the practice that the status of the worst element determines the final status, doesn't seem applicable although it may be regarded as a precautionary approach [3,7]. In case one species shows concentrations exceeding the regulatory levels, the (sub) region should be assessed as bad status for Descriptor 9; this judgment could easily be interpreted that consumption of all fish and seafood originating in that (sub) region would be dangerous.

Otherwise the Marine Strategy does not give specific rules, allowing Member States to adopt different criteria in order to use correctly health data for assessing environmental quality.

In Italy, a statistical range of acceptance was applied in order to avoid losing information, as reported below. Regarding the assessment of contaminants and the number of pollutants that exceeded law levels, Italy proposed that Good Environmental Status is achieved when concentrations are statistically lower than the thresholds identified by national and international legislation (Reg. 1881/2006 [23] and further updates [24,25,26], D.Lgs. 152/2006 [27]), estimated on samples of wild caught fish and aquaculture products (mussels) coming from national waters. Special attention was paid to assure traceability, selecting samples directly linked to the Italian (sub) regions laid down in the MSFD. The need arose since seafood is widely transported, its provenance may not be identifiable and human exposure monitoring programs often lack the necessary data to link the

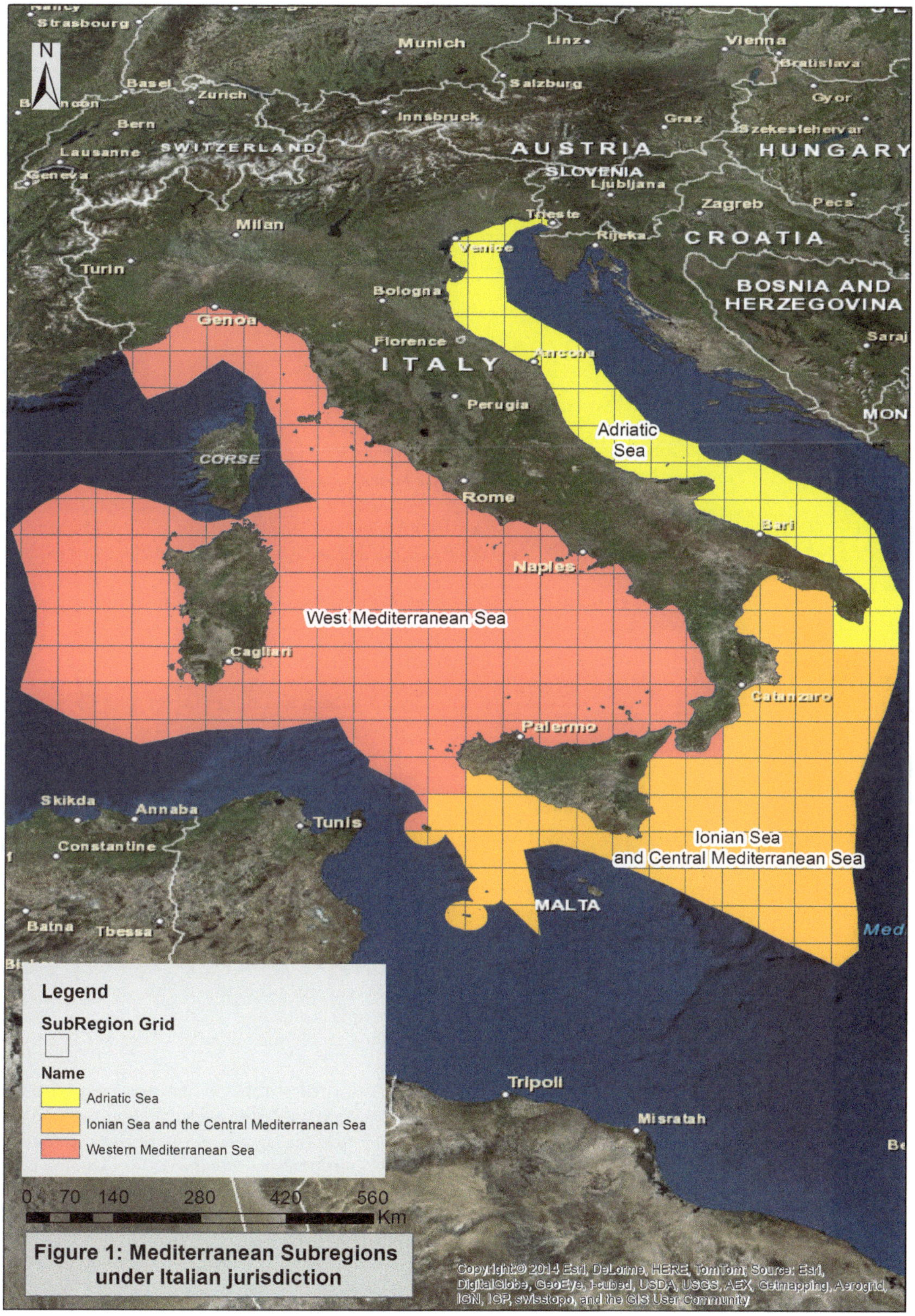

Figure 1: Mediterranean Subregions under Italian jurisdiction

Figure 1. Mediterranean Subregions under Italian jurisdiction.

samples to specific (sub) regions. In addition, Italy, as Task Group 9 suggests, took care to make a selection of species, choosing those more prone to biomagnify/bio-accumulate regulatory classes of contaminants, species representative of the different trophic levels or habitats, species representative for different subregions.

Table 1. Regulatory levels, reference legislation, code and foodstuff categories.

Category code	Legislation	Foodstuff	Regulatory levels
Cd 3.2.5	Reg.1881/2006/CE	Muscle meat of fish (footnote 24)	0,05 mg/kg w.w.
Cd 3.2.6	Reg.1881/2006/CE	Muscle meat of listened fish	0,10 mg/kg w.w.
Cd 3.2.8	Reg.1881/2006/CE	Crustaceans	0,50 mg/kg w.w.
Cd 3.2.9	Reg.1881/2006/CE	Bivalve molluscs	1,0 mg/kg w.w.
Cd 3.2.10	Reg.1881/2006/CE	Cephalopods	1,0 mg/kg w.w.
Hg 3.3.1	Reg.1881/2006/CE	Fishery products and muscle meat of fish (footnotes 24, 25, 26)	0,50 mg/kg w.w.
Hg 3.3.2	Reg.1881/2006/CE	Muscle meat of listened fish	1,0 mg/kg w.w.
Pb 3.1.5	Reg.1881/2006/CE	Muscle meat of fish (footnote 24)	0,3 mg/kg w.w.
Pb 3.1.6	Reg.1881/2006/CE	Crustaceans	0,50 mg/kg w.w.
Pb 3.1.7	Reg.1881/2006/CE	Bivalve molluscs	1,5 mg/kg w.w.
Pb 3.1.8	Reg.1881/2006/CE	Cephalopods	1,0 mg/kg w.w.
Dioxins 5.3	Reg.1259/2011/CE	Muscle meat of fish and Bivalve molluscs	3,5 pg/g w.w.
Sum dioxins and dioxin like PCBs 5.3	Reg.1259/2011/CE	Muscle meat fish and Bivalve molluscs	6,5 pg/g w.w.
Benzo(a)pyrene 6.1.4	Reg.1881/2006/CE	Muscle meat of fish (footnote 24)	2,0 μg/kg w.w.
Benzo(a)pyrene 6.1.5	Reg.1881/2006/CE	Crustaceans and Cephalopods	5,0 μg/kg w.w.
Benzo(a)pyrene 6.1.6	Reg.835/2011/CE	Bivalve molluscs	5 μg/kg w.w.
Sum PAH 6.1.6	Reg.835/2011/CE	Bivalve molluscs	30 μg/kg w.w.

Methodology

Marine Strategy Directive divided European marine waters in 4 regions: Baltic, North East Atlantic Ocean, Mediterranean Sea and Black Sea. Some of them, as Mediterranean Sea, had been subdivided in subregions and Commission Decision of 1 September 2010 [9] established that Member States should performed initial assessment in each subregions of jurisdiction.

Italian waters fall into Mediterranean region and Western Mediterranean Sea (WMS), Adriatic Sea (AS) and Ionian Sea and the Central Mediterranean Sea (ISCMS) subregions are under Italian authority (Fig. 1).

The regulatory framework is essentially composed of Regulation (EC) No 1881/2006 [23] and further updates [24,25,26] on contaminants levels, set taking into account their toxicity as well as their prevalence in the food chain, and of Regulations (EC) No 882/2004 [28], No 1883/2006 [29] and No 333/2007 [30] (and respective further updates) regarding sampling procedures, methods of analysis and analytical quality. The Regulations provide for consumers protection from chemical substances in food and set maximum levels for certain contaminants in foodstuffs in order to keep contaminants at levels which are toxicologically acceptable. Each regulated contaminant has a different threshold level, laid down according to the specific tissue analyzed. Descriptor 9 refers to these regulations, but also to other relevant national standards (in Italy i.e. D.Lgs. 152/2006 fixes mercury and lead levels in mussel tissues, concerning to water quality required for mussel farms).

Chemical contaminants monitored were those for which regulatory levels have been laid down for public health reasons: heavy metals (lead, cadmium, and mercury), polycyclic aromatic hydrocarbons (benzo(a)pyrene, benz(a)anthracene, benzo(b)fluoranthene and chrysene) and dioxins (including dioxin-like PCBs). Each contaminant was identified by category code reported in Reg. 1881/2006 [23], Reg. 835/2011[24] and Reg. 1259/2011 only for dioxins and dioxin-like PCBs [26], providing for specific values of edibility, relative to different species and analyzed tissues. In table 1, regulatory values, reference legislation, code and foodstuff categories are summarized.

It was noted that almost all the contaminants thought to be of concern for marine vertebrates are also of concern for human health, due to the transfer along the food web. Thus, the relationship between the degree of chemical contamination and health of marine vertebrates is of great relevance to contaminant-related human diseases [15]. This is likely to be an even greater area of interest in the coming years as the human demographic ageing proceeds across Europe, allowing longer-term accumulation of contaminants in human tissues.

For each contaminant, the number of excesses and their frequency on the total number of observed data were indicated.

Calculation of statistical range of acceptance

Based on the thresholds identified in the relevant national and international laws, statistical ranges of acceptance were calculated for the concentrations of regulated contaminants.

To calculate these ranges, it was utilized R software. The assumption used is that contaminants' concentrations follow lognormal distribution $Y \sim LN(\mu, \sigma)$ [31], [32], with parameter μ equal to the geometric mean of collected contaminant's data and σ equal to 20% of this value, as required for environmental data [33]. We have also tested this assumption, with Shapiro Normality test on logarithm of actually used data. In order to identify statistical ranges, it was assumed that threshold level (TL) is the 90th percentile of theoretical distribution and that 90th percentile of empirical data distribution must not exceed this level.

GIS elaborations

Useful data for the assessment were georeferenced in GIS (Software ArcGis 10.1) and exported into Feature Classes, taking

Table 2. Statistical outputs for Metals.

AS* (total cells 51)	Cd 3.2.5	Cd 3.2.6	Cd 3.2.8	Cd 3.2.9	Cd 3.2.10	Hg 3.3.1	Hg 3.3.2	Pb 3.1.5	Pb 3.1.7	Pb 3.1.8	Pb 3.1.6
% of spatial coverage	**27.5**	**11.8**	**5.9**	**47.1**	**9.8**	**82.4**	**11.8**	**33.3**	**49.0**	**9.8**	**5.9**
% of data within the limits	100	100	100	99.4	100.0	99.4	93.8	92.9	99.7	100	100
% of data outside the limits				0.6		0.6	6.2	7.1	0.3		
% of station within the limits	100	100	100	100	100	97.3	85.7	94.7	100	100	100
% of station outside the limits						2.7	14.3	5.3			
% of cells within the limits	100	100	100	100	100	95.2	83.3	94.1	100	100	100
% of cells outside the limits						4.8	16.7	5.9			
WMS (total cells 166)**	**Cd 3.2.5**	**Cd 3.2.6**	**Cd 3.2.8**	**Cd 3.2.9**	**Cd 3.2.10**	**Hg 3.3.1**	**Hg 3.3.2**	**Pb 3.1.5**	**Pb 3.1.7**	**Pb 3.1.8**	**Pb 3.1.6**
% of spatial coverage	**3.6**	**2.4**	-	**10.8**	-	**15.1**	**3.6**	**7.8**	**11.4**	-	-
% of data within the limits	85	75	-	100	-	88.3	87.5	87.2	96	-	-
% of data outside the limits	15	25	-		-	11.7	12.5	12.8	4	-	-
% of station within the limits	85.7	75	-	100	-	92.3	85.7	86.7	97.7	-	-
% of station outside the limits	14.3	25	-		-	7.7	14.3	13.3	2.3	-	-
% of cells within the limits	83.3	75	-	100	-	96	83.3	92.3	94.7	-	-
% of cells outside the limits	16.7	25	-		-	4	16.7	7.7	5.3	-	-
ISCMS* (total cells 77)**	**Cd 3.2.5**	**Cd 3.2.6**	**Cd 3.2.8**	**Cd 3.2.9**	**Cd 3.2.10**	**Hg 3.3.1**	**Hg 3.3.2**	**Pb 3.1.5**	**Pb 3.1.7**	**Pb 3.1.8**	**Pb 3.1.6**
% of spatial coverage	**5.2**	**5.2**	-	**5.2**	-	**7.8**	**9.1**	**7.8**	**3.9**	-	-
% of data within the limits	100	75	-	100	-	100	100	100	99.8	-	-
% of data outside the limits		25	-		-				0.2	-	-
% of station within the limits	100	60	-	100	-	100	100	100	100	-	-
% of station outside the limits		40	-		-					-	-
% of cells within the limits	100	50	-	100	-	100	100	100	100	-	-
% of cells outside the limits		50	-		-					-	-

* Adriatic Sea SubRegion;
** West Mediterranean Sea SubRegion;
*** Ionian Sea and Central Mediterranean Sea SubRegion.

Table 3. Statistical outputs for PAHs (Polycyclic Aromatic Hydrocarbons).

AS* (total cells 51)	Benzo(a)pyrene 6.1.4	Benzo(a)pyrene 6.1.5	Benzo(a)pyrene 6.1.6	Sum of PAHs 6.1.6
% of spatial coverage	2	-	29.4	29.4
% of data within the limits	100	-	100	100
% of data outside the limits		-		
% of station within the limits	100	-	100	100
% of station outside the limits		-		
% of cells within the limits	100	-	100	100
% of cells outside the limits		-		
WMS (total cells 166)**	**Benzo(a)pyrene 6.1.4**	**Benzo(a)pyrene 6.1.5**	**Benzo(a)pyrene 6.1.6**	**Sum of PAHs 6.1.6**
% of spatial coverage	3.6	1.2	3.6	4.8
% of data within the limits	100	100	94.3	98.7
% of data outside the limits			5.7	1.3
% of station within the limits	100	100	89.7	96.3
% of station outside the limits			10.3	
% of cells within the limits	100	100	66.7	88.9
% of cells outside the limits			33.3	
ISCMS* (total cells 77)**	**Benzo(a)pyrene 6.1.4**	**Benzo(a)pyrene 6.1.5**	**Benzo(a)pyrene 6.1.6**	**Sum of PAHs 6.1.6**
% of spatial coverage	-	-	2.6	2.6
% of data within the limits	-	-	85.7	94.4
% of data outside the limits	-	-	14.3	5.6
% of station within the limits	-	-	100	100
% of station outside the limits	-	-		
% of cells within the limits	-	-	100	100
% of cells outside the limits	-	-		

* Adriatic Sea SubRegion;
** West Mediterranean Sea SubRegion;
*** Ionian Sea and Central Mediterranean Sea SubRegion.

as reference system ETRS89. The Feature Classes have been created on the basis of regulatory codes. The data were processed both for a single year, in order to see the concentration trend in the subregion, and considering the whole period (2006–2012), as requested by MSFD. Polygonal grids were created to cover the entire subregion.

A critical aspect of the cartographic layers production was the definition of the reference grid (fishnet) by selecting:

- The starting point (lower left corner);
- the cell size(s);
- the number of rows and columns.

As in 2009 ISPRA was involved in the EuSeaMap project (Mapping European seabed habitats, [34]), in which a broad-scale modelled habitat map for the Western Mediterranean was produced together with the Spanish and French partners, the fishnet was created according to the grid used in this project.

GIS layers were created in geographic coordinates (ETRS 89), using angular measures in order to obtain a 1:1 cell size ratio. The chosen mesh was 50 km per side, coherently with the ecological differences (such as mobility, migratory patterns and feeding habits) of functional groups and habitats considered.

Value of "code" 1 and green colour's cell were assigned to data not exceeding the threshold levels. Value of "code" 2 and red colour's cell were assigned to data above reference thresholds.

The stations results which fall in the same cell and/or which have the same coordinates (e.g. data relating to the same station, but sampled in different years or related to differences species belonging to the same regulatory category) were integrated, as follows:

- if stations number with code 1 and 2 is different, the cell has code with the highest frequency.
- if stations number with code 1 is equal to stations number with code 2, the cell reaches code 2, in compliance with the precautionary principle [7];

So, in the cartographic representation, colour green had been assigned to cells with majority of station with code 1, while colour red to the cells with equal or majority of station with code 2.

Table 4. Statistical outputs for Dioxins and Dioxins like PCBs.

AS*(total cells 51)	Dioxins 5.3	Sum Dioxins and dioxin like 5.3
% of spatial coverage	-	56.9
% of data within the limits	-	100
% of data outside the limits	-	
% of station within the limits	-	100
% of station outside the limits	-	
% of cells within the limits	-	100
% of cells outside the limits	-	
WMS (total cells 166)**	**Dioxins 5.3**	**Sum Dioxins and dioxin like 5.3**
% of spatial coverage	1.2	4.2
% of data within the limits	100	92.9
% of data outside the limits		7.1
% of station within the limits	100	100
% of station outside the limits		
% of cells within the limits	100	100
% of cells outside the limits		
ISCMS* (total cells 77)**	**Dioxins 5.3**	**Sum Dioxins and dioxin like 5.3**
% of spatial coverage	1.3	1.3
% of data within the limits	87.5	100
% of data outside the limits	12.5	
% of station within the limits	85.7	100
% of station outside the limits	14.3	
% of cells within the limits	100	100
% of cells outside the limits		

* Adriatic Sea SubRegion;
** West Mediterranean Sea SubRegion;
*** Ionian Sea and Central Mediterranean Sea SubRegion.

For the three subregions, the percentage of spatial coverage for individual contaminants and the percentage of stations that exceed the limit (frequency of exceedances) were determined.

No review from the ethic committee was required as our research work did not involve any direct manipulation or disturbance of animals. Blood samples were collected by and under the permit of the Israel Nature and Parks Authority. Access to all sampling locations was not restricted and so no specific permissions were required.

Results and Discussion

Initial Assessment

The initial assessment examined substances for which maximum levels are established in fish, shellfish and other marine products destined for human consumption, defined a statistical range of acceptance, calculated the number and frequency of contaminants exceeding maximum regulatory levels and handed over these information to environmental quality assessment.

In the Mediterranean Sea approximately 15000 data were collected, concerning Metals (Hg, Cd, Pb), PAHs (benzo(a)pyrene and sum of benzo(a)pyrene, benz(a)anthracene, benzo(b)fluoranthene and chrysene) and dioxins (including dioxin-like PCBs).

Data related to Adriatic Sea were 7119 of which 5801 for Metals, 1086 for PAHs and 232 for HOCs.

In West Mediterranean Sea 1900 data were examined, of which 1344 for Metals, 513 for PAHs and 43 for HOCs. For Ionian Sea and Central Mediterranean Sea the number of analyzed data was equal to 5983, of which 5865 for Metals, 92 for PAHs and 26 for HOCs.

Data were elaborated and those unfitting in analytical quality (Annex III of Commission Regulation (EC) No 882/2004 [28]), in MSFD criteria and in GIS elaborations standards, were removed. Collected data concerned edible portion of different target species both from environmental and commercial value point of view, but also representing a coverage of the entire subregion as better as possible. Target species were principally coastal fish, tuna, mullet, mussels, shark, crustaceans, cephalopods, selected using Annex II of Task Group 9 Report [35]. For PAHs', bivalve molluscs were used as preferred species for initial assessment, being sedentary benthic species with a large lipid content [36,37].

Tables 2–4 show statistical outputs for each Subregion. Percentage of spatial coverage and percentage of data, stations and cells, within or outside regulatory limits, are reported. Regulatory levels (Tab. 1) are referred to categories established by Reg. 1881/2006 [23]. Limits of the sum of PAHs and benzo(a)pyrene in bivalve molluscs related to Reg. 835/2011/CE [24] while limits for s and dioxin like PCBs for muscle meat of fish and bivalve molluscs referred to Reg. 1259/2011/CE [26].

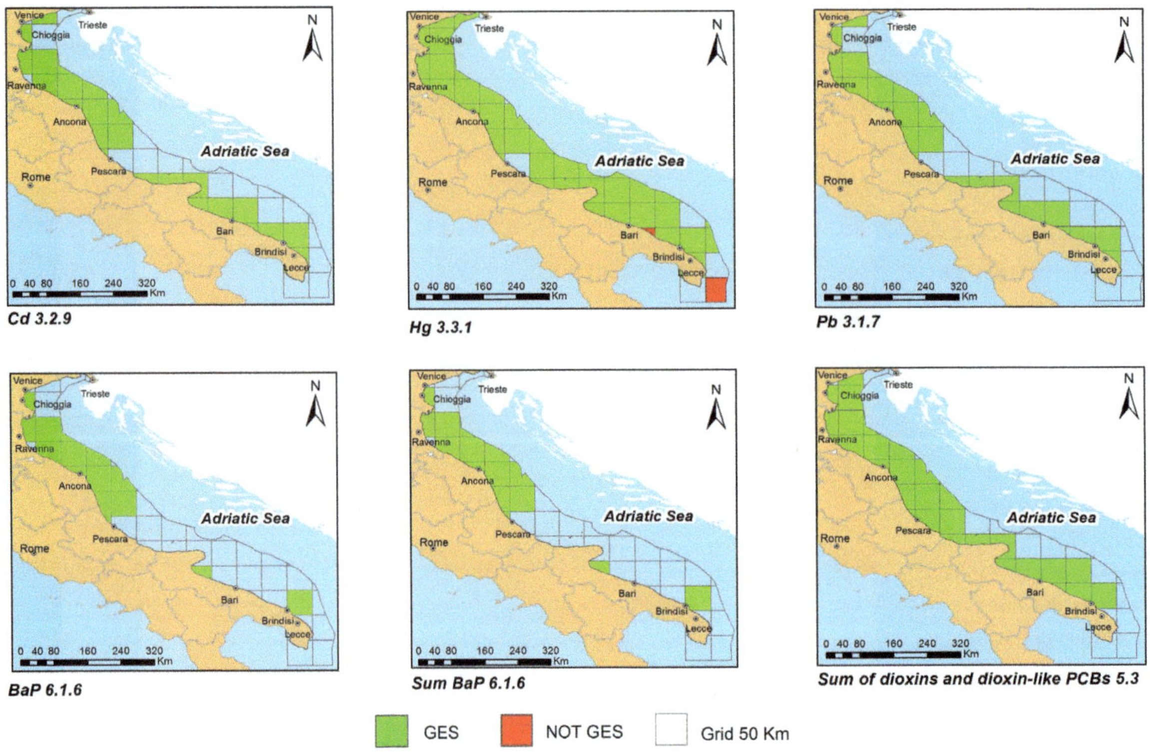

Figure 2. Results on Metals, PAH and Dioxins/Dioxin-like PCBs in Adriatic Sea Subregion (AS).

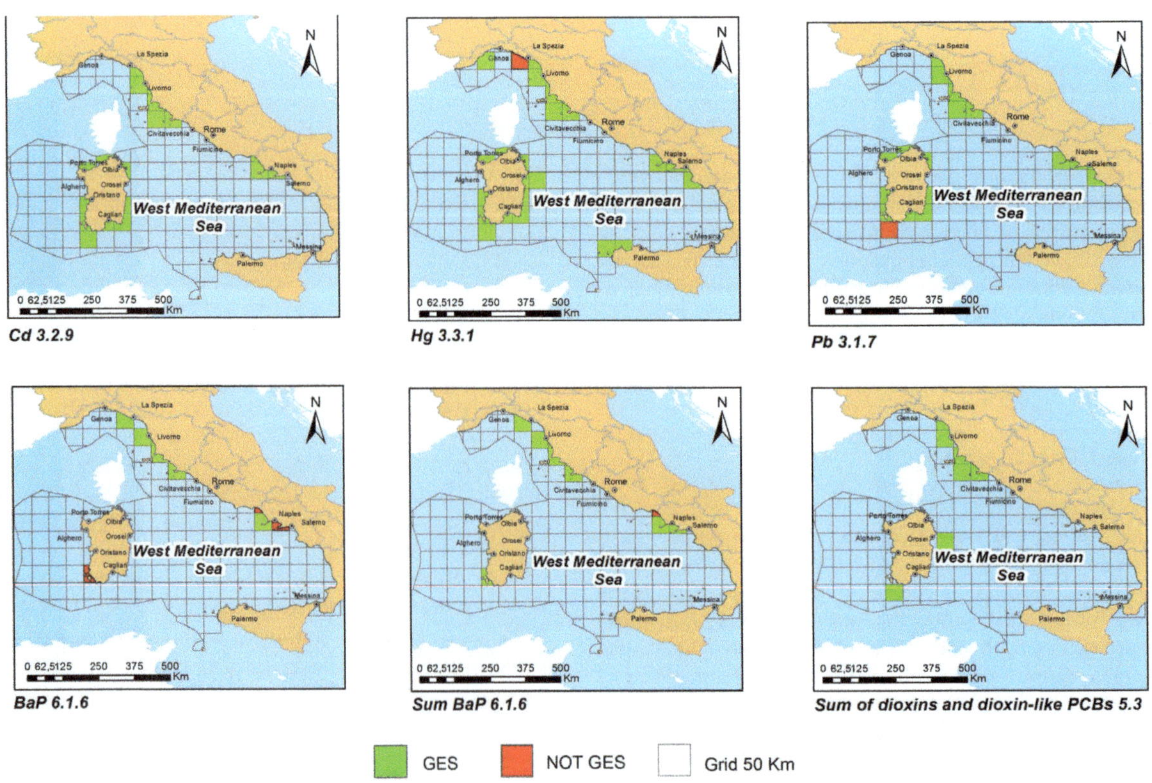

Figure 3. Results on Metals, PAH and Dioxins/Dioxin-like PCBs in Western Mediterranean (WMS).

Italy tried to hand over health information in terms of environmental assessment, without judging the outcome of a single sample, as requested by human consumption legislations, but integrating the results of all stations included into the same cell. For each regulatory category, different integration levels were performed: data within the same station; several stations falling in the same grid cell; same grid cell in different years.

The analyzed data showed that inorganic elements (metals) are usually more investigated than organic contaminants; indeed for metals the spatial coverage reaches 82% in AS subregion, while the maximum PAHs and sum Dioxins-dioxin like PCBs percentages are 29 and 57 respectively, still in AS subregion.

These data showed that the favourite organisms to assess the health risk are generally bivalve molluscs, as demonstrated by spatial coverage percentage of Cd 3.2.9, Hg 3.3.1(chiefly populated by molluscs), Pb 3.1.7 and BaP 6.1.6. Indeed, the mussel *Mytilus galloprovincialis* has all the desirable characteristics of a biological indicator and is the main species recommended to perform bio-monitoring activities; moreover, both the biology and ecology of the species are documented [38]. For these reasons mussel watch programs are widely diffused to evaluate environmental quality of marine water [38,39,40]. On the other hand, sedentary benthic species, having a high lipid content, are suggested as preferred species for PAHs' monitoring activities on human consumption [37,41].

Adriatic Sea (AS) subregion data referred to bivalve mollusc organisms, fishes (coastal fish, tuna, mullet, shark), crustaceans and cephalopods. In the Ionian Sea and Central Mediterranean Sea (ISCMS) subregion data distribution included mullet, nectobentonic fishes, bivalve molluscs, anchovies, sardines and seabream. Concerning Western Mediterranean (WMS) subregion, bivalve molluscs, crustaceans, mullet, seabream, seabass and flatfishes were studied.

Results on Metals, PAH and Dioxins/Dioxin-like compounds concerning to the seafood categories with greater spatial coverage are reported in figures 2 and 3. ISCMS results are not shown as the spatial coverage of data was very low compared to the number of subregion total cells.

Data were collected especially in coastal zones, where the main fishing activities take place; so the spatial coverage was often lacking in the open sea.

After integration, in all subregions, the environmental quality by metals was good, although some stations exceeded regulatory levels. In the AS subregion, Pb 3.1.5, and Hg 3.3.2 exceeded in muscle of twaite shad (*Alosa fallax*) and albacore tuna (*Thunnus alalunga*) while Hg 3.3.1 exceeded in blackbelly rosefish (*Helicolenus dactylopterus*), in thornback ray (*Raja clavata*) and in shellfish tissues. In the WMS subregion several stations exceeded regulatory levels concerning all codified categories of metals, except Cd 3.2.9. Indeed Pb 3.1.5–3.1.7 passed limits in fish muscle of seabream (*Dicentrarchus labrax*) and shells, Hg 3.3.1 in bivalves and Hg 3.3.2 in Atlantic bluefin tuna (*Thunnus thynnus*) and finally Cd 3.2.5–3.2.6 in fish muscle of seabream and in European anchovy (*Engraulis encrasicolus*). In ISCM subregion all data were below regulatory levels except Cd 3.2.6 in European anchovy, that covered only about 5% of the whole subregion.

These exceedances confirm the literature data regarding the different metal behaviour due to propensity to undergo biomagnification and to trophic level and foraging method/location of organisms. Thus, it is not surprising that tuna, that is at high trophic levels, had some excesses of mercury and that bottom feeding fishes had the excesses too. Animals living in close association with sediments, in which they bury and from where they mainly feed, are, in fact, more exposed to any sediment-associated contamination than other fish [42,43,44]. Another relevant point when evaluating the variability in data is the fish size. There is, in fact, a large body of data showing that mercury tissue concentrations increase with age/size of marine fauna [45,46,47]. Concerning Cd and Pb, the concentration ranges were quite narrow and similar, with exception of data from industrial sites in the coastal areas of ISCM and WMS subregions, suggesting that there is no trend in tissue residues of these metals based on taxonomic position or trophic status [48].

However, independently from the interaction complexity leading to a different toxic load among species, collected metals concentrations were in agreement with those reported in seafood from either the same environment [43,44,49] or other marine areas [50].

Concerning dioxins and dioxin-like compounds, in the AS and WMS subregions data were below the regulatory thresholds, while in the ISCMS only data referred to mussels (*Mytilus galloprovincialis*) showed some excesses; anyway the grid cells were all within the limits. The spatial cover for dioxins and dioxin-like compounds was higher in Adriatic subregion (56,9%), where the most part of mussel farming sites are located [51], while in the others subregions the coverage didn't reach 5% overall.

The monitoring of benzo(a)pyrene and the others congeners (benz(a)anthracene, benzo(b)fluoranthene and chrysene) was carried out in a systematic way in Adriatic Sea subregion, where data collected covered about 30% of the subregion and the environmental quality was good. Despite of that, in Western Mediterranean Sea and Ionian Sea and Central Mediterranean subregions the coverage percentage didn't go beyond 5% and some data in bivalve molluscs exceeded the regulatory thresholds for benzo(a)pyrene. It is well known that seasonal differences in biotic and abiotic factors regulate metabolic mechanisms and PAHs bioaccumulation and that bivalve molluscs are not able to metabolize PAHs, accumulating them in their organisms [49]. Nevertheless, several works described annual cyclic variation in the PAHs content of seafood, so that data from *Mytilus galloprovincialis* collected in winter exceeded regulatory levels [41,52,53].

Despite some exceedances, environmental quality of Italian subregions was good overall, regarding Metals, PAHs, Dioxins and Dioxin like compounds, taking into account coverage percentages too. Indeed, concentrations over thresholds were linked not only to the presence of the pollutant, but also to a whole range of biological (species, age, growth degree) and environmental (temperature, geochemical anomalies, salinity) factors which influence trace elements bioavailability [49,54,55,56]. Human health programmes effectively do not contemplate seasonal variation, age and sex which influence contaminants' bioavailability in fish and seafood. So higher levels of contaminants concentrations should not automatically be interpreted as a negative status.

Conclusion

The analysis of the collected data allowed to carry out the initial assessment on environmental quality of the 3 Italian subregions of the Mediterranean Sea. Results showed that the three subregions were overall in good status, since, for each substance or category, the greater part of the investigated cells had a concentration statistically lower than the acceptance range calculated assuming the regulatory levels as the 90 percentile of data distribution. However, it is needed to underline that data spatial coverage is limited, especially for synthetic compounds and, for this reason, results could have some limitations.

For this initial assessment Italy used data from health and environmental monitoring plans because existing national health monitoring programs didn't not cover all information needs for monitoring compliance to GES under Descriptor 9. It would be desirable improve monitoring activities considering Descriptor 9 in conjunction with requirements for Descriptor 8. Moreover it would be interesting to identify possible relations between contaminant levels in sediment and biota's, in order to explain the transfer of contaminants from sediment to biota and in order to establish a well-defined quantitative link between levels in fish and other seafood and levels of contaminants in marine environment. That might help to clearly demonstrate the transfer of them from marine environment to fish/fishery species and, at the end, to man.

So that researchers could best benefit from future monitoring programs (art.11 MSFD) and from information obtained, major adaptations would be needed regarding sampling plans' design, sampling procedures, matrix on which to perform the analysis as well as traceability to the location of catching or harvesting. In fact monitoring programs on human exposure use regulatory levels set for public health, but they often lack the necessary information to link the samples and results to specific subregions. In addition, these monitoring programs check out contaminants for which regulatory levels have been set, while other contaminants of relevance in fish and other seafood, distinguishing of a specific area, activity or input source, should be investigated to evaluate both spatial and temporal trends [35]. It would be opportune to set a limit also for total PAHs, organochlorine pesticides and organotin compounds. Infact, to date, the legislative Regulation 835/2011/CE [24] provides only a limit for the sum of 4 PAHs. Among toxic metals, only Pb, Cd and Hg have a limit set by E.C. Reg. 1881/2006 [23], while other metals also might become toxic when elevated doses are ingested [37,57].

Nevertheless there is still the need to improve the flow of information coming from institutions working in the health field, in order to validate the applied methodology and confirm the possibility to extend the use of health information arising from chemical contamination of seafood for the evaluation of the quality of the marine environment.

This shrewdness will be functional to fill part of the knowledge gaps that emerged from the initial assessment and nevertheless indispensable for a better definition of GES and Targets.

Acknowledgments

We are grateful to National Institute of Health, to Ministry of Health, to Experimental Zooprophylactic Institutes for provided data and to Prof. M.C. Fossi and coworkers for their expert advice and contribution that helped the completion of this work.

Author Contributions

Conceived and designed the experiments: CM. Performed the experiments: CM SL MTB. Analyzed the data: CM SL MTB MdA BDL. Contributed to the writing of the manuscript: CM SL MTB.

References

1. Ban N, Alder J (2008) How wild is the ocean? Assessing the intensity of anthropogenic marine activities in British Columbia, Canada. Aquatic Conserv.: Mar Freshw Ecosyst 18: 5–85.
2. Halpern BS, Walbridge S, Selkoe KA, Kappel CV, Micheli F (2008) A global map of human impact on marine ecosystems. Science 319: 948–952.
3. Borja A, Elliott M, Andersen JH, Cardoso A C, Carstensen J, et al. (2013) Good Environmental Status of marine ecosystems: What is it and how do we know when we have attained it? Mar Pollut Bull 76(1–2): 16–27.
4. Framework Directive 2008/56/EC (MSFD) on the Strategy for the Marine Environment.
5. Legislative Decree n. 190 of 13 October 2010.
6. Water Framework Directive (WFD) 2000/60/EC Directive.
7. Borja A, Elliott M, Carstensen J, Heiskanen A, van de Bund W (2010) Marine management – Towards an integrated implementation of the European Marine Strategy Framework and the Water Framework Directives. Mar Pollut Bull 60: 2175–2186.
8. http://ec.europa.eu/environment
9. Commission Decision 2010/477/EU of 1 September 2010.
10. Jakimska A, Konieczka P, Skóra K, Namieśnik J (2011) Bioaccumulation of metals in tissues of marine animals, part II: metal concentrations in animal tissues. Pol J Environ Stud 20(5): 1127–1146.
11. Cox GW (1997) Conservation Biology - 2nd ed. WCB.
12. Lang IA, Galloway TS, Scarlett A, Henley WE, Depledge M, et al. (2008) Association of urinary bisphenol A concentration with medical disorders and laboratory abnormalities in adults. JAMA 300(11): 1303–1310.
13. Melzer D, Rice NE, Lewis C, Henley WE, Galloway TS (2010) Association of urinary bisphenol a concentration with heart disease: evidence from NHANES 2003/06. PLoS One 5(1): 8673.
14. Gutierrez NL, Valencia SR, Branch TA, Agnew DJ, Baum JK, Bianchi PL, Cornejo-Donoso J, Costello C, Defeo O, Essington TE, Hilborn R, Hoggarth DD, LarsenAE, Ninnes C, Sainsbury K, Selden RL, Sistla S, Smith ADM, Stern-Pirlot A, Teck SJ, Thorson JT, Williams NE (2012) Eco-Label Conveys Reliable Information on Fish Stock Health to Seafood Consumers. PLoS One 7(8): e43765.
15. Fleming LE, Broad K, Clement A, Dewailly E, Elmira S, Knap A, Pomponi S A, Smith S, Solo Gabriele H, Walsh P (2006) Oceans and human health: Emerging public health risks in the marine environment. Mar Pollut Bull. 2006; 53(10–12): 545–560.
16. Kalay M, Ay Ö. and Canli M. (1999). Heavy metal concentrations in fish Tissues from the Northeast mediterrenean Sea. Bulletin of Environmental Contamination and Toxicology. 63: 673–681.
17. Gómez-Gutiérrez AI, Jover E, Bodineau L, Albaigés J, Bayona JM. (2006) Organic contaminant loads into the Western Mediterranean Sea: estimate of Ebro River inputs Chemosphere 65(2): 224–36.
18. El-Serehy HA, Aboulela H, Al-Misned F, Kaiser M, Al-Rasheid K, El-Din H E, (2012) Heavy metals contamination of a Mediterranean Coastal Ecosystem, Eastern Nile Delta, Egypt. Turk. J. Fish. Aquat. Sci. 12: 751–760.
19. Baumard P, Budzinski H, Garrigues P (May1998a) Polycyclic aromatic hydrocarbons in sediments and mussels of the western Mediterranean sea. Environ Toxicol Chem 17(5) 765–776.
20. Baumard P, Budzinski H, Michon Q, Garrigues P, Burgeot T, Bellocq J (July 1998b) Origin and Bioavailability of PAHs in the Mediterranean Sea from Mussel and Sediment Records. Estuar Coast Shelf S 47(1) 77–90.
21. Copat C, Bella F, Marine Castaing, Fallico R, Sciacca S, FerranteM, (2012) Heavy Metals Concentrations in Fish from Sicily (Mediterranean Sea) and Evaluation of Possible Health Risks to Consumers Bull Environ Contam Toxicol. 88(1): 78–83.
22. Miniero R, De Felip E, Magliulo M, Ferri F, Di Domenico A (2005) Selected persistent organic pollutants (POPs) in the Italian environment. Ann Ist Super Sanita 41(4): 487–492.
23. Commission Regulation (EU) No 1881/2006 of 19 December 2006.
24. Commission Regulation (EU) No 835/2011 of 19 August 2011.
25. Commission Regulation (EU) No 629/2008 of 2 July 2008.
26. Commission Regulation (EU) No 1259/2011 of 2 December 2011.
27. Legislative Decree 152/2006 Environment Regulation of April 29, 2006.
28. Commission Regulation (EU) No 882/2004 of 29 April 2004.
29. Commission Regulation (EU) No 1883/2006 of 19 December 2006.
30. Commission Regulation (EU) No 333/2007 of 28 March 2007.
31. Esman NA, Hammad YY (1977) Log-normality of environmental sampling data, Journal of Environmental Science and Health, Part A: Environmental Science and Engineering 12(1–2): 29–41.
32. Zou GY, Huo CY, Taleban J (2009) Simple confidence intervals for lognormal means and their differences with environmental applications - Environmetrics - Wiley Online Library.
33. Decreto Ministeriale Ambiente 14 aprile 2009, n. 56.
34. Cameron A, Askew N, (2011) EUSeaMap - Preparatory Action for development and assessment of a European broad-scale seabed habitat map final report. Available: http://jncc.gov.uk/euseamap
35. Swartenbroux F, Albajedo B, Angelidis M, Aulne M, Bartkevics V, et al. (2010) Marine Strategy Framework Directive – Task Group 9 Report Contaminants in fish and other seafood. EUR 24339 EN–2010. http://publications.jrc.ec.europa.eu/repository/handle/111111111/13669.
36. Perugini M, Visciano P, Giammarino A, Manera M, Di Nardo W, et al. (2007b) Polycyclic aromatic hydrocarbons in marine organisms from the Adriatic Sea, Italy. Chemosphere 66: 1904–1910.
37. Conti GO, Copat C, Ledda C, Fiore M, Fallico R, et al. (2012) Evaluation of heavy metals and Polycyclic Aromatic Hydrocarbons (PAHs) in *Mullus barbatus* from Sicily Channel and risk-based consumption limits. Bull Environ Contam Toxicol 88: 946–950.

38. Galgani F, Martínez-Gómez C, Giovanardi F, Romanelli G, Caixach J, et al. (2010) Assessment of polycyclic aromatic hydrocarbon concentrations in mussels (*Mytilus galloprovincialis*) from the Western basin of the Mediterranean Sea. Environ Monit Assess 172(1–4): 301–317.
39. Andral B, Galgani F, Tomasino C, Bouchoucha M, Blottiere C, et al. (2011) Chemical contamination baseline in the Western basin of the Mediterranean sea based on transplanted mussels. Arch Environ Contam Toxicol, 61(2): 261–271.
40. Scarpato A, Romanelli G, Galgani F, Andral B, Amici M, et al. (2010) Western Mediterranean coastal waters–Monitoring PCBs and pesticides accumulation in *Mytilus galloprovincialis* by active mussel watching: the Mytilos project. J Environ Monit 12(4): 924–935.
41. Perugini M, Visciano P, Manera M, Turno G, Lucisano A, et al. (2007a) Polycyclic aromatic hydrocarbons in marine organisms from the Gulf of Naples, Tyrrhenian Sea. J Agric Food Chem 55: 2049–2054.
42. Storelli MM, Giacominelli Stuffler R, Marcotrigiano GO (1998) Total mercury in muscle of benthic and pelagic fish from the South Adriatic Sea (Italy). Food Addit Contam 15: 876–883.
43. Storelli MM, Giacominelli Stuffler R, Storelli A, D'Addabbo R, Palermo C, et al. (2003) Survey of total mercury and methylmercury levels in edible fish from the Adriatic Sea. Food Addit Contam 20: 1114–1119.
44. Storelli MM, Giacominelli Stuffler R, Storelli A, Marcotrigiano GO, (2006) Cadmium and mercury in cephalopod molluscs: estimated weekly intake. Food Addit Contam 23: 25–30.
45. Joiris CR, Holsbeek L, Moatemri NL (1999) Total and methylmercury in sardines Sardinella aurita and Sardina pilchardus from Tunisia. Mar Pollut Bull 38: 188–192.
46. Storelli MM, Marcotrigiano GO (2000) Fish for human consumption: risk of contamination by mercury. Food Addit Contam 17: 1007–1011.
47. Storelli MM, Barone G, Piscitelli G, Marcotrigiano GO (2007) Mercury in fish: concentration versus fish-size and estimates intake dose. Food Addit Contam 6: 1–5.
48. Franca S, Vinagre C, Cacador I, Cabral HN (2005) Heavy metal concentrations in sediment, benthic invertebrates and fish in three salt marsh areas subjected to different pollution loads in the Tagus Estuary (Portugal). Mar Pollut Bull 50: 998–1003.
49. Martí-Cid R, Bocio A, Llobet JM, Domingo JL (2007) Intake of chemical contaminants through fish and seafood consumption by children of Catatonia, Spain: health risks. Food Chem Toxicol 45: 1968–1974.
50. Sivaperumal P, Sankar TV, Viswanathan Nair PG (2007) Heavy metal concentrations in fish, shellfish and fish products from internal markets of India vis-a-vis international standards. Food Chem 102: 612–620.
51. Prioli G, (2001) Censimento Nazionale sulla Molluschicoltura del Consorzio Unimar. 114 pp.
52. Marrone R, Mercogliano R, Palma G, Chirollo C, Smaldone G, et al. (2011) Presenza di idrocarburi policiclici aromatici in prodotti della piccola pesca del golfo di Napoli. AIVI online 1: 61–65.
53. Tornero V, d'Alcalà MR (2014) Contamination by hazardous substances in the Gulf of Naples nearby coastal areas: a review of sources, environmental levels and potential impacts in the MSFD perspective. Sci Total Environ 466–467: 820–840.
54. Jureša D, Blanuša M (2003) Mercury, arsenic, lead and cadmium in fish and shellfish from the Adritic Sea. Food Addit Contam 20(3): 241–246.
55. Storelli MM. (2008) Potential human health risks from metals (Hg, Cd, and Pb) and polychlorinated biphenyls (PCBs) via seafood consumption: Estimation of target hazard quotients (THQs) and toxic equivalents (TEQs). Food Chem. Toxicol. 46: 2782–2788.
56. Pastorelli AA, Baldini M, Stacchini P, Baldini G, Morelli Set al. (2012) Human exposure to lead, cadmium and mercury through fish and seafood product consumption in Italy: a pilot evaluation. Food Addit Contam Part A 29(12): 1913–1921.
57. Menzie CA, Ziccardi LM, Lowney YM, Fairbrother A, Shock SS, et al. (2009) Importance of considering the framework principles in risk assessment for metals. Environ Sci Technol 43: 8478–8482.

Marine Plastic Pollution in Waters around Australia: Characteristics, Concentrations, and Pathways

Julia Reisser[1,2,3]*, Jeremy Shaw[4], Chris Wilcox[3,5], Britta Denise Hardesty[3,5], Maira Proietti[6], Michele Thums[1,2,7], Charitha Pattiaratchi[1,2]

1 School of Environmental Systems Engineering, University of Western Australia, Perth, Western Australia, Australia, **2** Oceans Institute, University of Western Australia, Perth, Western Australia, Australia, **3** Wealth from Oceans Flagship, Commonwealth Scientific and Industrial Research Organisation, Floreat, Western Australia, Australia, **4** Centre for Microscopy, Characterisation and Analysis, University of Western Australia, Perth, Western Australia, Australia, **5** Marine and Atmospheric Research, Commonwealth Scientific and Industrial Research Organisation, Hobart, Tasmania, Australia, **6** Instituto de Oceanografia, Universidade Federal do Rio Grande, Rio Grande, Rio Grande do Sul, Brazil, **7** Australian Institute of Marine Science, Perth, Western Australia, Australia

Abstract

Plastics represent the vast majority of human-made debris present in the oceans. However, their characteristics, accumulation zones, and transport pathways remain poorly assessed. We characterised and estimated the concentration of marine plastics in waters around Australia using surface net tows, and inferred their potential pathways using particle-tracking models and real drifter trajectories. The 839 marine plastics recorded were predominantly small fragments ("microplastics", median length = 2.8 mm, mean length = 4.9 mm) resulting from the breakdown of larger objects made of polyethylene and polypropylene (e.g. packaging and fishing items). Mean sea surface plastic concentration was 4256.4 pieces km^{-2}, and after incorporating the effect of vertical wind mixing, this value increased to 8966.3 pieces km^{-2}. These plastics appear to be associated with a wide range of ocean currents that connect the sampled sites to their international and domestic sources, including populated areas of Australia's east coast. This study shows that plastic contamination levels in surface waters of Australia are similar to those in the Caribbean Sea and Gulf of Maine, but considerably lower than those found in the subtropical gyres and Mediterranean Sea. Microplastics such as the ones described here have the potential to affect organisms ranging from megafauna to small fish and zooplankton.

Editor: Graeme Clive Hays, University of Wales Swansea, United Kingdom

Funding: This project was funded by University of Western Australia (http://www.uwa.edu.au) and Commonwealth Scientific and Industrial Research (http://www.csiro.au). It has also been supported by Australia's Marine National Facility, Austral Fisheries, Australian Institute of Marine Science, CSIRO's Flagship postgraduate scholarship (JR), and the Shell social investment program (BDH and CW). The funders had no role in study design, data collection and analysis, decision to publish, or preparation of the manuscript.

Competing Interests: The authors have the following interests. This study was partly funded by Austral Fisheries and the Shell social investment program. There are no patents, products in development or marketed products to declare.

* E-mail: jureisser@gmail.com

Introduction

Plastics are a diverse group of materials derived from petrochemicals [1]. Their global production has grown exponentially from 1,700,000 tonnes in 1950 to 280,000,000 tonnes in 2011 [2]. The disposability of plastics, together with their low recycling rates, has contributed to a significant rise in the amount of waste produced globally [3]. For instance, in Australia, 1,433,046 tonnes of plastics were used in 2010–2011, of which only 20% was recycled. Moreover, around 37% of this plastic was for the manufacturing of single-use disposable packaging [4]. Plastics are transported from populated areas to the marine environment by rivers, wind, tides, rainwater, storm drains, sewage disposal, and even flood events. It can also reach the sea from vessels (e.g. fishing gear) and offshore installations [5]. Once in the oceans, they will either float at the ocean surface, or sink to the seafloor if made from polymers denser than seawater [6]. Buoyant plastics may be cast ashore by inshore currents or winds [7], or may enter the open ocean, where they tend to accumulate in convergence zones such as the ones formed by the five large-scale gyres (South and North Pacific, South and North Atlantic, and Indian [8–10]).

Marine plastics are known to undergo fragmentation into increasingly smaller pieces by photochemical, mechanical and biological processes [6,11]. Plastics are also directly manufactured in small sizes (<5mm), which may find their way into the oceans. These include virgin plastic pellets (pelletwatch.org; [12]), synthetic fibers from clothes [13], micro beads from cosmetics [14], and synthetic 'sandblasting' media [6]. There is increasing awareness that these small plastic particles (often called microplastics when smaller than 5 mm [6]) represent a significant proportion of the human-made debris present in the oceans. However, their at-sea spatial and temporal dynamics remain poorly assessed, mostly due to a lack of data on their characteristics and at-sea occurrence [15,16]. In Australia, the only published information on microplastics comes from a global study that recorded their occurrence in the sediments of Busselton beach (Western Australia) and Port Douglas (Queensland) [13]. Apart from this, our current knowledge on plastic contamination in the Australian marine environment is restricted to (1) beach litter cleanups that record mainly the occurrence of relatively large objects (e.g. [17–19]), (2) land-based surveys of marine megafauna impacted by marine debris (e.g. [17,20–22]), and (3) inferences based on plastic pollution reports from New Zealand (e.g. [23]).

The impacts of plastics on marine vertebrates, such as turtles, mammals and birds, have been well recognized since the 80's [24,25]. However, only recently has concern about the effects of small plastic particles on food webs and marine ecosystems been raised. More than half of modern plastics contain at least one hazardous ingredient [26] and those that end up in aquatic systems can become increasingly toxic by adsorbing persistent organic pollutants on their surface [27]. These concentrated toxins might then be delivered to animals via plastic ingestion and/or endocytosis [28,29] and transferred up their food webs [30–32]. This bio-magnification process is more likely to happen when plastics are small enough to be ingested by organisms that are close to the bottom of the ocean food web, such as planktivorous fish [33] and zooplankton [34]. For instance, it was inferred that small plastic particles found in the stomach contents of Southern Bluefin tuna captured close to Tasmania [35] were coming from the guts of their prey: myctophid fish [36]. In this scenario, plastic contaminants can be transferred to the affected organism and then biomagnified up the food chain. If this process is taking place, plastics can affect the health of food webs, which include humans as an apex predator.

Australia's acknowledgement of plastic threats to marine ecosystems is mostly limited to impacts from relatively large debris (e.g. abandoned fishing nets, plastic bags) on marine megafauna (e.g. turtles, mammals, birds) [37]. A first step towards a better understanding of the extent of marine plastic hazards to Australian organisms and environments is a better assessment of the occurrence and characteristics of plastic debris at-sea. To this end, we characterized (size, type, color, polymer) and estimated concentration (pieces km^{-2}) of plastics in waters around Australia using surface net tows. Additionally, potential pathways taken by the collected plastics were inferred using outputs of a dispersal model and trajectories of satellite-tracked drifting buoys.

Materials and Methods

Ethics Statement

Permits to conduct this field research were obtained from the Great Barrier Reef Marine Park Authority (GBRMPA: permit G11/34378.1). No other special permitting was required because sampling was limited to the collection of marine debris.

During seven transit voyages aboard Australian vessels (Figure 1), we undertook three consecutive 15-minute net tows (mean ± standard deviation tow length = 1.3±0.50 km) at 57 locations (hereafter called "net stations"), while the ship was travelling at a speed of 2 – 4 knots. These net tows sampled the air-sea interface, using a Neuston net (1.2×0.6 m mouth, 335 µm mesh) or a Manta net (1×0.17 m mouth, 333 µm mesh). After each net tow, the collected material was transferred to a container filled with seawater and examined for floating plastic pieces for at least an hour by a trained observer (J.R.). Each plastic piece was picked up with forceps and placed in a graduated dish to be counted, measured (length), photographed and classified into type (hard, soft, line, expanded polystyrene, pellet), and color. A random sample of 200 plastic pieces was selected for polymer composition analysis by Fourier transform infrared spectrometry (FT-IR; range = 500 – 4000 cm^{-1}). Polymer type was determined by comparing sample FT-IR spectra against known spectra from a database (Perkin-Elmer ATR of Polymers Library).

To estimate sea surface plastic concentrations (Cs, pieces km^{-2}), we first divided the number of plastic pieces found in the cod-end of each net tow by its towed area, which was estimated by multiplying net mouth width by tow length (determined from GPS position data). Mean Cs was then estimated for each of the 57 net stations by averaging the Cs of its three net tows. To our knowledge, this is the first study to take net tow replicates for marine plastic sampling. Apart from providing us measurements of Cs variability, our approach (i.e. execution of 3 short net tows instead of 1 long trawl) also avoided net clogging by gelatinous zooplankton.

Since buoyant plastics are vertically distributed due to wind-driven mixing, we also estimated depth-integrated plastic concentrations (Ci, pieces km^{-2}) by applying a one-dimensional column model [15]:

$$C_i = \frac{C_s}{1 - e^{-dw_b A_O^{-1}}}$$

Where:

d = immersion depth of the surface-towed net; equal to 0.17 m for the Manta net tows (full immersion of the net frame) and 0.3 m for the Neuston net tows (half of the frame immersed).

w_b = buoyant rise velocity of marine plastics; equal to 0.02 m s^{-1}. Preliminary experiments indicate that it ranges from 0.005 – 0.035 m s^{-1} [15].

A_o = near-surface turbulent (eddy) exchange coefficient, which was estimated by:

$$A_O = 1.5 u_{*w} k H_s$$

Where:

k = von Karman constant; equal to 0.4.

H_s = significant wave height (m).

u_{*w} = frictional velocity of water (m s^{-1}).

Both H_s and u_{*w} were taken from the ERA-Interim model [38]. There was a considerable similarity between wind fields of the ERA-Interim forecast model (U_{10}) and the wind speed measured by an anemometer (w) on five of our seven voyages (U_{10} = 0.85+1.04w, r^2 = 0.79, N = 39 net stations), indicating that the use of the model outputs is adequate.

To infer potential pathways taken by the collected plastics, we used two approaches: (1) application of the Australian Connectivity Interface Connie2 (csiro.au/connie2), and (2) trajectories of satellite-tracked buoys from the Global Drifter Program (aoml.noaa.gov/phod/dac). In our first approach, an area of 0.1° latitude by 0.1° longitude was created around each net station and particle-tracking models were run backwards in time. Particles were released within these areas over a 30-day period (25 particles per day), and subsequently tracked for a dispersal time equal to 45 days. These models were forced by averaged ocean current fields (2002 – 2006) of the month when the net station was sampled. Details of the particle tracking model, and the eddy-resolving/data-assimilating ocean general circulation model can be found in [39] and [40], respectively. In our second approach, an area of 4° latitude by 4° longitude was centered on each net station and drifters (drogued and un-drogued) that reached these regions were selected. The tracks starting from the drifter release point until they entered one of the net station areas were then plotted onto maps.

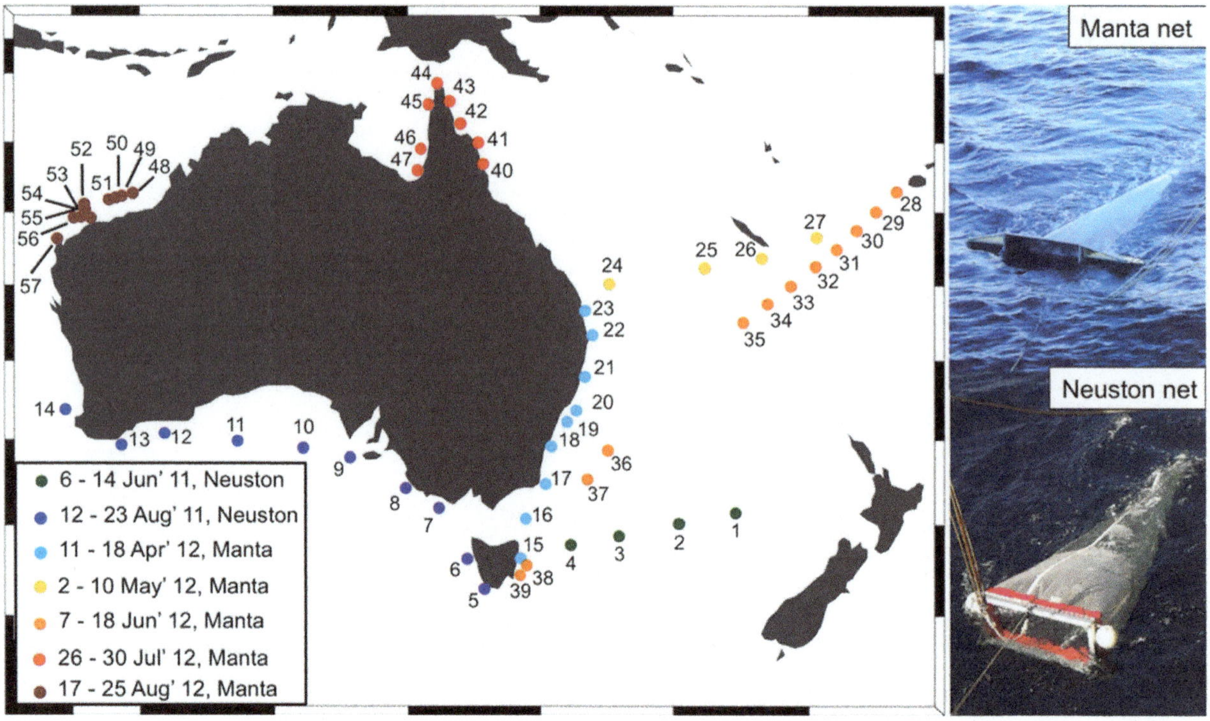

Figure 1. Location of the 57 net stations sampled during this study. Dot colors indicate the voyage when the net station was sampled and numbers follow the chronological order of sampling. Pictures of the two types of net used are shown in the right panel.

Results

We recorded 839 pieces of plastic, ranging in length from 0.4 to 82.6 mm (median = 2.8 mm, mean ± standard error = 4.9±0.27 mm, Figure 2). The majority of these plastic pieces had low circularity in their shape when compared to manufactured plastic particles (e.g. pellets and microbeads from cosmetics), suggesting they mostly resulted from the breakdown of larger items. The main plastic type was hard plastic (N = 633, median length = 2.4 mm, range = 0.7 – 57.0 mm) followed by soft plastic (N = 142, median length = 5.0 mm, range = 0.5 – 73.0 mm), plastic line (N = 54, median length = 10.3 mm, range = 2.0 – 82.6 mm), expanded polystyrene (N = 8, median length = 2.9 mm, range = 1.3 – 24.3 mm), and pellet (N = 2, both 4 mm). Most plastics were white/transparent (84.7%), but blue (8.3%) and other colors (7%) were also present. Of the 200 pieces subjected to FT-IR, 67.5% were made of polyethylene, 31% of polypropylene, 1% of expanded polystyrene, and 0.5% of ethylene vinyl acetate (Figure 3).

Approximately 80% of our net tows (136 out of 171), and 93% of our net stations (53 out of 57), had at least one piece of plastic (range: 0 – 68, median = 2, mean ± standard error = 4.9±0.63 pieces per net tow). Estimated sea surface plastic concentrations (*Cs*) for each net tow ranged from 0 to 48895.6 pieces km^{-2} (median = 1932.1 pieces km^{-2}, mean ± standard error = 4256.4±757.79 pieces km^{-2}) and the mean *Cs* of net stations varied between 0 and 23610.7 pieces km^{-2} (Figure 4, Table S1).

Relatively high mean *Cs* (>15500 pieces km^{-2}) were estimated only at low wind speeds (<7 m s^{-1}, Figure 5a). There was an inverse relationship between *Cs* and wind forcing ($b = -0.77$ in $Cs = a(u_{*w})^b$), which was relatively consistent with the biophysical model applied here (Figure 5b). When taking into account the effect of wind-mixing, net tow plastic concentrations increased by a mean factor of 2.8 (range: 1.04 – 10.0, median = 1.9). Hence, the amount of plastics collected by our net tows (*Cs*) represents anywhere between 10.0% and 96.1% (median = 52.7%, mean ± standard deviation = 50.0±24.47%) of the estimated total amount of plastic present in the water column (*Ci*, Figure 6).

Depth-integrated plastic concentration estimates (*Ci*) for each net tow ranged from 0 to 105438.6 pieces km^{-2} (median = 4363.7 pieces km^{-2}, mean ± standard error = 8966.3±1330.75 pieces km^{-2}) and the mean *Ci* of net stations ranged from 0 to 43194.5 pieces km^{-2} (Figure 7). In this scenario, plastic concentrations higher than 15500 pieces km^{-2} (red dots) were quite common, and those higher than 31500 pieces km^{-2} (dark red dots) were found close to populated areas (Brisbane and Fiji) as well as in some remote coastal regions (southwest Tasmania) and oceanic areas (Figure 7).

A wide range of pathways was taken by the virtual particles arriving at the net stations (Figure 8 and Maps S1). The routes taken by real drifters, from their release points to the net stations, showed similar patterns but covered larger areas due to their longer drifting time and wider range of release date (Figure 9 and Maps S2).

Discussion

We found that the surface waters around Australia are contaminated with small plastics that are mostly a by-product of the degradation of larger objects made of polyethylene and polypropylene. The high prevalence of plastic fragments smaller than 5 mm in Australian waters is consistent with other regions of the world's oceans, where microplastics were found to be the most abundant type of debris in all types of marine environment [8–10,13,41,42]. Plastic pollution levels were moderate when compared to concentrations in other marine areas [8–10,43,44]. Higher amounts of plastic were found close to cities on Australia's

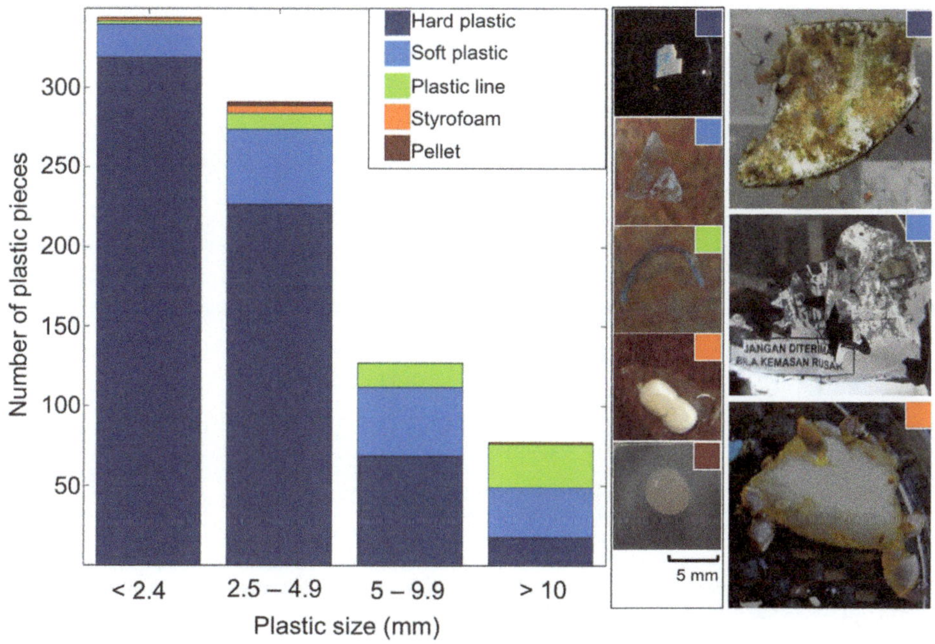

Figure 2. Size and types of marine plastics collected around Australia. Bars indicate the number of plastic pieces within each size category (<2.5, 2.5 – 4.9, 5 – 10, >10 mm) and colors show the amount of each plastic type within size categories. Examples of the types of plastic we collected are shown in the photos, including our biggest fragment of hard plastic (length = 57 mm, net station 32), soft plastic (length = 73 mm, net station 57, note the Indonesian words), and expanded polystyrene (Styrofoam cup fragment, length = 24.3 mm, net station 28).

east coast, as well as in remote locations (west Tasmania and North West Shelf). Recent studies reported toxicological effects of these small and contaminated plastics on a host of organisms, including large marine vertebrates [45] and fish [30–32,46]. As such, small plastics are a type of harmful marine debris, implying

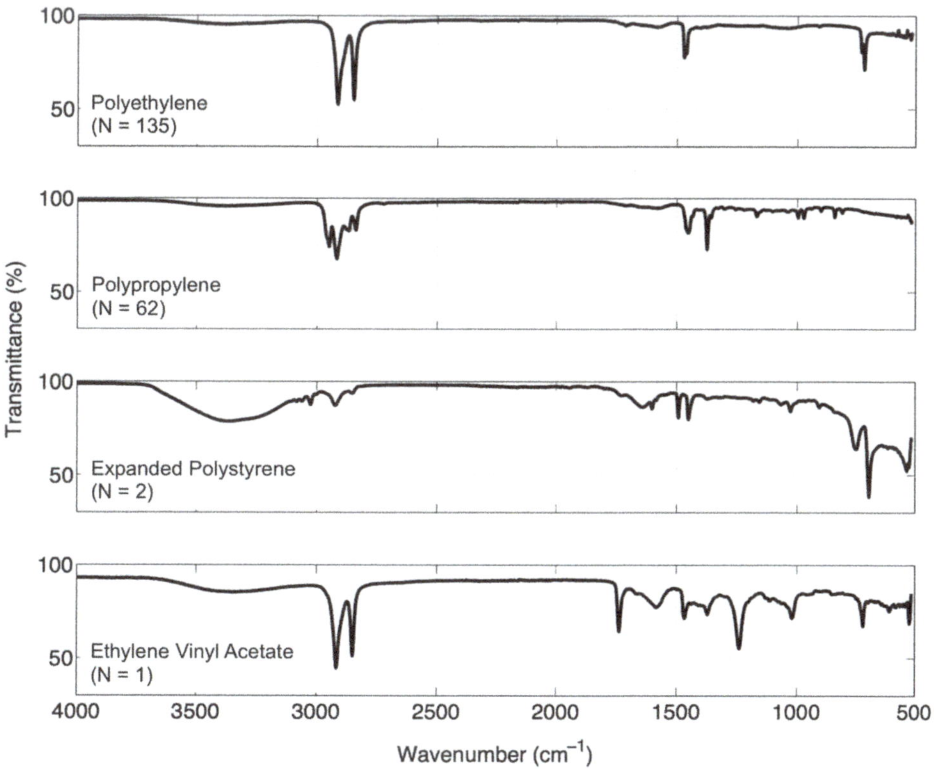

Figure 3. Mean infrared spectra of the plastic pieces within each polymer type.

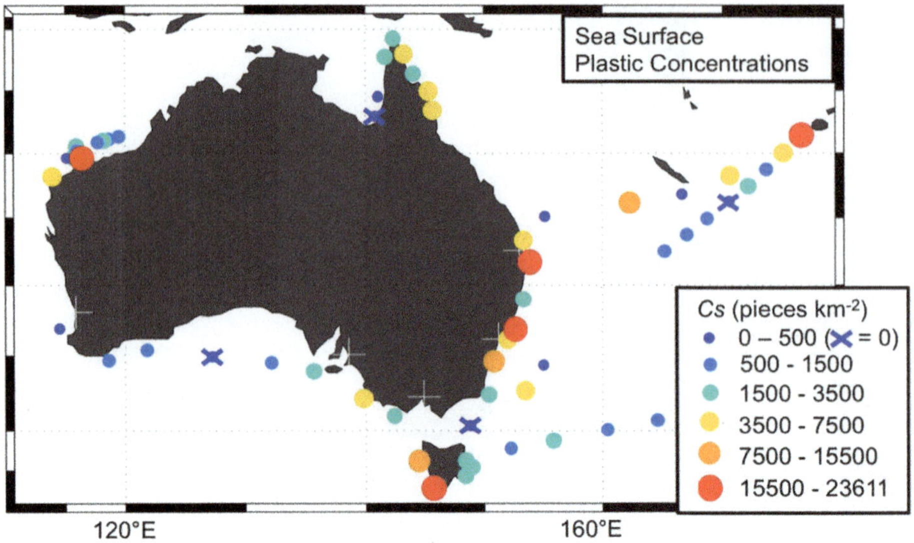

Figure 4. Mean sea surface plastic concentration (*Cs*) at the 57 net stations. White crosses indicate location of major Australian cities (population >1 million). From west to east: Perth, Adelaide, Melbourne, Sydney, and Brisbane.

that plastic hazards to Australian species and ecological communities are likely to be broader than those officially recognized.

Characteristics of marine plastics

Captured plastic particles ranged in size from 0.4 – 82.6 mm. The frequency distribution of different sized plastics, which was skewed towards smaller particles, provides evidence for the existence of smaller plastics. Current methods for assessing plastic pollution at the ocean surface rely on the use of nets, which omits plastic particles outside the collectible range of their mesh [47]. It will be critical for future investigations to develop efficient and reproducible techniques capable of detecting smaller buoyant

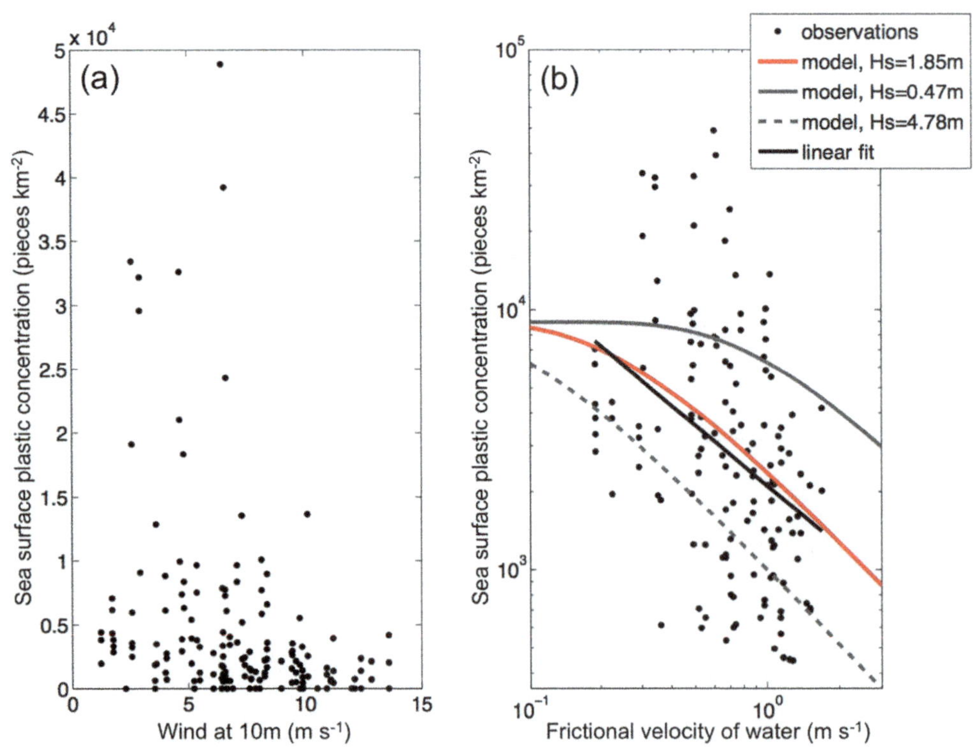

Figure 5. Sea surface plastic concentration (*Cs*) versus a) wind speed (U_{10}) and b) water friction velocity (u_{*w}). In (b) we also show the linear fit ($Cs = a\,(u_{*w})^b$) and theoretical model estimates for *Cs*, when depth-integrated plastic concentration (C_i) is equal to 8966 (mean *Ci* of the 171 net tows) and significant wave height (*Hs*) is equal to the mean (1.85 m), maximum (4.78 m) and minimum (0.47 m) values estimated for the 57 net stations.

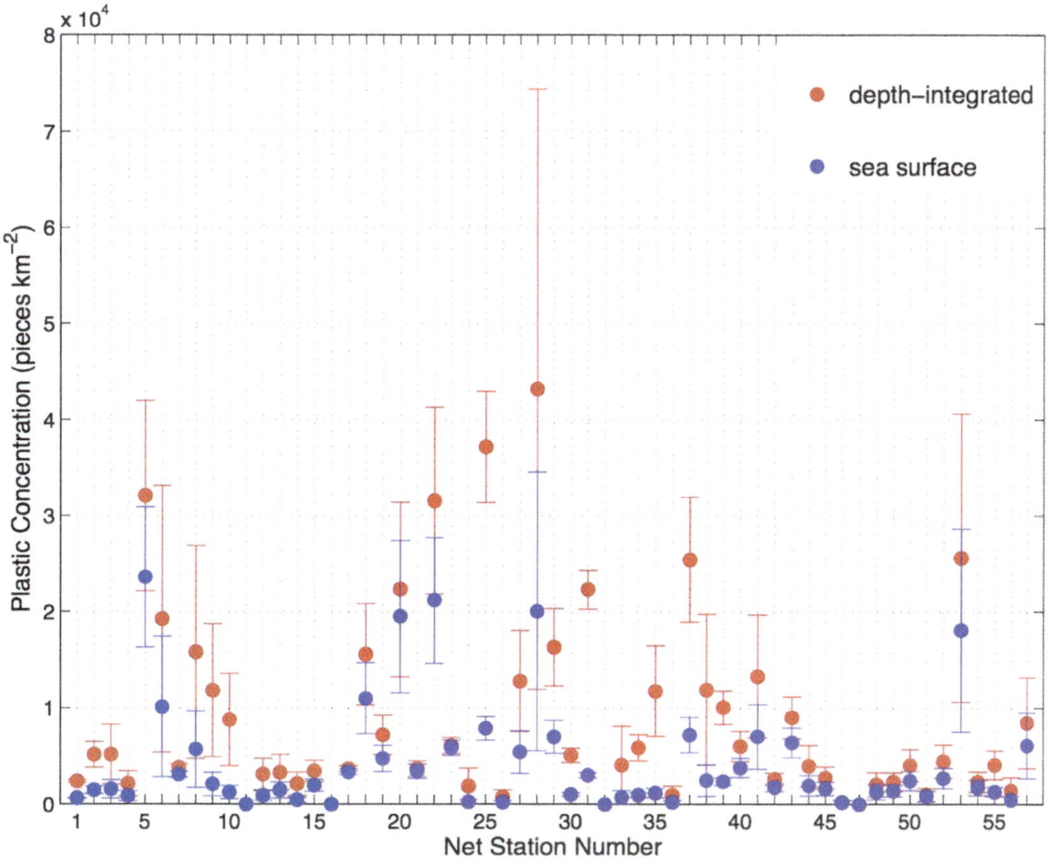

Figure 6. Mean and standard error of sea surface (*Cs*) and depth-integrated (*Ci*) plastic concentrations. Blue represents mean and standard error of *Cs* and red represents mean and standard error of *Ci*.

plastic particles (micro and nanoparticles). In addition, post processing techniques for sorting particles are also likely to miss small fragments [47]. An example of a new method with the potential to eliminate this limitation is the application of molecular mapping by reflectance micro-FT-IR spectroscopy, which does

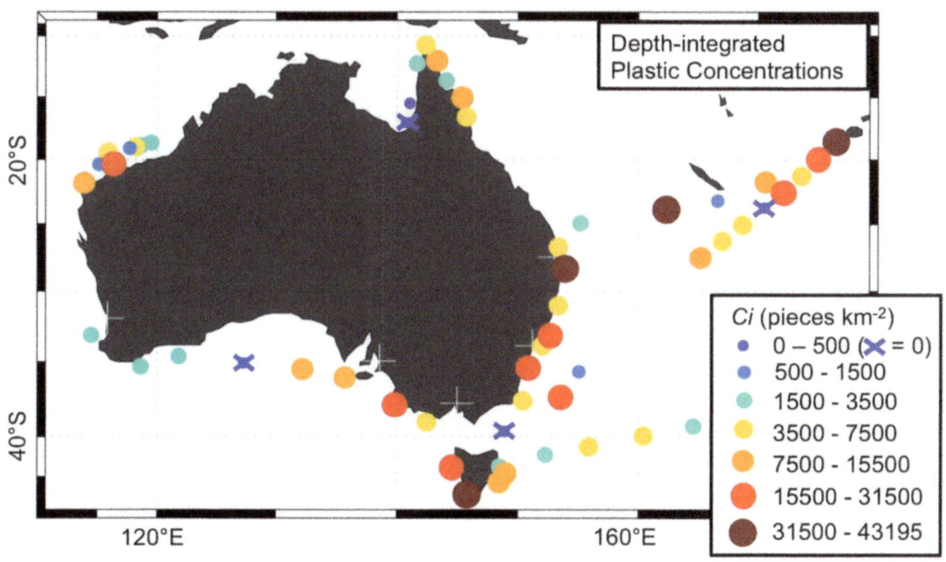

Figure 7. Mean depth-integrated plastic concentration (*Ci*) at the 57 net stations. White crosses indicate location of major Australian cities (population >1 million). From west to east: Perth, Adelaide, Melbourne, Sydney, and Brisbane.

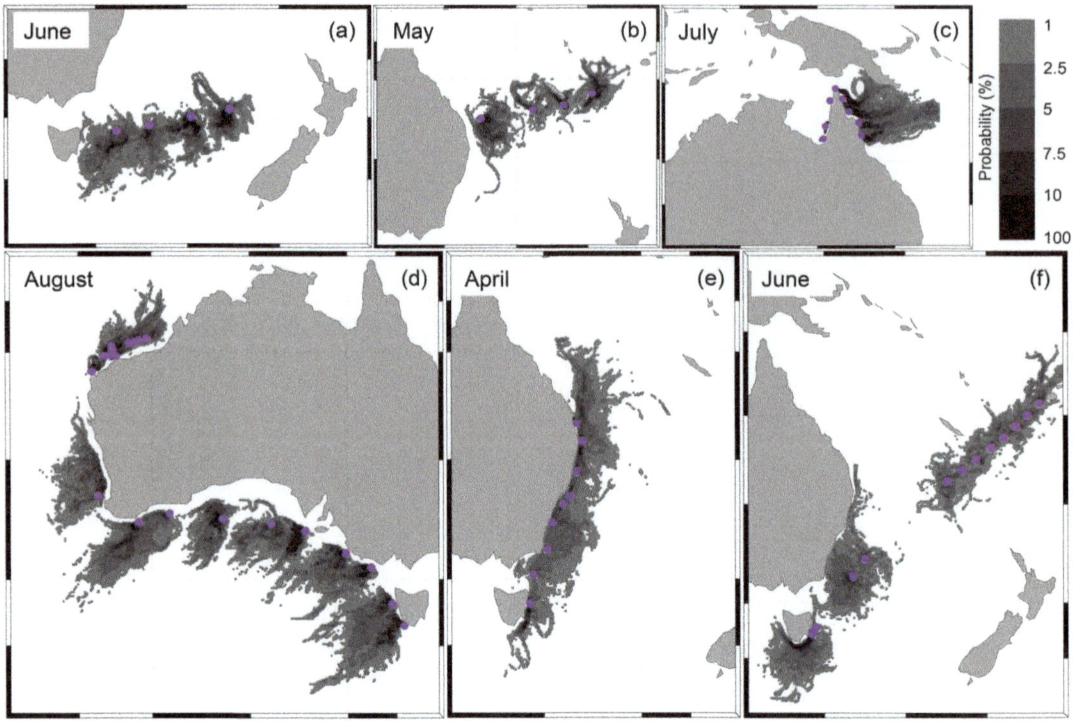

Figure 8. Cumulative probability distribution of virtual particles arriving at the 57 net stations. The month when the virtual particles (25 per day) were released is indicated in each panel. Backtracking dispersal time was equal to 45 days and arriving destinations (net stations) are marked with purple dots. See also Maps S1.

not rely on visual selection of plastic particles for characterization [48].

Hard plastics were by far the most common plastic type found (75.4%), but soft plastics (e.g. fragments of plastic wrappers) and lines (mostly fishing lines) were also relatively common (16.5% and 6.4%, respectively). It is interesting to note that soft plastics were more abundant in the larger size class (>2.4 mm). Our findings are consistent with recent studies documenting plastic pollution at the ocean surface, although explanations for variations in hard/soft plastic trends are not given [8,10,49]. Plastics gradually lose buoyancy in seawater as a result of biofilm formation [50]. We suggest that negative buoyancy due to biofouling occurs more quickly in soft/thin than in hard/thicker plastic fragments, resulting in a decline in the occurrence of soft plastics at the ocean surface, as they become smaller/older and begin to sink. Indirect evidence for this is that the proportion of soft plastics found in our coastal net stations was higher than that reported in open ocean settings further away from potential sources [49]. While a small number of experimental studies have confirmed that biofilms decrease the buoyancy of plastic items [50,51], none of them report the magnitude or speed of this process across different types of small fragments.

The plastics reported here were mostly white/transparent (84.7%) or blue (8.3%), which is consistent with reports from other investigations on buoyant marine plastics [49,52]. Depending on the feeding ecology of the affected animal, ingested plastic color proportions can differ from what is available in the environment [22]. For instance, ingested plastic color proportions in Australian shearwaters (*Ardenna pacifica* and *Ardenna tenuirostris*) are different from those reported by this study [20,21]. As these birds are known to use color vision to select their food [21,53], color can play a role in the ingestion risk associated with a certain plastic item. In contrast, the color proportion of plastics found in scats of fur seals (*Arctocephalus* spp.) at Macquarie Island (Australia) reflected what was available as flotsam in this environment [36]. These plastics are likely to be coming from the stomach contents of their main prey: the myctophid *Electrona subaspera*, which are pelagic small fish known to feed at night, selecting their food based on size rather than color [36].

The vast majority (98.5%) of the plastics detected were made of polyolefins (polyethylene and polypropylene), which is in agreement with what has been found for this size range of plastics in other marine regions around the world [47,49]. Polyethylene and polypropylene account for most of our global plastic production (38% and 24%, respectively [6]) and they are typically applied in the manufacturing of single-use disposable packaging. In addition to packaging, which reaches the oceans primarily from coastal areas, fishing equipment made of these polyolefins (e.g. fish crates, nets, ropes, fishing lines [17]) are also likely sources of the plastic particles registered here. Other types of polymers found in this study include two pieces of expanded polystyrene (Styrofoam), a type of plastic also used in packaging and fishing gear, and one fragment of ethylene vinyl acetate, which has several applications such as the making of shoe soles and foam mats.

Concentrations and sources

Our overall mean sea surface plastic concentration (Cs) was 4256.4 pieces km^{-2}, which is similar to mean values reported for other regions outside subtropical gyres, such the Caribbean Sea (mean $Cs = 1414$ pieces km^{-2}) and Gulf of Maine (mean $Cs = 1534$ pieces km^{-2}) [9]. Within subtropical gyres, Cs values tend to be higher but within the same order of magnitude: 20328 pieces km^{-2} in the North Atlantic Gyre [9], and 26898 pieces km^{-2} in the South Pacific Gyre [8]. The exception

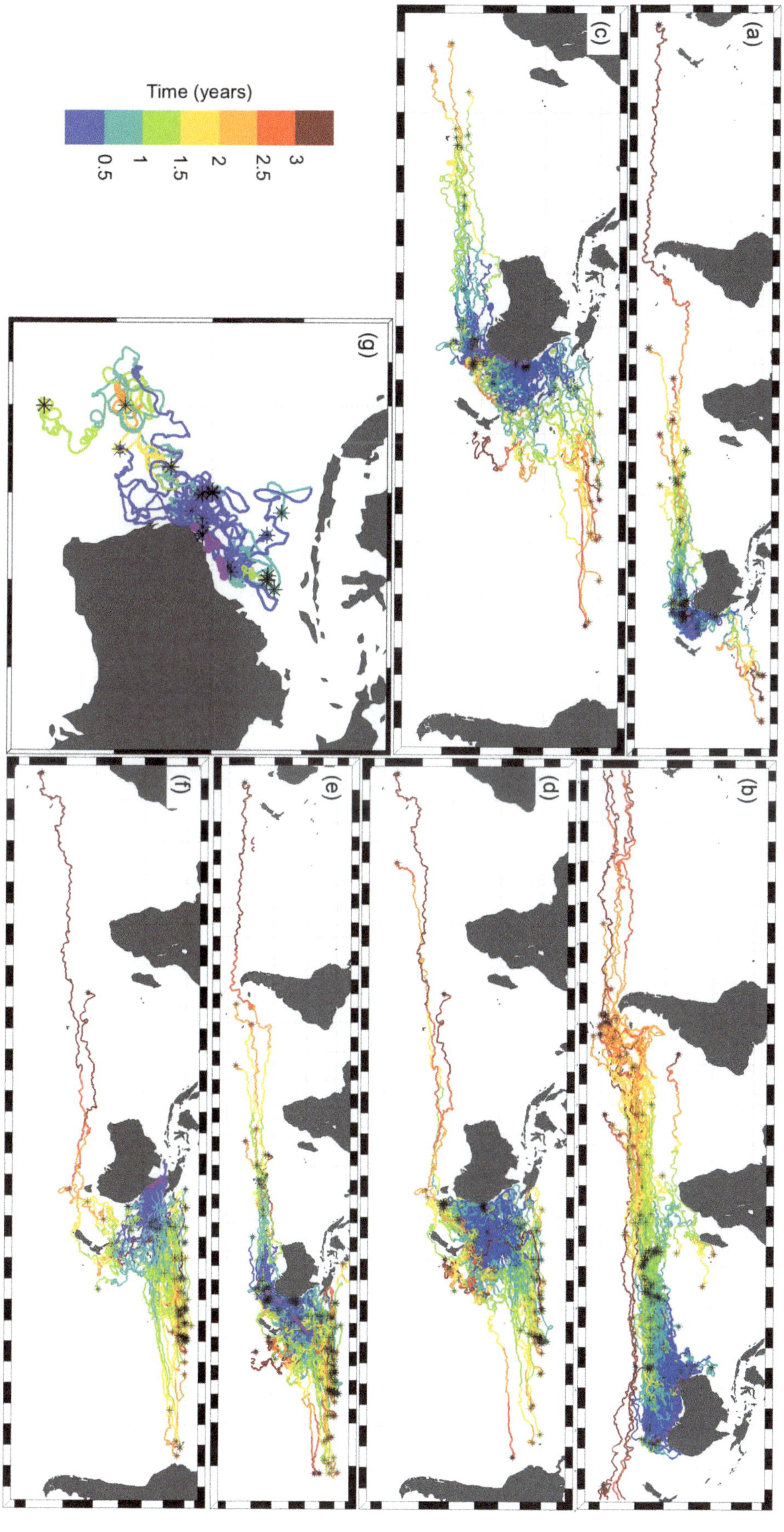
(a)
(b)
(c)
(d)
(e)
(f)
(g)
Time (years)
0.5
1
1.5
2
2.5
3

Figure 9. Real drifter pathways arriving at the 57 net stations. Purple dots indicate net station locations and asterisks indicate drifter release areas. See also Maps S2.

seems to be the subtropical waters of the North Pacific and Mediterranean, which present mean *Cs* values that are an order of magnitude higher than those reported here: 116000 pieces km^{-2} in the Mediterranean [43], 174000 pieces km^{-2} in Northwest Pacific [44], and 334271 pieces km^{-2} in Northeast Pacific [10]. The latter is also known as the "Great Pacific Garbage Patch" [10], which is the largest aggregator of floating marine particles [54].

Our findings show that the distribution of marine plastics is quite widespread (93% of our net stations had at least one plastic piece), patchy (i.e. high variability within and between net stations' *Cs*) and dynamic (*Cs* ranged from 10% to 91% of *Ci*). Therefore, better spatio-temporal data coverage is required in order to identify plastic pollution hotspots within Australian waters. However, our data already indicate some spatial patterns: we observed high plastic concentrations close to Sydney and Brisbane cities. This suggests that plastics along Australia's east coast are mostly associated with domestic inputs. Since high quantities of plastic were also found close to Viti Levu (Fiji), we hypothesize that part of the plastics coming from coastal areas remain in the vicinity of their sources for a long time, while fragmenting into smaller pieces. This suggestion of local retention of plastic debris is in agreement with findings of recent studies (e.g.[7]) and could be tested by developing high-resolution models able to simulate plastic transport in coastal environments.

While the relatively high concentrations of plastic found close to the East Australian coast (net stations 18–20, 22, 37) seem to originate from local sources of plastics, those found in southwest Tasmania/eastern South Australia (net stations 5, 6, 8), and the North West Shelf (net station 54) could be associated with international sources and/or maritime operations. The presence of internationally-based plastics is suggested by (1) a fragment with Indonesian words that was collected in North West Shelf (see Figure 2) and (2) beach surveys, which registered in South Australia plastic debris from South Africa and South America [19]. High plastic concentrations in the southern tip of Tasmania (net station 5) might be caused by convergence effects of the encounter of the East Australian and Zeehan coastal currents [19], whereas those found off the east coast (e.g. net station 37) could be associated with meso-scale eddies of the East Australian current [55].

Aside from this study and the one that developed the biophysical model we applied here [15], we are not aware of any investigation that quantitatively considers the effect of vertical mixing processes on plastic concentrations. This effect needs to be taken into account in future studies assessing at-sea plastic pollution to allow better comparisons between data collected under different sea states. An important step towards improved simulations of plastic distribution along the water column is to better quantify the buoyant rise velocity (w_b) of plastic particles from different oceanic and coastal surface waters. This variable has a considerable impact on the output of the model applied here. Furthermore, other environmental variables that were not taken into account in our one-dimensional column model (e.g. Langmuir circulation, breaking waves, mixed layer depth) could to be incorporated in this type of modeling.

Potential pathways

The model outputs and routes taken by real drifters showed that plastics we found could have moved via a wide range of routes. This is because our net stations are within regions that experience different hydrodynamics (e.g. North West Shelf, Great Australian Bright, Coral Sea, Tasman Sea) [40]. Plastics have the potential to reach the sampled sites by travelling with a range of currents, including: (1) Antarctic Circumpolar current [56], which can carry plastics from a wide area to several of our net stations, particularly those along the coast of Tasmania, south coast of Australia, and Tasman Sea (net stations 1–15, 38 and 39); (2) South Equatorial current in the Pacific Ocean [56,57], which can bring international plastics to the east coast of Australia (net stations 16–24, 40–45, 36, 37) and areas close to Fiji and New Caledonia (net stations 25–35); (3) East Australian current [55,56], which can carry plastics from domestic highly populated regions (e.g. Brisbane and Sydney) to the net stations along the coast of Tasmania (net station 5, 15, 38, 39), east coast of Australia (net stations 16–24, 36, 37) and the Tasman Sea (net stations 1–4); (4) Holloway, Leeuwin, South Australian, and Zeehan coastal current systems [58–61], which can bring plastics from international areas connected to the Indonesian Throughflow and Indian Gyre (e.g. Southeast Asia/Indonesia [16]), as well as from domestic populated areas, to the net stations of the North West Shelf (net stations 48–57), off Perth (net station 14), and along the south coast of Australia, Bass Strait, Tasman Sea, and coast of Tasmania (net stations 1–13, 15–17, 37–39); and (5) West Australia current [61], which could transport international marine plastics that accumulated in the Indian Gyre to the net stations in the North West Shelf (net stations 48–57) and off Perth (net station 14).

It is important to note that running models backwards and using drifter trajectories arriving at sampled locations can only provide an indication of the directions that the collected plastics could have taken. To precisely estimate plastic pathways is quite challenging, mostly because plastic source locations and quantities are still largely unknown. Moreover, there are still no methods to estimate the "age" (drifting time) of a certain plastic particle. For instance, only plastics with long drifting times (years) could have matched the long tracks of drifters. Another limitation of the real drifter approach is that the resulting pathway formed by all drifter tracks arriving at a certain region is not only dependent on the ocean current systems, but also on the locations where most of the drifters were released. For instance, sampled sites in the North West Shelf (net stations 48–57) had only a few drifters arriving at them. This is mostly due to the non-existence of drifters in the shallow waters of the Indonesian archipelago. The creation of a shallow-water drifter (e.g. [62]) release program in this area could bring crucial information to help inform marine plastic pathways and sources.

Final remarks

This investigation shows that the abundant and widespread small marine plastics around Australia are likely coming from a variety of domestic and international, land- and ocean-based sources. Even though marine plastic pollution is a global environmental issue, mostly caused by our massive production of plastic single-use disposable items, there are still no attempts to regulate plastic disposal on land at an international level [26]. Additionally, dumping of plastics into the oceans remains a significant issue owing the difficulties with regulation and enforcement [17,63]. We suggest further at-sea studies on the characterization, spatial distribution, and pathways of marine plastics in coastal and oceanic regions around Australia, as well as on marine plastic toxin loads and interactions between small plastic particles and organisms at all trophic levels of the food web. This would improve our current knowledge on the effects of plastic on the marine ecosystem as a whole.

Supporting Information

Table S1 Net tow data (N = 171). Columns indicate net station number, sampling date (day.month.year), location (degrees minutes), and sea surface plastic concentration (*Cs*; pieces per km^{-2}).

Maps S1 Cumulative probability distribution of virtual particles arriving at the 57 net stations. The month when the virtual particles (25 per day) were released is indicated in each panel. Backtracking dispersal time was equal to 45 days and arriving destinations (net stations) are marked with red dots.

Maps S2 Real drifter pathways arriving at the 57 net stations. Purple dots indicate net station locations and asterisks indicate drifter release areas.

Acknowledgments

We are thankful to CSIRO Marine National Facility, Australian Institute of Marine Science (AIMS), and Austral Fisheries for proving us with sea time onboard their vessels, as well as the staff and crew of RV Southern Surveyor, RV Solander, and Comac Enterprise for logistic support during the voyages. The authors also acknowledge the UWA School of Chemistry and Biochemistry, as well as the Centre for Microscopy, Characterisation and Analysis for their facilities, scientific and technical assistance. A special thank you to Don McKenzie, Mark Lewis, Oscar Del Borrello, Qamar Schuyler, Kathy Townsend, Stephen McCullum, Lisa Woodward, Alastair Graham, Sara Schofield, Tyrone Ridgway, Kim Brooks, Craig Steinberg, Gary Brinkman, Martin Exel, Andy Prendergast, Sebastian Holmes, Luana Lins, Dagmar Kubistin, Murphy Birnberg, and Cyprien Bosserelle for help with acquisition of data and/or data analysis.

Author Contributions

Conceived and designed the experiments: JR JS CW BDH MP MT CP. Performed the experiments: JR JS CW. Analyzed the data: JR CP. Contributed reagents/materials/analysis tools: JR JS CW BDH CP. Wrote the paper: JR JS MP MT.

References

1. Thompson RC, Swan SH, Moore CJ, vom Saal FS (2009) Our plastic age. Philosophical Transactions of the Royal Society B: Biological Sciences 364: 1973–1976.
2. PlasticsEurope (2012) Plastics - the Facts 2012. An analysis of European plastics production, demand and waste data for 2011. 38 p.
3. Hoornweg D, Bhada-Tata P (2012) What a waste: a global review of solid waste management. 98 p.
4. PACIA (2011) National plastics recycling survey (2010-11 financial year). 53 p.
5. Ryan PG, Moore CJ, van Franeker JA, Moloney CL (2009) Monitoring the abundance of plastic debris in the marine environment. Philosophical Transactions of the Royal Society B: Biological Sciences 364: 1999–2012.
6. Andrady AL (2011) Microplastics in the marine environment. Marine Pollution Bulletin 62: 1596–1605.
7. Thiel M, Hinojosa I, Miranda L, Pantoja J, Rivadeneira M, et al. (2013) Anthropogenic marine debris in the coastal environment: A multi-year comparison between coastal waters and local shores. Marine Pollution Bulletin 71: 307–316.
8. Eriksen M, Maximenko N, Thiel M, Cummins A, Lattin G, et al. (2013) Plastic pollution in the South Pacific subtropical gyre. Marine Pollution Bulletin 68: 71–76.
9. Law KL, Morét-Ferguson S, Maximenko NA, Proskurowski G, Peacock EE, et al. (2010) Plastic accumulation in the North Atlantic subtropical gyre. Science 329: 1185–1188.
10. Moore CJ, Moore SL, Leecaster MK, Weisberg SB (2001) A comparison of plastic and plankton in the North Pacific central gyre. Marine Pollution Bulletin 42: 1297–1300.
11. Davidson TM (2012) Boring crustaceans damage polystyrene floats under docks polluting marine waters with microplastic. Marine Pollution Bulletin 64: 1821–1828.
12. Mato Y, Isobe T, Takada H, Kanehiro H, Ohtake C, et al. (2001) Plastic resin pellets as a transport medium for toxic chemicals in the marine environment. Environmental Science & Technology 35: 318–324.
13. Browne MA, Crump P, Niven SJ, Teuten E, Tonkin A, et al. (2011) Accumulation of microplastic on shorelines woldwide: sources and sinks. Environmental Science & Technology 45: 9175–9179.
14. Fendall LS, Sewell MA (2009) Contributing to marine pollution by washing your face: microplastics in facial cleansers. Marine Pollution Bulletin 58: 1225–1228.
15. Kukulka T, Proskurowski G, Morét-Ferguson S, Meyer D, Law K (2012) The effect of wind mixing on the vertical distribution of buoyant plastic debris. Geophysical Research Letters 39: 1–6.
16. Lebreton LM, Greer S, Borrero J (2012) Numerical modelling of floating debris in the world's oceans. Marine Pollution Bulletin 64: 653–661.
17. Jones MM (1995) Fishing debris in the Australian marine environment. Marine Pollution Bulletin 30: 25–33.
18. Frost A, Cullen M (1997) Marine debris on northern New South Wales beaches (Australia): sources and the role of beach usage. Marine Pollution Bulletin 34: 348–352.
19. Edyvane K, Dalgetty A, Hone P, Higham J, Wace N (2004) Long-term marine litter monitoring in the remote Great Australian Bight, South Australia. Marine Pollution Bulletin 48: 1060–1075.
20. Carey MJ (2011) Intergenerational transfer of plastic debris by Short-tailed Shearwaters (*Ardenna tenuirostris*). Emu 111: 229–234.
21. Verlis K, Campbell M, Wilson S (2013) Ingestion of marine debris plastic by the wedge-tailed shearwater *Ardenna pacifica* in the Great Barrier Reef, Australia. Marine Pollution Bulletin 72: 244–249.
22. Schuyler Q, Hardesty BD, Wilcox C, Townsend K (2012) To eat or not to eat? Debris selectivity by marine turtles. PloS one 7: e40884.
23. Gregory MR (2009) Environmental implications of plastic debris in marine settings—entanglement, ingestion, smothering, hangers-on, hitch-hiking and alien invasions. Philosophical Transactions of the Royal Society B: Biological Sciences 364: 2013–2025.
24. Carr A (1987) Impact of nondegradable marine debris on the ecology and survival outlook of sea turtles. Marine Pollution Bulletin 18: 352–356.
25. de Stephanis R, Giménez J, Carpinelli E, Gutierrez-Exposito C, Cañadas A (2013) As main meal for sperm whales: plastics debris. Marine Pollution Bulletin 69: 206–214.
26. Rochman CM, Browne MA, Halpern BS, Hentschel BT, Hoh E, et al. (2013) Policy: classify plastic waste as hazardous. Nature 494: 169–171.
27. Rochman CM, Hoh E, Hentschel BT, Kaye S (2013) Long-Term field measurement of sorption of organic contaminants to five types of plastic pellets: implications for plastic marine debris. Environmental Science & Technology 47: 1646–1654.
28. von Moos N, Burkhardt-Holm P, Kohler A (2012) Uptake and effects of microplastics on cells and tissue of the blue mussel Mytilus edulis L. after an experimental exposure. Environmental Science & Technology 46: 11327–11335.
29. Teuten EL, Saquing JM, Knappe DR, Barlaz MA, Jonsson S, et al. (2009) Transport and release of chemicals from plastics to the environment and to wildlife. Philosophical Transactions of the Royal Society B: Biological Sciences 364: 2027–2045.
30. Basheer C, Lee HK, Tan KS (2004) Endocrine disrupting alkylphenols and bisphenol-A in coastal waters and supermarket seafood from Singapore. Marine Pollution Bulletin 48: 1161–1167.
31. Choy CA, Drazen JC (2013) Plastic for dinner? Observations of frequent debris ingestion by pelagic predatory fishes from the central North Pacific. Marine Ecology Progress Series 485: 155–163.
32. Gassel M, Harwani S, Park J-S, Jahn A (2013) Detection of nonylphenol and persistent organic pollutants in fish from the North Pacific Central Gyre. Marine Pollution Bulletin 64: 2374–2379.
33. Boerger CM, Lattin GL, Moore SL, Moore CJ (2010) Plastic ingestion by planktivorous fishes in the North Pacific Central Gyre. Marine Pollution Bulletin 60: 2275–2278.
34. Cole M, Lindeque P, Fileman E, Halsband C, Goodhead RM, et al. (2013) Microplastic ingestion by zooplankton. Environmental Science & Technology 47: 6646–6655.
35. Young JW, Lamb TD, Le D, Bradford RW, Whitelaw AW (1997) Feeding ecology and interannual variations in diet of southern bluefin tuna, *Thunnus maccoyii*, in relation to coastal and oceanic waters off eastern Tasmania, Australia. Environmental Biology of Fishes 50: 275–291.
36. Eriksson C, Burton H (2003) Origins and biological accumulation of small plastic particles in fur seals from Macquarie Island. AMBIO 32: 380–384.

37. Commonwealth of Australia (2009) Background paper for the threat abatement plan for the impacts of marine debris on vertebrate marine life. Available: http://www.environment.gov.au/biodiversity/threatened/publications/tap/marine-debris.html Accessed 2013 Jul 16.
38. Dee D, Uppala S, Simmons A, Berrisford P, Poli P, et al. (2011) The ERA-Interim reanalysis: configuration and performance of the data assimilation system. Quarterly Journal of the Royal Meteorological Society 137: 553–597.
39. Condie SA, Waring J, Mansbridge JV, Cahill ML (2005) Marine connectivity patterns around the Australian continent. Environmental Modelling & Software 20: 1149–1157.
40. Schiller A, Oke PR, Brassington G, Entel M, Fiedler R, et al. (2008) Eddy-resolving ocean circulation in the Asian–Australian region inferred from an ocean reanalysis effort. Progress in Oceanography 76: 334–365.
41. Browne MA, Galloway TS, Thompson RC (2010) Spatial patterns of plastic debris along estuarine shorelines. Environmental Science & Technology 44: 3404–3409.
42. Thompson RC, Olsen Y, Mitchell RP, Davis A, Rowland SJ, et al. (2004) Lost at sea: where is all the plastic? Science 304: 838–838.
43. Collignon A, Hecq J-H, Glagani F, Voisin P, Collard F, et al. (2012) Neustonic microplastic and zooplankton in the North Western Mediterranean Sea. Marine Pollution Bulletin 64: 861–864.
44. Yamashita R, Tanimura A (2007) Floating plastic in the Kuroshio current area, western North Pacific Ocean. Marine Pollution Bulletin 54: 485–488.
45. Fossi MC, Panti C, Guerranti C, Coppola D, Giannetti M, et al. (2012) Are baleen whales exposed to the threat of microplastics? A case study of the Mediterranean fin whale (*Balaenoptera physalus*). Marine Pollution Bulletin 64: 2374–2379.
46. Wright SL, Thompson RC, Galloway TS (2013) The physical impacts of microplastics on marine organisms: A review. Environmental Pollution 178: 483–492.
47. Hidalgo-Ruz V, Gutow L, Thompson RC, Thiel M (2012) Microplastics in the marine environment: a review of the methods used for identification and quantification. Environmental Science & Technology 46: 3060–3075.
48. Harrison JP, Ojeda JJ, Romero-González ME (2012) The applicability of reflectance micro-Fourier-transform infrared spectroscopy for the detection of synthetic microplastics in marine sediments. Science of the Total Environment 416: 455–463.
49. Morét-Ferguson S, Law KL, Proskurowski G, Murphy EK, Peacock EE, et al. (2010) The size, mass, and composition of plastic debris in the western North Atlantic Ocean. Marine Pollution Bulletin 60: 1873–1878.
50. Ye S, Andrady AL (1991) Fouling of floating plastic debris under Biscayne Bay exposure conditions. Marine Pollution Bulletin 22: 608–613.
51. Lobelle D, Cunliffe M (2011) Early microbial biofilm formation on marine plastic debris. Marine Pollution Bulletin 62: 197–200.
52. Carpenter EJ, Smith K (1972) Plastics on the Sargasso Sea Surface. Science 175: 1240–1241.
53. Bowmaker J (1980) Colour vision in birds and the role of oil droplets. Trends in Neurosciences 3: 196–199.
54. van Sebille E, England MH, Froyland G (2012) Origin, dynamics and evolution of ocean garbage patches from observed surface drifters. Environmental Research Letters 7: 044040.
55. Ridgway K, Dunn J (2003) Mesoscale structure of the mean East Australian Current System and its relationship with topography. Progress in Oceanography 56: 189–222.
56. Talley LD, Pickard GL, Emery WJ, Swift JH (2011) Descriptive physical oceanography: an introduction: Elsevier. 555 p.
57. Webb DJ (2000) Evidence for shallow zonal jets in the South Equatorial Current region of the southwest Pacific. Journal of Physical Oceanography 30: 706–720.
58. Ridgway K, Condie S (2004) The 5500-km-long boundary flow off western and southern Australia. Journal of Geophysical Research 109: 1–18.
59. Condie S, Andrewartha J (2008) Circulation and connectivity on the Australian North West Shelf. Continental Shelf Research 28: 1724–1739.
60. Sandery PA, Kämpf J (2007) Transport timescales for identifying seasonal variation in Bass Strait, south-eastern Australia. Estuarine, Coastal and Shelf Science 74: 684–696.
61. Pattiaratchi C, Woo M (2009) The mean state of the Leeuwin current system between North West Cape and Cape Leeuwin. Journal of the Royal Society of Western Australia 92: 221–241.
62. Ohlmann JC, White PF, Sybrandy AL, Niiler PP (2005) GPS-cellular drifter technology for coastal ocean observing systems. Journal of Atmospheric and Oceanic Technology 22: 1381–1388.
63. Rakestraw A (2012) Open oceans and marine debris: solutions for the ineffective enforcement of MARPOL Annex V. Hastings International and Comparative Law Review 35: 383–451.

14

Marine Litter Distribution and Density in European Seas, from the Shelves to Deep Basins

Christopher K. Pham[1,2]*, Eva Ramirez-Llodra[3,4], Claudia H. S. Alt[5], Teresa Amaro[6], Melanie Bergmann[7], Miquel Canals[8], Joan B. Company[3], Jaime Davies[9], Gerard Duineveld[10], François Galgani[11], Kerry L. Howell[9], Veerle A. I. Huvenne[12], Eduardo Isidro[1,2], Daniel O. B. Jones[12], Galderic Lastras[8], Telmo Morato[1,2], José Nuno Gomes-Pereira[1,2], Autun Purser[13], Heather Stewart[14], Inês Tojeira[15], Xavier Tubau[8], David Van Rooij[16], Paul A. Tyler[5]

1 Center of the Institute of Marine Research (IMAR) and Department of Oceanography and Fisheries, University of the Azores, Horta, Portugal, **2** Laboratory of Robotics and Systems in Engineering and Science (LARSyS), Lisbon, Portugal, **3** Institut de Ciències del Mar (ICM-CSIC), Barcelona, Spain, **4** Norwegian Institute for Water Research (NIVA), Marine Biology section, Oslo, Norway, **5** Ocean and Earth Science, University of Southampton, National Oceanography Centre, Southampton, United Kingdom, **6** Norwegian Institute for Water Research, Bergen, Norway, **7** Alfred-Wegener-Institut, Helmholtz-Zentrum für Polar- und Meeresforschung, Bremerhaven, Germany, **8** GRC Geociències Marines, Departament d'Estratigrafia, Paleontologia i Geociències Marines, Facultat de Geologia, Universitat de Barcelona, Campus de Pedralbes, Barcelona, Spain, **9** Marine Biology & Ecology Research Centre, Marine Institute, Plymouth University, Plymouth, United Kingdom, **10** Netherlands Institute for Sea Research (NIOZ), Texel, The Netherlands, **11** Institut Français de Recherche pour l'Exploitation de la Mer (IFREMER), Bastia, France, **12** National Oceanography Centre, University of Southampton Waterfront Campus, Southampton, United Kingdom, **13** OceanLab, Jacobs University Bremen, Bremen, Germany, **14** British Geological Survey, Murchison House, Edinburgh, United Kingdom, **15** Portuguese Task Group for the Extension of the Continental Shelf (EMEPC), Paço de Arcos, Portugal, **16** Renard Centre of Marine Geology (RCMG), Department of Geology and Soil Science, Ghent University, Gent, Belgium

Abstract

Anthropogenic litter is present in all marine habitats, from beaches to the most remote points in the oceans. On the seafloor, marine litter, particularly plastic, can accumulate in high densities with deleterious consequences for its inhabitants. Yet, because of the high cost involved with sampling the seafloor, no large-scale assessment of distribution patterns was available to date. Here, we present data on litter distribution and density collected during 588 video and trawl surveys across 32 sites in European waters. We found litter to be present in the deepest areas and at locations as remote from land as the Charlie-Gibbs Fracture Zone across the Mid-Atlantic Ridge. The highest litter density occurs in submarine canyons, whilst the lowest density can be found on continental shelves and on ocean ridges. Plastic was the most prevalent litter item found on the seafloor. Litter from fishing activities (derelict fishing lines and nets) was particularly common on seamounts, banks, mounds and ocean ridges. Our results highlight the extent of the problem and the need for action to prevent increasing accumulation of litter in marine environments.

Editor: Andrew Davies, Bangor University, United Kingdom

Funding: This research was supported by the European Community's Seventh Framework Programme (FP7/2007^2013) under the HERMIONE project, Grant agreement (GA) no. 226354. The authors would like to acknowledge further funds from the Condor project (supported by a grant from Iceland, Liechtenstein, Norway through the EEA Financial Mechanism (PT0040/2008)), Corazon (FCT/PTDC/MAR/72169/2006; COMPETE/QREN), CoralFISH (FP7 ENV/2007/1/21314 4), EC funded PERSEUS project (GA no. 287600), the ESF project BIOFUN (CTM2007-28739-E), the Spanish projects PROMETEO (CTM2007-66316-C02/MAR) and DOS MARES (CTM2010-21810-C03-01), la Caixa grant "Oasis del Mar", the Generalitat de Catalunya grant to excellence research group number 2009 SGR 1305, UK's Natural Environment Research Council (NERC) as part of the Ecosystems of the Mid-Atlantic Ridge at the Sub-Polar Front and Charlie-Gibbs Fracture Zone (ECOMAR) project, the Marine Environmental Mapping Programme (MAREMAP), the ERC (Starting Grant project CODEMAP, no 258482), the Joint Nature Conservation Committee (JNCC), the Lenfest Ocean Program (PEW Foundation), the Department for Business, Enterprise and Regulatory Reform through Strategic Environmental Assessment 7 (formerly the Department for Trade and Industry) and the Department for Environment, Food and Rural Affairs through their advisors, the Joint Nature Conservation Committee, the offshore Special Areas for Conservation programme, BELSPO and RBINS-OD Nature (Belgian Federal Government) for R/V Belgica shiptime. The footage from the HAUSGARTEN observatory was taken during expeditions ARK XVIII/1, ARK XX/1, ARK XXII/1, ARK XXIII/2 and ARK XXVI/2 of the German research icebreaker "Polarstern". The authors also acknowledge funds provided by FCT-IP/MEC to LARSyS Associated Laboratory and IMAR-University of the Azores (R&DU #531), Thematic Area E, through the Strategic Project (PEst-OE/EEI/LA0009/2011^2014, COMPETE, QREN) and by the Government of Azores FRCT multiannual funding. CKP was supported by the doctoral grant from the Portuguese Science Foundation (SFRH/BD/66404/2009; COMPETE/QREN). AP was supported by Statoil as part of the CORAMM project. MB would like to thank Antje Boetius for financial support through the DFG Leibniz programme. JNGP was supported by the doctoral grant (M3.1.2/F/062/2011) from the Regional Directorate for Science, Technology and Communications (DRCTC) of the Regional Government of the Azores. ERLL was supported by a CSIC-JAE-postdocotral grant with co-funding from the European Social Fund. Publication fees for this open access publication were supported by IFREMER. The funders had no role in study design, data collection and analysis, decision to publish, or preparation of the manuscript.

Competing Interests: The authors have declared that no competing interests exist.

* E-mail: phamchristopher@uac.pt

Introduction

Litter disposal and accumulation in the marine environment is one of the fastest growing threats for the world's oceans health. Marine litter is defined as "any persistent, manufactured or processed solid material discarded, disposed of or abandoned in the marine and coastal environment"[1]. The issue has been

Table 1. Sampling locations, date and methods for the collection of data on litter along with litter densities (mean number of items ha^{-1} and kg ha^{-1} ± standard errors).

Location	Year	Method	N° of samples	Mean depth (m)	Density (n ha^{-1})	Density (kg ha^{-1})	Area covered (ha)
ATLANTIC							
Continental slopes							
North Faroe-Shetland Channel	2006	TC	19	657	0.3±0.2	-	2.3
North-East Faroe-Shetland Channel	2006	TC	11	501	1.9±1.0	-	1.2
Continental shelf							
Norwegian Margin	2007	SUB	9	304	9.7±3.8	-	0.6
Submarine canyons							
Dangeard & Explorer Canyons	2007	TC	44	578	7.2±2.7	-	3
Nazaré Canyon	2007	ROV	13	3144	4.2±1.6	-	9
Lisbon Canyon	2007	ROV	1	1602	66.2	-	1
Setúbal Canyon	2007	ROV	1	2194	24.6	-	0.9
Cascais Canyon	2007	ROV	1	4574	10.6	-	1
Guilvinec Canyon	2008–2010	ROV	8	661	31.9±28.1	-	4.1
Whittard Canyon	2010	ROV-TC	11	2668	1.4±0.4	-	12.4
Seamounts, banks and mounds							
Anton Dohrn Seamount	2005–2009	TC	24	992	1.9±1.0	-	2.2
Condor Seamount	2010–2011	ROV	48	258	14.6±3.0	-	5.6
Josephine Seamount	2012	ROV	4	1455	5.7±3.3	-	0.9
Hatton Bank	2005–2011	ROV-TC	52	706	1.9±0.8	-	4
Rockall Bank	2005–2011	ROV-TC	29	702	0.7±0.5	-	2.4
Rosemary Bank	2006	TC	14	577	3.3±2.3	-	1.1
Pen Duick Alpha/Beta Mound	2009	ROV	7	534	2.5±1.7	-	1.1
Darwin Mounds	2011	ROV	7	1007	9.7±2.9	-	1.8
Ocean ridges							
North Charlie Gibbs Fracture Zone	-	ROV	24	2300	0.4±0.3	-	2.4
South Charlie Gibbs Fracture Zone	-	ROV	24	2600	2.9±1.4	-	2.4
Wyville-Thomson Ridge	2006	TC	15	670	10.9±4.3	-	1.2
MEDITERANEAN							
Continental slopes							
Calabrian Slope (Central Med.)	2009	Trawl	4	1400	-	0.6±0.4	18.9
Western Mediterranean Slope	2009	Trawl	8	1500	-	4±1.8	56
Crete-Rhodes Ridge (E. Med.)	2009	Trawl	8	1500	-	1.1±0.3	37.9
Blanes slope (NW Med.)	2009	Trawl	94	1387	-	1.2±0.4	407
Continental shelf							
Gulf of Lion (NW Med.)	2009	Trawl	52	85	0.4±0.1	-	276.4

Table 1. Cont.

Location	Year	Method	N° of samples	Mean depth (m)	Density (n ha^{-1})	Density (kg ha^{-1})	Area covered (ha)
Submarine canyons							
Blanes Canyon (NW Med.)	2009–2011*	ROV-Trawl	4 (13)	1496(1431)	32.1±11.9	0.7±0.2	2(33.9)
Gulf of Lion Canyons (NW Med.)	2009	Trawl	11	510	0.4±0.1	-	126.5
Deep basins							
Algero-Balearic Basin (W. Med.)	2009	Trawl	3	2883	-	1.8±1.5	16
Crete-Rhodes Ridge (E. Med.)	2009	Trawl	2	3000	-	1.2±0.3	2.8
Calabrian Basin (Central Med.)	2009	Trawl	3	2967	-	1.7±0.6	12.5
ARCTIC							
Continental slope							
HAUSGARTEN, station IV	1999–2011	TC-ROV	10	2450	13.6±7.9	-	72.2

*Numbers in parentheses refer to trawl surveys. ROV = remotely operated vehicle; TC = towed camera system; TRAWL = Otter Trawl or Maireta System; SUB = manned submersible.

highlighted by the United Nations Environment Program [1] and was included in the 11 Descriptors set by Europe's Marine Strategy Framework directive (2008/56/EC) (MSFD) [2]. The MSFD requires each Descriptor in all European marine waters not to deviate from the undisturbed state and reach Good Environmental Status (GES) by 2020.

With an estimated 6.4 million tonnes of litter entering the oceans each year [1], the adverse impacts of litter on the marine environment are not negligible. Besides the unquestionable aesthetic issue, litter can be mistaken for food items and be ingested by a wide variety of marine organisms [3–8]. Entanglement in derelict fishing gear is also a serious threat, particularly for mammals [9–11], turtles [12] and birds [13] but also for benthic biota such as corals [14,15]. High mortality of fish through "ghost fishing" is another consequence of derelict fishing gear in the marine environment [16]. Moreover, floating litter facilitates the transfer of non-native marine species (e.g. bryozoans, barnacles) to new habitats [17,18]. Barnes et al. [19] estimated that the dispersal of alien species through marine litter more than doubles the rate of natural dispersal processes, especially during an era of global change.

Although the type of litter found in the world's oceans is highly diverse, plastics are by far the most abundant material recorded [20–22]. Because of their persistence and hydrophobic nature, their impact on marine ecosystems is of great concern. Plastics are a source of toxic chemicals such as polychlorinated biphenyls (PCBs) and dioxins that can be lethal to marine fauna [23]. Furthermore, the degradation of plastics generates microplastics which, when ingested by organisms, can deliver contaminants across trophic levels [24–27].

Litter type, composition and density vary greatly among locations and litter has been found in all marine habitats, from surface water convergence in the pelagic realm (fronts) down to the deep sea where litter degradation is a much slower process [21]. The spatial distribution and accumulation of litter in the ocean is influenced by hydrography, geomorphological factors [21,28], prevailing winds and anthropogenic activities [29]. Hotspots of litter accumulation include shores close to populated areas, particularly beaches [30], but also submarine canyons, where litter originating from land accumulates in large quantities [28,31].

In Europe, much has been written on the abundance and distribution of litter on the coastline and in surface waters [32–41]. As more areas of Europe's seafloor are being explored, benthic litter is progressively being revealed to be more widespread than previously assumed [15,28,29,31,42–52]. The sources of litter accumulating on the seafloor are variable, depending upon interactions between distances from shore [31,45], oceanographic and hydrographic processes [47] and human activities such as commercial shipping [29] and leisure craft [43].

Early studies used trawling to quantify litter abundance on the seafloor [53], whilst more recent studies have demonstrated the potential of remotely operated vehicles (ROV), manned submersibles or towed cameras to study litter in the deep sea [15,31,43,47,54,55]. However, understanding spatial patterns in litter abundance and distribution in the deep sea is challenging, owing to the lack of standardization in the sampling and analytical methodologies used. Furthermore, the high cost of sampling in the deep sea has limited our ability to perform standardized surveys across large areas to understand fully the extent of this pollution issue.

The problem of marine litter on the deep seafloor was addressed by the EU-FP7 project HERMIONE, recognising the need to use the surveys conducted by all partners (although designed for other purposes) to gather data on litter in the deep sea. This paper

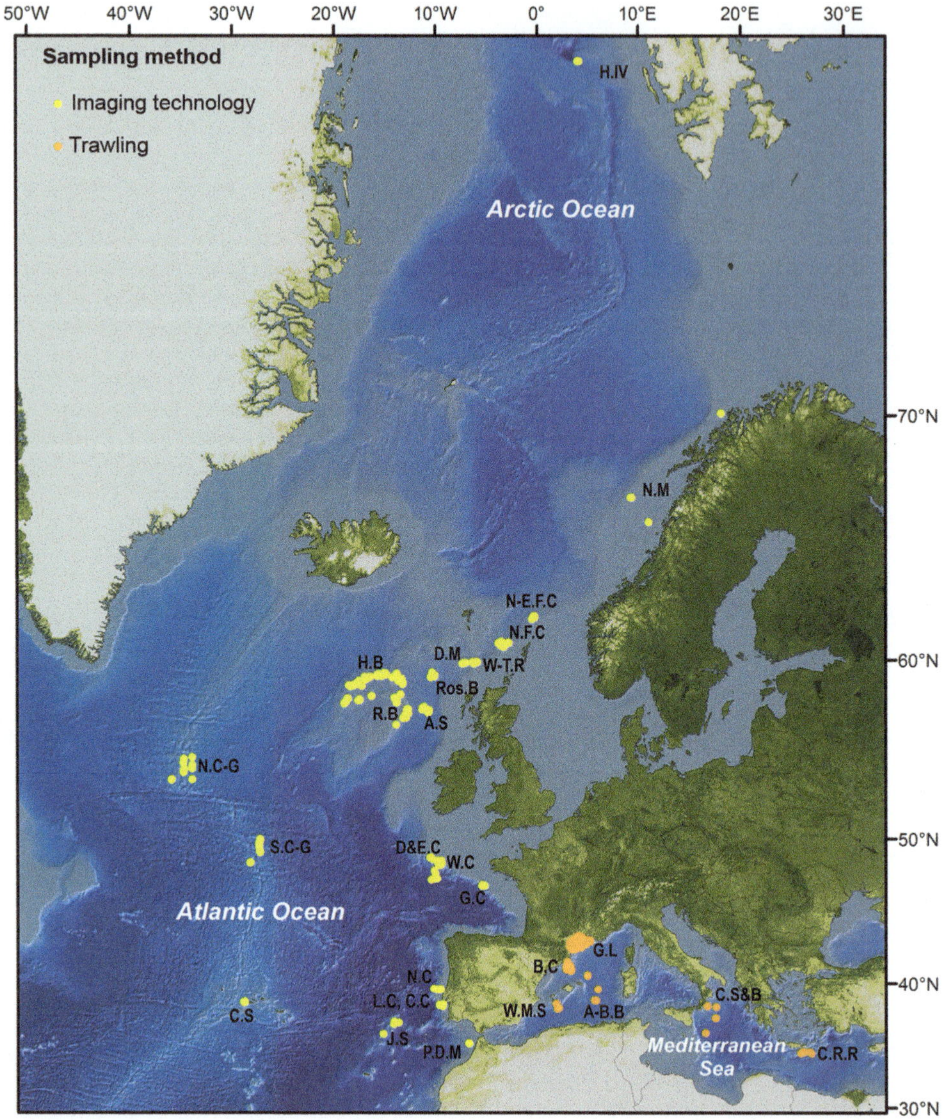

Figure 1. Locations of the study sites sampled with imaging technology (ROVs, manned submersible, towed camera systems) and trawling. A-B.B = Algero-Balearic Basin (W. Med.), A.S = Anton Dohrn Seamount, B.C = Blanes Canyon (NW Med.), C.C = Cascais Canyon, C.S = Condor Seamount, Calabrian Slope & Basin = C.S&B, Crete-Rhodes Ridge = C.R.R, D&E.C = Dangeard & Explorer Canyons, D.M = Darwin Mounds, G.L.C = Gulf of Lion canyons (NW Med.), G.L = Gulf of Lion, G.C = Guilvinec Canyon, H.B = Hatton Bank, H.IV = HAUSGARTEN, station IV, J.S = Josephine Seamount, L.C = Lisbon Canyon, N.C = Nazaré Canyon, N.C-G = North Charlie Gibbs Fracture Zone, N-E.F.C = North-East Faroe-Shetland Channel, N.F.C = North Faroe-Shetland Channel, N.W = Norwegian margin, P.D.M = Pen Duick Alpha/Beta Mound, R.B = Rockall Bank, Ros.B = Rosemary Bank, S.C = Setúbal Canyon, S.C-G = South Charlie Gibbs Fracture Zone, W.C = Whittard Canyon, W.M.S = Western Mediterranean slope, W-T.R = Wyville-Thomson Ridge.

presents the results on the distribution and densities of marine litter obtained during these surveys, with additional data provided by the UK's Mapping the Deep project as well as other previous projects. It provides a unique large-scale analysis of litter on the seafloor across different physiographic settings and depths.

Materials and Methods

Study areas

Data were gathered from surveys conducted during research cruises led by various European institutions between 1999 and 2011. A total of 32 sites in the northeastern Atlantic Ocean, Arctic Ocean and Mediterranean Sea were surveyed (Table 1; Figure 1). Surveyed sites were located on continental shelves and slopes, submarine canyons, seamounts, banks, mounds, ocean ridges and deep basins, at depths ranging from 35 to 4500 meters (Table 1).

Sampling methods

Sampling methods included both imaging technology (still photograph and video) and fishing trawls (Figure 1; Table 2). The Atlantic sites were surveyed uniquely using imaging technology, whilst sites located in the Mediterranean Sea were primarily investigated by trawling (except for some ROV transects in the Blanes submarine canyon). Video footage was collected by different ROVs (*Genesis*, *Isis*, *Liropus*, *Luso*, *Lynx*, *SP* and *Victor 6000*), manned submersible (*JAGO*, GEOMAR) and towed camera systems (Seatronics and the HD-video hopper video system). Still photographs were taken with the Ocean Floor

Table 2. Information on each platform used to collect video and photographs for the collection of data on litter densities and distribution on the seafloor of European waters.

Sampling platform	Name	Format	N° of samples	Total area surveyed (m^2)	Field of view (m)	References
Manned submersible	*Jago*	video	13	5561	1.5	[95]
ROVs	*Luso*	video	8	35587	3.6–4.4	[15]
	Sp	video	44	29749	2.3	[15]
	Isis	video	64	167308	2.0	[31]
	Genesis	video	20	86700	2.6	[96]
	Liropus	video	4	19867	3.0	[97]
	Lynx	video	19	3750	1.0	[98]
	Victor 6000	video	6	421840	10.0	[46]
Towed camera systems	*Seatronics*	video	194	158528	1.5	[99]
	HD video hopper system	video	6	21490	3.0	[100]
	Ocean Floor Observation System	photographs	2882	8570	0.8–11.6	[43]

Further technical information about each platform can be found in the indicated references.

Observation System (OFOS) at the HAUSGARTEN observatory, station IV. Technical details about each platform can be found elsewhere (see Table 2). Trawl samples were collected using two different gears: a net (GOC 73) with a 20 mm-diamond stretched mesh size at the cod-end [56] and an otter trawl Maireta System (OTMS), with a cod-end mesh size of 40 mm and an outer cover of 12 mm [29,57].

Analysis of image data

Protocols for video analysis varied slightly according to the platform used, but followed the same general outline. The entire footage was visualised and the number of litter items and depth recorded. Each litter item was classified into six different categories: plastic (all plastic with exception of fishing line and net), derelict fishing gear (fishing line or net), metal, glass, clinker (residue of burnt coal). Because of the low densities found at all sites, paper and cardboard, fabric, wood and unidentified items were grouped in the same category (other items). Although fishing lines and nets are mostly made of plastic, fishing gear was considered as a separate litter category because of our knowledge on its source and social implications and the particular impacts of this type of litter, such as ghost fishing and entanglement.

For each dive (sample), the area covered was calculated by multiplying the linear distance on the seafloor (off bottom footage were excluded from the analysis) by the average width of view of each of the platforms (Table 2).

For data derived from still photographs (OFOS), all images along each transect (taken at 30 s to 50 s-intervals) were analysed for the presence of litter items. Parallel laser points on the images allowed calculations of the area for each image; ranging between 0.8 and 11.6 m^2. For OFOS, each image was considered to be a separate sample, while for video data, each dive was considered a single sample.

Trawl data

Hauls in the Gulf of Lion (shelf and submarine canyons) were performed with a bottom trawl equipped with a GOC 73 net [56]. After trawling, litter items were counted and classified into the different categories (see above).

Trawling at the other Mediterranean sites was performed using an otter trawl Mareita System (OTMS). All litter items were separated and classified into different categories (see above) and weighed, after excess water and mud had been removed. The use of weight rather than number to quantify litter was based on the high abundance of broken plastics (from whole plastic bags to very small (<0.5 cm) pieces of plastics) and broken glass, which impeded the quantification of single items without overestimating abundances of certain categories over others [29].

Data analysis

For each sample (video and still photographs), litter density was estimated as items of litter hectare^{-1} (ha; 10,000 m^2) of seafloor surveyed. For trawl data where litter was measured in weight, litter density was estimated as kg of litter ha^{-1}. Sites were grouped into 6 different groups according to physiographic characteristics (Table 1); (1) continental shelves; (2) continental slopes (excluding submarine canyons); (3) submarine canyons; (4) seamounts, banks and mounds; (5) ocean ridges and (6) deep basins. Tests for investigating differences among litter densities across physiographic settings were done separately according to the unit in which litter density was estimated (number ha^{-1} or weight ha^{-1}). For both cases, the data were not normally distributed but variances were equal, therefore, the non-parametric Kruskal-Wallis rank sum test followed by a multiple comparison test (Dunn's pairwise comparison) were performed using the statistical package R. Variation in litter composition between physiographic settings were tested for significance using ANOSIM (Analysis of similarity) in PRIMER v6 software [58]. Bray-Curtis similarity [59] was calculated on log(x+1) transformation of the percentage contribution of litter type for each of the physiographic settings, across the entire data set. A similarity percentage analysis (SIMPER) was applied to identify the discriminating feature of the dissimilarities and similarities between physiographic settings.

Results

Litter density

Litter was found at all sites and all depths (from 35 m down to 4500 m) sampled. Most common litter items included plastic bags,

Figure 2. Litter items on the seafloor of European waters. A = Plastic bag entrapped by a small drop stone harbouring sponges (*Cladorhiza gelida, Caulophacus arcticus*), shrimps (*Bythocaris* sp.) and a crinoid (*Bathycrinus carpenterii*) recorded by an OFOS at the HAUSGARTEN observatory (Arctic) at 2500 m; B = Litter recovered within the net of a trawl in Blanes open slope at 1500 m during the PROMETO V cruise on board the R/V "García del Cid"; C = "Heineken" beer can in the upper Whittard canyon at 950 m water depth with the ROV Genesis; D = Plastic bag in Blanes Canyon at 896 m with the ROV "Liropus"; E = "Uncle Benn's Express Rice" packet at 967 m in Darwin Mound with the ROV "Lynx" (National Oceanography Centre, UK); F = Cargo net entangled in a cold-water coral colony at 950 m in Darwin Mound with the ROV "Lynx" (National Oceanography Centre, UK).

glass bottles and derelict fishing lines and nets (Figure 2). Locations with highest litter densities (>20 items ha^{-1}) included the Lisbon Canyon, the Blanes Canyon, the Guilvinec Canyon, and the Setúbal Canyon (Table 1; Figure 3). Sites with intermediate litter density (between 10 and 20 items ha^{-1}) were found on the Condor Seamount, the Wyville-Thomson Ridge, the continental slope of the HAUSGARTEN observatory and the Cascais Canyon (Figure 3). Low densities (between 2 and 10 items ha^{-1}) were recorded on the Darwin Mounds, off the Norwegian margin, in Dangeard and Explorer Canyons, on the Josephine Seamount, in the Nazaré Canyon, on the Rosemary Bank, south of the Charlie-Gibbs Fracture Zone and on the Pen Duick Alpha and Beta Mounds (Figure 3). The lowest litter density (<2 items ha^{-1}) was found on the Hatton Bank, the continental slope on the northern side of the Faroe-Shetland Channel, on the Anton Dohrn Seamount, in the Whittard Canyon, on the Rockall Bank, north of the Charlie-Gibbs Fracture Zone, and in the Gulf of Lion (in both the continental shelf and submarine canyons). Sites with higher litter density were found principally closer to shore (Figure 4), but there were exceptions, such as the samples from the Gulf of Lion where litter densities were low (Table 1).

The sites sampled by trawling in the Mediterranean revealed a relatively even distribution of litter but with a higher density on the continental slope, south of Palma de Mallorca (western Mediterranean) with a mean (±SE) of 4.0±1.8 kg of litter ha^{-1} as opposed to densities ranging between 0.7 and 1.8 kg of litter ha^{-1} at the other sites (Figure 5).

When grouping all sites into physiographic settings, there were significant differences in litter density (items ha^{-1}) between the various groups (Kruskal-Wallis $\chi^2 = 26.68$; $p<0.01$; DF = 4). Multiple comparisons tests indicated that litter density in submarine canyons was significantly higher than those from all other physiographic settings, reaching an average (± SE) of 9.3±2.9 items ha^{-1} (Figure 6a). Litter density on seamounts, mounds and banks was similar to the densities found on the continental slopes with mean (± SE) densities of 5.6±1.0 and 4.1±2.1 items ha^{-1}, respectively (Figure 6a). Mean (± SE) litter density for continental shelves and ocean ridges was 2.2±0.8 and 3.9±1.3 items ha^{-1}, respectively (Figure 6a). For Mediterranean sites, where litter density was quantified by weight rather than number of items, no significant differences were found in litter density between the three different physiographic settings (Kruskal-Wallis $\chi^2 = 3.88$; $p = 0.144$; DF = 2). However, litter density in deep basins was slightly higher (1.55±0.57 kg ha^{-1}) compared to continental slopes (1.36±0.34 kg ha^{-1}) and submarine canyons (0.71±0.25 kg ha^{-1}) (Figure 6b).

Litter composition

There was a high variability in the composition of litter across the different sites (Table 3). A total of 546 litter items were encountered throughout all sites surveyed with imaging technology. Plastic and derelict fishing gear were the most abundant litter items. Plastic represented 41% of the litter items, whilst derelict fishing gear accounted for 34% of the total. Clinker, glass and metal were least common (1, 4 and 7%, respectively). Items classified as "other items" accounted for 13% of the litter items encountered in sites surveyed by imaging technology and included wood, paper/cardboard, clothing, pottery, and unidentified material. Analysis of litter density from trawl surveys found plastic to be the most common litter type to be recovered (found in 98% of the trawls), followed by clinker (73%), fabric (48%), derelict fishing gear (33%), metal (31%) and glass (28%).

Results from ANOSIM showed that there were significant differences in litter composition between physiographic settings (1-way ANOSIM; Global $R = 0.32$; $p<0.001$), the analysis also showed some settings to be similar (Table S1). There were no significant differences between litter composition in submarine canyons and continental shelves ($R = 0.01$; $p = 0.58$). According to SIMPER analysis (Table S2), the similarity in composition between submarine canyons and continental shelves was mostly driven by plastic. Plastic was the dominant litter category for both settings (Figure 7). Litter composition on ocean ridges and on seamounts, banks and mounds did not show significant differences in litter composition ($R = 0.17$; $p = 0.06$), due to a predominance of derelict fishing gear (Figure 7). Finally, litter composition found on continental slopes was similar to deep basins ($R = -0.11$; $p = 0.87$). Clinker and plastic were the categories contributing most to the similarities between these two physiographic settings.

Discussion

The occurrence of litter on the seafloor has been far less investigated than in surface waters or on beaches, principally because of the high cost and the technical difficulties involved in sampling the seafloor at bathyal and abyssal depths [21,60].

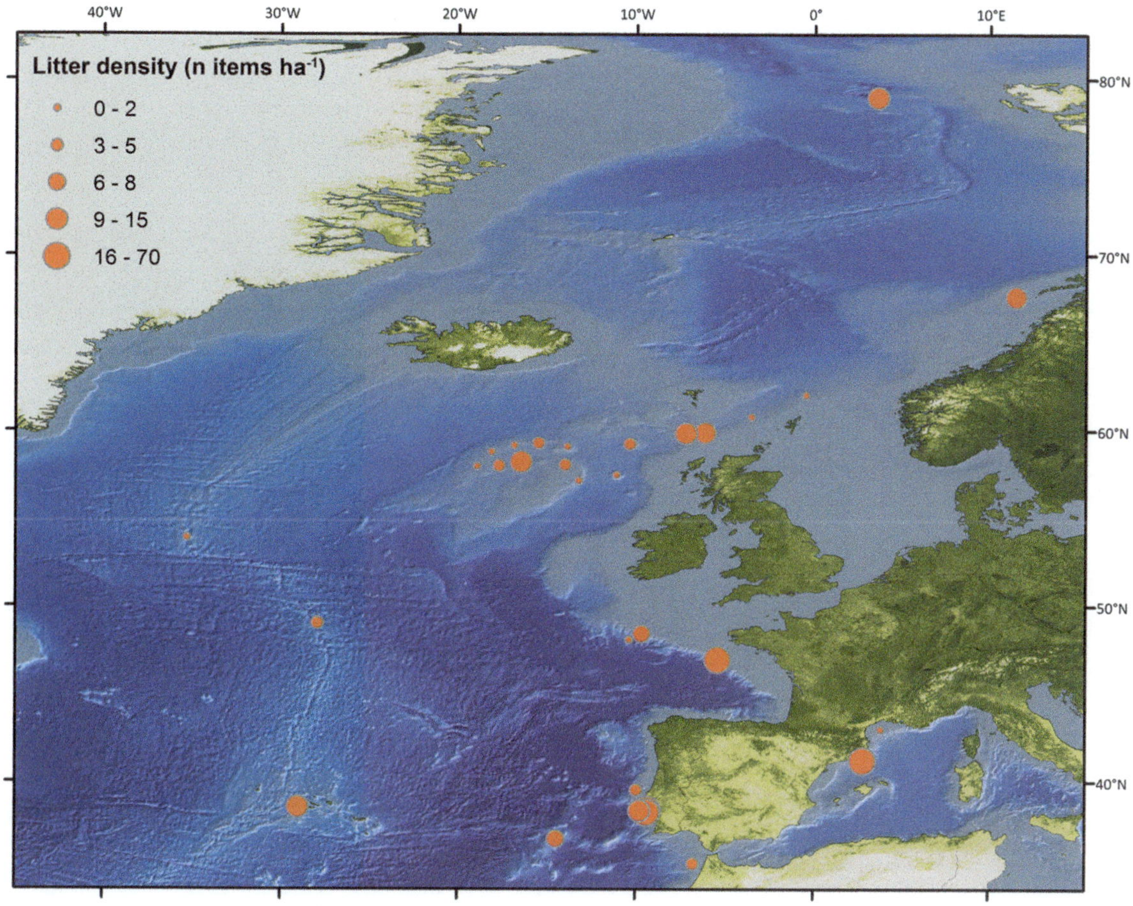

Figure 3. Litter densities (number of items ha^{-1}) in different locations across European waters obtained with ROVs, towed camera systems, manned submersible and trawls.

Considering such limitations and poor knowledge on litter accumulation in deep waters, every survey is of great value for obtaining information on litter density and distribution. In the present study, we integrated data collected during numerous cruises over a large regional scale into a single analysis, providing insight on the density and composition of litter across a wide variety of seafloor settings and over a large geographical area in European waters. Although standardisation of the data permitted comparisons between sites, dissimilarities in the sampling equipment implies that the results should be treated with caution. Furthermore, differences in the areas of the seafloor surveyed between locations may lead to overestimations or underestimations of the litter density. Also, studying litter from trawls introduces the issue of quantification units (number vs. weight), with no correct solution. When using number of items, certain litter categories may be overestimated such as plastic or glass that can break into many small pieces. As a counterpart, if weight is used, the abundance of litter type with different weights (e.g. heavy clinker vs. light plastic) cannot be compared. Ideally, both units for litter quantification will help to understand better trends, but the EU Marine Strategy Framework Directive stresses that for monitoring litter in the marine environment, number is mandatory whilst weight is only recommended [2].

Litter was found at all the locations surveyed, from sites close to population centres such as the Gulf of Lion or the Lisbon Canyon to as far as the South Charlie-Gibbs Fracture Zone on the Mid-Atlantic Ridge, located at about 2000 km from land. Litter was found from shallow waters (35 meters in Gulf of Lion) down to 4500 meters (Cascais Canyon). Such records were not surprising, as litter is known to be present in all seas and oceans of the planet, as remote as the Southern Ocean [21] and at depths as deep as 7216 m in the Ryuku trench, south of Japan [61]. The range of litter densities found on our study sites was within the same order of magnitude to the ones found on the seabed in other parts of the globe (North America [55,62,63], China [54], Japan [64,65]) and for other locations in Europe [28,44,45,47,48]. On the other hand, macro litter densities on the seabed were higher than reported for surface waters [32,66–69]. At the surface, floating litter tends to accumulate in frontal areas but eventually reaches the seabed when heavily covered by fouling organisms [70] or loaded with sediments. Contrary to a common notion that most plastic items float at the sea surface it has been estimated that 70% of the plastic sinks to the seafloor [23]. This results in macro litter accumulation on the seabed rather than in the open sea [21]. For example, on the seafloor of the Mediterranean Sea, our data showed much higher litter densities (0.4 to 48 litter items ha^{-1}) than that estimated to float at the surface (0.021 items ha^{-1}; [1]). Alternatively, floating litter may be transported for considerable distances and get washed ashore [71,72]. Litter density on the coastline is typically higher than on the seafloor given that there is an additional input of waste coming from inland sources (e.g. man-made drainage systems, recreational usage, rivers, winds, etc.) [71,73]. On European coasts, litter densities can exceed 30,000 litter items per linear km [1,41,74], while much higher densities

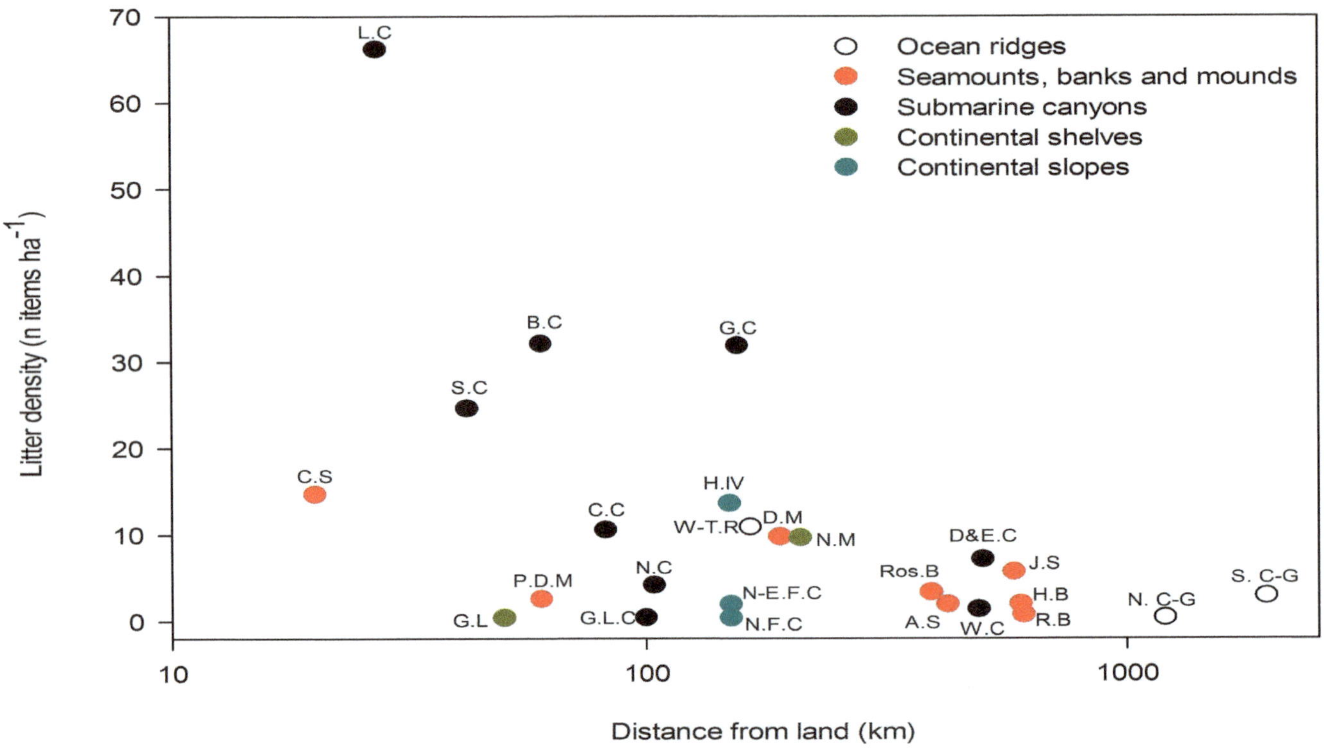

Figure 4. Litter densities (number of items ha^{-1}) in different locations across European waters according to their closest distances from land. *x* axis is in a Log_{10} scale. A.S = Anton Dohrn Seamount, B.C = Blanes Canyon (NW Med.), C.C = Cascais Canyon, C.S = Condor Seamount, D&E.C = Dangeard & Explorer Canyons, D.M = Darwin Mounds, G.L.C = Gulf of Lion canyons (NW Med.), G.L = Gulf of Lion, G.C = Guilvinec Canyon, H.B = Hatton Bank, H.IV = HAUSGARTEN, station IV, J.S= Josephine Seamount, L.C= Lisbon Canyon, N.C = Nazaré Canyon, N.C-G = North Charlie Gibbs Fracture Zone, N-E.F.C = North-East Faroe-Shetland Channel, N.F.C = North Faroe-Shetland Channel, N.W= Norwegian margin, P.D.M = Pen Duick Alpha/Beta Mound, R.B = Rockall Bank, Ros.B = Rosemary Bank, S.C = Setúbal Canyon, S.C-G = South Charlie Gibbs Fracture Zone, W.C = Whittard Canyon, W-T.R = Wyville-Thomson Ridge.

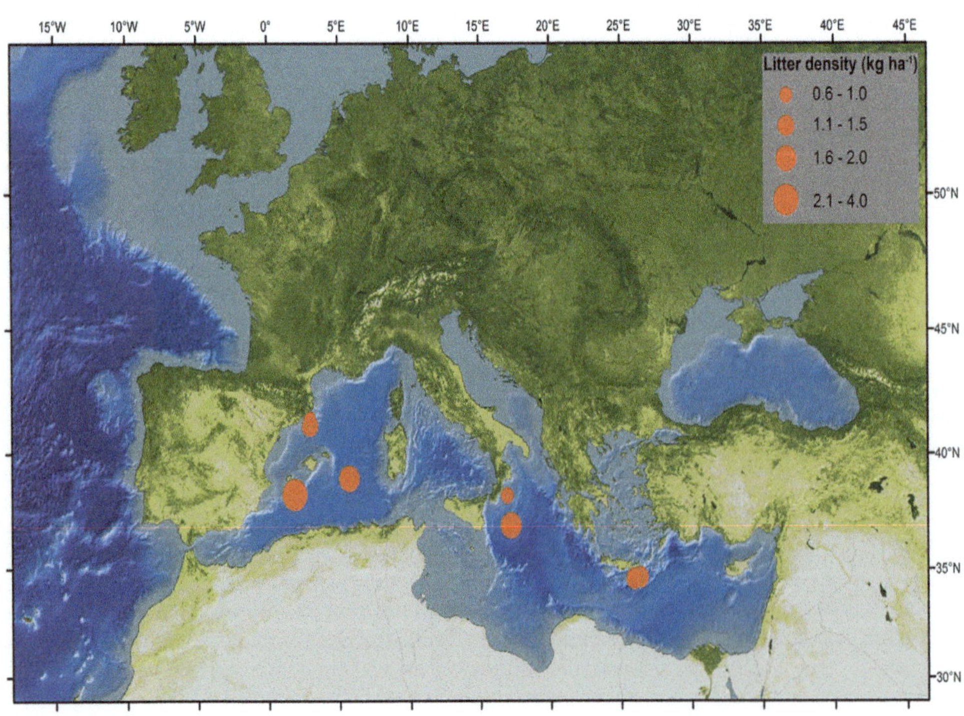

Figure 5. Litter densities (kg ha^{-1}) in different locations across the Mediterranean Sea obtained from trawl surveys.

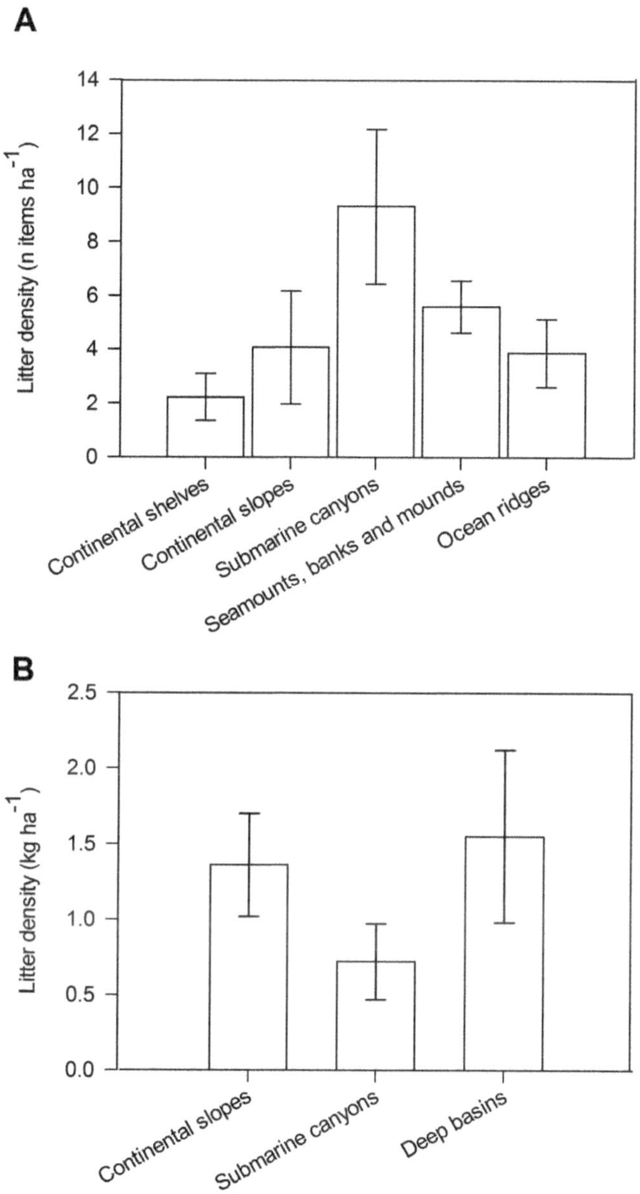

Figure 6. Mean litter density (± standard error) in A = number of items ha^{-1} and B = in kg of items ha^{-1}, across different physiographic settings in European waters.

have been reported for beaches in Indonesia [75] or on the beaches along Armaçao dos Buzios, Rio de Janeiro, Brazil [76]. However, comparisons between studies are challenging considering differences in the size of the litter items sampled and the sampling methodology used [77].

Our data showed a general increase in litter density in locations closer to the shore, a pattern previously reported for the French Mediterranean coast [47] and off California [55]. Nevertheless, low litter densities in some near-shore sites (e.g. Gulf of Lion or Faroe-Shetland channel) suggest that many other factors (such as geomorphology, hydrography and human activity) affect litter distribution and accumulation rates [29]. In the Gulf of Lion, Galgani et al. [47] suggested that low litter density on the shelf was caused by strong water flow from the Rhone River, transporting litter down south to deeper waters. A similar situation occurs in Monterey Bay where sediment and litter are being swept off the continental shelf down into Monterey Canyon [78]. Such phenomena may explain why continental shelves were the settings with overall lowest litter density, whilst submarine canyons had the highest litter concentration. Litter levels on seamounts, banks, mounds and ocean ridges were characterised by intermediate levels when compared to other physiographic settings. They are typically located far away from coastal areas where the main anthropogenic activities include fishing [79] and seabed mining [80,81]. The presence of litter on these settings is of concern because they harbor Vulnerable Marine Ecosystems (VMEs) (such as cold-water corals and hydrothermal vents) that have reduced capacity to recover from disturbance events and for which conservation is a global priority [82].

The types of accumulated litter can provide an indication on the human activities impacting a particular location. However, one must be cautious and consider the differences in the buoyancy and longevity of the different types of litter. For example, while some plastics sink to the seafloor, others float on the surface and are able to travel great distances before eventually sinking far from their initial dumping locations, following biofouling and degradation [23]. On the other hand, glass, metal and clinker will sink rapidly and are expected to be recovered from the seafloor close to sites where they were initially released. Cardboard and fabrics (of organic origin) will break down quickly, implying that such items will not reach the deep ocean with the frequency of more resistant materials such as plastic and negatively buoyant items such as glass, metal and clinker. Although it is difficult to determine the exact source of the litter observed on the seafloor, the dominant litter category can be used as an indicator to separate ocean and terrestrial sources [15,29,31,78]. Plastic (other than derelict fishing gear) was the most abundant litter category in submarine canyons, continental shelves and continental slopes. The predominance of plastics in submarine canyons reaffirms that litter accumulation in these habitats comes from coastal and land sources and that submarine canyons act as conduits for litter transport from continental shelves into deeper waters [21,28,29,31,47,78]. Therefore, submarine canyons can be considered to be accumulation zones of land-based marine litter in the deep sea. In fact, submarine canyons are areas where macrophyte detritus that originates from coastal areas accumulates in high quantities. This results in a localised increase of organic matter and high abundances of associated fauna, dominated by deposit and suspension-feeding invertebrates [83–85]. Since some deposit-feeders (e.g. holothurians) have been shown to select plastic fragments over sediment grains under laboratory conditions [7], the accumulation of plastics in submarine canyons could have detrimental effects for these ecologically important deep-sea organisms. Furthermore, plastic fragments contain a wide variety of persistent organic pollutants (POPs) that may accumulate in the consumer's tissues and can be transferred upwards in the trophic webs to predators, including humans [86].

Derelict fishing gear was the main litter item found on seamounts, banks, mounds and ocean ridges implying that, unlike submarine canyons, fishing activities are the major source of litter at those settings. Seamounts and banks are targeted by commercial fishing activities as they are often highly productive areas supporting dense aggregations of commercially valuable fish and shellfish [87]. At other locations where recreational [55,88] and commercial [28,54,62,89] fishing activities are intense, derelict fishing gear dominated the litter on the seabed. It was beyond the scope of this study to evaluate the impacts caused by derelict fishing gear, but numerous studies have shown diverse impacts including ghost fishing [16,90] and entanglement by sessile invertebrates such as corals [15], as well as causing damage to

Table 3. Composition of litter (%) in different locations on the seafloor of European waters.

Location	Derelict fishing gear	Glass	Metal	Plastic	Other items	Clinker
ATLANTIC						
Continental slopes						
North Faroe-Shetland Channel	100.0	0.0	0.0	0.0	0.0	0.0
North-East Faroe-Shetland Channel	100.0	0.0	0.0	0.0	0.0	0.0
Continental shelf						
Norwegian Margin	80.0	0.0	0.0	20.0	0.0	0.0
Submarine canyons						
Dangeard & Explorer Canyons	72.2	0.0	0.0	16.7	11.1	0.0
Nazaré Canyon	37.1	0.0	17.1	25.7	20.0	0.0
Lisbon Canyon	9.2	0.0	1.5	86.2	3.1	0.0
Setúbal Canyon	8.7	4.3	4.3	30.4	52.2	0.0
Cascais Canyon	9.1	0.0	0.0	54.5	36.4	0.0
Guilvinec Canyon	43.8	0.0	0.0	43.8	6.3	6.3
Whittard Canyon	28.6	7.1	14.3	42.9	0.0	7.1
Seamounts, banks and mounds						
Anton Dohrn Seamount	0.0	0.0	100.0	0.0	0.0	0.0
Condor Seamount	85.5	14.5	0.0	0.0	0.0	0.0
Josephine Seamount	42.9	28.6	14.3	0.0	14.3	0.0
Hatton Bank	87.5	0.0	12.5	0.0	0.0	0.0
Rockall Bank	33.3	0.0	66.7	0.0	0.0	0.0
Rosemary Bank	66.7	0.0	33.3	0.0	0.0	0.0
Pen Duick Alpha/Beta Mound	75.0	0.0	25.0	0.0	0.0	0.0
Darwin Mounds	10.0	0.0	15.0	60.0	15.0	0.0
Ocean ridges						
North Charlie Gibbs Fracture Zone	0.0	0.0	100.0	0.0	0.0	0.0
South Charlie Gibbs Fracture Zone	0.0	28.6	28.6	28.6	14.3	0.0
Wyville-Thomson Ridge	85.7	0.0	14.3	0.0	0.0	0.0
MEDITERANEAN						
Continental slopes						
Calabrian Slope (Central Med.)	13.2	0.0	8.4	36.2	26.6	15.5
Western Mediterranean Slope	21.6	0.6	0.2	12.1	0.6	64.9
Crete-Rhodes Ridge (E. Med.)	1.6	9.3	6.0	17.0	20.5	45.5
Blanes slope (NW Med.)	2.3	7.9	8.4	12.6	11.6	57.1
Continental shelf						
Gulf of Lion (NW Med.)	0.0	0.0	0.0	88.9	11.1	0.0
Submarine canyons						
Blanes Canyon (NW Med.)	3 (0.2)	3 (4.9)	6 (2.2)	78 (76.3)	9 (1.7)	0 (14.7)
Gulf of Lion Canyons (NW Med.)	0.0	0.0	0.0	67.3	32.7	0.0
Deep basins						
Algero-Balearic Basin (W. Med.)	16.5	0.8	29.6	14.0	2.1	37.0
Crete-Rhodes Ridge (E. Med.)	0.0	9.7	25.0	19.5	7.2	38.5
Calabrian Basin (Central Med.)	0.5	6.7	0.7	5.9	36.1	50.1
ARCTIC						
Continental slope						
HAUSGARTEN, station IV	2.5	2.5	2.5	60	32.5	0

*Numbers in parentheses refer to trawl surveys.

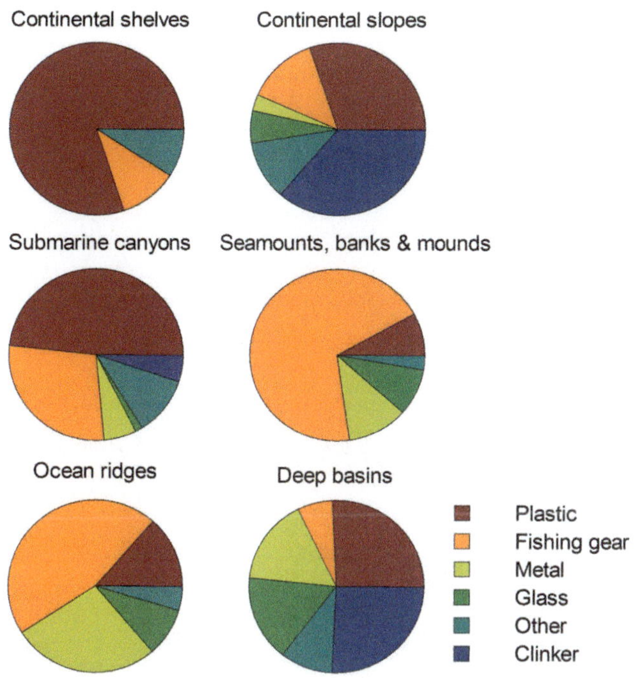

Figure 7. Litter composition in different physiographic settings across European waters.

fishing equipment [91]. Discarded trawl gear can also have a compounding effect by trapping more mobile litter resulting in a litter 'depot' that has a greater impact than single pieces of litter [31]. Since most fishing equipment (lines and nets) is made mostly of highly resistant plastics, such negative effects will likely persist for a long time. Sites located in deep basins and continental slopes were dominated by clinker. Clinker, the residue of burnt coal, was commonly dumped from steam ships from the late 18th century and well into the 20th century. In the Mediterranean Sea, its occurrence on the deep seafloor has been shown to coincide with such shipping routes [29]. However, it is important to acknowledge that in this study, deep basins and continental slopes were principally sampled by trawling and it is difficult to determine if the differences in litter composition with other physiographic settings are the results of differences in the sampling methodology, particularly since clinker is difficult to identify from underwater footage. Indeed, clinker was present in non-quantitative trawls undertaken at HAUSGARTEN (Bergmann, unpublished data), but could not be detected on images from the seafloor. Similarly, a high abundance of clinker was recovered from trawl surveys in Blanes Canyon that could not be identified in analysis of ROV footage from the same area (Table 3). Given that most of the clinker present on the seafloor was dumped over 100 years ago, sedimentation will have buried it, which would explain the differences in clinker quantification between images and trawl data. The deep seafloor is a passive accumulation area for litter, integrating information over long-time periods. If trawls are able to recover heavy clinker deposited on the seafloor over a century ago, these gears must be retrieving at the same time all of the lighter and most recent litter items, such as plastic for example, that have been accumulating only in the last 50 years. Overall, the composition of litter found on the seafloor showed some dissimilarity with the composition found on the coasts or in surface waters. Although plastics are dominant in all settings [70], some areas of the seafloor investigated here and elsewhere [28,44,45,54,78] harbour significant quantities of non-buoyant litter such as glass, metal and clinker, directly dumped from ships but that are seldom found in surface waters [41,68] or on the coasts [41,72]. The coasts and surface waters are a source of litter items for the open seas and all this litter, sooner or later, will sink to the seafloor where it accumulates.

The most common method used to provide data on benthic marine litter has been trawling, typically as a parallel objective to surveys directed to fish or benthic organism sampling [53]. With the recent development of optical methods fitted to platforms such as submersibles, ROV and drop-down systems, the use of underwater imaging technology has greatly increased our ability to quantify deep-sea litter. Both methods (imaging technology and trawling) have distinct assets for studying benthic litter that should be used in conjunction to best understand the dynamics of pollution on the seafloor. Video surveys can provide data for areas where topography is complex (e.g seamounts or canyon walls), habitats made by structure-building organisms (e.g. cold-water corals), or dynamic systems (e.g. hydrothermal vents and cold seeps), that cannot be accessed with a trawl [53]. Furthermore, imaging is a non-intrusive method that does not remove benthic organisms or damage the environment. On the other hand, a trawl has the advantages of recovering litter items of very small size (e.g. small plastic fragments) or that are buried in the sediments (e.g. clinker), which otherwise would not be detected through imaging technology. In addition, litter items collected with a trawl can be analysed in the laboratory to obtain further important information, such as state of degradation or colonisation by fouling organisms [92]. Such data will help understand sinking processes of plastic, facilitate the identification of their location of arrival into the ocean and provide information on the impacts of litter on marine organisms.

The large quantities of litter reaching the deep ocean floor is a major issue worldwide, yet little is known about its sources, patterns of distribution, abundance and, particularly, impacts on the habitats and associated fauna [1]. At present, density of litter in the deep sea is lower than found on some heavily polluted beaches [33,93], but unlike the coastal zone, only a tiny fraction of the (deep) seafloor has been surveyed to date. Furthermore, microplastic accumulation may become an important component of pollution in deep-sea ecosystems [94] that urgently needs to be evaluated. Our results for European waters show that litter sources are distinct across different physiographic settings and that their abundance is variable, most probably guided by a complex set of interactions between physiography, anthropogenic activities and hydrography. It is important that in the future, large-scale assessments are done in a standardised manner to understand fully the scale of the problem and set the necessary actions to prevent the accumulation of litter in the marine environment.

Supporting Information

Table S1 Results of analyses of similarity (ANOSIM) evaluating variation in the composition of litter among physiographic settings. RIDGE: ocean ridges; CANY: submarine canyons; SHELF: continental shelves; SLOPE: continental slopes; SBM: seamounts, banks and mounds; BASIN: deep basins.

Table S2 Similarity percentage analysis (SIMPER) of litter composition for each pooled physiographic settings (based on similarities revealed by ANOSIM) and the contribution of litter category to group similarity.

Acknowledgments

The authors would like to thank the captains, crews and scientific parties of all cruises for their help and support during the data collection. PT would like to thank Gideon Mordecai for analytical work and Doug Masson. Finally, the authors would like to thank Martin Thiel and two other anonymous reviewers, whose suggestions and comments greatly improved the manuscript. This is publication number 33575 of the Alfred-Wegener-Institut Helmholtz-Zentrum für Polar- und Meeresforschung.

Author Contributions

Conceived and designed the experiments: CKP ERL CHSA TA MB MC JBC JD GD FG KLH VAIH EI DOBJ GL TM JNGP AP HS IT XT DVR PT. Performed the experiments: CKP ERL CHSA TA MB MC JBC JD GD FG KLH VAIH EI DOBJ GL TM JNGP AP HS IT XT DVR PT. Analyzed the data: CKP. Wrote the paper: CKP.

References

1. UNEP (2009) Marine Litter: A Global Challenge. Nairobi. 232 p.
2. Galgani F, Hanke G, Werner S, De Vrees L (2013) Marine litter within the European Marine Strategy Framework Directive. ICES J. Mar. Sci. 70: 1055–1064.
3. Moser ML, Lee DS (1992) A 14-year survey of plastic ingestion by western north-atlantic seabirds. Colon. Waterbirds 15: 83–94.
4. Ryan PG (1988) Effects of ingested plastic on seabird feeding: Evidence from chickens. Mar. Pollut. Bull. 19: 125–128.
5. Bjorndal KA, Bolten AB, Lagueux CJ (1994) Ingestion of marine debris by juvenile sea turtles in coastal Florida habitats. Mar. Pollut. Bull. 28: 154–158.
6. Tomás J, Guitart R, Mateo R, Raga JA (2002) Marine debris ingestion in loggerhead sea turtles, *Caretta caretta*, from the Western Mediterranean. Mar. Pollut. Bull. 44: 211–216.
7. Graham ER, Thompson JT (2009) Deposit- and suspension-feeding sea cucumbers (Echinodermata) ingest plastic fragments. J. Exp. Mar. Biol. Ecol. 368: 22–29.
8. Carson HS (2013) The incidence of plastic ingestion by fishes: From the prey's perspective. Mar. Pollut. Bull. 74: 170–174.
9. Neilson JL, Straley JM, Gabriele CM, Hills S (2009) Non-lethal entanglement of humpback whales (*Megaptera novaeangliae*) in fishing gear in northern Southeast Alaska. J. Biogeogr. 36: 452–464.
10. Williams R, Ashe E, O'Hara PD (2011) Marine mammals and debris in coastal waters of British Columbia, Canada. Mar. Pollut. Bull. 62: 1303–1316.
11. Allen R, Jarvis D, Sayer S, Mills C (2012) Entanglement of grey seals *Halichoerus grypus* at a haul out site in Cornwall, UK. Mar. Pollut. Bull. 64: 2815–2819.
12. Carr A (1987) Impact of nondegradable marine debris on the ecology and survival outlook of sea turtles. Mar. Pollut. Bull. 18: 352–356.
13. Schrey E, Vauk GJM (1987) Records of entangled gannets (*Sula bassana*) at Helgoland, German Bight. Mar. Pollut. Bull. 18: 350–352.
14. Chiappone M, Dienes H, Swanson DW, Miller SL (2005) Impacts of lost fishing gear on coral reef sessile invertebrates in the Florida Keys National Marine Sanctuary. Biol. Conserv. 121: 221–230.
15. Pham CK, Gomes-Pereira JN, Isidro EJ, Santos RS, Morato T (2013) Abundance of litter on Condor seamount (Azores, Portugal, Northeast Atlantic). Deep-Sea Res. Part II-Top. Stud. Oceanogr. 98: 204–208.
16. Brown J, Macfadyen G (2007) Ghost fishing in European waters: Impacts and management responses. Mar. Pol. 31: 488–504.
17. Winston JE (1982) Drift plastic—An expanding niche for a marine invertebrate? Mar. Pollut. Bull. 13: 348–351.
18. Barnes DKA, Milner P (2005) Drifting plastic and its consequences for sessile organism dispersal in the Atlantic Ocean. Mar.Biol. 146: 815–825.
19. Barnes DKA (2002) Biodiversity: Invasions by marine life on plastic debris. Nature 416: 808–809.
20. Derraik JGB (2002) The pollution of the marine environment by plastic debris: a review. Mar. Pollut. Bull. 44: 842–852.
21. Barnes DKA, Galgani F, Thompson RC, Barlaz M (2009) Accumulation and fragmentation of plastic debris in global environments. Philos. Trans. R. Soc. Lond. B Biol. Sci. 364: 1985–1998.
22. Sheavly SB, Register KM (2007) Marine debris & plastics: environmental concerns, sources, impacts and solutions. J. Polym. Environ. 15: 301–305.
23. Engler RE (2012) The complex interaction between marine debris and toxic chemicals in the ocean. Environ. Sci. Technol. 46: 12302–12315.
24. Andrady AL (2011) Microplastics in the marine environment. Mar. Pollut. Bull. 62: 1596–1605.
25. Cole M, Lindeque P, Fileman E, Halsband C, Goodhead R, et al. (2013) Microplastic ingestion by zooplankton. Environ. Sci. Technol. 47: 6646–6655.
26. Farrell P, Nelson K (2013) Trophic level transfer of microplastic: *Mytilus edulis* (L.) to *Carcinus maenas* (L.). Environ. Pollut. 177: 1–3.
27. Murray F, Cowie PR (2011) Plastic contamination in the decapod crustacean *Nephrops norvegicus* (Linnaeus, 1758). Mar. Pollut. Bull. 62: 1207–1217.
28. Galgani F, Leaute JP, Moguedet P, Souplet A, Verin Y, et al. (2000) Litter on the sea floor along European coasts. Mar. Pollut. Bull. 40: 516–527.
29. Ramirez-Llodra E, Company JB, Sardà F, De Mol B, Coll M, et al. (2013) Effects of natural and anthropogenic processes in the distribution of marine litter in the deep Mediterranean Sea. Prog. Oceanogr. 118: 273–287.
30. Corcoran PL, Biesinger MC, Grifi M (2009) Plastics and beaches: A degrading relationship. Mar. Pollut. Bull. 58: 80–84.
31. Mordecai G, Tyler PA, Masson DG, Huvenne VAI (2011) Litter in submarine canyons off the west coast of Portugal. Deep Sea Res. Part II Top. Stud. Oceanogr. 58: 2489–2496.
32. Aliani S, Griffa A, Molcard A (2003) Floating debris in the Ligurian Sea, north-western Mediterranean. Mar. Pollut. Bull. 46: 1142–1149.
33. Ariza E, Jiménez JA, Sardá R (2008) Seasonal evolution of beach waste and litter during the bathing season on the Catalan coast. Waste Manag. 28: 2604–2613.
34. Gabrielides GP, Golik A, Loizides L, Marino MG, Bingel F, et al. (1991) Man-made garbage pollution on the Mediterranean coastline. Mar. Pollut. Bull. 23: 437–441.
35. Kornilios S, Drakopoulos PG, Dounas C (1998) Pelagic tar, dissolved/dispersed petroleum hydrocarbons and plastic distribution in the Cretan Sea, Greece. Mar. Pollut. Bull. 36: 989–993.
36. McCoy FW (1988) Floating megalitter in the eastern Mediterranean. Mar. Pollut. Bull. 19: 25–28.
37. Morris RJ (1980) Floating plastic debris in the Mediterranean. Mar. Pollut. Bull. 11: 125.
38. Shiber JG (1982) Plastic pellets on Spain's 'Costa del Sol' beaches. Mar. Pollut. Bull. 13: 409–412.
39. Thompson RC, Olsen Y, Mitchell RP, Davis A, Rowland SJ, et al. (2004) Lost at sea: Where is all the plastic? Science 304: 838.
40. Vauk GJM, Schrey E (1987) Litter pollution from ships in the German Bight. Mar. Pollut. Bull. 18: 316–319.
41. Van Cauwenberghe L, Claessens M, Vandegehuchte MB, Mees J, Janssen CR (2013) Assessment of marine debris on the Belgian Continental Shelf. Mar. Pollut. Bull. 73: 161–169.
42. Anastasopoulou A, Mytilineou C, Smith CJ, Papadopoulou KN (2013) Plastic debris ingested by deep-water fish of the Ionian Sea (Eastern Mediterranean). Deep-Sea Res. Part I Oceanogr. Res. Pap. 74: 11–13.
43. Bergmann M, Klages M (2012) Increase of litter at the Arctic deep-sea observatory HAUSGARTEN. Mar. Pollut. Bull. 64: 2734–2741.
44. Galgani F, Burgeot T, Bocquene G, Vincent F, Leaute JP, et al. (1995b) Distribution and abundance of debris on the continental shelf of the Bay of Biscay and in Seine Bay. Mar. Pollut. Bull. 30: 58–62.
45. Galgani F, Jaunet S, Campillo A, Guenegen X, His E (1995a) Distribution and abundance of debris on the continental shelf of the north-western Mediterranean Sea. Mar. Pollut. Bull. 30: 713–717.
46. Galgani F, Lecornu F (2004) Debris on the seafloor at "Hausgarten". Reports on Polar and Marine Research 488: 260–262.
47. Galgani F, Souplet A, Cadiou Y (1996) Accumulation of debris on the deep sea floor off the French Mediterranean coast. Mar. Ecol. Prog. Ser. 142: 225–234.
48. Galil BS, Golik A, Tuerkay M (1995) Litter at the bottom of the sea: A sea bed survey in the eastern Mediterranean. Mar. Pollut. Bull. 30: 22–24.
49. Katsanevakis S, Katsarou A (2004) Influences on the distribution of marine debris on the seafloor of shallow coastal areas in Greece (Eastern Mediterranean). Water, Air, & Soil Pollution 159: 325–337.
50. Kidd RB, Huggett QJ (1981) Rock debris on abyssal plains in the northeast Atlantic - a comparison of epibenthic sledge hauls and photographic surveys. Oceanol. Acta 4: 99–104.
51. Revill AS, Dunlin G (2003) The fishing capacity of gillnets lost on wrecks and on open ground in UK coastal waters. Fish Res. 64: 107–113.
52. Stefatos A, Charalampakis M, Papatheodorou G, Ferentinos G (1999) Marine debris on the seafloor of the Mediterranean Sea: examples from two enclosed gulfs in Western Greece. Mar. Pollut. Bull. 38: 389–393.
53. Spengler A, Costa MF (2008) Methods applied in studies of benthic marine debris. Mar. Pollut. Bull. 56: 226–230.
54. Lee D-I, Cho H-S, Jeong S-B (2006) Distribution characteristics of marine litter on the sea bed of the East China Sea and the South Sea of Korea. Est. Coast. Shelf Sci. 70: 187–194.
55. Watters DL, Yoklavich MM, Love MS, Schroeder DM (2010) Assessing marine debris in deep seafloor habitats off California. Mar. Pollut. Bull. 60: 131–138.
56. Fiorentini L, Dremière P-Y, Leonori I, Sala A, Palumbo V (1999) Efficiency of the bottom trawl used for the Mediterranean international trawl survey (MEDITS). Aquat. Living Resour. 12: 187–205.
57. Sardà F, Cartes JE, Company JB, Albiol A (1998) A Modified Commercial Trawl Used to Sample Deep-Sea Megabenthos. Fish. Sci. 64: 492–493.
58. Clarke KR, Clarke RK, Gorley RN (2006) Primer V6: User Manual - Tutorial: Plymouth Marine Laboratory.
59. Bray JR, Curtis JT (1957) An Ordination of the Upland Forest Communities of Southern Wisconsin. Ecol. Monogr. 27: 325–349.
60. Ramirez-Llodra E, Tyler PA, Baker MC, Bergstad OA, Clark MR, et al. (2011) Man and the last great wilderness: Human impact on the deep sea. PLoS ONE 6: e22588.

61. Miyake H, Shibata H, Furushima Y (2011) Deep-sea litter study using deep-sea observation tools. In: Omori K, Guo X, Yoshie N, Fujii N, Handoh IC et al., editors. Interdisciplinary Studies on Environmental Chemistry-Marine Environmental Modeling and Analysis: Terrapub. pp. 261–269.
62. Hess NA, Ribic CA, Vining I (1999) Benthic marine debris, with an emphasis on fishery-related items, surrounding Kodiak Island, Alaska, 1994–1996. Mar. Pollut. Bull. 38: 885–890.
63. Keller AA, Fruh EL, Johnson MM, Simon V, McGourty C (2010) Distribution and abundance of anthropogenic marine debris along the shelf and slope of the US West Coast. Mar. Pollut. Bull. 60: 692–700.
64. Kanehiro H, Tokai T, Matuda K (1996) The distribution of litter in fishing ground of Tokyo Gulf. Fish. Eng. 32: 211–217.
65. Kuriyama Y, Tokai T, Tabata K, Kanehiro H (2003) Distribution and composition of litter of Tokyo Gulf and its age analysis. Nippon Suisan Gakkaishi 69: 770–781.
66. Hinojosa IA, Thiel M (2009) Floating marine debris in fjords, gulfs and channels of southern Chile. Mar. Pollut. Bull. 58: 341–350.
67. Thiel M, Hinojosa I, Vasquez N, Macaya E (2003) Floating marine debris in coastal waters of the SE-Pacific (Chile). Mar. Pollut. Bull. 46: 224–231.
68. Thiel M, Hinojosa IA, Joschko T, Gutow L (2011) Spatio-temporal distribution of floating objects in the German Bight (North Sea). J. Sea Res. 65: 368–379.
69. Zhou P, Huang CG, Fang HD, Cai WX, Li DM, et al. (2011) The abundance, composition and sources of marine debris in coastal seawaters or beaches around the northern South China Sea (China). Mar. Pollut. Bull. 62: 1998–2007.
70. Derraik JGB (2002) The pollution of the marine environment by plastic debris: a review. Mar. Pollut. Bull. 44: 842–852.
71. Moore SL, Gregorio D, Carreon M, Weisberg SB, Leecaster MK (2001) Composition and distribution of beach debris in Orange County, California. Mar. Pollut. Bull. 42: 241–245.
72. Topcu EN, Tonay AM, Dede A, Ozturk AA, Ozturk B (2013) Origin and abundance of marine litter along sandy beaches of the Turkish Western Black Sea Coast. Mar. Environ. Res. 85: 21–28.
73. Silva-Iniguez L, Fischer DW (2003) Quantification and classification of marine litter on the municipal beach of Ensenada, Baja California, Mexico. Mar. Pollut. Bull. 46: 132–138.
74. Martinez-Ribes L, Basterretxea G, Palmer M, Tintore J (2007) Origin and abundance of beach debris in the Balearic Islands. Sci. Mar. 71: 305–314.
75. Willoughby NG, Sangkoyo H, Lakaseru BO (1997) Beach litter: an increasing and changing problem for Indonesia. Mar. Pollut. Bull. 34: 469–478.
76. Oigman-Pszczol SS, Creed JC (2007) Quantification and classification of marine litter on beaches along Armacao dos Buzios, Rio de Janeiro, Brazil. J. Coast. Res. 23: 421–428.
77. Ryan PG, Moore CJ, van Franeker JA, Moloney CL (2009) Monitoring the abundance of plastic debris in the marine environment. Philosophical Transactions of the Royal Society B-Biological Sciences 364: 1999–2012.
78. Schlining K, von Thun S, Kuhnz L, Schlining B, Lundsten L, et al. (2013) Debris in the deep: Using a 22-year video annotation database to survey marine litter in Monterey Canyon, central California, USA. Deep-Sea Res. Part I Oceanogr. Res. Pap. 79: 96–105.
79. Halpern BS, Selkoe KA, Micheli F, Kappel CV (2007) Evaluating and ranking the vulnerability of global marine ecosystems to anthropogenic threats. Conserv. Biol. 21: 1301–1315.
80. Halfar J, Fujita RM (2007) Danger of deep-sea mining. Science 316: 987.
81. He G, Ma W, Song C, Yang S, Zhu B, et al. (2011) Distribution characteristics of seamount cobalt-rich ferromanganese crusts and the determination of the size of areas for exploration and exploitation. Acta Oceanol. Sin. 30: 63–75.
82. Davies AJ, Roberts JM, Hall-Spencer J (2007) Preserving deep-sea natural heritage: Emerging issues in offshore conservation and management. Biol. Conserv. 138: 299–312.
83. Amaro T, Bianchelli S, Billett DSM, Cunha MR, Pusceddu A, et al. (2010) The trophic biology of the holothurian *Molpadia musculus*: implications for organic matter cycling and ecosystem functioning in a deep submarine canyon. Biogeosciences 7: 2419–2432.
84. Pagès F, Martín J, Palanques A, Puig P, Gili JM (2007) High occurrence of the elasipodid holothurian *Penilpidia ludwigi* (von Marenzeller, 1893) in bathyal sediment traps moored in a western Mediterranean submarine canyon. Deep-Sea Res. Part I Oceanogr. Res. Pap. 54: 2170–2180.
85. Vetter EW, Dayton PK (1998) Macrofaunal communities within and adjacent to a detritus-rich submarine canyon system. Deep Sea Res. Part II Top. Stud. Oceanogr. 45: 25–54.
86. Wright SL, Thompson RC, Galloway TS (2013) The physical impacts of microplastics on marine organisms: A review. Environ. Pollut. 178: 783–492.
87. Clark MR, Koslow JA (2008) Impacts of fisheries on seamounts. In: Pitcher TJ, Morato T, Hart PJB, Clark MR, Haggan N, et al., editors. Seamounts: Ecology, Fisheries & Conservation: Blackwell Publishing Ltd. pp. 413–441.
88. Moore SL, Allen MJ (2000) Distribution of anthropogenic and natural debris on the mainland shelf of the southern California Bight. Mar. Pollut. Bull. 40: 83–88.
89. Cho D-O (2011) Removing derelict fishing gear from the deep seabed of the East Sea. Mar. Pol. 35: 610–614.
90. Carr HA, Harris J (1997) Ghost-fishing gear: have fishing practices during the past few years reduced the impact? In: Coe J, Rogers D, editors. Marine Debris: Springer New York. pp. 141–151.
91. Nash AD (1992) Impacts of marine debris on subsistence fishermen: An exploratory study. Mar. Pollut. Bull. 24: 150–156.
92. Sanchez P, Maso M, Saez R, De Juan S, Muntadas A, et al. (2013) Baseline study of the distribution of marine debris on soft-bottom habitats associated with trawling grounds in the northern Mediterranean. Sci. Mar. 77: 247–255.
93. Martins J, Sobral P (2011) Plastic marine debris on the Portuguese coastline: A matter of size? Mar. Pollut. Bull. 62: 2649–2653.
94. Van Cauwenberghe L, Vanreusel A, Mees J, Janssen CR (2013) Microplastic pollution in deep-sea sediments. Environ. Pollut. 182: 495–499.
95. Purser A, Orejas C, Gori A, Tong RJ, Unnithan V, et al. (2013) Local variation in the distribution of benthic megafauna species associated with cold-water coral reefs on the Norwegian margin. Cont. Shelf Res. 54: 37–51.
96. Van Rooij D, Blamart D, De Mol L, Mienis F, Pirlet H, et al. (2011) Cold-water coral mounds on the Pen Duick Escarpment, Gulf of Cadiz: The MiCROSYSTEMS project approach. Mar. Geol. 282: 102–117.
97. Tubau X, Canals M, Lastras G,Company JB, Rayo X (2012) The PROMARES-OASIS DEL MAR shipboard party; Marine litter in the deep sections of the North Catalan submarine canyons from ROV video-inspection. 3rd Annual Hermione Meeting, Faro (Portugal), Abstr. Vol., p. 21.
98. Huvenne VAI (2011) Benthic habitats and the impact of human activities in Rockall Trough, on Rockall Bank and in Hatton Basin. National Oceanography Centre, Cruise Report No. 04, RRS James Cook Cruise 60, 133 pp.
99. Bullimore RD, Foster NL, Howell KL (2013) Coral-characterized benthic assemblages of the deep Northeast Atlantic: defining "Coral Gardens" to support future habitat mapping efforts. ICES J. Mar. Sci. 70: 511–522.
100. Lavaleye MSS (2011) CoralFISH-HERMIONE cruise report of Cruise 64PE345 with RV Pelagia Texel-Vigo, 28 Sept – 14 Oct 2011 to Belgica Mound Province (CoralFISH & HERMIONE) and Whittard Canyon (HERMIONE). NIOZ-cruise report. pp. 47.

15

Application of Empirical Wave Run-Up Formulas to the Polish Baltic Sea Coast

Dominik Paprotny*, Paweł Andrzejewski, Paweł Terefenko, Kazimierz Furmańczyk

Faculty of Geosciences, University of Szczecin, Szczecin, Poland

Abstract

Advanced, multidimensional models are typically applied when researching processes occurring in the nearshore. Relatively simple, empirical equations are commonly used in coastal engineering practice in order to estimate extreme wave run-up on beaches and coastal structures. However, they were mostly calibrated to the characteristics of oceanic coasts, which have different wave regime than a semi-enclosed basin like the Baltic Sea. In this paper we apply the formulas to the Polish Baltic Sea coast. The equations were adjusted to match local conditions in two test sites in Międzyzdroje and Dziwnówek, where beaches are under continuous video surveillance. Data from WAM wave model and coastal gauge stations were used, as well as precise measurements of the beaches' cross-sections. More than 600 run-up events spanning from June to December 2013 were analysed, including surges causing dune erosion. Extreme wave run-up $R_{2\%}$ was calculated and presented as a percentage value indicating what part of the beach was inundated. The method had a root-mean-square error of 6.1 and 6.5 percentage points depending on the test site. We consider it is a fast and computationally undemanding alternative to morphodynamic models. It will constitute a part of the *SatBałtyk Operating System-Shores*, delivering forecasts of wave run-up on the beaches for the entire Polish coastline.

Editor: Moncho Gomez-Gesteira, University of Vigo, Spain

Funding: This work was supported by the European Union through the Innovative Economy Operational Programme - project No. POIG.01.01.02-22-011/09-00 "The Satellite Monitoring of the Baltic Sea Environment (SatBałtyk)". The funders had no role in study design, data collection and analysis, decision to publish, or preparation of the manuscript.

Competing Interests: The authors have declared that no competing interests exist.

* Email: dominikpaprotny@op.pl

Introduction

Coastal zone represents a dynamic environment where land, sea and atmosphere meet. It is also a system where up to two-thirds of the world population lives and many other are visiting those areas frequently [1,2]. In order to sustainably manage recreational, industrial, agricultural and any other activities taking place near the shores it is necessary to understand the dynamic nature of both marine and terrestrial interactions at the coasts.

Small-scale elements of a depositional coast can experience rapid change in their morphology, mainly due to short-term sea level variations caused by storm surges [3–6]. In this perspective it is crucial for coastal managers to determine the impact of the sea on the coast especially in the aspect of potential hazard and risk to human activities as well as to the shore itself. This includes also the possibility of making predictions of those impacts.

Recently, an evaluation of risks, existing coastal management plans and civil engineering-produced protection schemes in nine European countries revealed that operational approaches to coastal risks and hazards had been virtually non-existent in most of those countries [7]. The biggest effort used to be channelled only to forecasting physical parameters (wind, waves, sea levels etc.), without considering their impact on the coast. One of the first steps to adopt an alternate approach to coastal hazards was the MICORE project in which nine prototype early warning systems based on open-source morphological XBeach model [8] were developed [9,10]. MICORE made it possible to test, with good results, the XBeach model at a variety of test sites along the European coasts, including the Polish southern Baltic coast [11].

Advanced, multidimensional models such as XBeach are typically applied when researching processes occurring in the nearshore. They mathematically explain typically complex and non-linear relationships between different variables and are indispensable in enhancing knowledge about coastal systems [12,13]. Models provide support in finding the best means to deal with as yet unforeseen events and their effects, both morphological and socio-economic. In Poland, efforts are being made to expand research on other parts of the coast. Currently three test sites are included in the *SatBałtyk Operating System-Shores* [14], the coastal-research component of a research project dedicated to the monitoring of the Baltic Sea [15]. Unfortunately, these models have a major drawback: they are computationally demanding, limiting the possibility to use them for detailed analysis of large parts of the shoreline.

But alongside those robust models some simpler methods were developed. At times when solving complex, non-linear equations was problematic to say the least, coastal engineers created empirical formulas. They were intended to improve the designs of coast protection structures. Wave run-up could be calculated and therefore enabled the engineers to design proper heights of embankments or seawalls [16,17]. In the 1980s scientists started field experiments trying to apply the equations to natural beaches, an attempt that proofed successful. Several new equations were also developed [18–24]. Yet unlike more sophisticated methods, they are not being used in operational forecasting, but are rather confined to engineering practice.

Nevertheless, the reliability of aforementioned methods, both the empirical equations and models, needs always to be verified when applied to a new area. One possibility is to conduct *in situ* measurements of the desired phenomena. Much more convenient, though, is to use an increasingly popular alternative in the form of video monitoring systems. This method was applied for coastal research in the early 1980s and developed ever since. Video cameras were used during several experiments, capturing wave run-ups or measuring sand bar movements [25,26]. Whereas traditional methods (*in situ* measurements, aerial and satellite imagery) are expensive and limited by weather conditions, video systems enable the researches to continuously monitor beach processes in the long-term [27,28]. Especially since the technique has been automated and turned from analogue to digital, it has become a standard procedure when calibrating models for wave run-up and rip current occurrence [23,27,29].

Research described in this paper was conceived while examining possibilities to expand the beach inundation module of *SatBałtyk Operating System-Shores* beyond the three test sites it currently includes. Preliminary analysis found that running XBeach model (1D variant) for an estimated 500 cross-sections of the coast (one for each kilometre of the coast) would be too time-consuming and therefore render the resulting forecasts useless. Moreover, it was designed specifically sandy coasts, which is not necessarily the case along the Baltic coast. Empirical equations of wave run-up on slopes were chosen as an alternative, for several reasons. Firstly, they are computationally undemanding, therefore creating a possibility to deliver forecasts without substantial delays. Secondly, they utilize the very same data as the system based on XBeach: wave parameters from WAM model, sea levels from M3D model and Maritime Offices' cross-sections of the shore.

It excepted that the formulas, once calibrated, will be built into an operational, forecasting system during 2014–2015 in support of the *SatBałtyk* project [14]. Each 1 km cross-section will contain data extracted from the latest available high-resolution digital elevation models, created using lidar technology and provided by the responsible maritime offices.

Additionally, *Satbałtyk* project expects its system and models to be used, apart from delivering forecasts covering up to several days, also to archive the results and calculate hindcasts. They will allow us to conduct seasonal and long-term analysis. Therefore, alongside the calibration of the method, we include elements of such an investigation for two test sites, on which this experiment is based.

Materials and Methods

Wave run-up is defined as "the landward extent of wave uprush measured vertically from the still water level" [24]. The earliest formulation of this is by Hunt [30], who gives the following equation:

$$\frac{R}{H_0} = \xi_0 \tag{1}$$

where R is the wave run-up, H_0 is the deep-water wave height and ξ_0 is the deep-water surf similarity parameter. The latter is given as follows [31]:

$$\xi_0 = \frac{\tan(\alpha)}{\sqrt{\frac{H}{L_0}}} \tag{2}$$

where α is the slope angle and L_0 is the linear theory deep-water wave length:

$$L_0 = \frac{gT^2}{2\pi} \tag{3}$$

where T is regular wave period. Hunt's formula can be used to calculate run-up from incidental regular waves. Later research reworked eq. (1) to include irregular waves and introduced additional empirical parameters based on field measurements and lab experiments [24]. Several very similar equations were developed, of which the most recent version from Mase [32] was chosen:

$$R_{2\%} = aH_{mo}\left(\frac{\tan(\alpha)}{\sqrt{\frac{H_{mo}}{L_{op}}}}\right)^b \tag{4}$$

where $R_{2\%}$ is wave run-up with a 2% probability of occurrence (i.e. 1 in 50 waves), H_{mo} is the significant deep-water wave height and L_{op} is the same as in eq. (3), except that T stands here for peak wave period (T_{p}). (The 2% run-up is commonly used in similar equations, though the reason for such a choice is not well explained. It seems to have a special statistical relation to significant wave height and originate from Dutch experiments carried out in the 1930s, but the report where these results were published apparently has gone missing [17]). Additionally, two empirical parameters can be found in eq. (4): a and b. Mase suggested $a = 1.86$ and $b = 0.71$, based on his laboratory tests [32]. Several field tests have been conducted by various authors, but they were mostly located on high-wave-energy coasts of the United States, the Netherlands, Portugal, Spain or Australia [23,28,33,34]. One study for the Baltic sea could be identified, to our best knowledge, albeit it did not include any adjustments of parameters [35]. As a result, the formula had to be calibrated on the Polish coast to ensure its accuracy. It should be also noted that other field-tested equations exist, developed by Hughes [22] and Stockdon [23]. Melby [24] tested those equations versus two incarnations of Hunt's formula using Stockdon's compilation of field experiments and found that Holman's [19] and Mase's equations had better statistical performance. Equation by Holman differs from Mase's only by the inclusion of a third parameter. Their performance, when calibrated, is virtually of the same quality, therefore Mase's variant was chosen as having less parameters to calibrate. This is even in spite of the fact that the main assumption behind the formula—a plain, impermeable slope—conforms more to flood protection structures than beaches, especially sandy ones.

Polish Baltic Sea coast is 500 km long, including 72 km along the Hel Peninsula. Approximately 80% is a dune coast, whereas the remaining part forms a cliff coast [36]. Vast majority of the coastline has a beach, except for river mouths, harbours, military bases and the swampy shores of Puck Bay. Research was carried

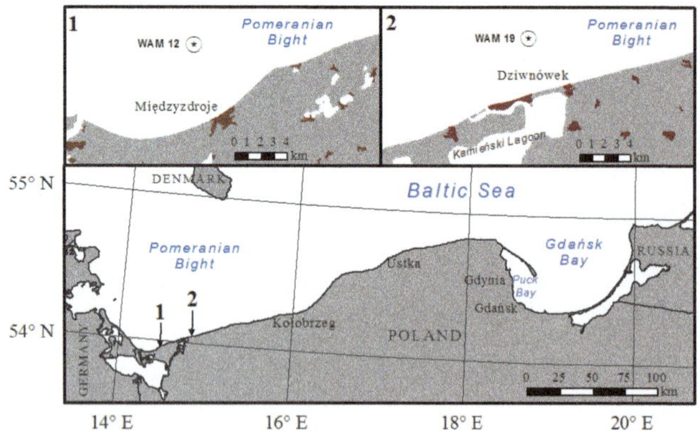

Figure 1. Polish Baltic Sea coast. Location of the test sites in Międzyzdroje (1) and Dziwnówek (2) is marked.

out in two locations in the western part of the coast, namely Dziwnówek and Międzyzdroje (Fig. 1).

Międzyzdroje is a town on Wolin Island, while Dziwnówek is a village located 30 km east of the former. Both are very popular summer resorts along the banks of the Pomeranian Bight. The bight is a shallow basin, where depths are generally below 20 meters. The coast in this area is oscillating, but in overall erosion is stronger than accumulation. It includes both cliffs and dunes, which are retreating, on average, by 10 cm per year [37].

The two selected cross-sections are located at about 414.2 and 386.2 km of the Maritime Office's kilometrage. Both shores include moderately developed dunes that reach 5–7 m above sea level. There are no coastal protection structures in those particular locations apart from some timber groynes in Dziwnówek. They are also under continuous video surveillance.

In Dziwnówek a video camera was mounted in 2009 in order to monitor a test site for the MICORE project [7]. Another camera was installed in Międzyzdroje in 2013 to support model validation activities in the SatBałtyk project [14]. They are both placed at the tops of observation towers used formerly by the military and border guards. The view from the cameras is presented in Fig. 2 (real-time images could also be found online [38]). The cameras were specifically adjusted for use as a close-range remote sensing device through a calibration processes described in literature [39,40].

The cross-sections under observation were surveyed with GPS-RTK devices while the cameras were operational, so that the exact location of each point along the profile could be unequivocally identified on all images from video monitoring. Measurements were conducted on July 11th, 2013 (Dziwnówek) and July 18th, 2013 (Międzyzdroje). In Dziwnówek the length of the profile, measured from the waterline at mean sea level to the base of the dune, is almost 33 m with an elevation difference of 2.32 m. For Międzyzdroje the values are 53 m and 1.88 m, respectively. Detailed data can be found in Table S1A in File S1.

A major issue is calculating the slope angle properly. The method assumes a single value, while in reality the beaches are more complex. The profiles shown in Fig. 3 are relatively smooth and do not include a longshore bar that typically appears near the waterline after storms. Presence of a sand bar can skew the results, but it did not appear throughout the course of the experiment. Due to the limitations of the measurement method used, the cross-sections do not include most of the underwater part of the foreshore. Approaches to the slope angle problem differ largely. For example, Douglass [20] suggested skipping the slope parameter altogether, based on his field experiments on the US Atlantic coast. On the opposite end, an 'effective slope' method was presented as the most accurate by Mayer and Kriebel [41], albeit it is most cumbersome to apply, as the slope angle is recalculated for every event between the wave breaking point and run-up on the slope (an iterative method). For the purpose of this study, the angle was calculated using linear regression based on all available measurements from the underwater part up to and including the dune base. Consequently, the angle amounts to 3.9° in Dziwnówek and 2.0° in Międzyzdroje. Notwithstanding, large disproportion between beach height and width (1:12 and 1:23, respectively) means that the assumption of a plain slope is still fairly accurate, at least as long the beach is not reshaped by a storm.

The final element of eq. (4) that had been acquired were data on waves, i.e. deep-water significant wave height and peak wave period. Since no actual measurements from buoys were available when the research described here was carried out, calculations from WAM wave model were used instead. This two-dimensional model of wave spectrum [42] was operationally implemented for

Figure 2. View from cameras in Międzyzdroje (left) and Dziwnówek (right). Cross-sections are marked in red.

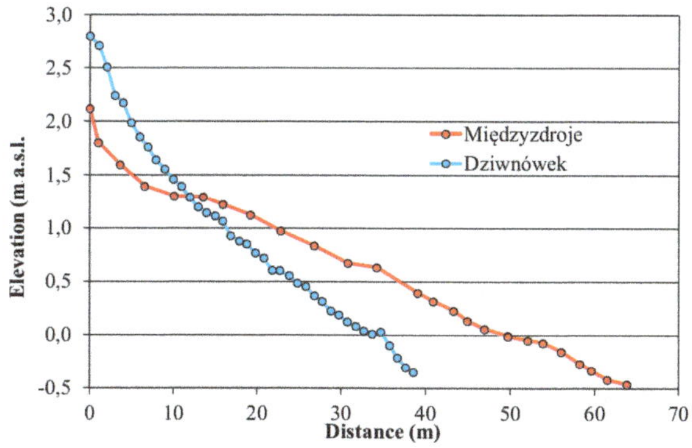

Figure 3. Cross-sections of two beaches in Międzyzdroje and Dziwnówek.

the Baltic Sea by the Interdisciplinary Centre for Mathematical and Computational Modelling [43]. It provides a 84-hour forecasts every 12 hours [44] (ICM 2014). The latest forecast for each event included in the calibration process was used. ICM's model has a 8 km grid; central points of those grids utilized in this study are marked in Fig. 1. WAM point no. 12 supplied information for Międzyzdroje, while point no. 19 for Dziwnówek.

One additional factor had to be included, namely sea level, because eq. (4) calculates run-up from still water level. Measurements were obtained from the Institute of Meteorology and Water Management, which publishes real-time data from tide gauges on its website [45]. Levels from gauges in Dziwnów and Świnoujście were used for the calibration process in Dziwnówek and Międzyzdroje, respectively. The Baltic is a non-tidal basin, which also marks a difference from other research, which was done hitherto on tidal coasts.

Wave run-up was extracted from two time periods during every day from June 20th up to December 8th, 2013. Such approach differs from other experiments [23], which analysed a few days or weeks in great detail. Here, almost half a year in less detail was chosen in order to ensure the method's validity over a longer period of time, given our ultimate goal of creating a forecasting tool. Only days with considerable wave heights or sea levels were analysed in one-hour periods, which was the temporal resolution of hydrologic data available. Such events occurred on 23th–24th of September, 24th–26th and 28th–29th of November, 2nd and 5th–7th of December. This last event was a fierce winter storm *Xavier*, which completely reshaped the beaches, creating a vast sandbar near the waterline and an indention towards the middle of the profile. In effect, calibration process was not continued after that date. During periods of calmer weather, wave run-ups from around 7 and 13 UTC were an analysed, with small alterations on some days due to availability of data. Thanks to this approach, small inundations of 0–20% would not completely dominate the calculations, even though they still represented the bulk of data acquired.

For each event that was selected, up to 15 minutes of footage were observed in order to extract the highest wave run-up on all images. Since there would be about 100 individual wave run-ups during this period of time the value obtained from the images is likely to correspond to the 2% run-up modelled by eq. 4 [33,46]. Measurements of the highest run-up during in the 15-min set of images sampled at 1 Hz were made along the profile lines mentioned above. Distance between the point of the highest run-up and the 'zero' point could therefore be obtained from the image. Thanks to GPS RTK measurements at different points along the profile line, the distance in pixels could be easily transformed to distance in the terrain. As a result, a percentage value of beach inundation could be obtained. The 'zero' point (0% extreme wave run-up) was located at the point of the profile that equals the mean sea level, defined as −0.08 m in Kronstadt-86 reference system or 500 cm at Polish gauge stations. On the other end of the scale, 100% run-up would mean that the extreme (2%) waves reached the dune base. The distance between those points constitutes the 'baseline' beach width. Therefore, modelled extreme beach inundation B could be written as:

$$B = \frac{R_{2\%} + W}{h} 100\% \tag{5}$$

where $R_{2\%}$ is the wave run-up from eq. (4), W is the sea level and h is the 'baseline' height of the beach. The inundation parameter could a have a negative value, since the sea level could be so low, that the waves reached below the mean sea level point. Therefore, a −10% value indicates that the beach is at the moment 10% longer than the 'baseline' width. On the other hand, maximum values of B were capped at 100%, since any higher value means that the waves reach outside the beach.

A total of 667 events were analysed: 367 in Międzyzdroje and 300 in Dziwnówek (the latter number is lower due to camera malfunctions). During those events, water levels varied from 426 cm to 596 cm; significant wave heights reached from 0.01 to 3.99 m and peak wave periods from 2.0 to 10.2 seconds. Large variety of values indicate that both storm surges and falls were included, as well as period of calm weather. The lowest measured beach inundation amounted to −22.5% in Międzyzdroje and −7.3% in Dziwnówek, while during the *Xavier* storm waves swashed the dunes in both locations (100% run-up). Histograms are presented in Fig. 4. Underlying data (except for sea levels) can be found in Table S1B in File S1.

Having obtained actual wave run-ups, empirical parameters a and b where adjusted, so that the linear regression line plotted using the data for each locality would pass through points of equal value of both actual and modelled run-up, i.e. constitute a 1:1 line. The correlation between the modelled and actual data was not

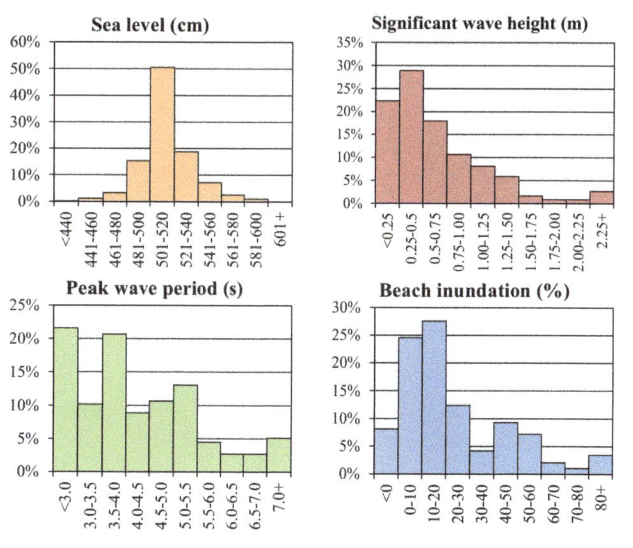

Figure 4. Histograms of parameters included in the study. Beach inundation pertains to actual (measured) wave run-up as % of the beaches' baseline width.

maximized here, but it decreased the formula's bias in estimating high run-ups.

A permit was required in order to conduct GPS measurements on the dunes. It was obtained from the Maritime Office in Szczecin. No other permissions were necessary. Video cameras used in this study were placed on the university's (Międzyzdroje) or private (Dziwnówek) parcels by kind consent of their respective owners.

Results

Results of the calibration process are presented in Fig. 5. Empirical parameters $a = 1.29$ and $b = 0.72$ where used, so that eq. (4) becomes:

$$R_{2\%} = 1.29 H_{mo} \left(\frac{\tan(\alpha)}{\sqrt{\frac{H_{mo}}{L_{op}}}} \right)^{0.72} \tag{6}$$

In general, the data concentrate along the regression line, with very few outlying elements. Root-mean-square error for data from Międzyzdroje is 6.1 percentage points, or 3.2 m of beach width. In Dziwnówek, the error is slightly larger, 6.5 percentage points, but because the beach is shorter, this value corresponds to 2.2 m of its width. The coefficient of determination is high and amounts to 0.94 in Międzyzdroje and 0.92 in Dziwnówek.

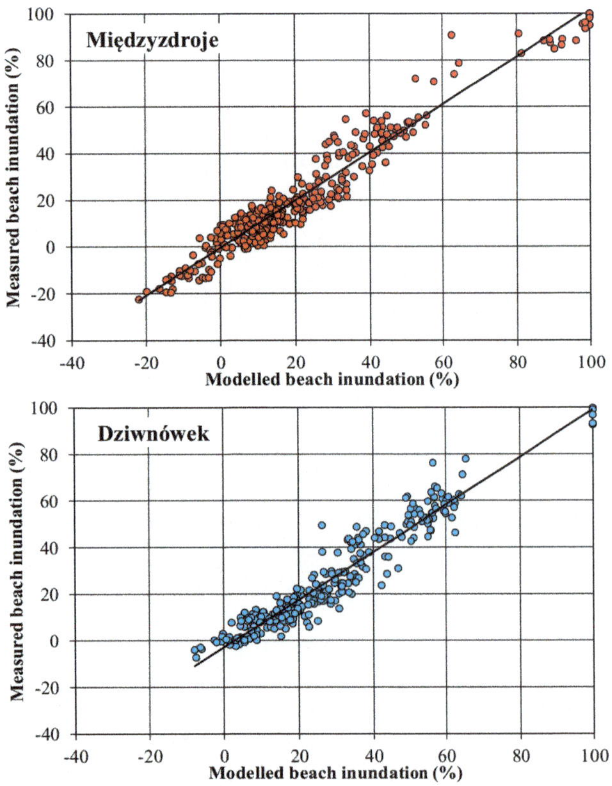

Figure 5. Measured and modelled beach inundation for Międzyzdroje and Dziwnówek.

In Międzyzdroje, the formula proved most accurate for small run-ups, up to 30%, while more energetic events where slightly underestimated. In Dziwnówek, it is the opposite. However, these are only slight deviations from the line of equality. On most occasions. Unfortunately, only a few cases of 60–80% run-ups where recorded, causing some uncertainty about the method's reliability for storm events, even though ca. 90% beach inundation was estimated correctly. It generally properly predicted waves reaching the dune base.

However, the impression of very good quality of the model is slightly misleading. An important factor regulating how far will the wave swash into the beach is the sea level. It contributed, on average, to about 22% of the actual beach inundation in Międzyzdroje and 32% in Dziwnówek. Since the data from tide gauges are fairly accurate (even though there is some considerable distance from the gauge station to Międzyzdroje), performance of 'bare' wave run-ups was calculated. Normalized RMS error amounted to 0.27 in Międzyzdroje (i. e. average error in estimating wave run-up was 27%) and 0.29 in Dziwnówek. Coefficient of determination was 0.85 and 0.81, respectively. This is a lower value than that recorded by the beach inundation indicator, but still a decent result.

Having obtained optimal parameters for the formula, wave run-up was recalculated using hourly data from WAM and M3D models. The latter is a hydrodynamic model providing, among other parameters, water levels for the Polish coast [47,48]. Fig. 6 presents a histogram of inundation from hourly data and its change over the course of 2013 (daily averages are presented for clarity of the graph). Run-ups below 30% constituted four-fifths of all events. The values for the beach in Dziwnówek were generally higher than in Międzyzdroje (20% against 12%). The reasons could be easily identified: not only Dziwnówek has a shorter and steeper beach, but also had higher waves ($H_{mo} = 0.53$ m against 0.47 m) and water levels (504 cm against 500 cm) during that period. The beach in Międzyzdroje was more inundated than Dziwnówek's only 5% of the time.

In Fig. 6 both calmer and more dynamic periods are visible. Especially standing out is the *Xavier* storm surge in early December. Only one event in February comes close. November and December observed the highest monthly means, with April and October on the opposite. In Dziwnówek, 18 days had average with inundation above 50%, whereas in Międzyzdroje the corresponding figure is 9 days. 2013 was not a particularly eventful year in terms of storms in general. Even *Xavier*, though generated strong winds and caused damages in some Western countries was at the end rather unimpressive in terms of coastal flooding. It could be estimated that it impacted the dunes in Dziwnówek for around one and a half day, and even less than that in Międzyzdroje. Naturally, the results for parts of the year could be not exactly accurate due to changes in morphology of the beaches, particularly after the December surges, where sandbars near the waterline appeared.

Discussion and Conclusions

The empirical parameters' values from eq. 6 do not differ much from other authors' findings. Parameter b, estimated at 0.72, is almost the same as Mase's lab-tests result of 0.71 [32] and Melby's approximation of 0.70 based on Stockdon's dataset of nine different field experiments [23,24]. Such a small difference between different studies could indicate that the value of around 0.7 has a universal application. On the other hand, a parameter's value (1.29) is significantly lower than Mase's 1.86 [32]. Other authors' estimates of this parameter ranged from 1.49 to 1.87 [24].

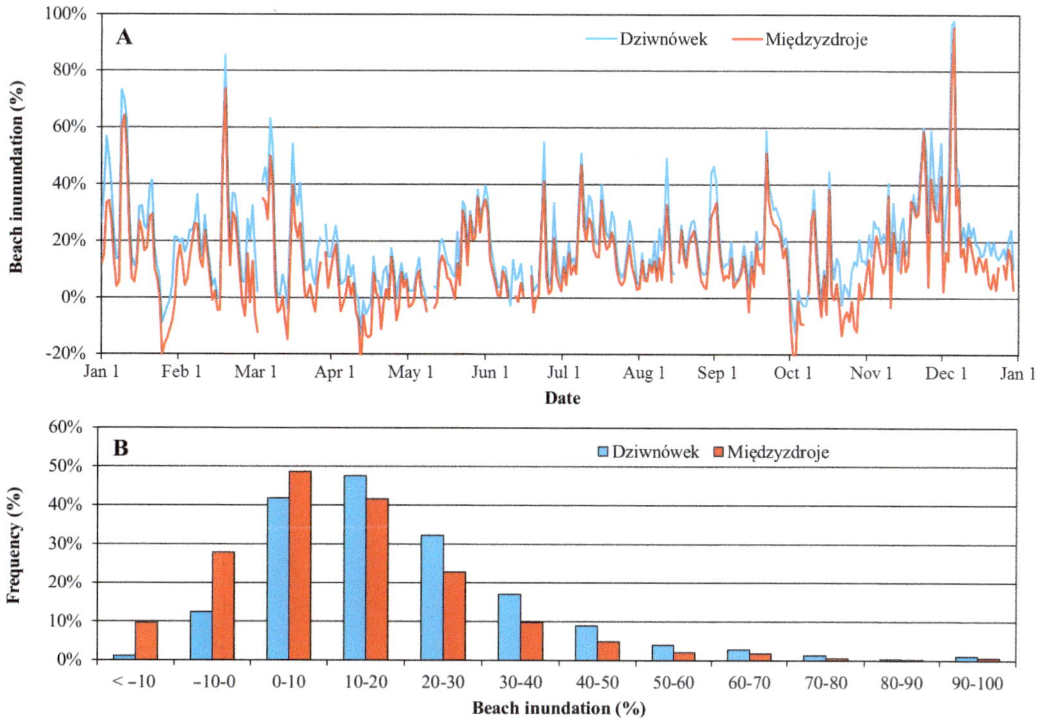

Figure 6. Extreme 2% beach inundation estimates for Międzyzdroje and Dziwnówek, hourly data from January 1st to December 31st, 2013. Daily averages (A) and histogram (B).

It is likely caused by a different research scope: Mase and others tested their methods using very large slope angles that are typical for hydrotechnic structures such as dykes. In contrast, Melby's estimate from field data is 1.10 [24]. It is likely caused by different environmental conditions, since the beaches used in that study were located along tidal coasts with high wave energies. Another source of the difference may be the permeability of the beaches, though authors working on the empirical equations generally did not give much consideration to this aspect apart from assuming that the slopes they tested were impermeable or permeable, therefore is hard to quantify the impact of this factor. Overall, bigger permeability lowers run-up [49] and the Polish beaches are definitely more permeable than those used in other studies.

Our results could be compared with other findings using Melby's Summary Performance Score. This statistic is an average of normalized RMS error, normalized bias and scatter index with an 0 to 1 scale, where 1 means no error [24]. Only wave run-up from still water level is taken into account here. In our case, the Performance score amounted to 0.78 in both locations. Application of Mase's equation adjusted by Melby to Stockdon's field data gave a result from 0.62 to 0.90 depending on the test site. Four out of nine beach had better score than 0.8, with an average of 0.84 [24]. Melby's findings were more accurate than our results, though it should be remembered that they were based on a much smaller number of run-up events (from 14 to 138, depending on the test site). Additional error could have come from the limitations of our dataset: the use of modelled wave data and inaccuracies in registering wave run-ups from images. Nevertheless, it can be concluded that the results of the calibration process were in line with other authors' findings.

Notwithstanding, some shortcomings of the method could be identified in the context of application for the whole Polish coast in *SatBałtyk Operational System-Shores*. First of all, the two test sites are both located on the western part of the coast, therefore not necessarily accurate enough for the middle and eastern section. Currently there is no possibility of validating the model there, since no suitable, scientific cameras are installed along the coast. What is more, the method has its potential inaccuracies originating in the heavy reliance on other models' output: modelled wave data are used for both calibration and operational forecasting. All this means that the method described here, though accurate for the test sites, should be subjected to further validation in other areas, before implemented to the entire Polish coast in an operational system.

When completed, the forecasting system could have several applications. For example, temporary constructions that are placed on the beach in the summertime could be removed in time to protect them from the effects of high wave run-up. Moreover, run-up predictions close to 100% would indicate potential dune/cliff erosion, an information useful for scientists interested in the subject of coastline change. As could be seen in Fig. 6 the method is also useful for analysing storm surges' development over time, as well as seasonal analysis. The topic will be investigated further in the future. More work is also expected on the calibration, too, pending availability of new data.

To sum up, the study proved that empirical wave run-up formulas could be applied to the Baltic Sea after calibration to local conditions. The method is accurate enough to be used in other work, especially lays foundations for an operational, forecasting system (which will naturally depend also on the quality of operational hydrologic models delivering data for calculating beach inundation). The errors recorded were not substantial despite applying a formula designed for very different conditions. This proves that this simple method, if properly used, can also be a valuable addition to more complex research.

Supporting Information

File S1 Contains: Table S1A. Cross-sections of the beaches. Table S1B. Extreme wave run-ups and wave data.

Author Contributions

Conceived and designed the experiments: DP KF. Performed the experiments: DP PA. Analyzed the data: DP KF. Contributed to the writing of the manuscript: DP PA PT.

Acknowledgments

The authors would like to thank Paulina Turczyk and Natalia Bugajny for participation in field surveys.

References

1. Trenhaile AS (1997) Coastal dynamics and landforms. Oxford: Clarendon Press. 366 pp.
2. Haslett SK (2009) Coastal Systems. New York: Routledge. 216 pp.
3. Johansson M, Boman H, Kahma KK, Launiainen J (2001) Trends in sea level variability in the Baltic Sea. Boreal Environment Research 6: 159–179.
4. Suursaar Ü, Kullans T, Ostmann M, Kõuts T (2003) Extreme sea level events in the coastal waters of western Estonia. Journal of Sea Research 49(4): 295–303. http://dx.doi.org/10.1016/S1385-1101(03)00022-4.
5. Suursaar Ü, Kullans T, Ostmann M, Saaremäe I, Kuik J, et al. (2006) Cyclone Gudrun in January 2005 and modeling its hydrodynamic consequences in the Estonian coastal waters. Boreal Environment Research 11: 143–169.
6. Kont A, Jaagus J, Aunap R, Ratas U, Rivis R (2008) Implications of sea level rise for Estonia. Journal of Coastal Research 24(2): 423–431. http://dx.doi.org/10.2112/07A-0015.1.
7. Ferreira Ó, Ciavola P, Armaroli C, Balouin Y, Benavente J, et al. (2009) Coastal Storm Risk Assessment in Europe: Examples from 9 study sites. Journal of Coastal Research SI 56: 1632–1636.
8. Roelvink D, Reniersc A, van Dongeren A, van Thiel de Vries J, et al. (2009) Modelling storm impacts on beaches, dunes and barrier islands. Coastal Engineering 56(11–12), 1133–1152. http://dx.doi.org/10.1016/j.coastaleng.2009.08.006.
9. Ciavola P, Ferreira Ó, Haerens P, Van Koningsveld M, Armaroli C (2011) Storm impacts along European coastlines. Part 2. Lessons learned from the MICORE Project. Environmental Science and Policy 14: 924–933.
10. Haerens P, Ciavola P, Ferreira Ó, van Dongeren A, Van Koningsveld M, et al. (2012) Online Operational Early Warning System Prototypes To Forecast Coastal Strom Impacts (CEWS). Coastal Engineering Proceedings 1(33): 1–8. http://dx.doi.org/10.9753/icce.v33.management.45
11. Bugajny N, Furmańczyk K, Dudzińska-Nowak J, Paplińska-Swerpel B (2013) Modelling morphological changes of beach and dune induced by storm on the Southern Baltic coast using XBeach (case study: Dziwnow Spit). Journal of Coastal Research SI 65: 672–677.
12. Dyke PPG, Moscardini AO, Robson EH, editors (1985) Offshore and Coastal Modeling. Berlin: Springer-Verlag. 404 pp.
13. Lakhan VC, Trenhaile AS (1989) Models and the coastal system. In: Lakhan VC, Trenhaile AS, editors. Applications in Coastal Modeling. Amsterdam: Elsevier. pp. 1–16.
14. Furmańczyk K, Andrzejewski P, Benedyczak R, Bugajny N, Cieszyński Ł, et al. (2014) Recording of selected effects and hazards caused by current and expected storm events in the Baltic Sea coastal zone. Journal of Coastal Research SI 70: 338–342.
15. Woźniak B, Bradtke K, Darecki M, Dera J, Dudzińska-Nowak J, et al. (2011) SatBałtyk – A Baltic environment al satellite remote sensing system – an ongoing project in Poland. Part 1: Assumptions, scope and operating range. Oceanologia 53(4): 897–924. http://dx.doi.org/10.5697/oc.53-4.925.
16. U.S. Army Corps of Engineers (1984) Shore Protection Manual. U.S. Army Engineer Waterways Experiment Station, Vicksburg, MS. 1070 pp.
17. Pullen T, Allsop NWH, Bruce T, Kortenhaus A, Schüttrumpf H, et al. (2007) EurOtop—Wave Overtopping of Sea Defences and Related Structures: Assessment Manual. Die Küste, Heft 73. 193 pp.
18. Guza RT, Thornton EB (1981) Wave set-up on a natural beach. Journal of Geophysical Research Oceans 86(C5): 4133–4137. http://dx.doi.org/10.1029/JC086iC05p04133.
19. Holman RA (1986) Extreme value statistics for wave run-up on a natural beach. Coastal Engineering 9(6): 527–544, http://dx.doi.org/10.1016/0378-3839(86)90002-5.
20. Douglass S (1992) Estimating Extreme Values of Runup on Beaches. Journal of Waterway, Port, Coastal, and Ocean Engineering 118(2): 220–224. http://dx.doi.org/10.1061/(ASCE)0733-950X(1992)118:2(220).
21. Massel SR, Pelinovsky EN (2001) Run-up of dispersive and breaking waves on beaches. Oceanologia 43(1): 61–97.
22. Hughes SA (2004) Estimation of wave run-up on smooth, impermeable slopes using the wave momentum flux parameter. Coastal Engineering 51(11): 1085–1104, http://dx.doi.org/10.1016/j.coastaleng.2004.07.026.
23. Stockdon HF, Holman RA, Howd PA, Sallenger AH (2006) Empirical parameterization of setup, swash, and runup. Coastal Engineering 53(7): 573–588. http://dx.doi.org/10.1016/j.coastaleng.2005.12.005.
24. Melby JA (2012) Wave Runup Prediction for Flood Hazard Assessment. ERDC/CHL TR-12-24 Technical Report. Vicksburg MS: Coastal and Hydraulics Laboratory, U.S. Army Engineer Research and Development Center. 126 pp.
25. Holman RA, Bowen AJ (1984) Longshore structure of infragravity wave motions. Journal of Geophysical Research: Oceans 89(C4): 6446–6452. http://dx.doi.org/10.1029/JC089iC04p06446.
26. Lippmann TC, Holman RA (1989) Quantification of sand bar morphology: A video technique based on wave dissipation. Journal of Geophysical Research: Oceans 94(C1): 995–1011. http://dx.doi.org/10.1029/JC094iC01p00995.
27. Davidson M, Van Koningsveld M, De Kruif A, Rawson J, Holman R, et al. (2007) The CoastView project: Developing video-derived Coastal State Indicators in support of coastal zone management. Coastal Engineering 54(6–7): 463–475. http://dx.doi.org/10.1016/j.coastaleng.2007.01.007.
28. Vousdoukas MI, Wziątek D, Almeida LP (2012) Coastal vulnerability assessment based on video wave run-up observations at a mesotidal, steep-sloped beach. Ocean Dynamics 62(1):123–137. http://dx.doi.org/10.1007/s10236-011-0480-x.
29. Aagaard T, Holm J (1989) Digitization of wave run-up using video records. Journal of Coastal Research 5(3): 547–551.
30. Hunt IA (1959) Design of seawalls and breakwaters. Journal of the Waterways and Harbors Division 85(WW3): 123–152.
31. Battjes JA (1974) Surf similarity. Proceedings of the 14th International Coastal Engineering Conference 1: 466–480.
32. Mase H (1989) Random wave runup height on gentle slope. Journal of Waterway, Port, Coastal, and Ocean Engineering 115(5): 649–661. http://dx.doi.org/10.1061/(ASCE)0733-950X(1989)115:5(649).
33. Power HE, Atkinson AL, Hammond T, Baldock TE (2013) Accuracy of wave runup formula on contrasting southeast Australian beaches. Proceedings of the Coasts and Ports 2013: 21st Australasian Coastal and Ocean Engineering Conference and the 14th Australasian Port and Harbour Conference: 1–6.
34. Díaz-Sánchez R, López-Gutiérrez JS, Lechuga A, Negro V (2014) Runup variability due to time dependence and stochasticity in the beach profiles: two extreme cases of the Spanish coast. Journal of Coastal Research SI 70: 1–6.
35. Hansen H, Larson M (2008) Implications of extreme waves and water levels in the southern Baltic Sea. Journal of Hydraulic Research (Supp. 2): 292–302. http://dx.doi.org/10.1080/00221686.2008.9521962.
36. Subotowicz W (1995) Transformation of the cliff coast in Poland. Journal of Coastal Research SI 22: 57–71.
37. Schwarzer K, Diesing M, Larson M, Niedermeyer R-O, Schumacher W, et al. (2003) Coastline evolution at different time scales – examples from the Pomeranian Bight, southern Baltic Sea. Marine Geology 194: 79–101. http://dx.doi.org/10.1016/S0025-3227(02)00700-4.
38. Remote Sensing and Marine Cartography Unit (2014) Kamera ZTiKM Dziwnówek – Zachód. Available: http://ztikm.szczecin.pl/kamera/. Accessed 2014 May 6.
39. Holland KT, Holman RA, Lippmann TC, Stanley J, Plant NG (1997) Practical use of video imagery in nearshore oceanographic field studies. IEEE Journal of Cceanic Engineering 22(1): 81–92. http://dx.doi.org/10.1109/48.557542.
40. Santos F, Pais-Barbosa J, Teodoro A, Gonçalves H, Baptista P, et al. (2012) Coastal Morphodynamic Features/Patterns Analysis Through A Video Based System And Image Processing. SPIE Proceedings 8538, Earth Resources and Environmental Remote Sensing/GIS Applications III. http://dx.doi.org/10.1117/12.974679.
41. Mayer RH, Kriebel DL (1994) Wave runup on composite-slope and concave beaches. Proceedings of the 24th Coastal Engineering Conference, ASCE: 2325–2339.
42. WAMDI Group (Hasselmann S, Hasselmann K, Bauer E, Janssen PAEM, Komen GJ, et al.) (1988) The WAM model – A third generation wave prediction model. Journal of Physical Oceanography 18(12): 1775–1810. http://dx.doi.org/10.1175/1520-0485(1988)018%3C1775:TWMTGO%3E2.0.CO;2.
43. Cieślikiewicz W, Paplińska-Swerpel B (2008) A 44-year hindcast of wind wave fields over the Baltic Sea. Coastal Engineering 55(11): 894–905. http://dx.doi.org/10.1016/j.coastaleng.2008.02.017.
44. Interdisciplinary Centre for Mathematical and Computational Modelling (2014) Numerical Weather Forecast. Available: http://www.meteo.pl/. Last accessed 2014 April 10.
45. Institute of Meteorology and Water Management (2014) Monitor – Bałtyk. Available: http://monitor.pogodynka.pl/baltyk/map. Accessed 2014 March 17.

46. Nielsen P, Hanslow DK (1991) Wave Runup Distributions on Natural Beaches. Journal of Coastal Research 7(4): 1139–1153. http://www.jstor.org/stable/4297933.
47. Jędrasik J, Cieślikiewicz W, Kowalewski M, Bradtke K, Jankowski A (2008) 44 Years Hindcast of the sea level and circulation in the Baltic Sea. Coastal Engineering 55(11): 849–860. http://dx.doi.org/10.1016/j.coastaleng.2008.02.026.
48. Kowalewski M, Kowalewska-Kalkowska H (2011) Performance of operationally calculated hydrodynamic forecasts during storm surges in the Pomeranian Bay and the Szczecin Lagoon. Boreal Environment Research 16(suppl. A): 27–41.
49. Kobayashi N, Cox DT, Wurjanto A (1991) Permeability Effects on Irregular Wave Runup and Reflection. Journal of Coastal Research 7(1): 127–136. http://www.jstor.org/stable/4297809.

16

Long-Term Changes in Nutrients and Mussel Stocks Are Related to Numbers of Breeding Eiders *Somateria mollissima* at a Large Baltic Colony

Karsten Laursen[1]*, Anders Pape Møller[2]

1 Institute of Bioscience, Aarhus University, Rønde, Denmark, 2 Laboratoire d'Ecologie, Systématique et Evolution, CNRS UMR 8079, Université Paris-Sud, Orsay, France

Abstract

Background: The Baltic/Wadden Sea eider *Somateria mollissima* flyway population is decreasing, and this trend is also reflected in the large eider colony at Christiansø situated in the Baltic Sea. This colony showed a 15-fold increase from 1925 until the mid-1990's, followed by a rapid decline in recent years, although the causes of this trend remain unknown. Most birds from the colony winter in the Wadden Sea, from which environmental data and information on the size of the main diet, the mussel *Mytilus edulis* stock exists. We hypothesised that changes in nutrients and water temperature in the Wadden Sea had an effect on the ecosystem affecting the size of mussel stocks, the principal food item for eiders, thereby influencing the number of breeding eider in the Christiansø colony.

Methodology/Principal Finding: A positive relationship between the amount of fertilizer used by farmers and the concentration of phosphorus in the Wadden Sea (with a time lag of one year) allowed analysis of the predictions concerning effects of nutrients for the period 1925–2010. There was (1) increasing amounts of fertilizer used in agriculture and this increased the amount of nutrients in the marine environment thereby increasing the mussel stocks in the Wadden Sea. (2) The number of eiders at Christiansø increased when the amount of fertilizer increased. Finally (3) the number of eiders in the colony at Christiansø increased with the amount of mussel stocks in the Wadden Sea.

Conclusions/Significance: The trend in the number of eiders at Christiansø is representative for the entire flyway population, and since nutrient reduction in the marine environment occurs in most parts of Northwest Europe, we hypothesize that this environmental candidate parameter is involved in the overall regulation of the Baltic/Wadden Sea eider population during recent decades.

Editor: Gabriele Sorci, CNRS, Université de Bourgogne, France

Funding: Financial support was received from the '15-June Fondation'. The funders had no role in study design, data collection and analysis, decision to publish, or preparation of the manuscript.

Competing Interests: The authors have declared that no competing interests exist.

* E-mail: kl@dmu.dk

Introduction

Leakage of nutrients from farmland to the marine environment has changed the benthic community by increasing primary and secondary production. For example, primary production in Kattegat doubled between the 1950's and the early 1990's, and the biomass of fish increased six-fold during the same period and eight-fold in the Baltic Sea during 1900–1970 [1,2]. However, the increase in nutrient load also had some undesirable effects in the marine ecosystem. Countries across Western Europe subsequently initiated programs in the 1970's and 1980's to reduce nutrient leakage from farmland, which led to reduced concentrations of nitrogen and phosphorus in the environment. In major rivers such as Elbe and the Rhine that receive water from large parts of European farmland and cities reduced concentrations of nutrients have been recorded since the late 1980's [3–5]. Continuous reductions in nutrients have also occurred in the North Sea and parts of the Baltic [6,7]. Reduced nutrient loads affect the marine ecosystem by changing community structure of phytoplankton and benthos, although it has been difficult to predict the consequences for the ecosystem due to high annual variation [8]. However, studies of single waterbird species show clear relationships between nutrients and reproduction, recruitment, survival and breeding numbers [9].

Here we focus on the factors determining the size of the Baltic/Wadden Sea flyway population of eider *Somateria mollissima* that has decreased during the last two decades [10]. This decline was first reported from the wintering grounds and later from breeding areas [11,12]. The reasons for the decline are poorly understood and epidemic outbreaks, parasite infestations, mass starvation, increased predation, and reduced survival of ducklings and adult females have all been suggested to be partly responsible for the decline [13–18]. However, no single study has so far shown that these factors account for the population decline, and a recent review [12] concluded that knowledge of the causes of declining survival of adult females is necessary for understanding the population decline.

The number of breeding eiders at Christiansø in the southern Baltic Sea (Fig. 1) is typical for several other colonies in the Baltic/

Wadden Sea flyway population by showing a steady increase during the 1930's–1980's with a peak in the mid 1990's, followed by rapidly declining numbers (Fig. 2)[12,19]. In addition eiders from this colony winter in the Danish waters (Kattegat and the Wadden Sea) as the vast majority of the flyway population [20]. The Christiansø colony has been monitored for 85 years and it is among the largest in the flyway breeding range counting up to 3,000 nesting females. The colony does not suffer from mammalian predation, and avian predation is limited due to close proximity to people. Thus female numbers returning to the colony may reflect conditions during the non-breeding season, i.e. ecological conditions at the wintering grounds. Female survival in the colony has not been estimated directly, but the annual number of breeders is supposed to reflect survival, since site fidelity of adult females is known to be almost 100% [21].

Eiders are capital breeders that build up most of their body reserves during winter for subsequent breeding [22]. When females arrive during spring they enter colonies and produce a clutch of 4–5 eggs that are incubated for about 26 days without the females leaving the nest to feed. Thus females must build up sufficient body reserves at the wintering grounds for breeding. Studies in the Wadden Sea show that wintering numbers of eiders are positively related to mussel stocks, and individuals feeding on mussels have better body condition than those taking other types of prey, which makes it attractive for eiders to feed on mussel beds [23–25]. Blue mussel *Mytilus edulis* can affect feeding conditions of eiders in at least two ways: (1) by increasing the flesh content and (2) by increasing the size of mussel stocks. First, the flesh content and reproduction in mussels are influenced by winter climate. During cold winters flesh content is large due to reduced respiration offering better foraging conditions for eiders than during mild winters [26–28]. Second, an increase in mussel stocks are based on large spatfall (production of eggs) in mussels, which are often initiated by cold winter climate making it possible for mussels to maintain large flesh content and produce large spatfall. Predation pressure is low in cold water and survival of mussel larvae during settling is high, giving rise to large cohorts 1–3 years after a cold winter. Thus cold winters maintain high flesh content in mussels that facilitate feeding conditions for eiders and increase their body condition for the subsequent breeding season. Cold winters increase mussel stock growth during the next 1–3 years and increase eider body conditions during the following years. However, flesh content and mussel growth depend on light, climatic conditions (temperature) and amount of nutrients that stimulate phytoplankton growth, the food of blue mussels [5,29,30]. Thus nutrients together with temperature are central variables in mussel stock growth and eider feeding conditions.

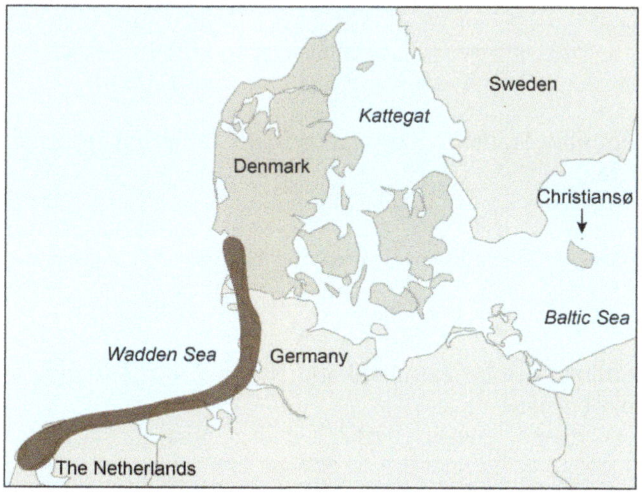

Figure 1. Eiders breeding at Christiansø are wintering in the Wadden Sea (toned brown) and in Kattegat. The eider colony at Christiansø in the Southern Baltic Sea was one of the largest colonies in the Baltic/Wadden Sea flyway population during the mid-1990's.

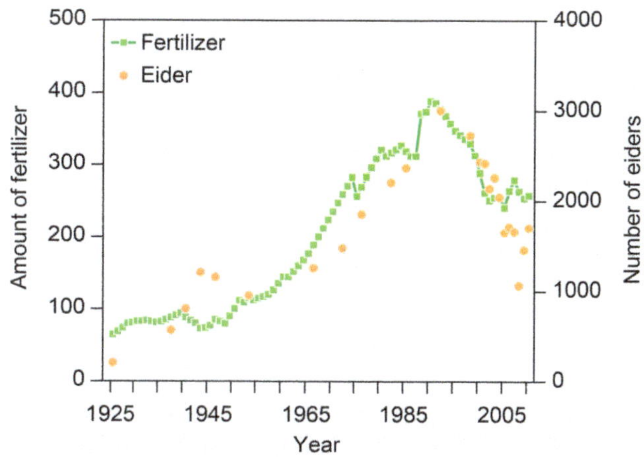

Figure 2. The number of breeding eiders at Christiansø and the amount of fertilizer used by Danish farmers (indexed as 1950 = 100) during 1925–2010.

The objectives of this study were to test four predictions concerning the effects of changes in nutrient availability as indexed by fertilizer use in agricultural areas together with climate conditions have affected the number of eiders breeding at Christiansø through an increase in mussel stocks. First, we tested whether the concentration of nitrogen and phosphorus in the Wadden Sea was explained by the amount of fertilizer used on farmland and winter precipitation. Second, we tested if the amount of fertilizer together with climate (water temperature) could explain the size of mussel stocks in the Wadden Sea. Third, we tested if the number of breeding eiders at Christiansø could be predicted by linear effects of fertilizer used by farmers together with an effect of climate (water temperature). Fourth, we tested if mussel stocks in the Wadden Sea could explain the number of breeding eiders at Christiansø. We conducted these analyses by relying on extensive sampling in the Danish waters by the national Danish environmental monitoring programme NOVANA and the TMAP-programme in the Wadden Sea [5,6].

The study was performed in the Baltic Sea at Christiansø, three rocky islands with a total area of 0.39 km^2, of which two are inhabited (Fig. 1). The islands are covered by grass, bushes and a few trees. The vast majority of eiders breed on the inhabited islands near buildings, in gardens and along paths. Pets such as dogs and cats are not allowed on the islands, making it possible for eiders to breed in close vicinity to people. This reduces predation by gulls during incubation. A detailed description is given by Franzmann [21] and Lyngs [31]. The colony is normally not influenced by ice cover in the Baltic Sea, which affects clutch size and breeding success of eiders in the north-eastern part of the Baltic Sea [32] and at other sites with ice cover [33].

Methods

Ethics Statement

Eider nests, eggs and chicks are protected in Denmark according to the Wildlife Management and Hunting Act from

1931. We use existing data for this study that were collected by local wardens and ornithologists, who had permission to count birds according to the national monitoring program. The eiders at Christiansø are breeding in gardens and close to public paths and are used to people close to the nests. Counts of nests are easily made without disturbance of the birds, and counts of eggs in nests are performed by carefully lifting the females off the nests.

Material

The number of eider nests was recorded since 1925 with annual intervals of 5.9 years (±0.4 SE) up to 1999 and thereafter annually until 2010. The number of breeding females was analysed in relation to the amount of fertilizer, mussel stocks and climate. In the analyses we included winter water temperature (mean °C for January–February) in the Wadden Sea measured in Marsdiep in the Dutch Wadden Sea [34], and updates. We used temperature measured in the Dutch Wadden Sea due to a long time series and because water temperature from this station reflects water temperature in the Danish waters and the western part of the Baltic Sea [35]. Information on precipitation during September-May was obtained from the Danish Meteorological Institute.

Information on nutrients as total nitrogen concentration (μg/l) was available from 1989 for the Danish waters, but Møller et al. [9] showed a close relationship between fertilizer (sum of artificial and natural fertilizers) use by Danish farmers and the concentration of nutrients in the Danish marine environment due to leakage of fertilizer from arable land to rivers and lakes and eventually the ocean. Information on fertilizer use by Danish farmers dates back to 1900 [36], and we use this as an indirect estimate of the amount of nutrients, which makes it possible to analyse the effects of nutrients for ecological variables for long time series. The annual amount of fertilizer use is estimated from autumn to spring the following year, e.g. the amount of fertilizer used in 1986 covers the amount used in autumn 1986 until spring 1987. Here the amount of fertilizer used in 1986/1987 is presented as 1986. From the Danish part of the Wadden Sea mussel stock data (metric tonnes) were available for 1987–2007 [25,37], and nitrogen and phosphorus concentration for 1989–2009 (data from the Danish Ministry of the Environment).

Statistical Procedures

We analyzed variation in and correlations among predictor variables to test for collinearity. We subsequently calculated variance inflation factors to determine whether collinearity posed a problem for the analyses. Furthermore, we tested for temporal trends in variables, and in case there was a significant temporal trend, we included year as a continuous predictor.

We made four major series of linear models. First, we tested whether N and P in the Danish Wadden Sea could be predicted by the amount of fertilizer used in agriculture by Danish farmers in year (i), year (i-1) and year (i-2). Time lags are used since fertilizer has to pass the soil to streams and lakes before it reaches coastal marine waters. These models all had predictors with variance inflation factors <5 [38].

Second, we tested if the amount of mussel stock in the Danish Wadden Sea could be predicted by the amount of fertilizer in year (i), year (i-1) and year (i-2), winter water temperature in year (i), year (i-1) and year (i-2) and the amount of mussel stock in year (i-1). We reduced these models by removing predictors relative to their associated P-values until all P-values were <0.10. These models all had predictors with variance inflation factors <10 and the final model <5 [38].

Third, we tested whether the abundance of breeding eiders at Christiansø could be predicted by the amount of fertilizer, winter water temperature and year. We included the amount of fertilizer in year (i), year (i-4) and (i-5) to account for effects of fertilizer on number of breeding females in the current year (i) and in year (i-4) and (i-5), to account for effects of fertilizer in the year of hatching for currently recruited females (we assume it takes 0–1 year for fertilizer to reach coastal waters and to stimulate mussel growth, and 3–4 years for eider ducklings to be mature [39]). Likewise, we included water temperature in year (i), year (i-4) and (i-5) to account for effects of temperature on number of breeding females in the current year (i) and effects of temperature in the year when recruiting females were hatched, i.e. year (i-4) and year (i-5). Finally, we included year as a continuous variable to account for temporal trends in variables other than fertilizer or temperature. We could only include variables in the tests that reflect these time lags in order to maintain variance inflation factors <10 and <5 in the final model [38].

Fourth, we tested if the number of breeding eiders in year (i) was predicted by the amount of mussel stock in year (i), year (i-1) and year (i-5). For justification for these time lags, see above. We only included variables that kept variance inflation factors <10 [38]. All analyses were made with JMP [40].

Results

Temporal Trends in Variables

The amount of fertilizer used by farmers increased from an index value of 64 in 1925 to about 100 in the late 1930's. In the early 1940's during the Second World War the use of fertilizer stabilized and later decreased, but in the 1950's it increased strongly again to reach a maximum index value of 377 in 1992, followed by a decrease to 250 in 2002 and later the amount stabilized until 2010 (Fig. 2). In the Wadden Sea the concentration of total nitrogen in spring increased during 1989–2000 from 755 μg/l to 2,141 μg/l after which it decreased to 1,297 μg/l in 2008. During the same period total phosphorus decreased from 55 μg/l in 1989 to 10–20 μg/l during 1990 to 2002 and <10 μg/l in 2010. Mussel stocks in the Wadden Sea increased from 40,000 metric tonnes in 1986 to 117,000 metric tonnes in 1994, while varying considerably during the following years, and it decreased to finally reach 12,000 metric tonnes in 2007. The breeding population of eiders at Christiansø increased from 200 females in 1925 to about 1,200 females in 1943. During the following years numbers declined, but increased to 3,000 females in 1992 followed by a decline to 1,060 females in 2008, and a small recent recovery (Fig. 2). All data are given in Supporting data (Table S1)

Summary Statistics, Correlations and Temporal Tends in Data

All predictor variables showed considerable variation during the study period (Table S2). There were strong correlations among several variables (Table S3) preventing their inclusion as predictors in the same analyses. All variables with the exception of water temperature showed significant temporal trends (Table S4).

Fertilizer Use on Farmland and N and P in the Marine Environment

Only concentration of total P during spring in the marine environment was predicted by fertilizer and only in the previous year (Fig. 3; $F = 9.26$, d.f. = 1, 18, $r^2 = 0.30$, $P = 0.007$, estimate (SE) = 0.0018 (0.0006)), while there was no similar effect for concentration of total N ($F = 0.04$, d.f. = 1, 18, $r^2 = -0.05$, $P = 0.85$) or for the different time lags (results not shown). This makes sense because fertilizer used in agriculture must first reach the marine environment through leaching and transport via soil,

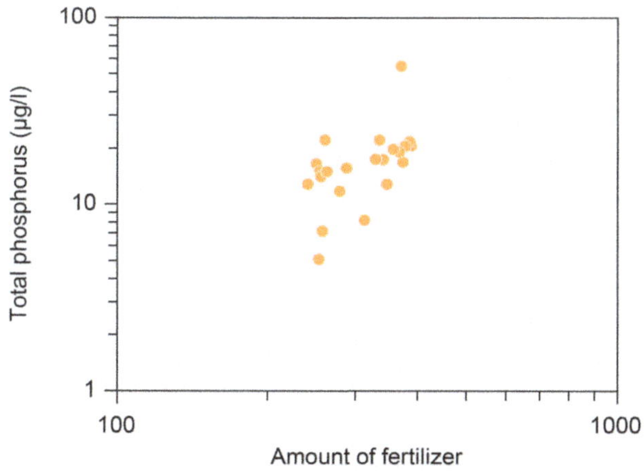

Figure 3. Total concentration of phosphate (µg/l) in the Danish Wadden Sea in year (i) in relation to fertilizer use in agriculture (1950 = 100) in the previous year (year (i-1)).

ditches and streams. Thus, P increases with amount of fertilizer use with a time lag of one year.

Mussel Stocks and Fertilizer Use

There was more mussel stock when fertilizer use was high in the current year ($F = 11.37$, d.f. = 1, 20, $r^2 = 0.33$, $P = 0.0030$, estimate (SE) = 0.005 (0.001)). No other time lag fitted the data better, nor did year enter as a significant predictor.

Number of Eiders in Relation to Fertilizer Use and Water Temperature

The abundance of eiders increased with fertilizer use in year (i) (Fig. 4; $F = 63.39$, d.f. = 1, 22, $r^2 = 0.73$, $P < 0.0001$, estimate (SE) = 6.650 (0.835)). There were no significant effects of year, fertilizer use with time lags or winter water temperature.

Number of Eiders in Relation to Mussel Stocks

The number of breeding eiders increased with the amount of mussel stock in year (i)(Fig. 5; $F = 26.54$, d.f. = 1, 8, $r^2 = 0.77$, $P = 0.0009$, estimate (SE) = 1003.65 (194.80)). This model accounted for 74% of the variance in abundance of eiders. There were no significant effects of year, mussel stock with time lag or winter water temperature.

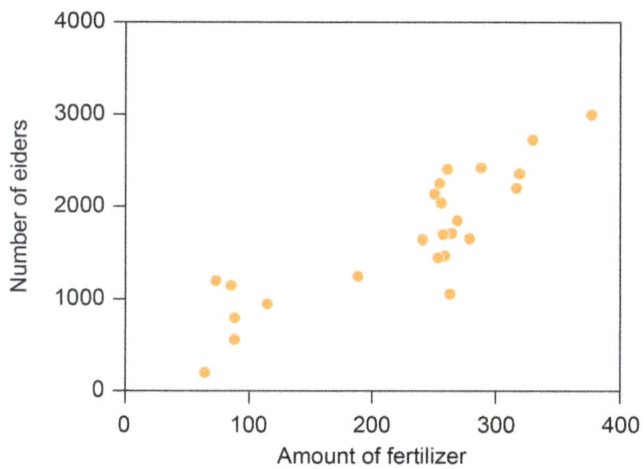

Figure 4. Number of breeding eiders at Christiansø in year (i) in relation to fertilizer use in Danish agriculture in the same year (year (i)) (1950 = 100).

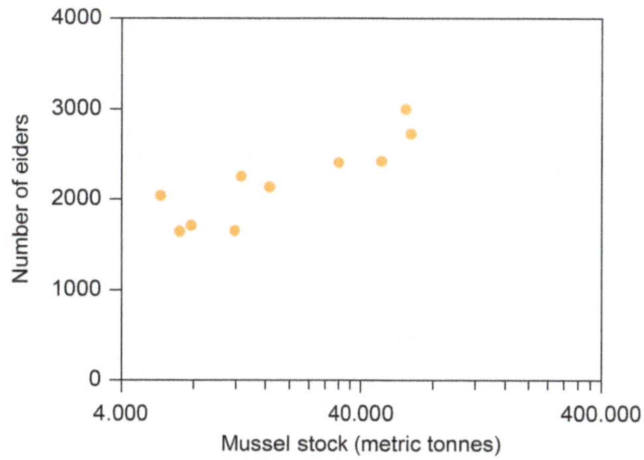

Figure 5. The number of breeding eiders at Christiansø in year (i) in relation to mussel stocks (metric tonnes) the same year (year (i)).

Discussion

The number of eiders breeding at Christiansø increased from 1925 until 1992–1995, followed by a steep decrease in recent years. Estimated breeding numbers in Sweden and Finland, where the vast majority of the eider population of the Baltic/Wadden Sea flyway breeds, show a similar trend from the 1980's until the first half of the 1990's with a subsequent decline in the second part of the 1990's [12]. The two Finnish colonies Söderskär and Aasla studied during four to five decades showed peaks in breeding numbers around 1985 and 1997, respectively, both with a steep subsequent decline [19]. Countrywide monitoring of breeding eiders in Denmark at ten-year intervals showed increasing numbers up to 1990 followed by stable numbers during 2000 and 2010 [41]. Annual, systematic counts of wintering eiders in the Wadden Sea showed stable numbers until 1995 followed by declines [42]. In the Danish waters, which constitute the main wintering grounds for this population of eiders, numbers declined by 54% between 1991 and 2000 [11,43], and for the flyway population as a whole a decline of 36% was recorded during the same period, although numbers partly recovered in recent years [12]. All these estimates from breeding and wintering grounds reveal a pronounced reduction (or stabilization in one country) in the number of eiders around 1995 with breeding numbers at Christiansø reflecting the general population trend.

However, the number of eiders on Christiansø does not reflect the total population size because non-breeding females in poor body condition probably spend the breeding season at sea, and they are thus not included in the data [12]. The number of non-breeding females is supposed to increase in periods with declining populations. This implies that the numbers of eiders in such periods probably are larger than those counted at the colony, and the population decline is smaller than reflected by the number of females in the colony. Finnish studies show that the age at first reproduction is younger when the number of eiders is increasing in a colony [39]. These two factors could have caused smaller numbers during the decline phase and larger numbers during the increase phase.

Fertilizer from agricultural areas ends up as nutrients in the marine environment where it stimulates the growth of mussel stocks. Although our data did not describe the precise time lag in the processes of transport of nutrients from farmland to the Wadden Sea, presumable due to short time series, we show a significant correlation between fertilizer use and the amount of phosphorus in the Wadden Sea and between the amount of fertilizer and mussel stocks. Furthermore, we found a strong positive correlation between mussel stocks and the number of breeding eiders at Christiansø. Finally, there was a significant correlation between mussel stocks in the Danish Wadden Sea and the number of breeding eiders at Christiansø.

Due to the close relationship found between nutrients and mussel stocks and between mussel stocks and the number of breeding eiders at Christiansø it is reasonable to assume that the reduced numbers of breeding females in the colony reflect individuals that have abandoned breeding (such females stay at sea and hence are not covered by the counts of incubating females), or individuals that have died due to reduced food availability. Females that have abandoned breeding or for other reasons have not appeared at the breeding colony have been reported from colonies in Finland [44] and on Iceland [45] in years with poor food availability. Emigration to other colonies is another explanation for the reduced number of breeders, although this explanation seems unlikely because countrywide surveys of breeding eiders in Denmark have shown stable numbers after 1995 with no increase in colonies adjacent to Christiansø [41]. In addition, adult females show an extremely high degree of philopatry close to 100% [21]. A similar degree of site fidelity by females has been reported from other colonies supporting the assumption that emigration of females does not occurred between colonies [46,47].

Data on mussel stocks from the Danish Wadden Sea were used in our statistical models, although eiders from Christiansø also spend part of the non-breeding season in other parts of the Wadden Sea, i.e. in Germany and the Netherlands [20,48]. However, Wadden Sea mussel stocks in Germany (both in Schleswig-Holstein and Lower Saxony) show the same pattern in abundance as in the Danish Wadden Sea with large stocks in 1998–2001 followed by low levels up to 2005, and small increases in more recent years [49]. Thus there is a common pattern in mussel stocks in major parts of the Wadden Sea, indicating that mussel stock data from the Danish part are representative of a larger region.

Eiders from Christiansø spend late winter and early spring in Kattegat, and a month before breeding they fly to the western Baltic Sea off the colony at Christiansø [20,48]. The amount of nutrients has also decreased in the Kattegat by more than 30% since 1991–1992 [6] as in the Danish Wadden Sea. A reduction in nutrients by 50% has occurred in coastal parts of the western Baltic Sea from the late 1980's to 2006 [7]. Despite reductions in nutrients in recent years, the sea off the eider colony at Christiansø has suffered from seasonal hypoxia and oxygen depletion during summer and autumn, and this has reduced or eliminated most of the bottom fauna especially large, slow-growing species such as mussels in favour of small, rapidly reproducing organisms [7]. Studies in the Baltic Sea have shown that foraging during the pre-breeding period is important for eiders [50]. Thus hypoxia at the pre-breeding grounds and the reduction in nutrients at the wintering grounds has both reduced the abundance of mussel for eiders from Christiansø.

In conclusion, marine areas where eiders from Christiansø occur during the non-breeding season have experienced reductions in nutrient levels for more than two decades. Because low nutrient levels reduce mussel stocks the population of breeding eiders at Christiansø has declined. The population trend for the eider colony at Christiansø is representative for the general trend of the Baltic/Wadden Sea flyway population. Reductions in nutrients in the marine environment occur in several geographical areas [51] and in most parts of North-western Europe. Therefore, we hypothesize that this environmental variable account for the overall reduction in the total eider population of the Baltic and Wadden Seas. Hence, our findings may provide a general explanation for the trends in eider numbers in large parts of Western Europe.

Supporting Information

Table S1 Data used for the statistical analyses. Numbers of eider at Christiansø, the amount of fertilizer used by Danish farmers (indexed 1950 = 100), water temperature (C°) during January-February (data from the Dutch Wadden Sea), mussel stocks (metric tonnes) and total nitrogen (µg/l) and total phosphorus (µg/l) in both spring and autumn in the Danish Wadden Sea.

Acknowledgments

We thank the Christiansø Field Station for access to eider data, and Chris M. Waltho for comments and suggestions on an early draft of the manuscript. The outline of this paper was formulated at the NOWAC meeting in connection with the Third Pan-European Duck Symposium in Jindrichuv Hradec, Czech Republic in April 2012.

Author Contributions

Conceived and designed the experiments: KL APM. Performed the experiments: KL. Analyzed the data: KL APM. Contributed reagents/materials/analysis tools: KL APM. Wrote the paper: KL APM. Obtain permission for use of data: KL.

References

1. Nielsen E, Richardson K (1996) Can changes in the fisheries yield in the Kattegat be linked to changes in primary productions? ICES J Mar Sci 53: 988–994.
2. Nixon SW, Buckley BA (2002) 'A strikingly rich zone': Nutrient enrichment and secondary production in coastal marine ecosystems. Estuaries 25: 782–796.
3. Stanners D, Bourdea P (1995) Europe's environment, the Dobirs assessment. European Environment Agency, Copenhagen.
4. Lenhart H, Pätsch J (2001) Daily nutrient loads of the European continental rivers for the years 1977–1998. Berichte aus dem Zentrum für Meeres- und Klimaforschung, Reihe B: Ozeanographie 40: 1–146.
5. Van Beusekom JEE, Loebl M, Martens P (2009) Distant riverine nutrient supply and local temperature drive the long-term phytoplankton development in a temperate costal basin. J Sea Res 61: 26–33.
6. Hjort M, Josefson AB (2010) Marine områder 2008. NOVANA, tilstand og udvikling i miljø- og naturkvalitet. Faglig rapport fra DMU nr. 760. (In Danish).
7. BSEP (2009) Eutrophication in the Baltic Sea – and integrated thematic assessment of the effects of nutrient enrichment in the Baltic Sea region. Baltic Sea Environment Proceedings No.115B.
8. Phillippart CJM, Beukema JJ, Cadée GC, Dekker R, Goedhart PW, et al. (2007) Impacts of nutrients reduction on coastal communities. Ecosystems 10: 95–118.

9. Møller AP, Flensted-Jensen E, Mardal W (2007) Agriculture, fertilizers and life history of a coastal seabird. J Anim Ecol 76: 515–525.
10. Delany S, Scott D (2006) Waterbird Population Estimates. Wetlands International, Wageningen, The Netherlands.
11. Desholm M, Christensen TK, Scheiffarth G, Hario M, Ens B, et al. (2002) Status of the Baltic/Wadden Sea population of Common Eider *Somateria mollissima*. Wildfowl 53: 167–203.
12. Ekroos J, Fox AD, Christensen TK, Petersen IK, Kilpi M, et al. (2012) Declines amongst breeding Eider *Somateria mollissima* numbers in the Baltic/Wadden Sea flyway. Ornis Fenn 89: 81–90.
13. Christensen TK, Bregnballe T, Andersen TH, Dietz HH (1997) Outbreaks of Pasteurellosis among wintering and breeding common Eiders *Somateria mollissima* in Denmark. Wildlife Biol 3: 125–128.
14. Pedersen K, Dietz HH, Jørgensen JC, Christensen TK, Bregnballe T, et al. (2003) *Pasteurella multocida* from outbreaks of avian cholera in wild and captive birds in Denmark. J Wildl Dis 39: 808–816.
15. Camphuysen CJ, Berrvoets CM, Cremers HJWM, Dekinga A, Dekker R, et al. (2002) Mass mortality of the Common Eiders (*Somateria mollissima*) in the Dutch Wadden Sea, winter 1999/2000: starvation in a commercially exploited wetland of internationally importance. Biol Cons 1006: 303–317.
16. Kats R (2007) Common Eiders *Somateria mollissima* in the Netherlands: The rise and fall of breeding and wintering populations in relation to the stock of shellfish. Rijksuniversiteit Groningen, The Netherlands.
17. Kilpi M, Öst M (2002) The effects of White-tailed Sea Eagle predation on breeding Eider females off Tvärminne, Western Gulf of Finland. Suomen Riista 48: 27–33. (English summary).
18. Hario M, Selin K (2002) Cohort-specific differences in reproductive output in a declining Common Eider *Somateria mollissima* population. In: Noer H, Nehls G ,(eds.), Population processes in the Common Eider *Somateria mollissima*. Dan Rev Game Biol 16. 33–45.
19. Rönkä M, Saari L, Hario M, Hänninen J, Lehikoinen E (2001) Breeding success and breeding population trends of waterfowl: implications for monitoring. Wildlife Biol 17: 225–239.
20. Noer H (1991) Distribution and movements of Eider *Somateria mollissima* populations wintering in Danish waters, analysed from ringing recoveries. Dan Rev Game Biol 14 (3): 1–32.
21. Franzmann N-E (1980) Ederfuglenes (*Somateria m. mollissima*) ynglebiotop og populationsdynamik på Christians Ø. Licentiat afhandling. Københavns Universitet
22. Meijer T, Drent R (1999) Re-examination of the capital and income dichotomy in breeding birds. Ibis 141: 399–414.
23. Nehls G, Ketzenberg C (2002) Do common Eiders *Somateria mollissima* exhaust their food resources? A study on natural mussel *Mytilus edulis* beds in the Wadden Sea. Dan Rev Game Biol 16: 47–61.
24. Laursen K, Frikke J (2008) Hunting from motorboats displaces Wadden Sea Eiders *Somateria mollissima* from their favoured feeding distribution. Wildlife Biol 14: 423–433.
25. Laursen K, Asferg K, Frikke J, Sunde P (2009) Mussel fishery affects diet and reduces body condition of Eiders *Somateria mollissima* in the Wadden Sea. J Sea Res 62: 22–30.
26. Beukema JJ, Essink K, Michaelis H, Zwarts L (1993) Year-to-year variability in the biomass of macrobenthic animals on tidal flats of the Wadden Sea: how predictable is food sources for birds? Neth J Sea Res 31: 319–330.
27. Honkroop PJC, Beukema JJ (1997) Loss of body mass in winter in three intertidal bivalve species: an experimental and observational study of the interacting effects between water temperature, feeding time and feeding behaviour. J Exp Mar Biol Ecol 212: 277–297.
28. Waldeck P, Larsson K (2013) Effects of winter water temperature on mass loss in Baltic blue mussels: Implications for foraging sea ducks. J Exp Mar Biol Ecol 444: 24–30.
29. Colijn F, Cadée CC (2003) Is phytoplankton growth in the Wadden Sea light or nitrogen limited? J Sea Res 49: 83–93.
30. Carstensen J, Sànchez-Camachos M, Duarte CM, Krause-Jensen D, Marbà N (2011) Connecting the Dots: Responses of coastal ecosystems to changing nutrients concentrations. Environ Sci Technol 45: 9122–9132.
31. Lyngs P (1992) Ynglefuglene på Græsholm 1925–90. Dansk Orn Foren Tidsskr 94: 12–18. (In Danish with an English summary).
32. Lehikoinen A, Kilpi M, Öst M (2006) Winter climate affects subsequent breeding success of common eiders. Global Change Biol 12: 1355–1365.
33. Chaulk GK, Mahoney ML (2012) Does spring ice cover influence nest initiation date and clutch size in common eiders? Polar Biol 35: 645–653.
34. Van Aken HM (2001) 140 years of daily observations in a tidal inlet (Marsdiep). ICES Mar Sci Symp 219: 359–361.
35. Mackenzie BR, Schiedek D (2007) Long-term sea surface temperature baselines-time series, spatial covariation and implications for biological processes. J Marine Systems 68: 405–420.
36. Duus K, Zinglersen E (2000) Geografistatistik 2000. Geografforlaget, Copenhagen.
37. Kristensen PS, Hoffmann E (2000) Fishery for blue mussels in Danish waters 1989–1999. DFU report no. 72-00, 38 pp. Danish Fishery Research Institute, Charlottenlund. (In Danish with an English summary).
38. McClave JT, Sincich T (2003) Statistics. 9th Edition. New Jersey: Prentice-Hall.
39. Hario M, Rintala J (2009) Age of first breeding in the Common Eider *Somateria m. mollissima* population in the northern Baltic Sea. Ornis Fenn 86: 81–88.
40. SAS (2010) SAS Enterprice Guide 5.1. SAS Institute, USA.
41. Christensen TK, Bregnballe T (2012) Status of the Danish breeding population of Eiders *Somateria mollissima* 2010. Dansk Orn Foren Tidsskr 105: 195–205.
42. Laursen K, Blew J, Eskildsen K, Hälterlein B, Kleefstra R, et al. (2010) Migratory Waterbirds in the Wadden Sea 1987–2008. Wadden Sea Ecosystem No. 30. Common Wadden Sea Secretariat, Joint Monitoring Group of Migratory Birds in the Wadden Sea, Wilhelmshaven, Germany.
43. Petersen IK, Pihl S, Hounisen JP, Holm TE, Clausen P, et al. (2006) Landsdækkende optællinger af vandfugle januar-februar 2004. Faglig rapport fra DMU nr. 606. Danmarks Miljøundersøgelser, Miljøministeriet. (In Danish).
44. Öst M, Smith BD, Kilpi M (2008) Social and maternal factors affecting duckling survival in Eiders *Somateria mollissima*. J Anim Ecol 77: 315–325.
45. Jónsson JE, Gardarsson A, Gill JA, Petursdóttir UK, Petersen A, et al. (2013) Relationships between Long-term Demograpy and Weather in a Sub-Arctic Population of Common Eider. PLoS One 8(6): e67093. doi:10.1371/journal.pone.0067093.
46. Reed A (1975) Migration, homing and mortality of breeding females Eiders *Somateria mollissima dresseri* of the St. Lawrence estuary, Quebec. Ornis Scand 6: 41–47.
47. Swennen C (1990) Dispersal and migratory movements of Eiders *Somateria mollissima* breeding in the Netherlands. Ornis Scand 21: 12–27.
48. Bønløkke JJ, Madsen J, Thorup K, Pedersen KT, Bjerrum M, et al. (2006) Dansk Trækfugleatlas. Rhodos, Humlebæk. (In Danish with an English summary).
49. Nehls G, Witte S, Büttger H, Dankers N, Jansen J, et al. (2009) Beds of blue mussels and Pacific oysters. Quality Status Report 2009 (Marencic H, de Vlas H, eds.), Thematic Report No. 11: 1–29. Wadden Sea Ecosystem No. 25. Common Wadden Sea Secretariat, Trilateral Monitoring and Assessment Group. Wilhelmshaven. Germany.
50. Rigou Y, Guillemette M (2010) Foraging effort and pre-laying strategy in breeding common eiders. Waterbirds 33: 314–322.
51. Møller AP (2013) Biological consequences of global change for birds. Integr Zool 8: 150–158.

Burrows of the Semi-Terrestrial Crab *Ucides cordatus* Enhance CO_2 Release in a North Brazilian Mangrove Forest

Nathalie Pülmanns[1]*, Karen Diele[2], Ulf Mehlig[3], Inga Nordhaus[1]

1 Department of Ecology, Leibniz Center for Tropical Marine Ecology (ZMT), Bremen, Germany, **2** School of Life, Sport and Social Sciences, Edinburgh Napier University, Edinburgh, United Kingdom, **3** Instituto de Estudos Costeiros, Universidade Federal do Pará - Campus de Bragança, Bragança, PA, Brazil

Abstract

Ucides cordatus is an abundant mangrove crab in Brazil constructing burrows of up to 2 m depth. Sediment around burrows may oxidize during low tides. This increase in sediment-air contact area may enhance carbon degradation processes. We hypothesized that 1) the sediment CO_2 efflux rate is greater with burrows than without and 2) the reduction potential in radial profiles in the sediment surrounding the burrows decreases gradually, until approximating non-bioturbated conditions. Sampling was conducted during the North Brazilian wet season at neap tides. CO_2 efflux rates of inhabited burrows and plain sediment were measured with a CO_2/H_2O gas analyzer connected to a respiration chamber. Sediment redox potential, pH and temperature were measured in the sediment surrounding the burrows at horizontal distances of 2, 5, 8 and 15 cm at four sediment depths (1, 10, 30 and 50 cm) and rH values were calculated. Sediment cores (50 cm length) were taken to measure the same parameters for plain sediment. CO_2 efflux rates of plain sediment and individual crab burrows with entrance diameters of 7 cm were 0.7–1.3 $\mu mol\ m^{-2}\ s^{-1}$ and 0.2–0.4 $\mu mol\ burrows^{-1}\ s^{-1}$, respectively. CO_2 released from a *Rhizophora mangle* dominated forest with an average of 1.7 *U. cordatus* $burrows^{-1}\ m^{-2}$ yielded 1.0–1.7 $\mu mol\ m^{-2}\ s^{-1}$, depending on the month and burrow entrance diameter. Laboratory experiments revealed that 20–60% of the CO_2 released by burrows originated from crab respiration. Temporal changes in the reduction potential in the sediment surrounding the burrows did not influence the CO_2 release from burrows. More oxidized conditions of plain sediment over time may explain the increase in CO_2 release until the end of the wet season. CO_2 released by *U. cordatus* and their burrows may be a significant pathway of CO_2 export from mangrove sediments and should be considered in mangrove carbon budget estimates.

Editor: Navnith K.P. Kumaran, Agharkar Research Institute, India

Funding: N.P. and I.N. were funded by the Leibniz Center for Tropical Marine Ecology (project 6201, http://www.zmt-bremen.de). K.D. received funding from the MASTS pooling initiative (The Marine Alliance for Science and Technology for Scotland, http://www.masts.ac.uk/) and its support is gratefully acknowledged. MASTS is funded by the Scottish Funding Council (grant reference HR09011) and contributing institutions. The European Commission funded the contributions by U.M. within this study via the international cooperation through the "Coastal Research Network on Environmental Changes" (CREC, http://www.forst.tu-dresden.de/CREC/) as part of its 7th 825 Framework Programme. The CREC project is classified as a Marie Curie Action (FP7-PEOPLE-2009-IRSES). The funders had no role in study design, data collection and analysis, decision to publish, or preparation of the manuscript.

Competing Interests: The authors have declared that no competing interests exist.

* Email: nathalie.puelmanns@gmail.com

Introduction

In mangrove ecosystems, semi-terrestrial burrowing crabs are important bioturbators affecting the biogeochemistry of the sediment and carbon cycling [1–5]. Their bioturbation activity includes constructing and maintaining of burrows leading to the reworking of sediment particles and organic material. Organic material, such as litter, is also reworked by their feeding, ingestion and defecation and can become buried in the sediment leading to carbon storage [1,6–8]. Crabs accumulate carbon as body biomass by feeding on mangrove litter which otherwise would have been washed away by the tide [9–14]. However, crabs may also account for carbon depletion in the sediment. Their burrows present an extension of the sediment surface area [1,15], thus, at low tide sediment surrounding the burrows can become oxidized by atmospheric oxygen. Atmospheric oxygen enters only a few millimeters into the mangrove sediment [8,16,17] where oxic reduction can take place. During burrow maintenance, crabs may mix more oxidized sediment inside their burrows with surrounding less oxidized sediment. The resulting increase in reduction potential affects the sediment surrounding the burrows up to a distance of several centimeters [18,19]. This may alter the type of microbial carbon oxidation from mainly anaerobic (e.g. sulfate reduction or methanogenesis) to partially aerobic decomposition processes (e.g. aerobic respiration or iron reduction) [1,8,20,21], which can result in higher CO_2 efflux rates [22] and consequently to carbon loss from the sediment [6]. Thus, bioturbation by crabs and associated changes in the microbial decomposition process possibly influence CO_2 release from the sediment.

To better evaluate the role of burrowing crabs in carbon cycling, it is important to quantify the amount of carbon stored and released by the crabs and their burrows. Bouillon et al. [23]

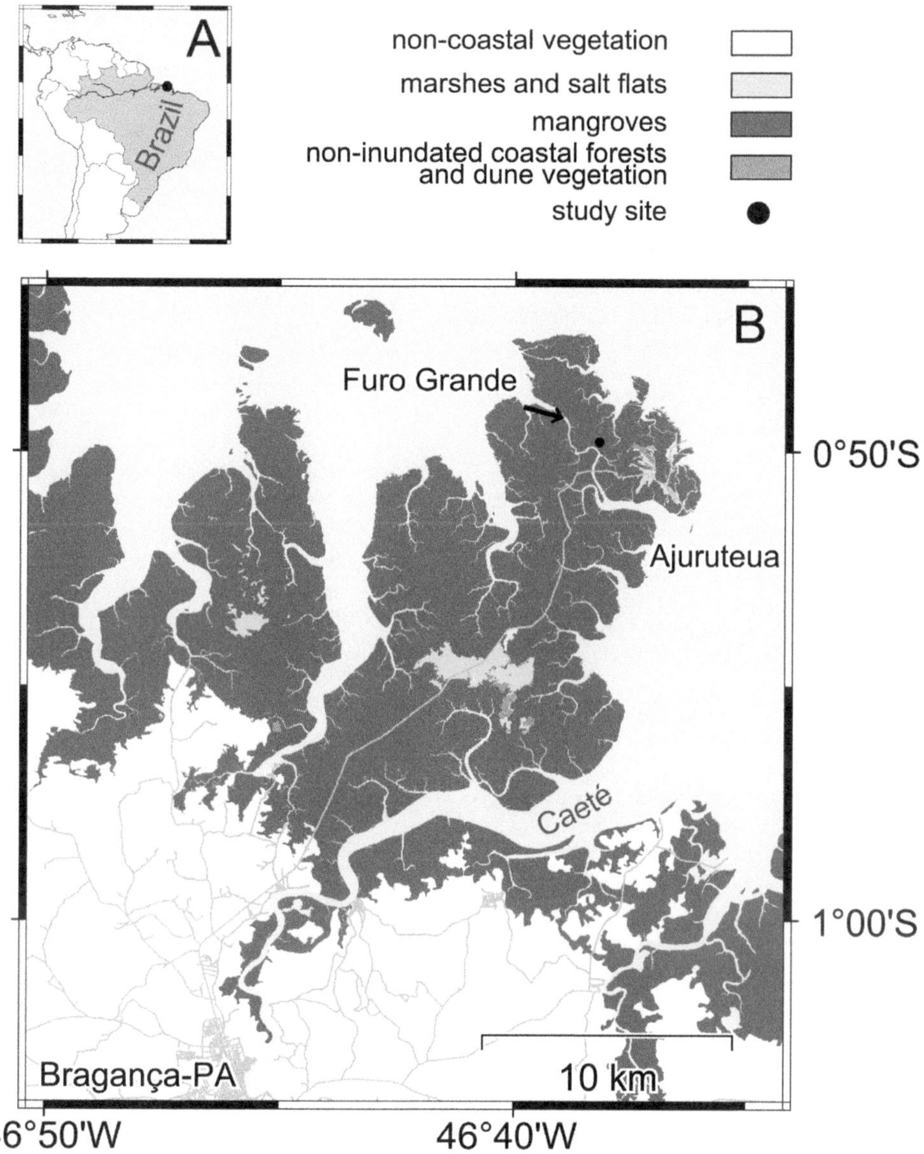

Figure 1. Study area on the Ajuruteua peninsula, Northern Brazil. The study was performed in Brazil (A) near the city of Bragança (46°38′W 0°50′S, B). The study site is marked by a black dot and is located at the tidal channel Furo Grande (B).

assume that the CO_2 released by burrows, animal respiration and roots is currently underestimated and may be an uncertain factor within the global carbon budget estimates for mangrove forests. Studies investigating the CO_2 release by crab burrows or the reduction potential in the sediment surrounding the burrows have only been conducted for the relatively small fiddler crabs [18,19,24]. Inhabited fiddler crab burrows increased the CO_2 release from mangrove sediment approximately 2 to 5 times compared to plain sediment [19,24]. 58% of the released CO_2 originated from the sediment surrounding the burrows and the remainder from the crab inside the burrow [24]. Studies characterizing the reduction potential in sediment surrounding fiddler crab burrows in salt marsh and mangrove sediments found a 1.5–4 cm thick oxidized zone, where iron reduction replaced the usually predominant sulfate reduction [18,19]. This zone also presented the highest CO_2 production rates. The rates decreased with increasing distance from the burrow, indicating that the change in reduction potential can influence carbon decomposition processes [18]. The availability and quality of carbon as well as the temperature can further influence carbon decomposition processes and thereby CO_2 efflux from the sediments [8,21].

Araújo Jr. et al. [25] investigated the bioturbation activity of the large and abundant crab *Ucides cordatus* (Ucididae), which, together with much smaller fiddler crabs, influence with their burrows the microtopography of the sediment in Brazilian mangrove forests. Araújo Jr. et al. [25] concluded from measurements of the reduction potential in sediment cores from bioturbated and non-bioturbated sites that *U. cordatus* burrows oxidized the sediment. However, the study did not provide information on the supposed gradient and the horizontal range of the reduction potential in the sediment surrounding the burrows, likewise, no CO_2 efflux rates were measured.

In the present study the effects of bioturbation by *U. cordatus* on carbon dynamics and sediment reduction potentials were evaluated by (i) quantifying the CO_2 efflux rate of individual burrows compared to plain sediment and (ii) measuring the reduction potential in radial profiles in the sediment around burrows at different depths in a *Rhizophora mangle* forest, respectively. We hypothesized that (i) sediment with *U. cordatus* burrows has higher CO_2 efflux rates than sediment without burrows and (ii) reduction potentials in radial profiles of the sediment surrounding the burrows down to 50 cm depth decrease gradually from the burrow wall, approximating conditions of non-bioturbated sediment.

Materials and Methods

The artisanally harvested study species *Ucides cordatus* is listed as endangered in the Brazilian legislation due to the risk of overfishing. All research activities were allowed by a research permit granted by SISBIO (Sistema de Autorização e Informação em Biodiversidade, authorization number 30007-1). Field work was performed in the extractive reserve "Reserva Extrativista Marinha de Caeté-Taperaçu".

Study area

The study area is located on the Ajuruteua peninsula, the western bank of the Caeté river estuary in North Brazil (Figure 1). Field work was performed in the high intertidal zone near the tidal channel Furo Grande (46°38′W 0°50′S) (Figure 1, B). The region has semidiurnal tides with amplitudes of 2 to 5 m. During spring tide days the study site is flooded twice a day, while during neap tide days tidal heights are too low to inundate the whole mangrove forest [26,27]. The forest is dominated by the mangrove tree *Rhizophora mangle* (Rhizophoraceae). Other mangrove tree species are also present including *Avicennia germinans* (Acanthaceae) and *Laguncularia racemosa* (Combretaceae) [26].

Mean annual temperature in 2012 was 26.3 ± 1°C (mean ± standard deviation, [28]). Total precipitation recorded at the weather station in Tracuateua (50 km from the study area) was 1552 mm (Figure 2). In Northern Brazil, the wet season generally occurs from January to August and the dry season (monthly precipitation <100 mm) from September to December [28]. Surface water salinity of the tidal channel Furo Grande varied between 22 and 37 throughout the year 2012 (unpublished data).

In the study area *U. cordatus* is the only large burrowing crab (max. carapace width of 9.9 cm in the Caeté estuary, [29]). An average density of 1.7 ind. m^{-2} was reported for the Furo Grande area in *R. mangle* dominated forests [30]. *U. cordatus* prefers to construct its burrow around the root system of *R. mangle*, because large roots structure the sediment and thereby increase sediment stability and provide shelter against predators [31]. Burrows of *U. cordatus* can have various forms. They can be simple burrows with one opening and one corridor, U-shaped with more than one opening, or even more complex structures with a maximum of three openings [32]. In northeast Brazil simple burrows are the most common type (more than 85% of all investigated burrows, n = 735) [32]. In the Furo Grande area simple burrows are also the most common burrow type (unpublished observation). Such burrows have the following morphology: Their corridors initially descend with a slight slope and then bend vertically downwards until forming a terminal burrow chamber ([14], Figure 3). At our study site at the Furo Grande total burrow corridor lengths (initial horizontal and vertical part) ranged from 73 to 219 cm with an average of 123 ± 32 cm (mean ± standard deviation, n = 30, own data). The inner walls of the burrows of *U. cordatus* have no conspicuous inner mucus layer such as known from Polychaetes (e.g. reviewed by Kristensen and Kostka [33]). The burrows of the other crab species, for example sympatric fiddler crabs, also seem to be unlined (own observation, [34]).

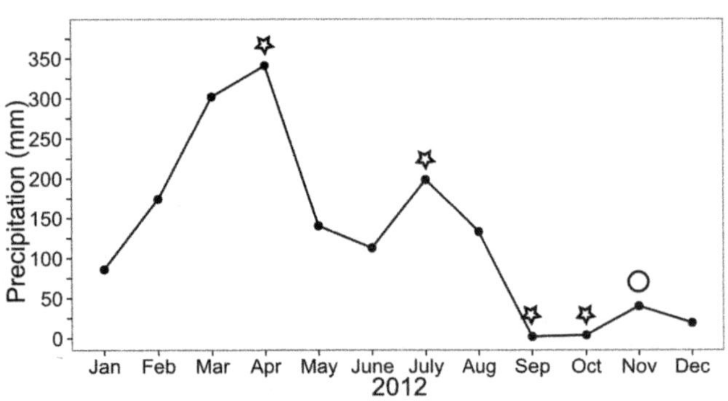

Figure 2. Total monthly precipitation in 2012. Precipitation (mm) was recorded by the weather station in Tracuateua (50 km southwest from the study site), Pará, Brazil [28]. The asterisks mark the sampling months (CO_2 efflux rate and rH of burrows and control sediment). The circle marks the only control sampling (CO_2 efflux rate and rH), which had to be conducted two neap tide cycles later than the corresponding sampling in October 2012 due to technical reasons.

CO_2 efflux rates

CO_2 efflux rates of both burrows and plain sediment were measured in April, July, September and October 2012 around neap tides during the daytime. Due to technical reasons, measurements of sediment with burrows and plain sediment were performed during two succeeding neap tide cycles. In October 2012, the sampling of plain sediment was performed two neap tide cycles later.

U. cordatus burrows with only few *R. mangle* stilt roots nearby were chosen for the measurements to minimize the effect of root respiration. Furthermore, freshly excavated and blackish sediment had to be in front of their entrances indicating the presence of a live crab and ensuring sampling of recently bioturbated sediment. The diameter of the burrow entrances and if possible the current

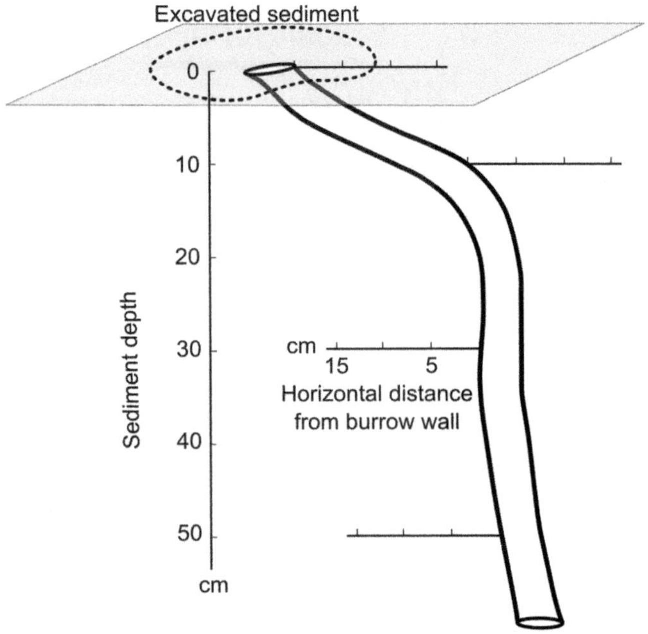

Figure 3. Vertical and horizontal sampling scheme of the sediment surrounding *U. cordatus* burrows for the rH measurements.

depths from the sediment surface to the water surface inside the burrow were recorded with a measuring tape before taking measurements. If the water surface inside the burrows was too low to measure, measurement was delayed to the rH measurements, when parts of the burrow walls were removed to facilitate access. The area of the burrow wall exposed to air was estimated based on the entrance diameter and the distance to the water surface.

For measuring the CO_2 efflux rate of a burrow, a PVC collar with 20 cm diameter was positioned around the entrance and slightly pushed several centimeters into the sediment. To avoid any influence of CO_2 released during and after the installation of the collar, the measurements were delayed for 1 h. A respiration chamber connected to a CO_2/H_2O infrared gas analyzer (LI-8100A, LI-COR, Biosciences) was then placed on top of the PVC collar. Since the respiration chamber was opaque no CO_2 was consumed through photosynthesis during measurements. The CO_2 efflux was recorded over 2 min and the measurement repeated five times. Between replicates, the chamber was opened for 25 s to release the accumulated CO_2. Most probably, these measurements included the respiration of the crab inside the burrow. The contribution of crab respiration to the measured CO_2 efflux rate was estimated by a separate experiment (see below). Sediment temperature was measured outside the collar at 2 cm sediment depth with a thermocouple (OMEGA Engineering).

As a control, the CO_2 efflux rate of plain sediment without *U. cordatus* burrows and at the most with very few fiddler crab burrows was performed, following the same measurement protocol as for the sediment with burrows. However, four replicates were taken, because preliminary tests had shown that variance of measured values for plain sediment was smaller than for sediment with *U. cordatus* burrows.

The CO_2 efflux rates were calculated for a linear flux rate [35] taking into account the surface sediment temperature and surface area inside the collar. For the measurement with *U. cordatus* burrows, the surface area inside the collar also included the burrow wall surface area. It also had to be considered that the relationship between burrow opening area and surface sediment area inside the collar did not represent the average relationship of burrow opening area to plain sediment area at the study site due to specific selection of single burrows and the small collar area. Under natural conditions, the proportion of burrow to plain area is lower (1.7 ind. m^{-2}, [30]). Therefore, the calculated CO_2 efflux rates inside collars with a crab burrow could not be directly compared with rates of plain sediment. To nevertheless allow for a comparison between the CO_2 efflux rates of plain sediment with sediment containing burrows at a given burrow density, the following calculations were performed: First, the CO_2 efflux rates of crab burrows without the surface sediment were estimated. To do this, the corresponding monthly mean CO_2 efflux rates of the plain sediment measurements were subtracted from the calculated CO_2 efflux rate of the burrows, considering the areal proportions of each term as follows:

$$F_b = (F_t \times A_t - F_s \times A_s) / A_b$$

F is the CO_2 efflux rate and A the area, which are both linked to plain sediment s, burrow b, total measured area t or total CO_2 efflux t. The total measured area refers to the surface area inside the PVC collar. Mean control CO_2 efflux rates of each month were used for F_s. From this calculation values for F_b (µmol m^{-2} s^{-1}) were obtained. From these values we derived CO_2 efflux rate of individual burrows (µmol $burrows^{-1}$ s^{-1}) with entrance diameters of 5, 6 and 7 cm corresponding to the range of burrow diameters sampled in the field. CO_2 efflux rates for mangrove sediment based on the CO_2 efflux rate of plain sediment and individual burrows were then obtained with the help of size-specific crab burrow density data. Two burrow densities were chosen, one represents the mean *U. cordatus* burrow density in a *R. mangle* dominated forests with 1.7 ind. m^{-2} for the Furo Grande area [30]. To simplify the calculation, this density of crabs can be related to 1.7 $burrows^{-1}$ m^{-2} since most *U. cordatus* burrows have only one opening. The other chosen burrow density was 11.9 $burrows^{-1}$ m^{-2} (unpublished data) representing the increased burrow density around dense *R. mangle* stilt roots.

Respiration of *U. cordatus*

Measuring the CO_2 efflux rate of a burrow without a live crab was not feasible as removal of the crab would have strongly disturbed the sediment. To nevertheless allow an estimation of the proportion of the CO_2 efflux rate contributed by the crabs, their respiration rates were determined in an additional experiment. Adult male *U. cordatus* were caught nearby the study site. The carapace width (cm) of each animal was measured and the wet mass (g) was calculated by the regression formula for male *U. cordatus* provided by Diele [36]:

$$\text{Wet mass} = 0.4489 \times \text{carapace width}^{2.9533}$$

In the laboratory, crabs were placed in a PVC basin (20 cm diameter, 6 cm height) with the respiration chamber mounted on top. Under natural conditions crabs rest in air and water (own observation). Therefore, measurements were performed with different crabs in a dry basin (1. treatment: respiration in air, $n = 14$) and in a basin containing seawater (2. treatment: respiration in seawater, water level 4 cm, salinity 32, $n = 14$). CO_2 released by each crab was measured five times for 10 min with breaks of 25 s in-between to open the chamber and release the accumulated CO_2. Crabs were not acclimated inside the chamber prior to measurements. Therefore, the five consecutive measurements presented a gradient of crab respiration from a least acclimated state to a more acclimated one. CO_2 respiration rates of crabs (µmol CO_2 kg^{-1} s^{-1}) were calculated as described above, but related to wet mass instead of surface area in the calculation.

In the next step the contribution of crab respiration to the CO_2 efflux rate was estimated for individual burrows with a burrow entrance diameter (BED) of 5, 6 and 7 cm. Since the inhabitants of the burrows used for CO_2 measurements were not captured, their carapace width was estimated from the BED by a least squared regression formula based on data collected by Piou et al. [31]:

$$\text{Carapace width} = (\text{BED} - 0.32) \div 0.97 \; (R^2 = 0.5, n = 310)$$

The wet mass was then estimated from the calculated carapace width with the formula for male *U. cordatus* by Diele [36] as indicated above. With the estimated wet mass of crabs inhabiting burrows with entrance diameters of 5, 6, and 7 cm, respectively, the respiration rate of individual crabs (µmol CO_2 $crab^{-1}$ s^{-1}, in air and in seawater) was calculated based on the first and the last measurement of the five consecutive respiration measurements. These two measurements represent a "worst" (crab least acclimated, resting in air) and a "best" case scenario (crab with the longest acclimatization time and resting in water), since preliminary tests had shown that crabs resting in air had higher CO_2 efflux rates compared to crabs resting in water.

Reduction potential

After the CO_2 efflux rate measurements in the field were performed, reduction potentials of both burrow walls and of non-bioturbated sediment were measured at the selected sites on the same day. Redox potential (±1.0 mV), pH (±0.1) and temperature (±0.1°C) were measured in the sediment excavated by the crabs outside the burrow and in the upper 2 cm of the burrow water with a Sartorius ORP (redox) combination electrode and a WTW Sentix 41 pH-electrode connected to a WTW portable meter (Multi 340i), respectively. The above mentioned parameters were also measured in the sediment surrounding the burrow (from here on referred to as "burrow wall sediment" - BWS) at sediment depths of 1, 10, 30 and 50 cm. To be able to measure the reduction potential in deeper sediment depths, one half of the burrow was carefully opened. When the desired measuring depth inside the burrows was below the water table, the water was manually removed with a small vial before taking the measurement. Measurements beyond 50 cm sediment depth were technically too difficult due to the rapid refilling of the burrow with water from below. At each sediment depth, the electrodes were inserted perpendicularly to the burrow wall, thereby subsequent readings at horizontal distances of 2, 5, 8 and 15 cm from the BWS were obtained (Figure 3). As reduction potentials in the sediment around fiddler crab burrows have been observed as being influenced at a range of 1–4 cm [18,19] and *U. cordatus* burrows are larger, we choose a larger maximal horizontal distance of 15 cm to insure that the change in reduction potential were within the measured radius.

In addition to the burrows, non-bioturbated sediment (referred to as the control) was also measured for comparison. Sediment cores were taken at least 15 cm away from burrows and stilt roots to sample areas where the influence of burrows and roots on the surrounding sediment was most likely low. A peat sampler (Eijkelkamp) of 50 cm length and 6 cm diameter was used to obtain the cores. Redox potential, pH and temperature of the sediment core were measured at depths of 1, 10, 30 and 50 cm.

Redox potential, temperature and pH values of each measurement were used to calculate the rH values [37]. These values are an indicator of the reduction force of a redox system and range between 0 (strongly reducing conditions) and 42 (strongly oxidizing conditions).

Statistical analyses

The statistical analyses were carried out following protocols for data exploration and analysis of Zuur et al. [38,39] using the statistical programming environment R [40] with the "nlme" [41] and the "ggplot2" [42] package. Data were checked for outliers (Cook distance) and these were, if necessary, removed. For information on data sets used for the analyses see Data S1.

Burrow entrance diameters, depth from the surface sediment to the water surface inside burrows, and rH values of burrow water and excavated sediment were each separately analyzed with a one way ANOVA for differences over time. When the trend over time was not linear, time was considered as a factor and a Tukey post-hoc test was applied to detect differences throughout the wet season.

Gaussian linear mixed-effects (LME) models were used for the following analyses, as they can account for inner variation between sediment cores or burrows [38,43,44]. The CO_2 efflux rates of burrows (F_b) and controls were analyzed with a LME model to test for differences over time. The random intercept of the model allowed for heterogeneity among burrows and between the sampling sites of plain sediment. To find the optimal set of fixed terms, a backwards model selection was used based on the maximum likelihood ratio test (ML) and/or on the Akaike Information Criterion (AIC). The validity of the model was checked by examination of diagnostic plots of model residuals. Final models were checked for homogeneity between residuals versus fitted values and covariates. Independence was examined by plotting residuals versus time. The final model was generated and evaluated with the restricted maximum likelihood estimation (REML) and these models are presented in Tables S1, S2, S3, S4, S5 and S6. All following LME models were analyzed as described above. Estimated values from statistical analyses are presented as mean ± standard error (se). Respiration rates of crabs were analyzed with a LME model to test for difference between treatments, number of measurements (from least acclimated to acclimated) and their respective interaction terms. The random intercept accounted for heterogeneity between individual crabs. No function for the temporal correlation structure was considered in the repeated measurements of the same crab, to be able to detect those differences between repeated measurements.

The rH data for the BWS were analyzed with a LME model to test for differences among sediment depths, horizontal distances from the burrow wall, time and their interaction terms. The random intercept accounted for heterogeneity among individual burrows. Control rH values were tested with a LME model for differences between sediment depths, time and their interaction term. The random term allowed for heterogeneity between sediment cores. Since general trends over time for the control and burrow rH data had already been analyzed in the two LME models and to keep the following models simple, control and burrow rH data were analyzed separately for each month. Monthly data sets of rH values from burrows and controls were analyzed in four LME models to test for differences among horizontal sampling levels, sediment depths and their interaction terms. BWS values of 2 (BWS2), 5 (BWS5), 8 (BWS8) and 15 cm (BWS15) and the control (>15 cm) represented the five horizontal sampling levels. BWS values and the control value were set as categorical covariates, because the exact horizontal distances of the control cores to the neighboring below ground burrow corridors was unknown. The random term allowed for heterogeneity among sampling locations (sediment core or burrow).

A Pearson correlation was applied to measure the relationship between surface rH values and CO_2 efflux rates of the plain sediment. Because the CO_2 efflux rate of a burrow could be associated to several rH values in the burrow wall sediment (from horizontal distances of 2 cm in several sediment depths) no correlation was performed.

Results

CO_2 efflux rates

Burrow entrance diameters did not significantly differ over time ($F = 3.3$, $df = 1$, $p = 0.07$, Table 1) insuring that burrows of different samplings were comparable to each other. Depths from the sediment surface to the water surface inside burrows changed over time ($F = 9.9$, $df = 1$, $p < 0.001$) and were largest in October (Tukey post-hoc: $p \leq 0.002$, Table 1). The mean burrow wall area exposed to air was 0.04 ± 0.02 m^2 (mean ± standard deviation, n = 78) and ranged from 0 (burrow completely filled with water) to 0.1 m^2.

Examining the CO_2 efflux rates of burrows (F_b) for differences over time showed that the LME model which included the term time was not better than the model without this term (Likelihood Ratio/L. Ratio = 1.1, $df = 1$, $p = 0.3$, Table S1) indicating that measured values did not differ over time. By contrast, control sediment CO_2 efflux rates increased over time (Table 1); removing

Table 1. Field data of measured and estimated parameters.

	2012							
	April		July		September		October	
	Burrow entrance diameter (cm)							
Mean ± se (n)	6.6±0.3	(24)	5.9±0.2	(24)	6.2±0.1	(19)	5.8±0.3	(11)
Minimum	4.1		3.4		5.3		4.6	
Maximum	9.6		7.4		7.4		7.4	
	Burrow water level from the surface (cm)							
Mean ± se (n)	16.2±0.9	(24)	21.7±2.9	(24)	18.3±1.3	(19)	33.8±2.2	(11)
Minimum	7		5		10		27	
Maximum	26		50		29		50	
	CO_2 efflux rate of plain sediment (μmol m^{-2} s^{-1})							
Mean ± se (n)	0.7±0.07	(88)	0.9±0.07	(96)	1.3±0.07	(93)	1.3±0.06	(96)
	CO_2 efflux rate of individual burrows (μmol $burrow^{-1}$ s^{-1})							
BED: 5 cm, mean ± se (n)	0.1±0.01	(116)	0.2±0.02	(119)	0.1±0.01	(95)	0.1±0.01	(55)
BED: 6 cm, mean ± se (n)	0.2±0.01	(116)	0.3±0.03	(119)	0.1±0.01	(95)	0.2±0.02	(55)
BED: 7 cm, mean ± se (n)	0.3±0.02	(116)	0.4±0.04	(119)	0.2±0.01	(95)	0.2±0.03	(55)
	Reduction potential of burrow water							
Mean ± se (n)	16.2±0.9	(22)	16.8±0.5	(22)	18.0±0.9	(16)	16.2±0.7	(9)
	Reduction potential of excavated sediment							
Mean ± se (n)	15.5±0.8	(25)	16.1±0.7	(24)	16.0±0.7	(17)	17.5±0.9	(12)

Burrow entrance diameter, burrow water level from the surface, CO_2 efflux rate of plain sediment and inhabited burrows with specific diameter, reduction potential of burrow water and excavated sediment.
Abbreviation: BED = burrow entrance diameter (in cm), n = sample size, se = standard error.

the term time from the corresponding LME model led to a less suitable model, so the factor was retained (L. Ratio = 15.3, df = 1, $p<0.001$, Table S2). Calculated CO_2 efflux rates of individual burrows ranged from 0.1 µmol burrows^{-1} s^{-1} for burrows with a BED of 5 cm to 0.4 µmol burrows^{-1} s^{-1} for burrows with a BED of 7 cm (Table 1).

Respiration of *U. cordatus*

Mean carapace width and derived wet masses of crabs are given in Table 2. A LME model of the crab respiration data with the interaction term treatment×number of observations was significantly better than a model without this interaction (L. Ratio = 12.5, df = 1, $p<0.001$, Table S3). This indicates that crab respiration rates varied considerably between individuals of one treatment, but decreased during the five consecutive measurements. Crabs resting in seawater had the lowest respiration rates (Table 2). During the measurements, the least acclimated crabs produced foam in front of their mouthparts, whereas more acclimated crabs did not foam and stopped respiring for a few seconds up to several minutes. Respiration rates of individual crabs with a wet mass corresponding to burrow entrance diameters of 5, 6 and 7 cm, respectively, are presented for the "worst" (crab least acclimated, resting in air) and "best" case scenarios (crab with the longest acclimatization time and resting in water) in Table 2. The crabs' contribution to the CO_2 efflux rates of individual burrows was 55–100% in the "worst" case and 20–60% in the "best" case scenario (Table 1, Table 2).

Estimates for CO_2 release of mangrove sediment

Calculated CO_2 efflux rate estimates of mangrove sediment with 1.7 and 11.9 burrows^{-1} m^{-2} increased over time. The reason for this was an increase in CO_2 efflux rates of plain sediment over time due to seasonality, while individual burrow CO_2 efflux rates remained constant over time (Figure 4). Estimates of CO_2 efflux rate of sediment with *U. cordatus* burrows (Table 3, Figure 4) are 15 to 500% (considering 1.7 and 11.9 burrows^{-1} m^{-2}, respectively) higher than for plain sediment (Table 3). Despite the small contribution of burrow openings to total sediment surface area (< 5%), *U. cordatus* burrows, especially in densely rooted sediment, can increase the CO_2 efflux rate of plain sediment by 92 to 500% (Table 3, Figure 4).

Reduction potential

The rH values of both burrow water (F = 0.6, df = 1, $p = 0.4$, Table 1) and excavated sediment (F = 2.3, df = 1, $p = 0.1$, Table 1) did not differ over time. The final model of the BWS rH data indicates that rH values decreased with sediment depth and horizontal distance as well as over time (Figure 5); a LME model of the BWS rH data with a three way interaction term was not significantly better than a model without this interaction (L. Ratio = 4.4, df = 9, $p = 0.9$, Table S4). Furthermore, the interaction terms horizontal depth×time (L. Ratio = 1.0, df = 3, $p = 0.8$, Table S4) and sediment depth×time (L. Ratio = 3.2, df = 3, $p = 0.4$, Table S4) were eliminated. The model with the interaction term horizontal distance×sediment depth was better than without interaction (L. Ratio = 17.2, df = 9, $p = 0.045$, Table S4). However, we continued with the simpler (using fewer degrees of freedom) and slightly better model (AIC of model with the interaction term: 4389.1 and without interaction term: 4388.4) by excluding the interaction term. Finally, comparing models where each main term was excluded once with the original model containing all three main terms showed that the latter was best (sediment depth: L. Ratio = 555.0, df = 3, $p<0.001$; horizontal distance: L. Ratio = 35.6, df = 3, $p<0.001$; time: L. Ratio = 20.7, df = 1, $p<0.001$, Table S4).

The model of the control rH data with the interaction term sediment depth×time was significantly better (L. Ratio = 10.1, df = 1, $p = 0.0015$, Table S5) than the model without this term indicating that control rH values decreased with sediment depth, but had different depth-rH curves over time (Figure 5). Further, the surface rH values positively correlated with the CO_2 efflux rates of plain sediment (correlation coefficient = 0.18, $p<0.001$).

In the next step, BWS and control rH data were compared separately for each sampling campaign. The comparison of the four models including and excluding the interaction term sediment depth×time showed that the interaction term significantly improved the four models (April: L. Ratio = 18.0, df = 4, $p = 0.001$, July: L. Ratio = 33.2, df = 4, $p<0.001$, September: L. Ratio = 38.3, df = 4, $p<0.001$, October: L. Ratio = 44.0, df = 4, $p<0.001$, Table S6). Hence, at any point of time rH values decreased with increasing sediment depth, and depth-rH curves differed between horizontal sampling levels, in the way that rH values decreased mostly with increasing horizontal distances (Figure 5). Measured pH values are not shown in detail; they ranged from 3.5 to 7.1 for plain sediment and from 4.3 to 7.8 for burrow wall sediment.

Discussion

Measured CO_2 efflux rates of individual burrows are composed of three elements: The respiration of the inhabiting crab itself, the CO_2 released by the burrow wall sediment and the CO_2 released by the water inside the burrow.

The crabs' contribution to the CO_2 efflux rate of burrows

In our study measurements of CO_2 exhaled by isolated crabs showed that high CO_2 production by respiration was related to stress. Crabs in our study, particularly those kept without seawater in the PVC basin, were probably not fully acclimated during the 50 min respiration measurements, because CO_2 respiration of crabs resting in air were always higher in the five measurements compared to crabs resting in water (Table 2). Some crabs exhaled four or fewer times in 10 min (see minimum values in Table 2) suggesting reasonable acclimatization.

During field measurements, crabs were sometimes seen inside their burrows when the respiration chamber was opened between measurements. The crabs did not appear to be stressed by the chamber, since they moved slowly and did not produce foam, as they do when they are handled or otherwise stressed [45]. Therefore, respiration rates of *U. cordatus* measured in the laboratory for the "worst" case scenario (crabs with no acclimatization time and resting in air) are probably overestimating the crabs' respiration under field conditions. Respiration rates for the "best" case scenario (crabs with longest acclimatization time resting in water) on the other hand, probably resemble field conditions more closely. Our data thus suggest that crabs contributed about 20–60% to the CO_2 efflux rates of individual burrows, which leaves the burrow CO_2 efflux rates without crabs to range between 0.04–0.32 µmol burrows^{-1} s^{-1} for burrows of 5 to 7 cm entrance diameter (Table 1). Respiration rates of an acclimated *U. cordatus* specimen (250 g wet mass, in seawater) are 0.19±0.04 µmol CO_2 crab^{-1} s^{-1} and are similar to rates recorded for a large sesarmid crab (*Neoepisesarma versicolor*, 250 g wet mass, in seawater) of 0.24 µmol CO_2 crab^{-1} s^{-1} [46].

Table 2. Data of *U. cordatus* respiration measurements.

	Crab in air		Crab in seawater	
	Carapace width (cm)			
Mean ± se (n)	7.3±0.1	(14)	7.0±0.1	(14)
Median	7.4		7.0	
Minimum	6.5		6.2	
Maximum	7.8		7.5	
	Wet mass (g)			
Mean ± se (n)	161.6±6.3	(14)	140.4±5.9	(14)
Median	162.4		137.7	
Minimum	113		98.2	
Maximum	193.5		172.4	
	Crab respiration (μmol CO_2 kg^{-1} s^{-1})			
Mean ± se (n) of measurement 1	2.3±0.4	(13)	1.4±0.2	(13)
Mean ± se (n) of measurement 2	1.4±0.2	(13)	1.0±0.2	(12)
Mean ± se (n) of measurement 3	1.1±0.1	(13)	0.9±0.2	(12)
Mean ± se (n) of measurement 4	0.9±0.1	(12)	0.7±0.1	(12)
Mean ± se (n) of measurement 5	0.9±0.1	(12)	0.8±0.2	(12)
Minimum of all measurement (n)	0.4	(63)	0.04	(61)
Maximum of all measurement (n)	4.1	(63)	2.4	(61)
	Crab respiration - worst case (μmol CO_2 $crab^{-1}$ s^{-1})			
BED: 5 cm, WM: 28 g, mean ± se (n)	0.11±0.02	(13)	0.07±0.01	(13)
BED: 6 cm, WM 55 g, mean ± se (n)	0.19±0.03	(13)	0.12±0.02	(13)
BED: 7 cm, WM 93 g, mean ± se (n)	0.30±0.05	(13)	0.19±0.02	(13)
	Crab respiration - best case (μmol CO_2 $crab^{-1}$ s^{-1})			
BED: 5 cm, WM: 28 g, mean ± se (n)	0.04±0.004	(12)	0.04±0.01	(12)
BED: 6 cm, WM 55 g, mean ± se (n)	0.07±0.01	(12)	0.06±0.01	(12)
BED: 7 cm, WM 93 g, mean ± se (n)	0.12±0.01	(12)	0.10±0.02	(12)

Carapace width, estimated wet mass after Diele [36], average respiration rates during the five measurements and respiration rate of individual crabs of a specific wet mass inhabiting burrows with different burrow entrance diameters for the "worst" case scenario (crab not acclimated, values calculated from measurement 1) and for the "best" case scenario (crab with longest acclimatization time, values calculated from measurement 5).
Abbreviation: BED = burrow entrance diameter (in cm), n = sample size, se = standard error, WM = wet mass.

CO_2 efflux rates of burrows and sediment

CO_2 efflux rates of inhabited *U. cordatus* burrows were at least 20 times higher (Table 1) than estimated rates from inhabited fiddler crab burrows (0.005 μmol $burrows^{-1}$ s^{-1}) measured in a Tanzanian mangrove forest [24]. Higher CO_2 efflux rates of *U. cordatus* burrows are probably caused by the larger body size and higher respiration of the crab and the corresponding larger burrow size (and depth) compared with *Uca* spp. and their burrows (average wet mass of *Uca* spp.: 3.2 g, 0.002 μmol CO_2 $crab^{-1}$ s^{-1}, average burrow entrance diameter not reported, [24]). Mean CO_2 efflux rates of plain sediment in our study ranged from 0.7 to 1.3 μmol m^{-2} s^{-1} and were similar to other mangrove sediments [20,22], e.g. to values measured by Lovelock [47] for sites in the Caribbean, Australia and New Zealand, which ranged from −0.25 to 2.97 μmol m^{-2} s^{-1}. In line with our first hypothesis, CO_2 efflux rates of mangrove sediment containing burrows inhabited by *U. cordatus* were 15 to 500% higher than for plain sediment depending on burrow density and month (Table 3). In a Tanzanian mangrove forest, Kristensen et al. [24] recorded CO_2 efflux rates of sediment with inhabited *Uca* spp. burrows in areas without obvious above ground roots. Measured rates ranged between 0.68 μmol m^{-2} s^{-1} (68±17 $burrows^{-1}$ m^{-2}) and 1.38 μmol m^{-2} s^{-1} (237±53 $burrows^{-1}$ m^{-2}) and were similar to ours with 1.0±0.02 until 1.7±0.05 μmol m^{-2} s^{-1} based on an average *U. cordatus* burrow density of 1.7 $burrows^{-1}$ m^{-2} (Table 3). Although CO_2 release of individual fiddler crab burrows is much lower than that of *U. cordatus*, the burrow density of fiddler crabs at the Tanzanian study site was much higher (68–626 $burrows^{-1}$ m^{-2}, [24]) than burrow densities of *U. cordatus* recorded near our study site (1.7 $burrows^{-1}$ m^{-2} at an average, Diele et al. [30]). Thus, a high density of burrows may compensate for a lower per-burrow CO_2 efflux rate.

In addition to *U. cordatus* burrows, *Uca* spp. burrows were also present at our study site (28 to 105 $burrows^{-1}$ m^{-2} with an average of 57±18 $burrows^{-1}$ m^{-2}, mean±standard deviation, n = 42, with each sample area 1 m^2, unpublished data). Future investigations should focus on the entire crab community of a mangrove site, involving all burrowing crabs, to obtain more precise estimates of overall CO_2 efflux rates from the mangrove sediment.

The amount of CO_2 released by sediment or burrows depends on the prevalence of organic carbon in the sediment and its oxidation pathways by microorganisms [48–50]. In a study on the Ajuruteua peninsula the organic matter content and mean cell

Table 3. CO_2 efflux rates of mangrove sediment.

	2012							
	April		**July**		**September**		**October**	
Density: 1.7 burrows^{-1} m^{-2}	**CO_2 efflux rate of mangrove sediment (μmol m^{-2} s^{-1})**							
BED: 5 cm, mean ± se (n)	1.0±0.02	(116)	1.2±0.03	(119)	1.5±0.01	(95)	1.5±0.02	(55)
BED: 6 cm, mean ± se (n)	1.1±0.02	(116)	1.3±0.05	(119)	1.6±0.02	(95)	1.6±0.03	(55)
BED: 7 cm, mean ± se (n)	1.2±0.03	(116)	1.5±0.06	(119)	1.7±0.02	(95)	1.7±0.05	(55)
Density: 11.9 burrows^{-1} m^{-2}	**CO_2 efflux rate of mangrove sediment (μmol m^{-2} s^{-1})**							
BED: 5 cm, mean ± se (n)	2.4±0.12	(116)	3.2±0.22	(119)	2.5±0.08	(95)	2.7±0.17	(55)
BED: 6 cm, mean ± se (n)	3.2±0.17	(116)	4.2±0.32	(119)	3.0±0.12	(95)	3.4±0.24	(55)
BED: 7 cm, mean ± se (n)	4.1±0.23	(116)	5.4±0.44	(119)	3.6±0.16	(95)	4.1±0.33	(55)
Density: 1.7 burrows^{-1} m^{-2}	**Increase of CO_2 efflux rate to plain sediment (%)**							
BED: 5 cm (AP: 0.3%)	43		33		15		15	
BED: 6 cm (AP: 0.5%)	57		44		23		23	
BED: 7 cm (AP: 0.7%)	71		67		31		31	
Density: 11.9 burrows^{-1} m^{-2}	**Increase of CO_2 efflux rate to plain sediment (%)**							
BED: 5 cm (AP: 2.3%)	243		256		92		108	
BED: 6 cm (AP: 3.4%)	357		367		131		162	
BED: 7 cm (AP: 4.6%)	486		500		177		215	

Rates and the increase in % compared to plain sediment for average burrow densities of 1.7 burrows^{-1} m^{-2} for *Rhizophora mangle* dominated areas at the Furo Grande [30] and 11.9 burrows^{-1} m^{-2} for densely rooted areas around *R. mangle* trees.

Abbreviation: AP = areal proportion of the burrow opening per m^2, BED = burrow entrance diameter (in cm), n = sample size, se = standard error, WM = wet mass.

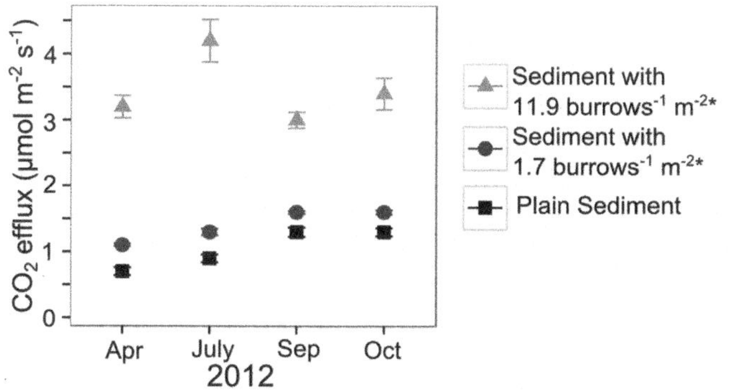

Figure 4. CO_2 efflux rates (µmol m^{-2} s^{-1}) of plain sediment and sediment with a burrow density of 1.7 and 11.9 burrows^{-1} m^{-2}. *For these estimates all burrow entrance diameters were set at 6 cm.

numbers of bacteria were measured in the BWS of *U. cordatus* burrows (70 cm depth) and were similar to corresponding values of the surface sediment [51]. Therefore, microbial activity in BWS may be more similar to surface sediment than to non-bioturbated sediment at 70 cm sediment depth. However, CO_2 released from burrows may not exclusively originate from carbon oxidation in the BWS, but also from the carbon stock in deeper sediment layers which may be released via the burrows [20,24]. Furthermore, the amount of CO_2 released from the water accumulated inside the burrow is unknown. The water may contain CO_2 originating from bacterial, meio- and macrofaunal respiration and from tidal input. How the bicarbonate system of the burrow water is constituted is also an unknown, e.g. how such systems react to increasing/decreasing CO_2 concentrations and whether changing water levels inside burrows influence the burrow water CO_2 gas exchange. CO_2 dissolved in water may also be partly exported by the groundwater during the ebb tide [52].

Apart from microbial decomposition in the sediment, CO_2 is produced by root respiration. Information about root respiration of mangrove tree species is scarce; root respiration rates are reported to be low for *R. mangle* trees. Kristensen et al. [24] estimated the amount of CO_2 released by pneumatophores of *Sonneratia alba* and *Avicennia marina* in a Tanzanian mangrove forest, yielding an equivalent or higher efflux rate than sediment with burrows. Due to technical reasons we had to avoid measuring near aboveground aerial roots. However, underground root respiration may have contributed to the CO_2 released by burrows and, equally, by surface sediment without burrows. As *U. cordatus* prefers areas with *R. mangle* roots when constructing its burrows [31], their burrows possibly function as an exhalant channel for CO_2 produced by root respiration. To test this hypothesis, further work on CO_2 release must overcome technical restrictions and focus on areas where crabs and mangrove tree roots occur closely together.

Temporal and spatial patterns of reduction potentials

Reduction potentials of the burrow water and the excavated sediment were similar during the wet season and relatively oxidized compared to the BWS and control rH values in deeper sediment depths. This indicates that burrow water is regularly mixed with more oxidized water or oxygen by, for example, tidal flushing or groundwater flow [53–55]. Further, excavated sediment oxidizes at the surface when in contact with atmospheric oxygen. Thus, bioturbation by crabs significantly oxidizes previously reduced sediment.

In contrast to the stable values in the burrow water and the excavated sediment, rH values in the BWS changed over time. In April, BWS had the lowest rH values. At 1 and 10 cm sediment depth rH values were even lower compared to plain sediment. The rising water column inside the burrows (Table 1) and waterlogging of the sediment caused by the high precipitation rates in this month may have led to anoxic conditions along the burrow wall. In contrast, during the last sampling at the end of the wet season rain showers, temporally filling up burrows became less frequent and water levels sank to 50 cm below the surface (Table 1). The longer exposure of BWS to air at the end of the wet season led to more oxidized conditions at 30–50 cm sediment depths compared with plain sediment. Additionally, from July onwards BWS rH and control rH values increased with decreasing precipitation rates (Figure 2, Figure 5).

The spatial pattern of the *U. cordatus* BWS rH values indicated a wider oxidized zone, reaching up to 15 cm away from the burrow, compared to the *Uca sp.* burrows, where the oxidized zone reached only 1.5–4 cm into the surrounding sediment [18,19]. This difference is most likely due to the much larger size and depth of *U. cordatus* burrows, leading to improved solute exchange conditions between the BWS and tidal water, groundwater and atmospheric air compared to the smaller fiddler crab burrows. Why rH values for the up to 15 cm horizontal distance from the burrow wall did not approximate control values remains unclear. However, an approximation may have been masked by the high variance of the rH values due to nearby below ground burrow galleries or roots [56], which could not be seen during the measurements. Since rH values in the BWS were not always more oxidized compared to plain sediment we reject our second hypothesis. The reduction potential of the sediment is further affected by different environmental and biotic factors such as water logging, tree roots and age of the forest stand [1,18,19,48,49,57].

Temporal pattern in CO_2 release

CO_2 efflux rates of plain surface sediment increased over time. Possibly surface sediment became more oxidized towards the end of the wet season and a change in the microbial decomposition process may thus have resulted in the different CO_2 efflux rates [22]. Another possible reason could be that CO_2, which is normally released over the sediment surface, may have been dissolved after rainfalls and stored in rain water filling sediment pore space. This would have increased the sediment CO_2 concentrations, while less CO_2 would have been released from the sediment, as observed in our study at the beginning of the wet season. Corresponding results have been obtained by studies performed in a temperate pinewood forest [58,59]. In contrast to the CO_2 efflux rates of plain surface sediment, the CO_2 efflux rates of individual burrows did not change over time, despite simultaneously varying rH values. The reasons for this are not clear.

Conclusions

Until now, the ecological role of *Ucides cordatus* has been mostly discussed in terms of carbon retention through litter feeding. The results of the present study emphasize the crabs' importance for carbon export via CO_2 release. Respiration by *U. cordatus* and CO_2 release by its burrows increase the CO_2 efflux rate compared to plain sediment by 15–71% assuming a conservative average density of 1.7 burrows^{-1} m^{-2} for *Rhizophora mangle* dominated forests at the Caeté estuary in North Brazil. This increase in CO_2 efflux rate is particularly high considering

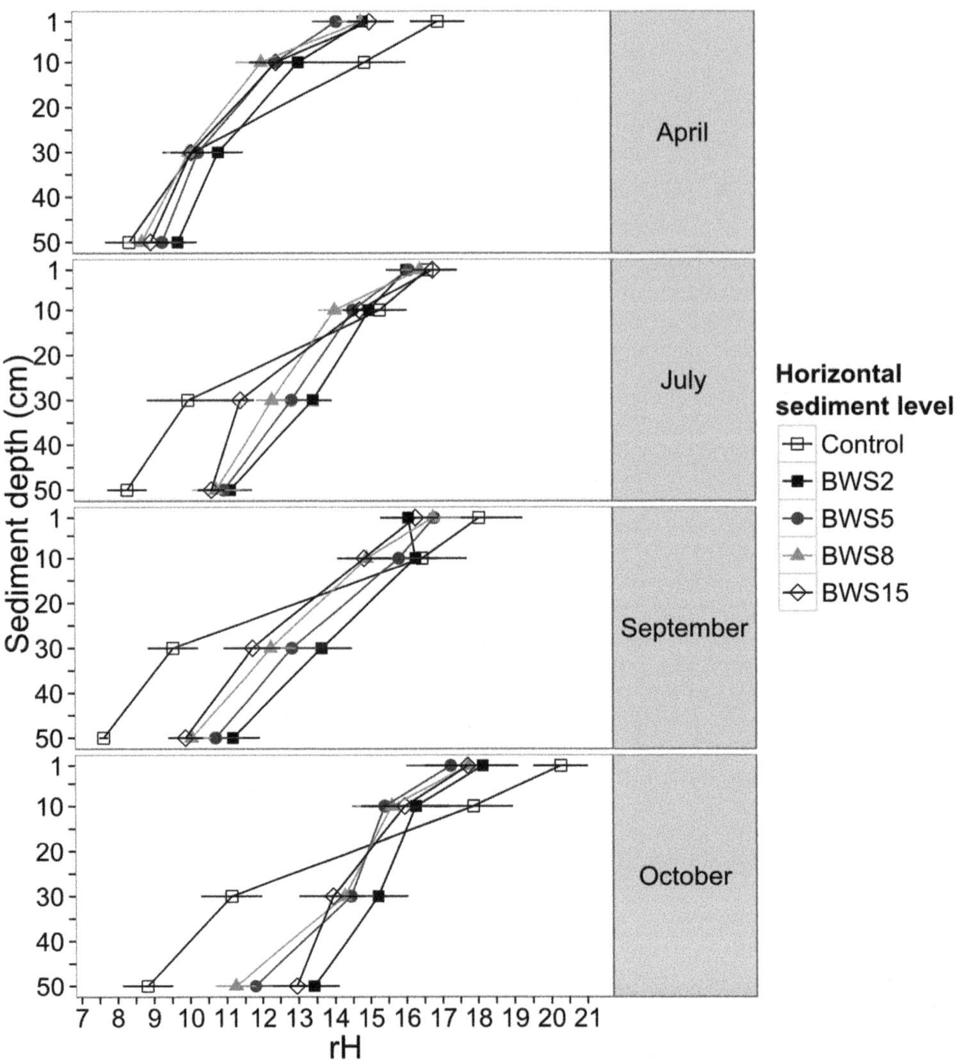

Figure 5. Burrow and plain sediment rH values for each sampling campaign in the year 2012. Mean sediment rH values ± standard error (se) plotted against sediment depth (cm) (sample size (n) for each month: April: n = 308, July: n = 363, September: n = 282 and October: n = 240). Different grey shades and symbols represent the different horizontal sampling distances from the burrow wall sediment (BWS, BWS2 = 2 cm, BWS5 = 5 cm, BWS8 = 8 cm, BWS15 = 15 cm and control >15 cm).

the relatively low proportion (0.3–0.7%) for which crab burrow entrances of *U. cordatus* account for per sediment area. Furthermore, burrows lead to changes in the reduction potential in the sediment and may thereby indirectly influence microbial decomposition pathways and CO_2 release. This study further demonstrates that the reduction potential around the burrow walls underlies spatio-temporal variations which need to be considered to improve estimates of CO_2 release from mangrove sediments. Future studies should disentangle carbon release and carbon storage by crabs and determine whether these two processes are in balance. Moreover, we underline the importance of including CO_2 efflux rates of burrows and respiration rates of crabs into carbon budget estimates of mangrove ecosystems.

Acknowledgments

We thank the Instituto do Estudos Costeiros (IECOS) of the Federal University of Pará (UFPa), Brazil for the support during field and laboratory work. Further we would like to express our thanks to the fishermen and students helping us during this time. We are also grateful to Cyril Piou for providing additional data.

Author Contributions

Conceived and designed the experiments: NP KD UM IN. Performed the experiments: NP. Analyzed the data: NP. Contributed reagents/materials/analysis tools: NP UM. Contributed to the writing of the manuscript: NP KD UM IN.

References

1. Kristensen E (2008) Mangrove crabs as ecosystem engineers; with emphasis on sediment processes. J Sea Res 59: 30–43.
2. Smith III TJ, Boto KG, Frusher SD, Giddins RL (1991) Keystone species and mangrove forest dynamics: the influence of burrowing by crabs on soil nutrient status and forest productivity. Estuar Coast Shelf Sci 33: 419–432.
3. Smith NF, Wilcox C, Lessmann JM (2009) Fiddler crab burrowing affects growth and production of the white mangrove (*Laguncularia racemosa*) in a restored Florida coastal marsh. Mar Biol 156: 2255–2266.
4. Lee SY (1998) Ecological role of grapsid crabs in mangrove ecosystems: a review. Mar Freshw Res 49: 335–343.
5. Lee SY (2008) Mangrove macrobenthos: Assemblages, services, and linkages. J Sea Res 59: 16–29.
6. Andreetta A, Fusi M, Cameldi I, Cimò F, Carnicelli S, et al. (2014) Mangrove carbon sink. Do burrowing crabs contribute to sediment carbon storage? Evidence from a Kenyan mangrove system. J Sea Res 85: 524–533.
7. Kristensen E, Penha-Lopes G, Delefosse M, Valdemarsen T, Quintana CO, et al. (2012) What is bioturbation? The need for a precise definition for fauna in aquatic sciences. Mar Ecol Prog Ser 446: 285–302.
8. Kristensen E (2000) Organic matter diagenesis at the oxic/anoxic interface in coastal marine sediments, with emphasis on the role of burrowing animals. Hydrobiologia 426: 1–24.
9. Robertson AI (1986) Leaf-burying crabs: Their influence on energy flow and export from mixed mangrove forests (*Rhizophora* spp.) in northeastern Australia. J Exp Mar Bio Ecol 102: 237–248.
10. Robertson AI, Daniel PA (1989) The influence of crabs on litter processing in high intertidal mangrove forests in tropical Australia. Oecologia 78: 191–198.
11. Emmerson WD, McGwynne LE (1992) Feeding and assimilation of mangrove leaves by the crab *Sesarma meinerti* de Man in relation to leaf-litter production in Mgazana, a warm-temperate southern African mangrove swamp. J Exp Mar Bio Ecol 157: 41–53.
12. Twilley RR, Pozo M, Garcia VH, Rivera-Monroy VH, Zambrano R, et al. (1997) Litter dynamics in riverine mangrove forests in the Guayas River estuary, Ecuador. Oecologia 111: 109–122.
13. Nordhaus I, Wolff M, Diele K (2006) Litter processing and population food intake of the mangrove crab *Ucides cordatus* in a high intertidal forest in northern Brazil. Estuar Coast Shelf Sci 67: 239–250.
14. Schories D, Barletta-Bergan A, Barletta M, Krumme U, Mehlig U, et al. (2003) The keystone role of leaf-removing crabs in mangrove forests of North Brazil. Wetl Ecol Manag 11: 243–255.
15. Katz LC (1980) Effects of burrowing by the fiddler crab, *Uca pugnax* (Smith). Estuar Coast Mar Sci 11: 233–237.
16. Revsbech NP, Sørensen J, Blackburn TH, Lomholt JP (1980) Distribution of oxygen in marine sediments measured with microelectrodes. Limnol Oceanogr 25: 403–411.
17. Michaels RE, Zieman JC (2013) Fiddler crab (*Uca* spp.) burrows have little effect on surrounding sediment oxygen concentrations. J Exp Mar Bio Ecol 448: 104–113.
18. Gribsholt B, Kostka JE, Kristensen E (2003) Impact of fiddler crabs and plant roots on sediment biogeochemistry in a Georgia saltmarsh. Mar Ecol Prog Ser 259: 237–251.
19. Nielsen OI, Kristensen E, Macintosh DJ (2003) Impact of fiddler crabs (*Uca* spp.) on rates and pathways of benthic mineralization in deposited mangrove shrimp pond waste. J Exp Mar Bio Ecol 289: 59–81.
20. Kristensen E, Bouillon S, Dittmar T, Marchand C (2008) Organic carbon dynamics in mangrove ecosystems: A review. Aquat Bot 89: 201–219.
21. Canfield DE (1994) Factors influencing organic carbon preservation in marine sediments. Chem Geol 114: 315–329.
22. Leopold A, Marchand C, Deborde J, Chaduteau C, Allenbach M (2013) Influence of mangrove zonation on CO_2 fluxes at the sediment–air interface (New Caledonia). Geoderma 202–203: 62–70.
23. Bouillon S, Borges A V., Castañeda-Moya E, Diele K, Dittmar T, et al. (2008) Mangrove production and carbon sinks: A revision of global budget estimates. Global Biogeochem Cycles 22: 1–12.
24. Kristensen E, Flindt MR, Ulomi S, Borges AV, Abril G, et al. (2008) Emission of CO_2 and CH_4 to the atmosphere by sediments and open waters in two Tanzanian mangrove forests. Mar Ecol Prog Ser 370: 53–67.
25. Araújo Jr JMC, Otero XL, Marques AGB, Nóbrega GN, Silva JRF, et al. (2012) Selective geochemistry of iron in mangrove soils in a semiarid tropical climate: effects of the burrowing activity of the crabs *Ucides cordatus* and *Uca maracoani*. Geo-Marine Lett 32: 289–300.
26. Krause G, Schories D, Glaser M, Diele K (2001) Spatial patterns of mangrove ecosystems: the bragantinian mangroves of northern Brazil (Bragança, Pará). Ecotropica 7: 93–107.
27. Krause G (2010) The geography of the Bragança coastal region. In: Saint-Paul U, Schneider H, editors. Mangrove Dynamics and Management in North Brazil. Springer Berlin Heidelberg, Vol. 211. 19–34.
28. INMET (2013) Available: http://www.inmet.gov.br/portal/index.php?r=estacoes/estacoesConvencionais. Accessed 2013 Jan 15.
29. Diele K, Koch V (2010) Growth and mortality of the exploited mangrove crab *Ucides cordatus* (Ucididae) in N-Brazil. J Exp Mar Bio Ecol 395: 171–180.
30. Diele K, Koch V, Saint-Paul U (2005) Population structure, catch composition and CPUE of the artisanally harvested mangrove crab *Ucides cordatus* (Ocypodidae) in the Caeté estuary, North Brazil: Indications for overfishing? Aquat Living Resour 18: 169–178.
31. Piou C, Berger U, Feller IC (2009) Spatial structure of a leaf-removing crab population in a mangrove of North-Brazil. Wetl Ecol Manag 17: 93–106.
32. Araújo MSLC de, Calado TCS (2011) Burrows architecture of the crab *Ucides cordatus* (LINNAEUS, 1763) (Crustacea, Decapoda, Ucididae) in a mangrove swamp of Brazil. Trop Oceanogr 39: 155–165.
33. Kristensen E, Kostka JE (2005) Macrofaunal Burrows and Irrigation in Marine Sediment: Microbiological and Biogeochemical. In: Kristensen E, Haese RR, Kostka JE, editors. Interactions between Macro– and Microorganisms in Marine Sediments. Washington, DC: American Geophysical Union. 125–157.
34. Curran HA, Martin AJ (2003) Complex decapod burrows and ecological relationships in modern and Pleistocene intertidal carbonate environments, San Salvador Island, Bahamas. Palaeogeogr Palaeoclimatol Palaeoecol 192: 229–245.
35. LI-COR Biosciences (2010) LI-8100 Instruction Manual. Rev. 1.0.
36. Diele K (2000) Life history and population structure of the exploited mangrove crab *Ucides cordatus* (Linnaeus, 1763) (Decapoda: Brachyura) in the Caeté estuary, North ZMT Contribution 9, University Bremen, Germany.
37. Pöpel F (2000) Lehrbuch für Abwassertechnik und Gewässerschutz. Müller Jur.Vlg.C.F.
38. Zuur AF, Ieno EN, Walker N, Saveliev AA, Smith GM (2009) Mixed Effects Models and Extensions in Ecology with R. New York: Springer.
39. Zuur AF, Ieno EN, Elphick CS (2010) A protocol for data exploration to avoid common statistical problems. Methods Ecol Evol 1: 3–14.
40. R Core Team (2012) R: A Language and Environment for Statistical Computing. Vienna, Austria.
41. Pinheiro JC, Bates DM, DebRoy S, Sarkar D, R Core Team (2012) nlme: Linear and nonlinear mixed effects models: 1–102.
42. Wickham H (2009) ggplot2: Elegant graphics for data analysis.
43. Pinheiro JC, Bates DM (2000) Mixed effects models in S and S-Plus. New York: Springer.
44. Zuur AF, Ieno EN, Smith GM (2007) Analysing Ecological Data. New York: Springer.
45. Diele K, Koch V (2010) Comparative population dynamics and life histories of North Brazilian mangrove ccabs, genera *Uca* and *Ucides* (Ocypodoidea). In: Saint-Paul U, Schneider H, editors. Mangrove Dynamics and Management in North Brazil. Springer Berlin Heidelberg, Vol. 211. 275–285.
46. Thongtham N, Kristensen E (2005) Carbon and nitrogen balance of leaf-eating sesarmid crabs (*Neoepisesarma versicolor*) offered different food sources. Estuar Coast Shelf Sci 65: 213–222.
47. Lovelock CE (2008) Soil respiration and belowground carbon allocation in mangrove forests. Ecosystems 11: 342–354.
48. Alongi DM, Tirendi F, Trott LA, Xuan TT (2000) Benthic decomposition rates and pathways in plantations of the mangrove *Rhizophora apiculata* in the Mekong delta, Vietnam. Mar Ecol Prog Ser 194: 87–101.
49. Alongi DM, Sasekumar A, Tirendi F, Dixon P (1998) The influence of stand age on benthic decomposition and recycling of organic matter in managed mangrove forests of Malaysia. J Exp Mar Bio Ecol 225: 197–218.
50. Kristensen E, Andersen FØ, Holmboe N, Holmer M, Thongtham N (2000) Carbon and nitrogen mineralization in sediments of the Bangrong mangrove area, Phuket, Thailand. Aquat Microb Ecol 22: 199–213.
51. Nordhaus I, Wolff M (2007) Feeding ecology of the mangrove crab *Ucides cordatus* (Ocypodidae): food choice, food quality and assimilation efficiency. Mar Biol 151: 1665–1681.
52. Alongi DM, Carvalho NA, Amaral AL, da Costa A, Trott L, et al. (2012) Uncoupled surface and below-ground soil respiration in mangroves: implications for estimates of dissolved inorganic carbon export. Biogeochemistry 109: 151–162.
53. Xin P, Jin G, Li L, Barry DA (2009) Effects of crab burrows on pore water flows in salt marshes. Adv Water Resour 32: 439–449.
54. Susilo A, Ridd P V., Thomas S (2005) Comparison between tidally driven groundwater flow and flushing of animal burrows in tropical mangrove swamps. Wetl Ecol Manag 13: 377–388.

55. Heron SF, Ridd PV (2008) The tidal flushing of multiple-loop animal burrows. Estuar Coast Shelf Sci 78: 135–144.
56. Ferreira TO, Otero XL, de Souza Junior VS, Vidal-Torrado P, Macías F, et al. (2010) Spatial patterns of soil attributes and components in a mangrove system in Southeast Brazil (São Paulo). J Soils Sediments 10: 995–1006.
57. Marchand C, Baltzer F, Lallier-Vergès E, Albéric P (2004) Pore-water chemistry in mangrove sediments: relationship with species composition and developmental stages (French Guiana). Mar Geol 208: 361–381.
58. Maier M, Schack-Kirchner H, Hildebrand EE, Schindler D (2011) Soil CO_2 efflux vs. soil respiration: Implications for flux models. Agric For Meteorol 151: 1723–1730.
59. Maier M, Schack-Kirchner H, Hildebrand EE, Holst J (2010) Pore-space CO_2 dynamics in a deep, well-aerated soil. Eur J Soil Sci 61: 877–887.

Egg Size Effects across Multiple Life-History Stages in the Marine Annelid *Hydroides diramphus*

Richard M. Allen*¤a, Dustin Marshall¤b

School of Biological Sciences, University of Queensland, St Lucia, Queensland, Australia

Abstract

The optimal balance of reproductive effort between offspring size and number depends on the fitness of offspring size in a particular environment. The variable environments offspring experience, both among and within life-history stages, are likely to alter the offspring size/fitness relationship and favor different offspring sizes. Hence, the many environments experienced throughout complex life-histories present mothers with a significant challenge to optimally allocate their reproductive effort. In a marine annelid, we tested the relationship between egg size and performance across multiple life-history stages, including: fertilization, larval development, and post-metamorphosis survival and size in the field. We found evidence of conflicting effects of egg size on performance: larger eggs had higher fertilization under sperm-limited conditions, were slightly faster to develop pre-feeding, and were larger post-metamorphosis; however, smaller eggs had higher fertilization when sperm was abundant, and faster planktonic development; and egg size did not affect post-metamorphic survival. The results indicate that egg size effects are conflicting in *H. diramphus* depending on the environments within and among life-history stages. We suggest that offspring size in this species may be a compromise between the overall costs and benefits of egg sizes in each stage and that performance in any one stage is not maximized.

Editor: Peter P. Fong, Gettysburg College, United States of America

Funding: This project was funded by Australian Research Council (DP0556552) to Dustin J. Marshall. R.M.A. was supported by an Australian Post-Graduate Award. The funders had no role in study design, data collection and analysis, decision to publish, or preparation of the manuscript.

Competing Interests: The authors have declared that no competing interests exist.

* Email: richard.allen@uqconnect.edu.au

¤a Current address: Department of Oceanography, Dalhousie University, Halifax, Nova Scotia, Canada
¤b Current address: School of Biological Sciences, Monash University, Clayton, Victoria, Australia

Introduction

Understanding the evolutionary ecology of offspring size is a central component of life-history theory. Larger offspring typically have higher fitness (e.g., in plants [1], [2], invertebrates [3], [4], and vertebrates [5], [6], [7]), but because the resources available for reproduction are limited, mothers must maximize their own fitness by balancing the benefits of producing larger, fitter offspring with the costs of decreased fecundity [8], [9]. A-single, optimal offspring size however is unlikely for any given species, because the strength and direction of the offspring size-fitness relationship is mediated by the environmental conditions experienced by the offspring [9], [10], [11], [12], [13], [14], [15], [16]. In some environments, offspring size effects on performance are positive and strong, and so selection favors mothers that increase their own fitness by producing larger offspring [14]. In other environments, selection can be positive but weak, absent, or negative; in these environments, selection will favor mothers that produce smaller, but more numerous offspring [17], [18], [19]. Such environmentally-driven variation in the relationship between offspring size and offspring fitness presents mothers with a significant challenge to optimally allocate their reproductive effort, and can lead to significant variation in offspring size within a population [20].

The challenges of optimally provisioning offspring are exacerbated in species with complex life-histories. Individuals can experience very different environments as propagules, larvae, and adults; and, the different environments encountered by offspring within each life-history stage may favor a different offspring size [21] (but see [22], [23], [24]; and [25] for other traits). More generally, complex life-cycles can allow organisms to adapt independently to the ecologically distinct environments of each stage [26], but offspring size is not a reversible trait: once an offspring is independent of its mother, the mother cannot reclaim any unused resources, nor supplement any deficiencies. For example, there is conflicting selection on seed size in oak seedlings: larger seeds have higher fitness during early ontogeny (such as germination rate, and seedling growth and survival), but suffer higher post-dispersal mortality due to size-selective predation [23]. Similar examples are found in salmon, frogs and marine invertebrates [18], [21], [24]. The optimal offspring size for a given population will therefore be a compromise between offspring size effects across the entire life-history, but may be constrained by highly sensitive stages or those under the most intense selection [24], [27].

Critical to the argument that conflicting pressures across multiple life-history stages can shape an optimal offspring size is the assumption that initial differences in offspring size can persist to affect later life-history stages [28]. In species with a short and non-feeding larval stage, we see evidence that offspring size and maternal effects can persist into adult life, and evidence of variable or conflicting effects on offspring size [23], [28], [29]. However, offspring size and other maternal effects can dissipate over the course of development and contribute little to adult fitness (e.g., [30], [31], [32]). Failure to detect these effects in later stages is often attributed to the environment contributing more to subsequent adult phenotypes than maternal effects [33], [34]. This appears to be particularly true for species with feeding developmental stages, because variation in the nutritional environment experienced by the offspring can correlate with adult nutrition and phenotypes, and overwhelm egg size effects [35]. Furthermore, small offspring may supplement deficiencies in maternal provisioning by prolonging development or accelerating growth/feeding rates to acquire more exogenous resources and achieve comparable size at maturity to larger offspring ([3], [17], [36], [37] [38] [39]. Although a feeding larval stage is common in most animal taxa (e.g., most insects, amphibians, fish, and marine invertebrates), few studies have determined the persistence of offspring size effects in species with this type of life-history. In part, this may reflect logistical difficulties in tracking individuals with complex life-cycles across different stages, so studies that examine conflicting selection on offspring size are exceedingly rare in species with feeding larvae.

Here, we examine the effects of natural variation in egg size on performance across multiple life-history stages in *Hydroides diramphus*. *Hydroides diramphus* has a complex life-cycle consisting of egg, embryonic, larval, and post-metamorphic stages. Experimentally, we test to see if egg size effects can persist across multiple life-history stages, and in doing so, we identify the potential range of life-history stages that may be influenced egg size-performance relationships. Additionally, we examine whether egg size influences proxies of offspring performance among each stage of these stages to describe when different offspring sizes should be favoured.

Materials and Methods

General Methods

Study species and site. *Hydroides diramphus* is a polychaete tubeworm found in benthic marine assemblages and is a non-native species in Australia. All collections and field experiments were conducted between May and September 2008 from the floating docks at the Scarborough Marina, Redcliffe, Queensland, Australia (27° 10' 45" S, 153° 06' 18" E). The floating docks are privately owned and managed by the Moreton Bay Boat Club, and permission was granted by club management for the collection of specimens and complete experiments. Worms were collected and transported to the laboratory in thermally insulated tanks for spawning and larval culture.

In *H. diramphus* the sexes are separate, and fertilization occurs externally. Generally, the fertilization environment is highly variable in natural populations and is strongly influenced by proximity to mates [40]. Many marine invertebrates spawn synchronously, and at high densities fertilization can be as high as 100% in some species [41], [42], [43]. In *H. diramphus*, individuals can touch or be millimeters apart at high densities, but also be meters apart at low population densities (Allen personal observations). Therefore, natural fertilization in *H. diramphus* likely occurs across a range of sperm concentrations from very dense when female and male tubes overlap to approaching zero where mates are far apart. Importantly, the fertilization environment can be a selective pressure on egg size in marine invertebrates [27]. Because larger eggs present a larger target to sperm, under low-sperm concentrations, larger eggs have higher fertilization success; but at high-sperm concentrations, large eggs can suffer polyspermy: a fatal condition that results from eggs being fertilized by multiple sperm [27], [46].

Once fertilized, *H. diramphus* embryos develop for 12–14 hrs and hatch into obligate feeding (planktotrophic) larvae. Under natural conditions, mortality is generally high during embryonic and larval development, and only a small fraction of the number of offspring produced by mothers typically recruit back to adult populations [45]. In other taxa, size-dependent mortality during development has been documented (e.g., [46], [47]), but in marine invertebrates, size dependent mortality has not been explored. More generally, the specific sources of mortality remain largely unknown in marine invertebrates, but larval mortality is usually reported as an instantaneous rate function of the time spent in the plankton [45], [48]. Importantly, the relationship between offspring size and development time has strong theoretical foundations derived from allometric and kinetic principles (e.g., [49]) and variable developmental relationships among marine taxa have been demonstrated by empirical studies [4]. Hence, the offspring size-development time relationship is a good proxy for probability of planktonic survival in marine invertebrate larvae [50].

Once *H. diramphus* larvae acquire enough resources during planktotrophic development, larvae become competent to settle and metamorphose into the adult environment. In the post-metamorphic environment, egg size may also be under selection, because other studies on marine invertebrates with different life-histories have shown that offspring size (as eggs or larvae) can influence growth, competitive ability, survival and reproductive output [4], [51], [52].

Obtaining gametes and larval culture techniques. In the laboratory, gametes were obtained using a strip spawning technique [53]. Worms were gently removed from tubes and placed in 30 mm diameter petri dishes with <5 ml of sterilized seawater (seawater microwaved at 1000 watts for 3 min past boiling and allowed to return to ambient temperature). Reproductively mature individuals released gametes immediately. Fertilization techniques varied with experiments (see below for details), but generally, eggs were placed in a beaker with a few drops of fresh sperm obtained from males immediately prior to fertilization. To ensure maximum fertilization and avoid sperm concentration selecting for particular egg sizes, the drops of sperm were added in a stepwise manner at 10–15 min intervals so that concentration of sperm gradually increased, but also allowed time for the polyspermic block to take place. At 22°C, trochophore larvae hatch from developing embryos at approximately 12–14 hours. Trochophore larvae were reared at concentrations of 50 larvae ml^{-1}, and fed *Isocrysis galbana* (Tahitian strain) in standing cultures at concentrations of 100,000 cells ml^{-1}. Fresh, sterilized seawater and phytoplankton were replaced every second day. Competency to settle and metamorphose was reached between 6 and 18 days after fertilization, and was determined by morphological examination of larvae [54].

Measurements. Egg size was used to estimate maternal investment. We recognize that egg size does not correlate perfectly with maternal investment in all species [55]; however, among annelids at least, egg size does correlate with the energy content of eggs [56]. Furthermore, three additional selective mechanisms may operate on egg size independently of energy content: 1) mothers face a physical constraint of brood-space [4]; 2) fertilization success is determined by physical egg size (outlined above, [27], [57]; and 3) egg size is a trait typically related with development mode and other life-history traits under selection in marine invertebrates and other taxa [3], [4], [6].

To measure eggs and larvae, samples were fixed in 5 ml seawater with a few drops of 12% formalin. Eggs and larvae were photographed with a compound microscope and camera (PixeLINK Capture SE Ver. 1.0). We measured the two-dimensional cross-sectional area of eggs and pre-feeding trocophore larvae (lateral view). Competent larvae were also measured as the cross-sectional area along a longitudinal orientation. The calcareous tubes that *Hydroides* secrete and inhabit are enlarged as juveniles develop to accommodate growth. Hence, tube length is a good indicator of worm size and growth in *Hydroides* species [58], [59]. The relationship between tube length and worm size is strongest after 9 days in similar *Hydroides* species [58], so at 10 days in the field, worms were returned to the laboratory and tube length was measured under a dissecting microscope. Maternal worms were measured out of their tubes as length from the most anterior part of the collar to the most posterior tip of the abdomen. All traits on digital images were measured with Image-Pro express version 5.1.

A natural source of egg size variation: the relationship between egg size and maternal size. The first goal of the study was to describe the existing amount of variation in egg size for the *H. diramphus*. There are many potential sources of variation in offspring size that are important to many aspects of life-history theory and ecological relevance, and are correlated with maternal traits [60]. One of the more common maternal effects on offspring size is the relationship between maternal size and offspring size [4], [60]. To describe the standing variation in egg size, and investigate whether maternal size is a natural source of egg size variation, we examined the relationship between maternal size and egg size. We spawned and measured 26 females, and sampled 25 eggs from each mother. The average egg size per mother was used as the unit of replication to estimate the relationship between maternal size and egg size.

Egg size effects during early development

The aim of the following experiment was to investigate how egg size and the fertilization environment mediate the maternal-offspring size relationship and affects fertilization success. A pilot study revealed the relationship between sperm concentration and fertilization success conformed to the fertilization curve found in other marine invertebrates (e.g., urchins, [57]: and polychaetes, [61]). We found fertilization success peaked at $\sim 23 \times 10^6$ sperm ml^{-1} with 88% of eggs fertilized, and at a sperm concentration of ~ 23 sperm ml^{-1}, fertilization success was 3.5% (Allen RM unpublished data). We used these two sperm concentrations to represent the extremes in selection pressure due to fertilization success. Sperm concentrations were determined by three replicate counts of sperm samples using a haemocytometer. In polychaetes, unfertilized eggs may cleave and polyspermic embryos may also continue to develop, but neither unfertilized nor polyspermic embryos develop past hatching. Hence, larval size of newly hatched larvae was the response variable of interest and a proxy for the size of eggs that were successfully fertilized (because larger eggs produce larger larvae; see *Results: Egg size effects during early development*). Twenty-three mothers were spawned, and the broods split into two separate petri dishes; then the fertilized the eggs from each mother at either the 'high' or 'low' sperm concentrations. At hatching, we measured the size of larvae from each mother and sperm concentration. Maternal size was also recorded as a proxy for egg size, and each mother was the unit of replication.

The aim of the next experiment was to investigate the effects of egg size on early development time, using time-lapse photography to measure early cleavage rates. We placed a sample of fertilized embryos in a 30 mm diameter petri dish under a dissector microscope. One image was taken every minute for 90 minutes. We followed the development of individually measured eggs from first cleavage (2-cell stage) to second cleavage (4-cell stage) to estimate early development time. We did not measure from time of fertilization to subsequent stages because we could not account for variation in exact moment each egg was fertilized, and we did not measure past the 4-cell stage because differentiating further stages was unreliable at the image resolution.

We then tested whether egg size affected the size of larvae. Twenty-six mothers were spawned and a sample of 25 eggs from each brood measured; then the remaining eggs fertilized in separate petri dishes (one dish per brood). As described in the *Obtaining gametes and larval culture techniques* section, we avoided selecting against particular egg size by fertilizing eggs in the stepwise manner described above. We allowed the embryos to develop to hatching, where we then fixed and measured a sample of 25 newly hatched larvae from each dish.

Egg size effects during planktonic development

Measuring a single egg and following that individual through multiple life-history stages is logistically very difficult to replicate in large numbers in most free-spawning marine invertebrates, because of their small size, and sensitivity to laboratory culture. Hence, instead of using maternal and egg size as covariates, we grouped mothers into 'large' and 'small' size classes that would enable the manipulation of egg size at the level of culture jar to investigate maternal size and egg size effects during planktonic development. To achieve this, we spawned 60 mothers, and ranked them roughly from largest to smallest. We then omitted the six mothers closest to the median in order to reduce any overlap in egg sizes. The eggs from large and small mothers were pooled within each size treatment, then immediately split into experimental jars. By grouping mothers into size classes and removing intermediately sized mothers, we improved the reliability that the large maternal size class contained large eggs. Approximately, 75% of the variation in egg sizes among replicate jars at the beginning of the experiments that used this procedure (see *Results*).

The goal of the next experiment was to investigate the effect of maternally derived egg size variation on planktonic development time to settlement. Like most species with a planktotrophic larval stage, *H. diramphus* larvae must pass through an obligate feeding stage where sufficient resources are required to settle and metamorphose. As described above, eggs from 60 mothers were sorted into size classes and split into 20 experimental jars (10×600 ml beakers of each size class) containing 250 ml of sterilized seawater. We fertilized the eggs in each jar with sperm pooled from five males; again, using step-wise fertilization to avoid selecting for particular egg sizes (*Obtaining gametes and larval culture techniques*). The embryos developed to hatching, after which the water was changed, and the concentration of larvae adjusted to ~ 50 larvae ml^{-1} across all jars. Phytoplankton was added to each jar (*Isocrysis* 100,000 cells ml^{-1}) and the larvae were allowed to develop. To ensure all larvae remained in the culture during water changes, we placed a 25 μm cylindrical-filter into the culture vessel, and from inside the filter, we pippetted, then discarded, the waste water. The front of the filter was thoroughly rinsed to retain all larvae within the culture. Settlement was measured as the proportion of larvae that had settled every three days for fifteen days, at which time almost all larvae settled (see *Results*). In some circumstances, marine invertebrate larvae may have a prolonged planktonic period because suitable sites for sedentary adult life are rare, or food limitation slows larval development (e.g., [62], [63]). Both instances could affect the size

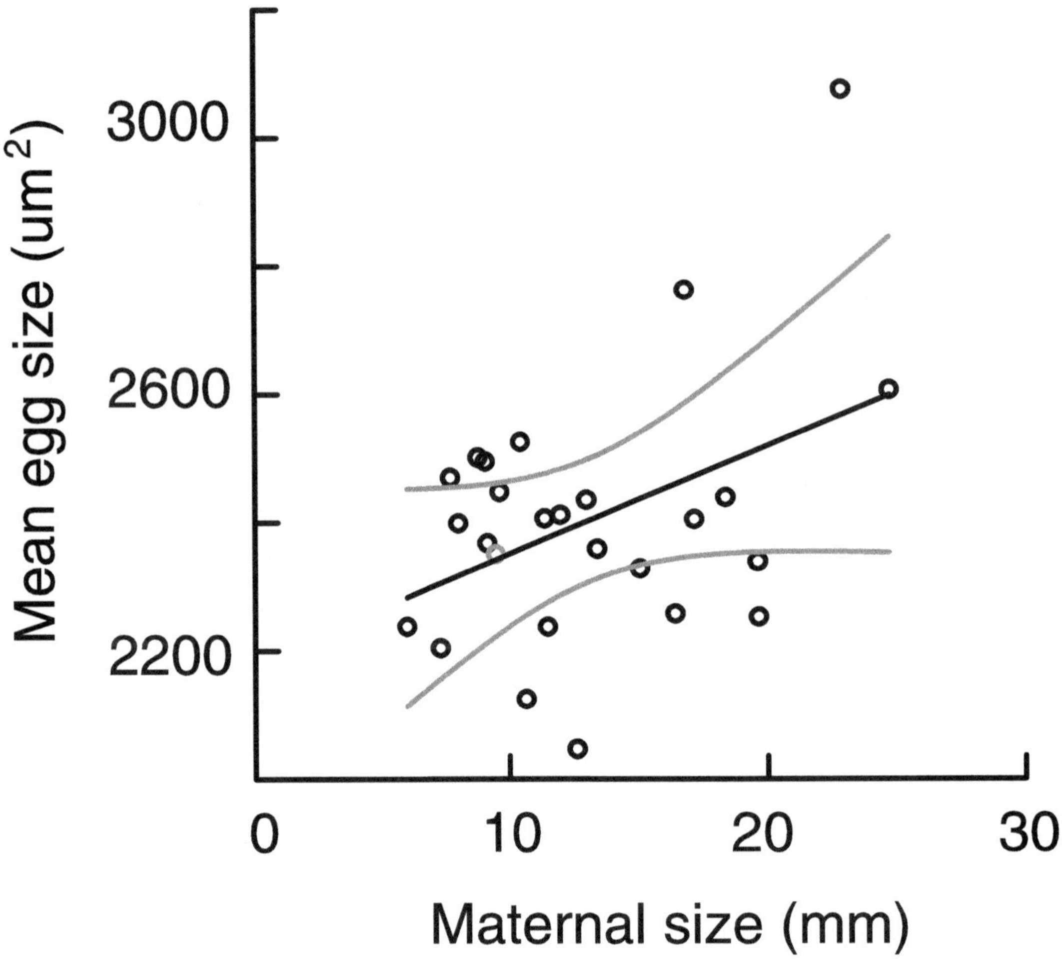

Figure 1. The relationship between maternal worm length and average egg area per mother (*P*=0.034, R^2=0.175) in *Hydroides dirampus.* Solid lines represent significant line of best fit; dotted lines represent 95% confidence intervals.

of larvae at settlement (e.g., [62], [63]). Ample food was provided (at a concentration above what larvae can consume before food was replaced) and *H. dirampus* larvae readily settle on surfaces with a biofilm, so settlement represents a 'best case' scenario where larvae can settle and metamorphose at a time that suits them.

Egg size effects on post-metamorphic performance

To examine the effect of egg size on performance after metamorphosis, we repeated the above procedures for spawning, fertilization, larval feeding and sorting of eggs into two size classes, but in this experiment eggs were split into 16 replicate jars (8 jars of 'large' and 'small' eggs each).

After 10 days, a high proportion of larvae were competent to settle, so the larvae from each experimental jar were transferred into corresponding petri dishes for settlement (60 mm diameter; 16 jars, 16 dishes). Dishes were pre-roughened and conditioned in natural seawater for at least three days prior to settlement, because many marine invertebrates prefer to settle on a roughened surface with a biofilm [64]. Once larvae were added, dishes were left for 48 hrs to allow larvae to settle and metamorphose. Once settled, ~20 larvae per dish were marked by circling individuals with a graphite pencil. Any extra settlers or unsettled larvae were haphazardly culled and discarded. A 6 mm hole was drilled in the centre of each dish, and the dishes were transported to the field where they were fixed to a 500×500 mm backing panel and hung one meter below the floating docks. Dishes were left in the field for 10 days, after which dishes were returned to the lab, and scored for survival and tube-length was measured. All surviving worms were measured, and the average length of worms per dish was the unit of replication.

Data analysis

All data were analyzed using Systat ver. 11 and checked for the assumptions of each statistical test. Where data sets were not normally distributed, they were transformed as indicated for each test below. Where no transformation is indicated, data did not require transformation. Data are available in supplementary File S1.

We analyzed the relationship between maternal body size and average egg size; maternal size and coefficient of variation in egg size; egg size and larval size; and egg size and early development time with linear regression. We also checked whether sorting eggs into size class based on maternal size were significantly different with a one-way ANOVA.

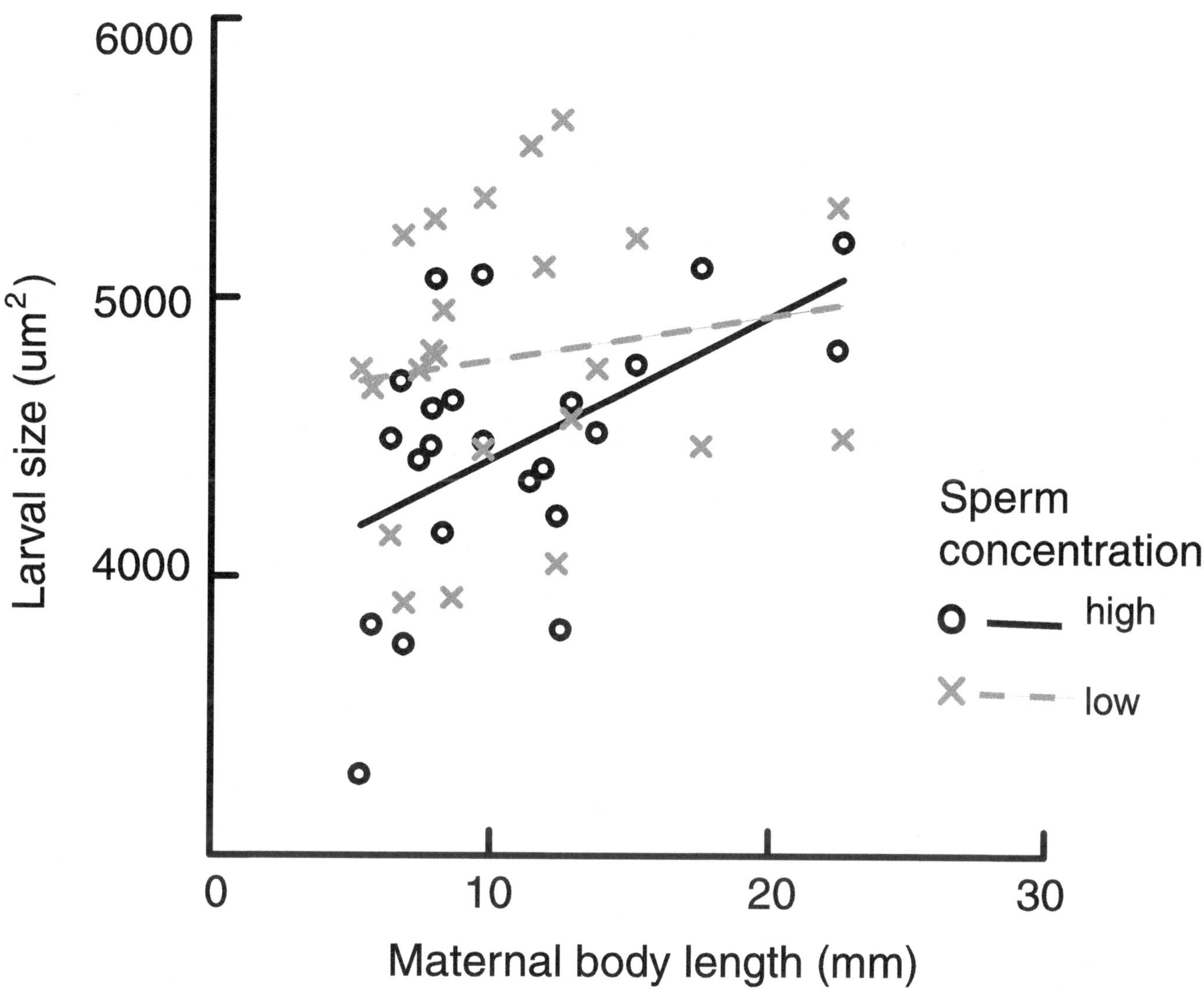

Figure 2. The effect of sperm concentration and maternal size on the size of larvae that complete development to hatching in *Hydroides diramphus*. Low and high sperm concentrations were ~23 sperm ml^{-1} (× and dotted line) and ~23×10^{6} sperm ml^{-1} (○ and solid line), respectively.

The effect of sperm concentration on the size of larvae that hatched was analyzed with ANCOVA where sperm concentration was the categorical variable and maternal size (as a proxy for egg size) was the covariate. Average larval size per dish was the response variable.

We analyzed the effect of larval size on planktonic development time with repeated measures ANOVA, where larval size was a fixed factor, and jar was the unit of replication. The proportion of larvae settled was the response variable, transformed by $\log_{10}$ to normalize the distribution. The Greenhouse-Geisser eigenvalue was low, and therefore the Greenhouse-Geisser adjusted *P*-value was reported as the test of significance because it is a more conservative test [65].

The effect of maternal size class on post-metamorphic size and survival was analyzed with a one-way ANOVA. The post-metamorphic size data distribution was skewed, and was square-root transformed for analysis. The proportional survival data were also not normally distributed, and were arc-sin square-root transformed for analysis.

Results

Relationships between maternal size and offspring size

Mothers in the study were on average 13.1 mm in length (±5 mm SE; CV =0.382). Among mothers, the mean egg size was 2403.22 μm^2 (measured as 2-D surface area; ±202.3 μm^2 SE). The coefficient of variation in egg size was 0.084 among all mothers (based on average egg size per mother), and the average CV within each mother was 0.101. Maternal body length was positively correlated with average egg area ($F_{1,\ 24}=5.083$, $P=0.034$, $R^2=0.175$). On average, doubling the length of mothers from 10 to 20 mm resulted in an approximately 11.4% increase in the area of the eggs produced by mothers (Figure 1).

Egg size effects during early development

The sperm environment in which eggs were fertilized significantly affected the size of newly hatched larvae, where larvae from eggs fertilized under low-sperm conditions had an average cross-sectional area that was ~10% larger than the larvae from eggs fertilized under high-sperm conditions ($F_{1,\ 42}=4.255$, $P=0.045$;

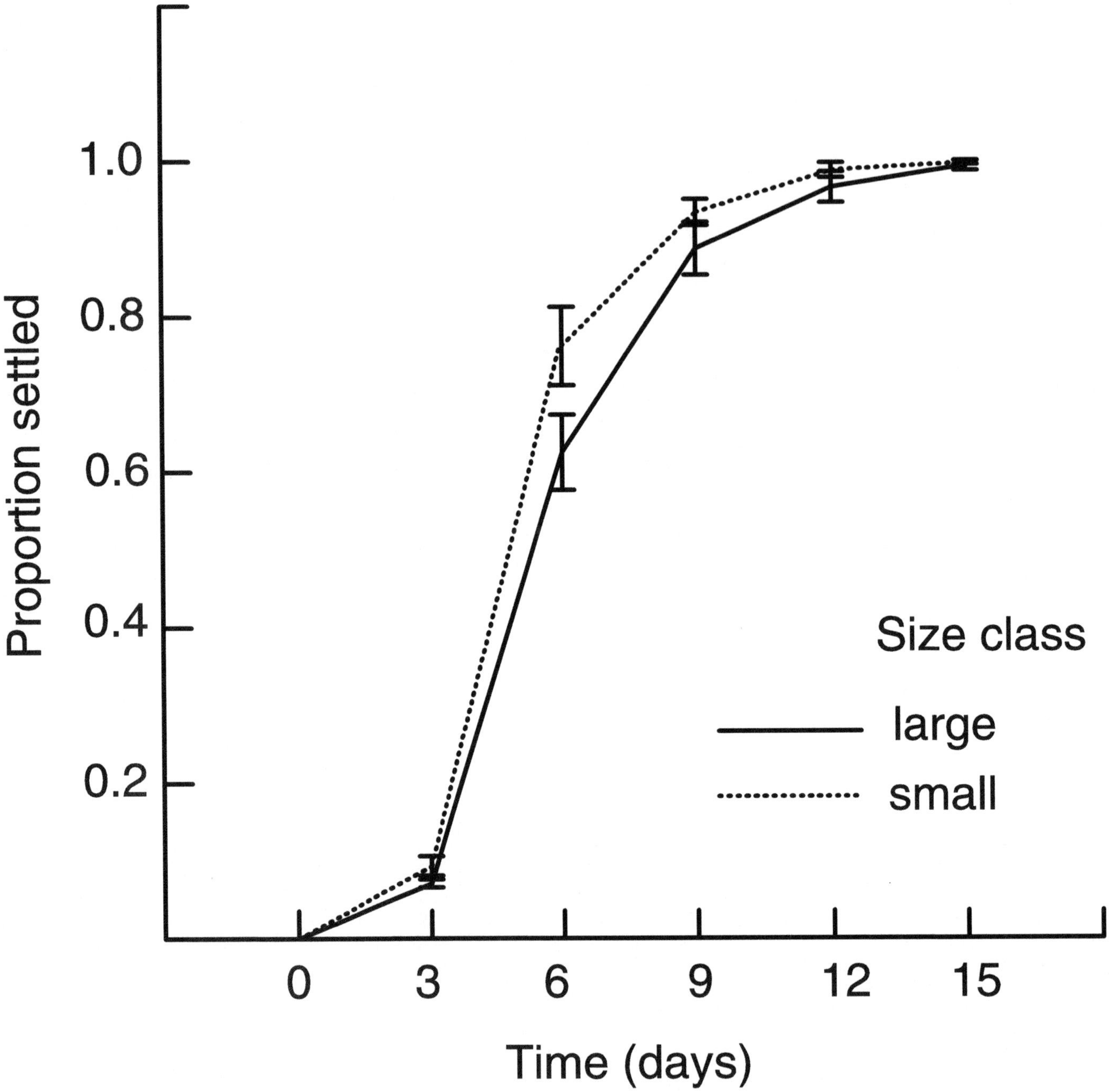

Figure 3. The relationship for larval settlement over fifteen days under conditions where *Hydroides dirampus* larvae were fed at concentrations of 100,000 cells ml^{-1} *Isocrysis*. Two treatments of larval size were applied: solid line represents the proportion of settlement of large larvae and the dashed line represents the proportion of settlement of small larvae. Error bars are ± SE.

Figure 2). Larval size was positively correlated with initial egg size (see below), therefore, the differences in larval size between treatments is likely to reflect different fertilization success due to egg size, where larger eggs are fertilized more frequently under low sperm concentrations. Maternal length was positively related to larval area, independent of sperm concentration ($F_{1,\ 42} = 5.390$, $P = 0.025$; Figure 2). There was no interaction between sperm concentration and maternal size ($F_{1,\ 42} = 1.50$, $P = 0.227$).

Once fertilized, larger eggs developed significantly faster than smaller eggs ($F_{1,\ 69} = 5.252$ $P = 0.025$): on average, eggs that were 50 µm in diameter took ~8 min longer to develop from the 2-cell stage to the 4-cell stage, than eggs that were 60 µm in diameter. This effect of egg size on cleavage rates was weak: only 7% of the variation in development time could be explained by egg size ($R^2 = 0.071$).

Larger eggs developed into larger larvae at hatching ($F_{1,\ 23} = 9.665$, $P = 0.005$, $R^2 = 0.296$); although, the eggs from one replicate did not develop into larvae. The overall effect was significant, where on average, an increase in egg area of 25% resulted in a 4.7% increase in the cross-sectional area of the newly hatched trocophore larvae.

Table 1. Repeated measures analysis testing the effect of larval size on the proportion of *Hydroides diramphus* settlement over time.

Source	df	MS	*F*	*P*
Between Jars				
Larval size	1.000	0.006	5.523	0.030
Error	18.000	0.001		
Within Jars (across time)				
Time	4.000	0.274	413.555	<0.001†
Time × larval size	4.000	0.002	2.403	0.114†
Error	72.000	0.001		

†Greenhouse-Geisser adjusted *P*-values: $\varepsilon = 0.429$.
Larvae were fed a constant food source of *Isocrysis*.

Egg size effects during planktonic development

We checked that using maternal size class was an effective way of manipulating egg size variation, and found it was more reliable than maternal size as a covariate: eggs in the large size class were 15% larger in volume than eggs in the small size class ($F_{1,14} = 42.73$, $P<0.001$, $R^2 = 0.753$).

Larval settlement increased over time (Figure 3, Table 1). Initially, settlement was slow with about 10% of larvae settling after three days. Average settlement increased to 70% between days three and six, and by day twelve almost all larvae had settled. The rate of settlement was also significantly affected by larval size, where larvae from smaller eggs settled sooner, and this effect was consistent across all time periods (no interaction between larval size and time on settlement; Table 1).

Egg size effects during post-metamorphosis

After 10 days under field conditions, juveniles in the large size class (coming from large eggs) were on average longer in length than juveniles from the small size class ('large' 1.79 mm±0.23 SE and 'small' 0.92 mm±0.14 SE; $F_{1,14} = 11.569$, $P = 0.004$; Figure 4a). Overall survival of post-metamorphic juveniles was 44% after 10 days in the field, but there was no effect of size class on survival ($F_{1,\ 14} = 1.411$, $P = 0.255$; Figure 4b).

Discussion

The results show *Hydroides diramphus* egg size effects can endure post-settlement and influence performance in early and late life-history stages. But, the effects of egg size on performance were variable both within and among life-history stages. On average, larger eggs had higher fertilization under sperm-limited conditions, were slightly faster to develop prior to feeding, and were larger post-metamorphosis. Considering these positive egg size-dependent performances, it might be expected that larger egg sizes should be favoured. However, smaller eggs had higher fertilization success when sperm was abundant, and a shorter planktonic development period overall, but egg size did not affect

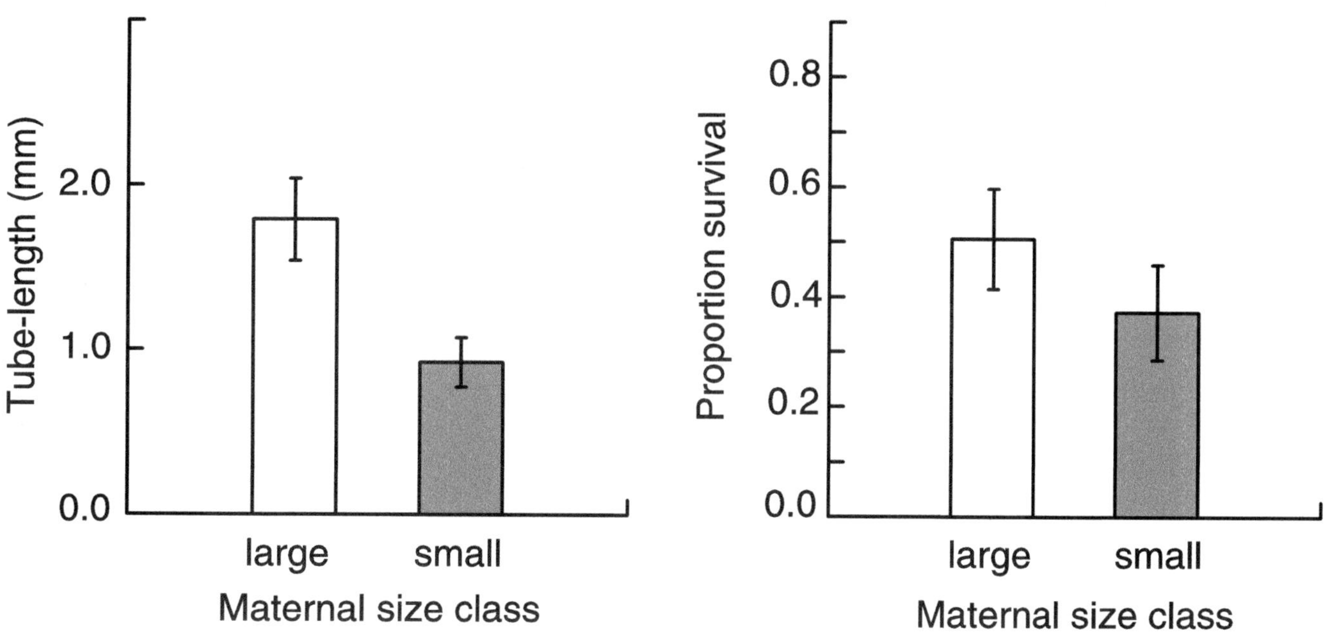

Figure 4. The effect of maternal size class on post-performance after 10 days under field conditions, for the tube worm *Hydroides diramphus*. (A) Post-metamorphic length of worms. (B) Post-metamorphic survival. Data presented untransformed. Error bars are ± SE.

post-metamorphic survival. These negative and neutral effects of increased egg size suggest that in some conditions smaller egg sizes should be favored. Further, in a previous study on *H. diramphus*, egg size has been shown to interact in a highly complex and context-dependent manner with larval nutritional and post-metamorphic density [35]. Hence, egg size effects in *H. diramphus* demonstrate considerable conflicting performance effects depending on the environment and life-history stage that makes an optimal egg size difficult to predict.

High variability in offspring size effects are reported generally (see reviews: [3], [6]), and more specifically in marine invertebrates with similar performance proxies as *H. diramphus* [28], [29], [66]; and reviewed by [4], [60]. During fertilization in broadcast-spawners, egg size had been repeatedly shown to be under strong selection depending on local sperm concentrations [27], [43], [44]. Typically, larger eggs have higher fertilization success in sperm limited environments [4], and the results of the current study confirm this trend.

Development through sensitive embryonic and larval stages is also considered a major source of fitness variation in marine invertebrates, where longer developments times are thought to increase the probability of mortality [8]. Generally, larger embryos and smaller larvae that feed spend longer developing, although this trend is not ubiquitous [4]. The results of the present study demonstrate a negative relationship between egg size and egg cleavage rates, but differ from the general trend for the larval period, as larvae from smaller eggs were faster to settle. Classic life-history theory for marine invertebrates suggest that the relationship between the pre-feeding and feeding development periods should be inversely related [8], but this rests on the assumption that larvae settle at a similar sizes and differences in egg size do not persist post-metamorphosis. The results show that initial egg size was positively related to post-metamorphic size; hence the results for each developmental period and post-metamorphic size suggest the relationship among these traits and offspring size may depend on complicated tradeoffs. Two independent studies on egg size effects in the sea urchin *Strongylocentrous droebachiensis* showed evidence that such a tradeoff may exist. Sinervo and McEdward [36] found larvae from small eggs had longer planktonic periods, and metamorphosed at the same size as larvae from large eggs. But, Hart [67] found no effect of egg size on planktonic period, yet did show a small positive effect of egg size on size at metamorphosis. Explanations for why the results among two different studies on the same species are inconsistent are not known, but may be due to some other mediating environmental factor. Within a single experiment, *H. diramphus* has demonstrated that egg size effects are not straightforward and depended on complex interactions among the larval nutrition and post-metamorphic environments [35]. Other studies have shown that smaller offspring may use compensatory growth mechanisms to achieve comparable sizes at metamorphosis or maturity as larger offspring by prolonging development or accelerating growth/feeding rates [3], [17], [37], [38], [39]. Overall the dynamic and interdependent relationships among egg size and subsequent performance at later stages, demonstrate the need to assess the benefits of egg size and when egg size benefits are conflicting under the context of variable environments across the entire life history of the study organism.

Acknowledgments

Thanks to J. David Aguirre for comments on the manuscript; and the Moreton Bay Boat Club for access to the floating docks.

Author Contributions

Conceived and designed the experiments: RMA DJM. Performed the experiments: RMA. Analyzed the data: RMA. Contributed reagents/materials/analysis tools: RMA DJM. Contributed to the writing of the manuscript: RMA DJM.

References

1. Stanton ML (1984) Seed variation in wild radish - effect of seed size on components of seedling and adult fitness. Ecology 65: 1105–1112.
2. Roach DA, Wulff RD (1987) Maternal effects in plants. Annual Review of Ecology and Systematics 18: 209–235.
3. Fox CW, Czesak ME (2000) Evolutionary ecology of progeny size in arthropods. Annual Review of Entomology 45: 341–369.
4. Marshall DJ, Keough MJ (2008) The evolutionary ecology of offspring size in marine invertebrates. Advances in Marine Biology 53: 1–60.
5. Sinervo B (1990) The evolution of maternal investment in lizards - an experimental and comparative analysis of egg size and its effects on offspring performance. Evolution 44: 279–294.
6. Bernardo J (1996) The particular maternal effect of propagule size, especially egg size: Patterns, models, quality of evidence and interpretations. American Zoologist 36: 216–236.
7. Einum S (2003) Atlantic salmon growth in strongly food-limited environments: Effects of egg size and paternal phenotype? Environmental Biology of Fishes 67: 263–268.
8. Vance RR (1973) On reproductive strategies in marine benthic invertebrates. American Naturalist 107: 339–352.
9. Smith CC, Fretwell SD (1974) Optimal balance between size and number of offspring. American Naturalist 108: 499–506.
10. Brockelman WY (1975) Competition, fitness of offspring, and optimal clutch size. American Naturalist 109: 677–699.
11. Parker GA, Begon M (1986) Optimal egg size and clutch size - effects of environment and maternal phenotype. American Naturalist 128: 573–592.
12. McGinley MA, Temme DH, Geber MA (1987) Parental investment in offspring in variable environments - theoretical and empirical considerations. American Naturalist 130: 370–398.
13. Sibly R, Calow P, Smith RH (1988) Optimal size of seasonal breeders. Journal of Theoretical Biology 133: 13–21.
14. Fox CW (1997) The ecology of body size in a seed beetle, *Stator limbatus*: Persistence of environmental variation across generations? Evolution 51: 1005–1010.
15. Marshall DJ, Cook CN, Emlet RB (2006) Offspring size effects mediate competitive interactions in a colonial marine invertebrate. Ecology 87: 214–225.
16. Allen RM, Buckley YM, Marshall DJ (2008) Offspring size plasticity in response to intraspecific competition: An adaptive maternal effect across life-history stages. American Naturalist 171: 225–237.
17. Berven KA, Chadra BG (1988) The relationship among egg size, density and food level on larval development in the wood frog (*Rana sylvatica*). Oecologia 75: 67–72.
18. Kaplan RH (1992) Greater maternal investment can decrease offspring survival in the frog *Bombina orientalis*. Ecology 73: 280–288.
19. Moran AL, Emlet RB (2001) Offspring size and performance in variable environments: Field studies on a marine snail. Ecology 82: 1597–1612.
20. Johnston TA, Leggett WC (2002) Maternal and environmental gradients in the egg size of an iteroparous fish. Ecology 83: 1777–1791.
21. Hendry AP, Day T, Cooper AB (2001) Optimal size and number of propagules: Allowance for discrete stages and effects of maternal size on reproductive output and offspring fitness. American Naturalist 157: 387–407.
22. Hendry AP, Day T (2003) Revisiting the positive correlation between female size and egg size. Evolutionary Ecology Research 5: 421–429.
23. Gomez JM (2004) Bigger is not always better: Conflicting selective pressures on seed size in *Quercus ilex*. Evolution 58: 71–80.
24. Marshall DJ, Morgan SG (2011) Ecological and evolutionary consequences of linked life-history stages in the sea. Current Biology 21:R718–R725.
25. Schluter D, Price TD, Rowe L (1991) Conflicting selection pressures and life-history trade-offs. Proceedings of the Royal Society of London Series B-Biological Sciences 246: 11–17.
26. Moran NA (1994) Adaptation and constraint in the complex life-cycles of animals. Annual Review of Ecology and Systematics 25: 573–600.

27. Levitan DR (1993) The importance of sperm limitation to the evolution of egg size in marine invertebrates. American Naturalist 141: 517–536.
28. Crean AJ, Monro K, Marshall DJ (2011) Fitness consequences of larval traits persist across the metamorphic boundary. Evolution 65: 3079–3089.
29. Rius M, Turon X, Dias GM, Marshall DJ (2010) Propagule size effects across multiple life-history stages in a marine invertebrate. Functional Ecology 24: 685–693.
30. Heath DD, Fox CW, Heath JW (1999) Maternal effects on offspring size: Variation through early development of chinook salmon. Evolution 53: 1605–1611.
31. Lindholm AK, Hunt J, Brooks R (2006) Where do all the maternal effects go? Variation in offspring body size through ontogeny in the live-bearing fish *Poecilia parae*. Biology Letters 2: 586–589.
32. Wilson AJ, Reale D (2006) Ontogeny of additive and maternal genetic effects: Lessons from domestic mammals. American Naturalist 167:E23–E38.
33. Kinghorn BP (1983) A review of quantitative genetics in fish breeding. Aquaculture 31: 283–304.
34. Mousseau TA, Dingle H (1991) Maternal effects in insect life histories. Annual Review of Entomology 36: 511–534.
35. Allen RM, Marshall DJ (2013) Phenotypic links among life-history stages are complex and context-dependent in a marine invertebrate: interactions among offspring size larval nutrition and postmetamorphic density. Functional Ecology 27: 1358–1366.
36. Sinervo B, McEdward LA (1988) Developmental consequences of an evolutionary change in egg size - an experimental test. Evolution 42: 885–899.
37. Fox CW (1994) Maternal and genetic influences on egg size and larval performance in a seed beetle (*Callosobruchus maculatus*) - Multigenerational transmission of a maternal effect. Heredity 73: 509–517.
38. Podolsky RD, Moran AL (2006) Integrating function across marine life cycles. Integrative and Comparative Biology 46: 577–586.
39. Auer SK, Arendt JD, Chandramouli R, Reznick DN (2010) Juvenile compensatory growth has negative consequences for reproduction in Trinidadian guppies (*Poecilia reticulata*). Ecology Letters 13: 998–1007.
40. Yund PO (1990) An in situ measurement of sperm dispersal in a colonial marine hydroid. Journal of Experimental Zoology 253: 102–106.
41. Pennington JT (1985) The ecology of fertilization of echinoid eggs: the consequences of sperm dilution, adult aggregation and synchronous spawning. Biological Bulletin 169: 417–430.
42. Sewell MA, Levitan DR (1992) Fertilization success during a natural spawning of the dendrochirote sea-cucumber *Cucumaria miniata*. Bulletin of Marine Science 51: 161–166.
43. Levitan DR, Petersen C (1995) Sperm limitation in the sea. Trends in Ecology & Evolution 10: 228–231.
44. Marshall DJ, Styan CA, Keough MJ (2002) Sperm environment affects offspring quality in broadcast spawning marine invertebrates. Ecology Letters 5: 173–176.
45. Morgan SG (1995) Life and death in the plankton: larval mortality and adaptation. Pages 279–321 in LA McEdward, editor. Ecology of Marine Invertebrate Larvae. CRC Press, Inc., Boca RatonBoca Raton.
46. McGurk MD (1986) Natural mortality of marine pelagic fish eggs and larvae – role of spatial patchiness. Marine Ecology Progress Series 34: 227–242.
47. Kiflawi M (2006) On optimal propagule size and developmental time. Oikos 113: 168–173.
48. Strathmann RR (1985) Feeding and nonfeeding larval development and life-history evolution in marine invertebrates. Annual Review of Ecology and Systematics 16: 339–361.
49. Gillooly JF, Charnov EL, West GB, Savage VM, Brown JH (2002) Effects of size and temperature on developmental time. Nature. 417 70–73.
50. McEdward LR, Miner BG (2003) Fecundity-time models of reproductive strategies in marine benthic invertebrates: fitness differences under fluctuating environmental conditions. Marine Ecology Progress Series 256: 111–121.
51. Strathmann RR (1977) Egg size, larval development, and juvenile size in benthic marine invertebrates. American Naturalist 111: 373–376.
52. Marshall DJ, Bolton TF, Keough MJ (2003) Offspring size affects the post-metamorphic performance of a colonial marine invertebrate. Ecology 84: 3131–3137.
53. Allen RM, Marshall DJ (2010) The larval legacy: Cascading effects of recruit phenotype on post-recruitment interactions. Oikos 119: 1977–1983.
54. Scheltema RS, Williams IP, Shaw MA, Loudon C (1981) Gregarious settlement by the larvae of *Hydroides dianthus* (Polychaeta, Serpulidae). Marine Ecology-Progress Series 5: 69–74.
55. Moran AL, McAlister JS (2009) Egg size as a life history character of marine invertebrates: Is it all it's cracked up to be? Biological Bulletin 216: 226–242.
56. Jaeckle WB (1995) Variation in the size, energy content and biochemical composition of invertebrate eggs: Correlates to the mode of larval development. Pages 49–78 in LA McEdward, editor. Ecology of Marine Invertebrate Larvae. CRC Press, Inc., Boca Raton.
57. Vogel H, Czihak G, Chang P, Wolf W (1982) Fertilization kinetics of sea urchin eggs. Mathematical Biosciences 58: 189–216.
58. Qian P, Pechenik JA (1998) Effects of larval starvation and delayed metamorphosis on juvenile survival and growth of the tube-dwelling polychaete *Hydroides elegans* (Haswell). Journal of Experimental Marine Biology and Ecology 227: 169–185.
59. Miles CM, Wayne ML (2009) Life history trade-offs and response to selection on egg size in the polychaete worm *Hydroides elegans*. Genetica 135: 289–298.
60. Marshall DJ, Allen RM, Crean AJ (2008) The ecological and evolutionary importance of maternal effects in the sea. Oceanography and Marine Biology: An Annual Review 46: 203–250.
61. Hollows CF, Johnston EL, Marshall DJ (2007) Copper reduces fertilisation success and exacerbates Allee effects in the field. Marine Ecology Progress Series 333: 51–60.
62. Toonen RJ, Pawlik JR (2001) Settlement of the gregarious tube worm *Hydroides dianthus* (Polychaeta: Serpulidae). II. Testing the desperate larva hypothesis. Marine Ecology Progress Series 224: 115–131.
63. Phillips NE (2002) Effects of nutrition-mediated larval condition on juvenile performance in a marine mussel. Ecology 83: 2562–2574.
64. Unabia CRC, Hadfield MG (1999) Role of bacteria in larval settlement and metamorphosis of the polychaete *Hydroides elegans*. Marine Biology 133: 55–64.
65. Quinn GP, Keough MJ (2002) Experimental design and data analysis for biologists. Cambridge University Press, Cambridge.
66. Dias GM, Marshall DJ (2010) Does the relationship between offspring size and performance change across the life-history? Oikos 119: 154–162.
67. Hart MW (1995) What are the costs of small egg size for a marine invertebrate with feeding planktonic larvae? American Naturalist 146: 415–426.

19

Virio- and Bacterioplankton Microscale Distributions at the Sediment-Water Interface

Lisa M. Dann[1]*, James G. Mitchell[1], Peter G. Speck[1], Kelly Newton[1], Thomas Jeffries[2], James Paterson[1]

1 School of Biological Sciences, Flinders University, Adelaide, South Australia, Australia, **2** School of the Environment, University of Technology Sydney, Sydney, New South Wales, Australia

Abstract

The marine sediment-water interface is an important location for microbially controlled nutrient and gas exchange processes. While microbial distributions on the sediment side of the interface are well established in many locations, the distributions of microbes on the water side of the interface are less well known. Here, we measured that distribution for marine virio- and bacterioplankton with a new two-dimensional technique. Our results revealed higher heterogeneity in sediment-water interface biomass distributions than previously reported with a greater than 45– and 2500-fold change cm^{-1} found within bacterial and viral subpopulations compared to previous maxima of 1.5- and 1.4-fold cm^{-1} in bacteria and viruses in the same environments. The 45-fold and 2500-fold changes were due to patches of elevated and patches of reduced viral and bacterial abundance. The bacterial and viral hotspots were found over single and multiple sample points and the two groups often coincided whilst the coldspots only occurred over single sample points and the bacterial and viral abundances showed no correlation. The total mean abundances of viruses strongly correlated with bacteria ($r = 0.90$, $p < 0.0001$, $n = 12$) for all three microplates ($n = 1350$). Spatial autocorrelation analysis via Moran's I and Geary's C revealed non-random distributions in bacterial subpopulations and random distributions in viral subpopulations. The variable distributions of viral and bacterial abundance over centimetre-scale distances suggest that competition and the likelihood of viral infection are higher in the small volumes important for individual cell encounters than bulk measurements indicate. We conclude that large scale measurements are not an accurate measurement of the conditions under which microbial dynamics exist. The high variability we report indicates that few microbes experience the 'average' concentrations that are frequently measured.

Editor: A. M. Ibekwe, U. S. Salinity Lab, United States of America

Funding: Funding provided to JGM by the Australian Research Council. The funders had no role in study design, data collection and analysis, decision to publish, or preparation of the manuscript.

Competing Interests: The authors have declared that no competing interests exist.

* Email: lisa.dann@flinders.edu.au

Introduction

Marine ecosystems are dependent on the microbial loop to carry out critical marine biogeochemical processes such as carbon and nitrogen fixation [1–8]. These marine biogeochemical processes often occur at or near cell surfaces with interactions occurring over micrometres to centimetres [9–17]. However, marine microbe distributions are commonly measured in large volumes and then that value is extrapolated over larger areas or volumes in what is sometimes termed a 'mean field' approach [12,18–24].

There is growing awareness of deficiencies in this mean field approach for productivity, biomass and process estimates [12,25–28]. Studies on patch dynamics in terrestrial and aquatic ecosystems show most organisms have heterogeneous spatial distributions. This patchiness would be missed if the sampling resolution used was not appropriate for the size of the organism studied [29–32]. For instance, Brentnall *et al.* [33] found that the mean field production of planktonic communities greatly underestimated actual production due to small scale patchiness. The little work on picoplankton distributions indicates that there is microscale spatial heterogeneity of bacteria and viruses, and that 'hotspots' and 'coldspots' are common, with bacterial and viral abundance changing 5-fold and 1.4-fold cm^{-1} [15,34].

Hotspots are areas of high microbial abundance which are believed to result from interactions between microbes and organic matter [9,12,13,15,35]. These viral and bacterial hotspots were often found to be due to bacterial response to nutrient patches, as marine bacteria have been found to accumulate around areas of high nutrient concentration via chemotaxis [9,31,36–40], This aggregation of bacteria then results in bacterial hotspots that are often followed by increased viral production via lysis due to close host proximity [15]. In contrast, coldspots are areas of low microbial abundance that are believed to be a result of lysis or grazing events in the case of lowered bacterial abundance and grazing or attachment in the case of lowered viral abundance [17].

The presence of these hotspots and coldspots are the units of microscale patchiness. The lower size limit of patches is controlled by the Batchelor scale, which is the smallest scale that nutrient gradients can occur before being dispersed by diffusion [41,42]. Chemotactic bacteria can exploit ephemeral nutrient gradients above the Batchelor scale. This exploitation leads to hotspots in bacterial abundance [37,38]. Patchiness in bacterial and viral distributions are also dependent on the season and time of sampling [13,39]. For instance, previous studies have shown bacterial abundance and activity were higher in the late afternoon [13]. This study ensured sampling was taken at the same time of day and within the same season to avoid confounding from these low frequency signals.

Although hotspots are a ubiquitous feature of microbial distributions they are often vaguely defined in studies, primarily being classified as patches of high microbial abundance that 'exceed' or are 'elevated above' background variation across one or two sampling points [12,13,43]. The methods for determining background are not given and the amount of elevation is rarely stated and differs considerably between studies making comparative analysis difficult. For instance, in the case of Seymour *et al.* [12] the data shown for hotspots suggests that elevation above background is less than one order of magnitude whereas in Seymour *et al.* [43] a conservative value of ≥4-fold higher is considered as a hotspot. In addition to this, coldspots have also lacked a definitive characterisation despite being observed in the microscale distributions of microbial communities previously [12,15,17].

As well as hotspots and coldspots, surface gradients are a common pattern seen in microbial distributions, and are seen when bacterial and viral abundance is highest at an interface and then dissipates into the water column as the distance from the interface increases. These surface gradients arise from the sinking of organic matter and its incorporation into the benthos followed by degradation and transformation by microbes leading to high nutrient concentrations and hence high bacterial and viral concentrations directly above the sediment-water interface [35]. The microscale distributions of microbial communities on the sediment side of the interface are well established, whilst the distributions of microbes on the water side of the interface are less well known and where this has been investigated the focus has been largely on bacteria [12,15,34]. However, of the studies analysing the water side, surface gradients have been seen in the distributions of bacterial and viral subpopulations above coral colonies [14], in the water column above an anchialine sinkhole [44] and in coastal [12] and estuary environments [15,35]. Viral studies show the greatest level of heterogeneity is found 1–2 cm above the sediment-water interface with abundance at St Kilda mangroves in South Australia varying by more than 2-fold, over a 15 cm sampling distance, about 1.13-fold cm^{-1}, with the highest concentrations found 1.5 cm from the sediment-water interface [15,35]. Similar patterns were seen in the vertical profiles in Seymour *et al.* [35] which showed high viral and bacterial abundance closest to the sediment-water interface with approximate increases of 1.4- and 1.5-fold cm^{-1} in total virus and total bacteria. However, due to one-dimensional sampling and lack of replication, determining whether these changes in abundance were single or multiple point maxima or surface gradients was unknown.

The null hypothesis tested in this study was that patchiness has no effect on mean biomass estimates and therefore the mean field approach is an accurate estimate of biomass. To test this hypothesis, two-dimensional sampling with extensive replication was used which builds upon previous work by Seuront *et al.* [45], Seymour *et al.* [13–15,35,43] and Waters *et al.* [17]. Here, a new sampler with 9 mm resolution provided replication via the collection of 8 vertical profiles per sampler or 24 profiles per environment, which is much larger than the previous maxima of 5 single vertical profiles per environment [35]. This replication allowed a detailed picture of microbial distributions immediately above the sediment water interface. The value of this design is that it shows how much the 'mean field' approach underestimates absolute concentration and misses the resources, competition and viral exposure gradients that microbes experience.

Materials and Methods

Study site

Samples were collected from Saint Kilda mangroves and Port Noarlunga in South Australia. These environments were chosen to build on previous work using these sites [12,13,15,35]. The site at Port Noarlunga is located near the mouth of the Onkaparinga River and is bordered by sand dunes covered in remnant vegetation. Previous studies have described the water at Port Noarlunga as oligotrophic [12]. Samples were collected from Port Noarlunga (35°16′S, 138°47′E) on the 12th of July 2011 at 11:20 am; the temperature of the water was 12.3°C with a pH of 9.4, a dissolved O_2 level of 14.3 ppm and a conductivity value of 40.7 ppt.

The St Kilda site was characterised by hypersaline lagoons less than 1 m deep in dense mangrove forests. Previous studies have shown St Kilda mangroves host a microbial community that is characterised by highly productive microbial mats and sulphur-oxidising bacteria [12,13,35,46]. Samples were collected from St Kilda (34°74′S, 138°54′E) on the 6th of July 2011 at 10:00 am; the temperature of the water was 11.5°C with a pH of 8.5, a dissolved O_2 level of 14.3 ppm and a conductivity value of 41.5 ppt.

Environmental conditions data was collected using a Hydrolab Datasonde 4a sensor. Permission to access the sampling site from the St Kilda mangrove trail site was provided by the National Parks and Wildlife Rangers of the City of Salisbury Council. Specific permission to access the Noarlunga sampling site was not required. The field studies did not involve endangered or protected species.

Sampling device (and supplementary methods)

Microplate triplicates were used to measure the microscale spatial distribution of marine viral and bacterial populations. Corning Costar TC-treated cell culture cluster microplates were used. They were 8 cm×12.3 cm, nonpyrogenic, polystyrene and sterile, consisting of 96 flat bottom 8 mm×8 mm wells which held 360 µl per well. The centre to centre distance between wells was 9 mm and the well perimeters were 7.2 cm at the ends and 10.8 cm on the sides resulting in a collection area of about 77.8 cm^2. This allowed the collection of 8 vertical profiles per microplate, which each consisted of 12 sampling points (Figure S1). Microplate triplicates gave 24 vertical profiles. The sampler was placed vertically against the respective surface and removed in a vertical motion with a 16 cm×16 cm glass plate used as a cover to minimise mixing and turbulence when collecting samples.

Sample collection and preservation

All samples were transferred into 2 ml cryovials containing 4 µl of glutaraldehyde (0.5% final concentration) using a pipette and stored at 4°C in the dark for 15 minutes. Samples were then snap frozen in liquid nitrogen and stored in a −80°C laboratory freezer [47]. Flow cytometric analysis was performed within three weeks to avoid deterioration [47].

FCM sample preparation and analysis

Many previous studies have used flow cytometry for the enumeration of bacteria and viruses due to its rapidity and accuracy [48,49]. Here we use the same flow cytometry technique as previous studies [12–15,34,35,43] and built on these studies by adding extensive replication. Briefly, samples were defrosted and prepared for flow cytometric viral and bacterial enumeration by diluting each sample 1:10 with Tris-EDTA buffer (pH 8.0, 0.2 µm filtered, 10 mM Tris, 1 mM EDTA) and staining with a nucleic acid-specific dye, SYBR Green I (1:500 dilution commercial stock; Molecular Probes). Samples were then incubated at 80°C in the dark for 10 minutes to optimise viral counts [35,47–51]. Reference beads (1 µm diameter, Molecular Probes) were added as an internal concentration and size standard with a final concentration of approximately 10^5 beads ml^{-1} in each sample. Measured flow cytometry parameters were normalised to the fluorescence and concentration of these beads [4,47–51].

Flow cytometry was conducted on a FACSCanto II flow cytometer (BD) equipped with a blue (488 nm, 20 mW, air-cooled), red (633 nm, 17 mW) and violet (405 nm, 30 mW) laser. Forward-angle light scatter (FSC), right-angle light scatter (SSC) and green fluorescence (SYBR I) were acquired for each sample and a phosphate-buffered saline (PBS) solution was used as a sheath fluid. The flow cytometer settings were normalised to fluorescence and bead concentration [35,50]. Each sample was run for two minutes at a medium flow rate setting for Noarlunga samples and a low flow rate setting for St Kilda samples due to more suspended particulate matter being present in St Kilda samples. In each flow cytometry session, triplicate blank control samples were analysed which consisted of 500 µl of 0.2 µl filtered Tris-EDTA buffer stained with 12.5 µl of SYBR Green I to eliminate any background noise that may have been created during the sample preparation or from flow cytometer artifacts.

Sample processing

The resulting cytograms, density plots and histograms acquired from flow cytometry were exported as listmode files and analysed using Win Midi 2.9 (Joseph Trotter) to enumerate the bacterial and viral populations present [35]. Viral and bacterial populations were discriminated via side-scatter (SSC), which indicates cell size; and SYBR Green fluorescence, which is indicative of nucleic acid content [47,48,51].

Data analysis and representation

For this study, rank abundance graphs were used to discriminate between hotspots, coldspots and background values. The background was determined according to Weibe [52] where the median value of the dataset is used as the background due to inclusion or exclusion of hotspots or coldspots not overly affecting the values obtained.

Surfer 10 (Golden Software, Inc.) was used to create two dimensional contour plots of the spatial distribution of viral and bacterial populations. A minimum contour interval value of at least 1000 was chosen when constructing plots due to this being larger than the maximum FACSCanto II flow cytometer error observed. This error refers to the background noise seen within triplicate blank control samples during flow cytometric analysis. This value is conservative as the maximum machine error was less than 36 events μl^{-1}.

Correlations were determined via Pearson's coefficient with the α of 0.05 being reduced by sequential Bonferroni [53]. Spatial autocorrelation statistics were used to determine whether correlations were present two-dimensionally and multi-directionally between sample points that were proximate and had similar values. From this, values were derived that illustrated the spatial autocorrelation present in the dataset. The two most common geospatial statistical tests are Moran's I and Geary's C statistics and both of these were employed as they allow statistical testing on smaller datasets [17].

Moran's I Statistic. Moran's *I* spatial autocorrelation statistics test within CrimeStat 3.3 (courtesy of Ned, Levine software) was used to determine the level of randomness within each population. The equation is as follows:

$$I = \frac{N\sum i\sum j W_{ij}(X_i-\overline{X})(X_j-\overline{X})}{\left(\sum i\sum j W_{ij}\right)\sum i(X_i-\overline{X})^2}$$

Where N is the sample number, X_i and X_j represent the variable values at a specific locations, i and j (where $i \neq j$), $\overline{X}$ is the mean of the variable and W_{ij} is the weight applied to the i and j comparison. A weighted Moran's I test was used which gave a weight value of 1 to sample points adjacent and a weight value of 0 to sample points not adjacent.

The Moran's I statistical test [54] is a global statistical test used to determine spatial autocorrelation within a set of values. It is multi-directional, being able to use vertical, horizontal and diagonal directional analysis to consider all position correlations. The Moran's I value ranges from +1 to −1, with +1 indicating perfect clustering where high values are proximate and low values are proximate, −1 indicating perfect dispersion where high values are found far apart and low values are found far apart and zero being indicative of a random distribution. Critical cut-off values for most statistical tests are dependent on the scale and values used and are often well established through previous literature using the same or similar types of studies. However, in the instance of Moran's I values for small-scale spatial studies of bacteria and viruses there is only 1 study to compare which I values are indicative of significance. Waters et al. [17] looked at phytoplankton distributions on a 2 cm and 4 cm scale. From looking at the I values obtained in the Waters et al. [17] study, it appears I values between 0.08 and 0.18 indicate a considerable level of clustering, as opposed to a random distribution, which is far less than the general clustering I value of +1.

Moran Correlograms. Moran correlograms were constructed using the Moran's I statistic, which is applied to pairs of sample values separated by a lag distance, which in this case was the distance between each sampling well (0.9 cm). Geostatistical analysis requires each lag distance to have ≥30 pairs of sample points to enable statistical reliability; therefore sample intervals with less than this were not included in the correlograms [17,55]. The standard error obtained via Moran's I analysis was used to determine if correlograms and lag distance values were significant. Output values were used to determine whether spatial autocorrelation or independence was present at each distance interval from the sediment-water interface.

Geary's C Statistic. Geary's C statistics [56] test within CrimeStat 3.3 (courtesy of the Ned, Levine software) is a global statistic that can be used in conjunction with Moran's I as it is more sensitive to local clustering and is able to identify deterministic patterning of extreme values and non-spatially related chance phenomena within the dataset, which cannot be identified using Moran's I alone. Geary's C is similar to Moran's I, however it calculates the spatial autocorrelation of a dataset by the deviation in intensity of each sample value's location compared to one another, whereas in Moran's I spatial autocorrelation is calculated by the cross-product of the deviations from the mean within the dataset. The equation is as follows:

$$C = \frac{(N-1)\left[\sum i \sum j W_{ij}(X_i - X_j)^2\right]}{2(\sum i \sum j W_{ij}) \sum i (X_i - \overline{X})^2}$$

All terms are the same as Moran's I. For Geary's C, the values range from 0 to approximately 2, however there is no definitive upper limit [57]. A value of 1 indicates spatial independence, whilst values <1 indicate positive spatial autocorrelation and values >1 indicate negative spatial autocorrelation. Therefore this statistical test is inversely related to the Moran's I test [58]. The Moran's I and Geary's C can be used in conjunction to identify deterministic patterning of extreme values [57].

Results

FCM analysis

Flow cytometric analysis revealed three distinct bacterial subpopulations and two distinct viral subpopulations at both sites which were comparable to those identified previously in marine systems [12,15,48]. Subpopulation discrimination was based on the presence of aggregated regions on biparametric cytograms of side scatter and SYBR green fluorescence and discrete peaks in monoparametric histograms of SYBR green fluorescence (Figure. 1A–D). Bacterial subpopulations were differentiated and separated into high-DNA (HDNA) and low-DNA (LDNA) groups (Figure. 1A–D). The HDNA 1 bacterial population is the intermediate bacterial population that contains bacteria with a size and nucleic acid content that is intermediate to the LDNA and HDNA 2 populations. The two viral subpopulations were identified and separated into two virus-like particle subpopulations (VLP 1 and VLP 2).

At Noarlunga, the total mean abundance of the VLP 1 subpopulation was 3.7×10^6 cells ml^{-1} ($95\%CI = 0.3\times10^6$ cells ml^{-1}, $n = 269$), whereas it was 1.3×10^6 cells ml^{-1} for the VLP 2 subpopulation ($95\%CI = 0.1\times10^6$ cells ml^{-1}, $n = 269$) (Figure. 1A and 1B). Hence, the total mean abundance of VLP 2 was 2.8-fold less than the total mean abundance of VLP 1. At St Kilda, the total mean abundance of VLP 2 was also less than VLP 1, but to a greater extent as the total mean abundance of VLP 1 was 75×10^5 cells ml^{-1} ($95\%CI = 5.1\times10^5$ cells ml^{-1}, $n = 209$) compared to 6.1×10^5 cells ml^{-1} ($95\%CI = 0.45\times10^5$ cells ml^{-1}, $n = 209$) for VLP 2 (Figure. 1C and 1D). Hence, the total mean abundance of VLP 2 was 12.3-fold less than the total mean abundance of VLP 1.

For the bacteria, the LDNA subpopulation at Noarlunga was less abundant than the HDNA 1 and the HDNA 2 subpopulations. The total mean abundance for LDNA was 0.4×10^5 cells ml^{-1} ($95\%CI = 0.03\times10^5$ cells ml^{-1}, $n = 269$) compared to total mean abundances of 3.1×10^5 cells ml^{-1} ($95\%CI = 0.2\times10^5$ cells ml^{-1}, $n = 269$) for HDNA 1 and 2.7×10^5 cells ml^{-1} ($95\%CI = 0.2\times10^5$ cells ml^{-1}, $n = 269$) for HDNA 2 (Figure. 1A and 1B). Hence, the total mean abundance of LDNA was up to 7-fold less than the total mean abundance of HDNA 1 and HDNA 2. Whereas at St Kilda, HDNA 1 was the least abundant bacterial subpopulation (Figure. 1C and 1D) with total mean abundances of 3.0×10^5 cells ml^{-1} ($95\%CI = 0.32\times10^5$ cells ml^{-1}, $n = 209$) for LDNA, 2.1×10^5 cells ml^{-1} ($95\%CI = 0.47\times10^5$ cells ml^{-1}, $n = 209$) for HDNA 1 and 5.9×10^5 cells ml^{-1} ($95\%CI = 1.7\times10^5$ cells ml^{-1}, $n = 209$) for HDNA 2. Hence, the total mean abundance of HDNA 1 was 2-fold less than the total mean abundances of LDNA and HDNA 2 (Table S1 and S2).

Vertical profiles

At Noarlunga, the mean of 24 vertical profiles had an r value of 0.9 ($p<0.0001$, $n = 12$) and showed almost identical patterns between the total mean bacterial and total mean viral population (Figure. 2A), as well as the total mean LDNA, HDNA 1 and HDNA 2 subpopulations (Figure. S2A), the total mean VLP 1 and VLP 2 subpopulations (Figure. S2B) and the mean vertical profile of VLP 1 and LDNA from microplate 1, which all had an r value of 0.99 ($p<0.0001$, $n = 12$) (Figure. S3). In addition, 145 of the 240 possible bacterial and viral single vertical profile subpopulation correlations were significantly correlated with a p-value generally less than 0.0001, such as the single vertical profile of VLP 1 and LDNA from microplate 1 which had an r value of 0.98 ($p<0.0001$, $n = 12$) (Figure. 3B). However, 95 of the single vertical profiles were not significantly correlated, such as the single vertical profile of VLP 1 and LDNA from microplate 3 which had an r value of −0.33 ($p = 0.12$, $n = 12$) (Figure. 3A). At St Kilda, the mean of 24 vertical profiles was not significantly correlated (Figure. 2B). However, the total mean LDNA, HDNA 1 and HDNA 2 populations were correlated ($r = 0.98$, $p<0.0001$, $n = 12$), as well as the total VLP 1 and total VLP 2 populations ($r = 0.64$, $p = 0.02$, $n = 12$). Of the 240 possible bacterial and viral single vertical profile subpopulation correlations, only 64 were significantly correlated with a p-value generally less than 0.0001, whilst 176 were not correlated.

Determination of hotspots and coldspots

To determine the presence of hotspots and coldspots in the viral and bacterial distributions, rank abundance graphs were created (Figure. 4A–D). In some instances only one linear trend was seen within the background values and in others there were breaks in the fits that indicated two or three separate linear trends. St Kilda showed primarily one linear trend, for instance in the VLP 1 and HDNA 1 subpopulations (Figure. 4A–B) whilst Noarlunga showed primarily two linear trends within their background values, for instance in the VLP 1 and HDNA 2 subpopulations (Figure. 4C–D). The hotspots were identified as the sample points that exceeded this linear fit (Figure. 4A–D). Exponent values for these hotspots ranged from −0.20 to −1.1 for the bacteria and −0.14 to −0.63 for the viruses at Noarlunga, whilst exponent values at St Kilda ranged from −0.27 to −1.4 for the bacteria and −0.23 to −0.66 for the viruses (Figure. 4A–D). Lastly, the coldspot sample values were identified as sample points that fit a linear trend but exhibited steeper slopes than what were seen in the background values due to large differences between adjacent sample points (Figure. 4A–D).

Two dimensional distributions

Two-dimensional contour plots of both sampling sites showed heterogeneous distributions with patches or hotspots of high abundance that occurred over single or multiple sample points as well as gaps or coldspots of low abundance that occurred only over single sample points (Figure. 5A–D).

At Noarlunga, high levels of heterogeneity were seen, with the VLP 2 subpopulation showing the highest level of heterogeneity of the two viral subpopulations. It was due to the presence of coldspots in the VLP 2 subpopulation that resulted in a 2585-fold change in abundance over 0.9 cm, with a coldspot of 6.64×10^2 particles ml^{-1} below a background of 1.07×10^6 particles ml^{-1} (Figure. 5A). For the bacterial subpopulations, the highest level of heterogeneity was seen in the LDNA subpopulation with a maximum hotspot of 2.59×10^5 cells ml^{-1} above a background of 3.47×10^4 cells ml^{-1} resulting in a 12.9-fold change in abundance over 0.9 cm (Figure. 5B).

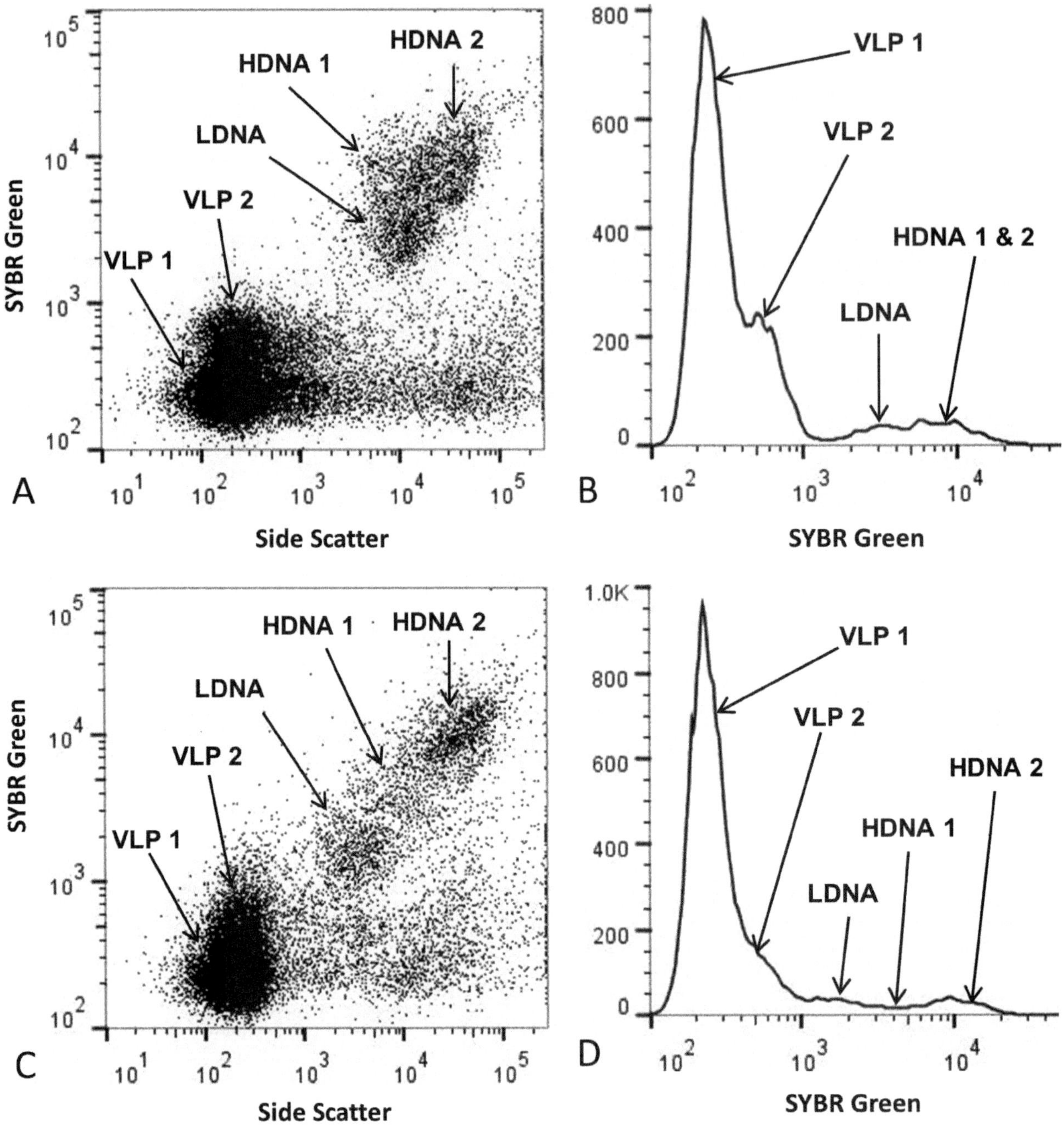

Figure 1. Identification of bacterial and viral subpopulations via flow cytometry. Flow cytometric cytograms of side-scatter light versus green fluorescence (SYBR Green) and histograms of green fluorescence (SYBR Green). **Noarlunga: A** cytogram **B** histogram; **St Kilda: C** cytogram **D** histogram showing two distinct viral populations (VLP 1 and VLP 2) and three distinct bacterial populations (LDNA, HDNA 1 and HDNA 2).

At St Kilda, the highest viral heterogeneity was present in the VLP 1 subpopulation which had a maximum hotspot of 1.62×10^7 particles ml^{-1} above a background of 6.12×10^6 cells ml^{-1} which resulted in a 10.52-fold change in heterogeneity seen over 0.9 cm (Figure. 5C). For the bacterial subpopulations at St Kilda a 45.2-fold change in heterogeneity was seen in the HDNA 2 subpopulation due a maximum hotspot of 1.45×10^7 cells ml^{-1} above a background of 3.7×10^5 cells ml^{-1} (Figure. 5D). There were a range of heterogeneities over 0.9 cm (Figure. S4) indicating a variety of intensities for hotspots and coldspots. Representative adjacent orthogonal gradient distributions as rank abundances show the distribution of abundance values for selected bacterial and viral populations at St Kilda and Noarlunga (Figure. S4).

Spatial autocorrelation

Moran's I. At Noarlunga, the Moran's I values for all of the bacterial subpopulations were non-significant. Whereas at St Kilda, significant Moran's I values were found in 33% of the bacterial subpopulations; the LDNA subpopulation of microplate 2 and 3 and the HDNA 2 subpopulation and total bacteria of microplate 2. These significant Moran's I values ranged from 0.039–0.070, with all the non-significant Moran's I values being

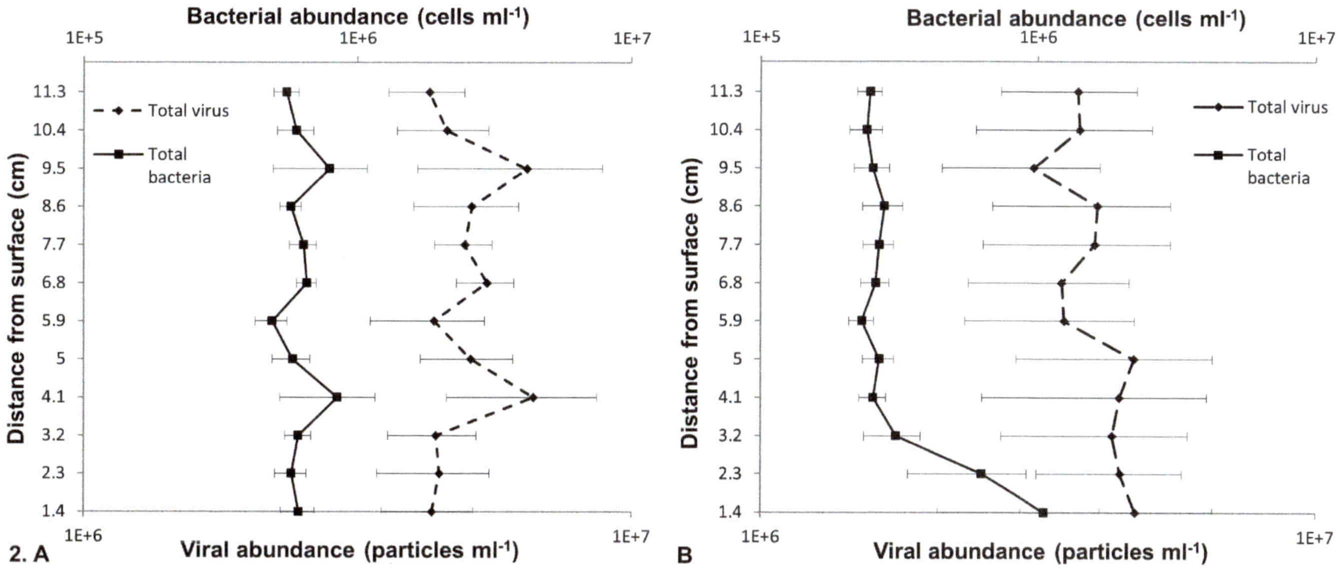

Figure 2. Comparison of total mean bacterial and viral populations at Noarlunga via vertical depth profiles. Total mean bacterial and total mean viral population abundance within all three microplates. **A** Noarlunga, (n = 1350). **B** St Kilda, (n = 1045). Error bars represent the 95% confidence intervals obtained from all three replicates (n = 12).

below these values (≤0.036). For these bacterial subpopulations z scores under the assumption of randomisation were much larger than the cut-off point of +1.96, with z scores ranging from 2.5 to 3.8. All the viral subpopulations at Noarlunga and St Kilda had non-significant Moran's I values.

All of the significant Moran's I values for the bacterial subpopulations were positive (Table S3 and S4). This trend is evident in the 2 dimensional contour plots of HDNA 2 and LDNA at St Kilda which show the presence of sediment-water interface surface gradients (Figure 6A–B). Statistical significance was not found for negative Moran's *I* values (Table S3 and S4). In geographic data, statistically significant positive values are more common than statistically significant negative values.

Moran correlograms. Moran correlograms for the bacterial subpopulations that had significant Moran's I values showed a general trend of positive to negative spatial autocorrelation as the

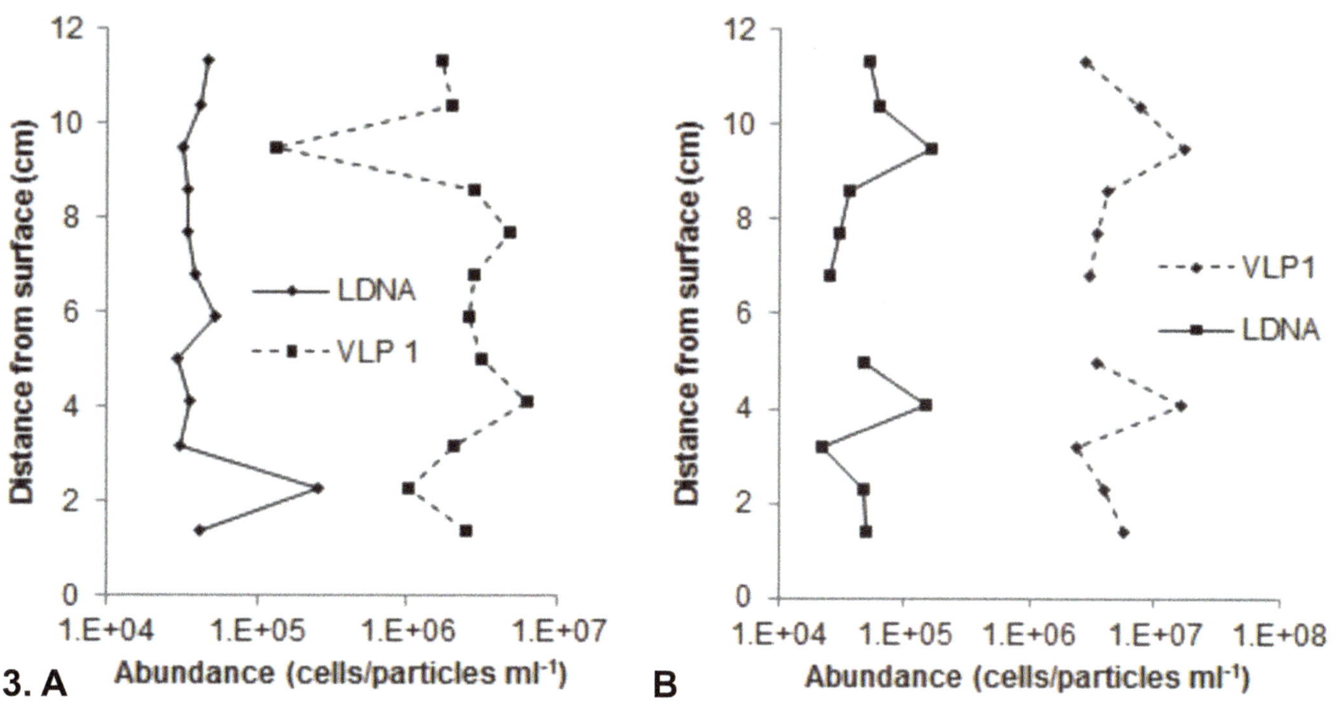

Figure 3. Single vertical profile of VLP 1 and LDNA populations at Noarlunga. A Microplate three showing little to no association. **B** Microplate one showing association. Gap in profile indicates a missing data point.

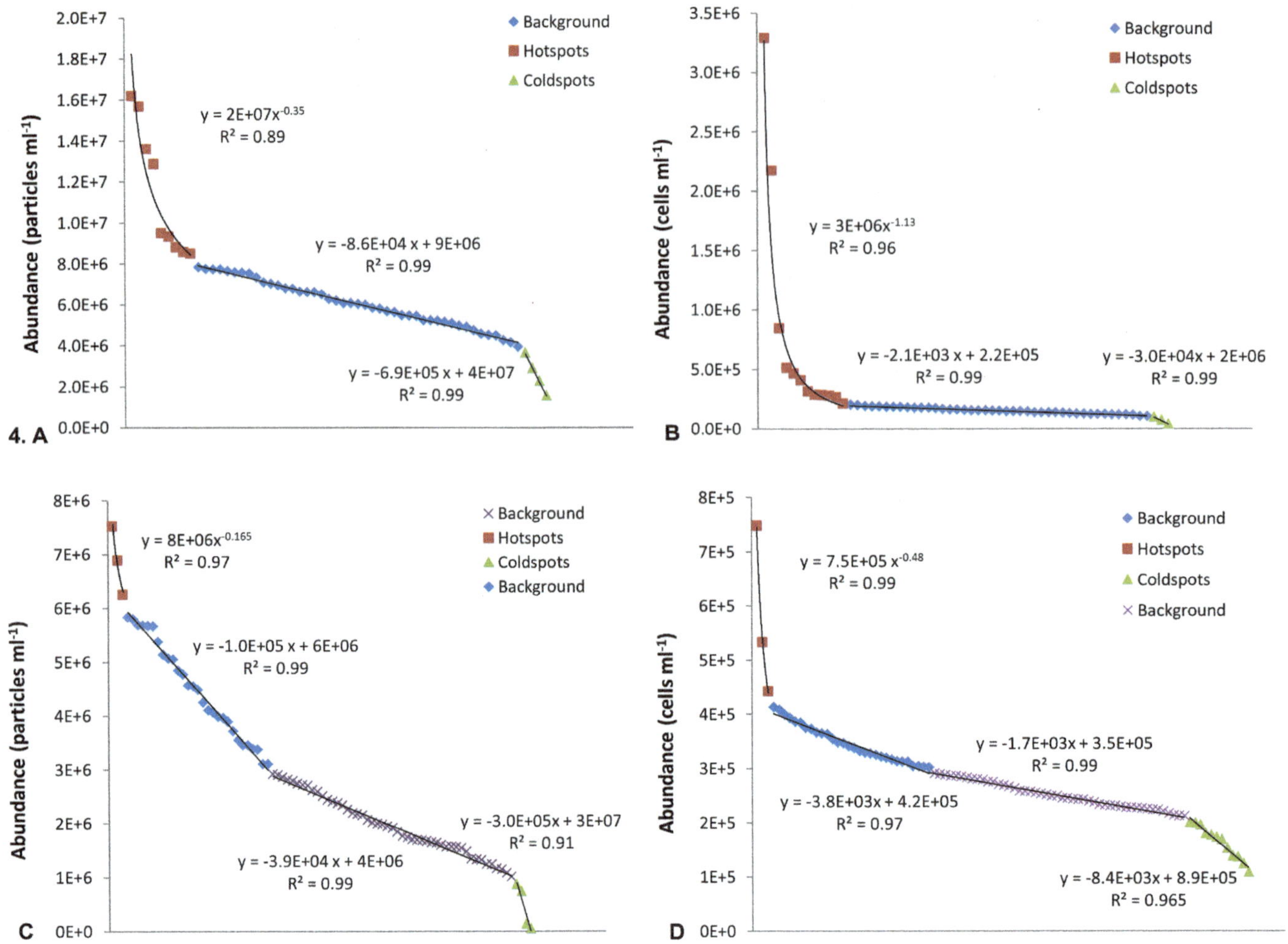

Figure 4. Rank abundance graphs used to differentiate hotspot and coldspot values from background values. One linear trend within the background values were characteristic of St Kilda whilst two linear trends were primarily seen at Noarlunga. **St Kilda: A** VLP 1, microplate 2. **B** HDNA 1, microplate 2. **Noarlunga: C** VLP 1, microplate 3. **D** HDNA 2, microplate 2.

distance from the sediment-water interface increased. On closer inspection, the correlograms for these bacterial subpopulations showed positive spatial correlation in the sample points located within 5.9 cm of the sediment-water interface and spatial independence as this distance from the sediment-water interface increased at >5.9 cm (Figure. 7A–D). This is indicative of surface gradients, which are shown in figures 6A–B. Moran correlograms for the LDNA, HDNA 1 and total bacteria in microplate 1, and HDNA 1 in microplate 2 at St Kilda also followed this trend despite having non-significant Moran's I values. Geary's C statistical testing was used to further analyse these subpopulations.

At Noarlunga and St Kilda, all of the viral subpopulation correlograms were non-significant and showed no general trend, with their distributions alternating between positive and negative spatial autocorrelation throughout the sampling distance indicating the presence of hotspots and coldspots.

Geary's C. At Noarlunga, significant Geary's C values were obtained from all of the bacterial subpopulations in microplate 1, whilst all of the bacterial subpopulations in microplate 2 and 3 were non-significant. Geary's C values were between 0.87 and 1.02; however, significance was only seen in C values of 0.87 (Table S3 and S4).

In contrast, at St Kilda significant Geary's C values were seen in 75% of the bacterial subpopulations; all of the subpopulations in microplate 1 and 2, as well as LDNA within microplate 3. Significant C values were between 0.73 and 0.92 with C values between 0.96 and 0.99 showing non-significance (Table S3 and S4).

At Noarlunga, significant Geary's C values were found in the total virus and VLP 2 subpopulation in microplate 1 whilst all other viral subpopulations were non-significant. Significance was seen in C values of 0.87, whilst C values of 0.98–1.02 showed non-significance. At St Kilda, all of the viral subpopulations had non-significant Geary's C values. C values were higher than what was seen in the bacteria at St Kilda, being between 0.96 and 1.02, which was similar to the C values obtained for the viruses at Noarlunga (Table S3 and S4).

Using Geary's C to look in more detail at the randomly distributed subpopulations that showed positive to negative spatial autocorrelation within the Moran correlograms revealed they had significant Geary's C values (Table S3 and S4). These subpopulations were the bacterial subpopulations in microplate 1, as well as the HDNA 1 subpopulation in microplate 2 at St Kilda and the bacterial subpopulations and VLP 2 and total virus populations in microplate 1 at Noarlunga.

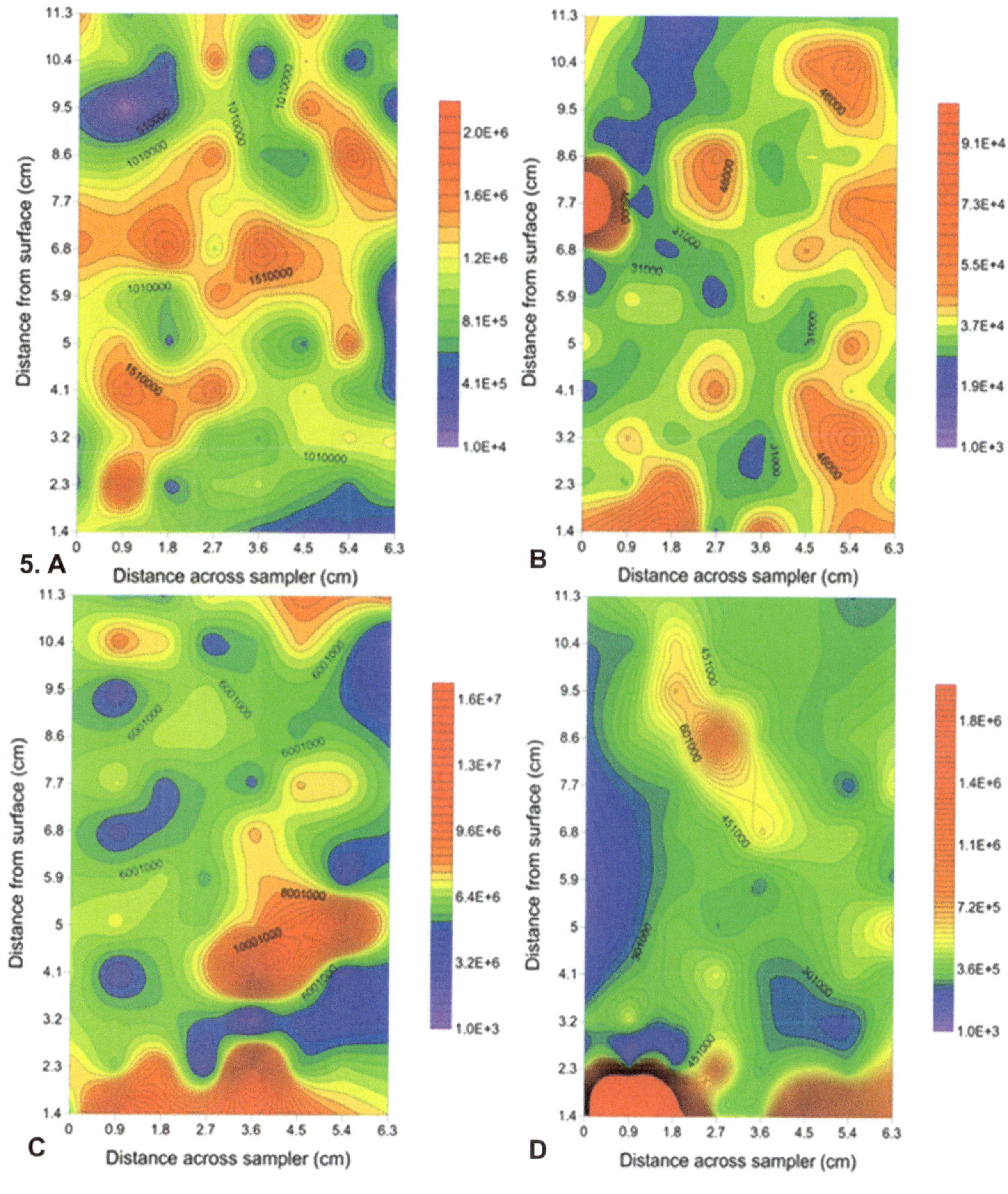

Figure 5. Two-dimensional contour plots showing the highest change in heterogeneity due to the presence of hotspots and coldspots within bacterial and viral subpopulations. Hotspots and coldspots seen across a distance of 6.3 cm×11.3 cm using Surfer 10 (Golden Software, Inc.). Noarlunga: A VLP 2 showing a 2585-fold change in abundance over 0.9 cm. B LDNA showing a 12.9-fold change in abundance over 0.9 cm. St Kilda: C VLP 1 showing a maximum 10.52-fold change in heterogeneity seen over 0.9 cm. D HDNA 2 showing a maximum 45.2-fold change in heterogeneity seen over 0.9 cm. There were a range of heterogeneities over 0.9 cm (Fig. S4) indicating a variety of intensities for hotspots and coldspots. Abundance levels are indicated by a colour intensity scale in units of cells/particles ml^{-1}. Solid red circles indicate areas of abundance higher than the maximum contour level selected. A minimum contour interval value of at least 1000 was chosen based on maximum machine error. The faint gridlines show sample interval.

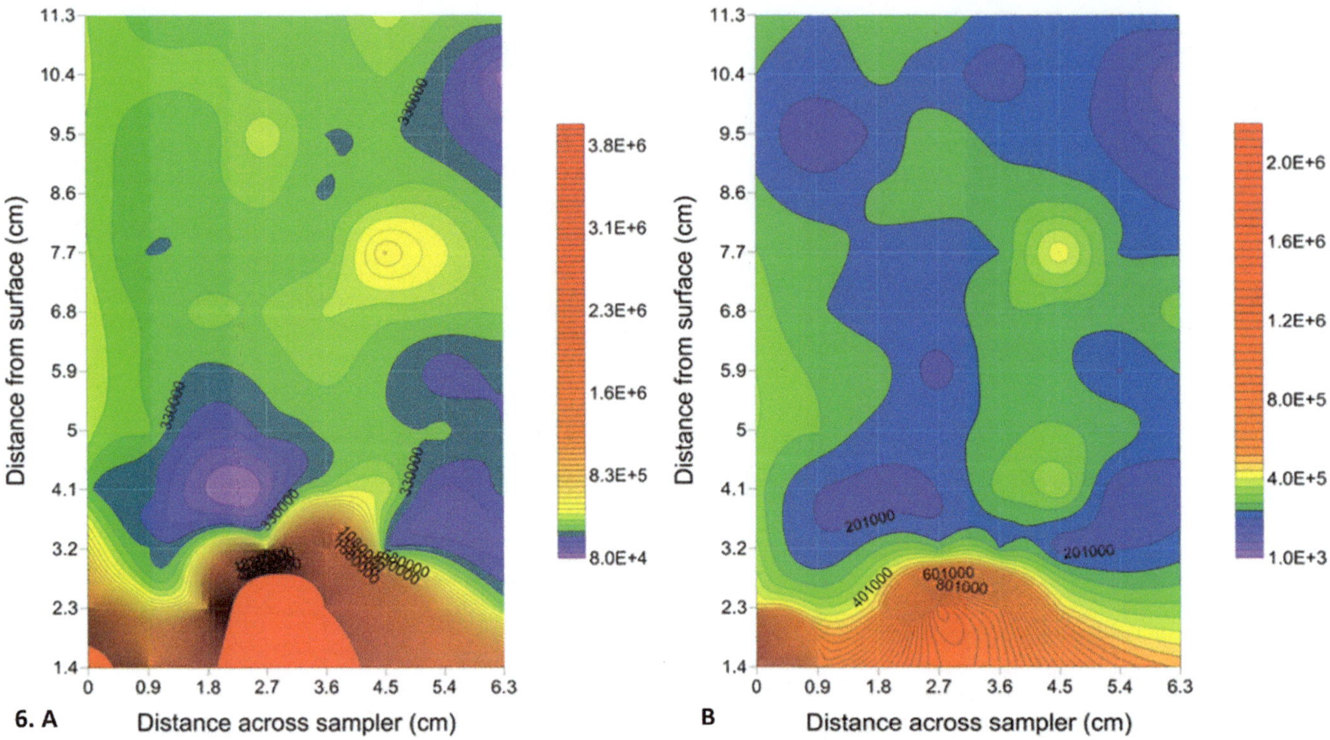

Figure 6. Two-dimensional contour plots showing surface gradients within bacterial subpopulations at St Kilda. Surface gradients seen across a distance of 6.3 cm×11.3 cm using Surfer 10 (Golden Software, Inc.). Microbial abundance levels are indicated by a colour intensity scale in units of cells ml^{-1}. **A** HDNA 2, microplate 2. **B** LDNA, microplate 2. Solid red circles indicate areas of abundance higher than the maximum contour level selected. A minimum contour interval value of at least 1000 was chosen based on maximum machine error. The faint gridlines show sample interval.

Discussion

Bacterioplankton and virioplankton distributions

Surface gradients. Strong surface gradients were found at St Kilda, which are well established along with characteristic low dissolved oxygen levels [35]. These strong gradients rising from the sediment surface showed up to a 15-fold change in abundance over the 11.3 cm vertical sampling distance collected above the sediment-water interface, The presence of these gradients is perhaps due to the strong input of organic matter from the mangroves [46]. In support of this, Seymour *et al.* [35] found microscale microbial distributions were highly correlated to nutrient concentration gradients. Seymour *et al.* [15] found the shear velocities at St Kilda at the time of sampling were not high enough for the resuspension of microbes. Given the protected nature of the environment, resuspension may be minimal compared to exposed sites and would not play a role in the surface gradients observed.

Bacteria and viruses immediately above the sediment are exposed to a variety of biological and non-biological parameters that will affect their distribution [35]. At both sampling sites clear abundance gradients were often seen in relation to distance from the sediment-water interface. The remineralisation of organic nutrients is known to cause gradients with dissolved and particulate matter, which usually show decreasing concentrations as the distance from the sediment-water surface increases [35,59].

Hotspots and coldspots. Previous studies could not determine whether a high or low point in a one dimensional profile was a gradient, a multiple point hotspot/coldspot or a single point maximum/minimum [35]. In this study, two-dimensional sampling revealed the presence of single and multiple point hotspots and single point coldspots in bacterial and viral populations at Noarlunga and St Kilda. In previous studies, local increases in abundance of up to 2 times above background over a distance of 1 cm have been considered hotspots [13,14,35,43], whereas this study showed hotspots with increases of up to 2585 and 45.2 times cm^{-1} in viral and bacterial subpopulations above the sediment-water interface. These hotspots were high in bacteria and viruses, consistent with bacteriophage infection being dependent on host density [15]. The correlation of these bacterial and viral hotspots suggests a repeated viral infection and lysis cycle for at least some of the bacterial species present [12,13,15], which, in turn, suggests that groups of picoplankton, at least close to the sediment interface remain close together for extended periods. Through entrainment in the overlying water these near-sediment hotspots may act to inoculate the overlying water column. With viral hotspots showing increases of up to 2585 times cm^{-1} this would contribute substantially to the abundance of viral particles hence playing a role in viral production and succession. In the case of coldspots, correlations were seen with viral subpopulations but not with bacterial subpopulations. There are many possible dynamics at play here. For instance, this may indicate these coldspots are capturing the stable bacterial community numbers, which have not yet been exposed to viral communities or perhaps areas in which compatible viral particles were not present hence leading to viral decay or extensive grazing as the viruses would not have been able to protect themselves from external sources by infecting the bacteria present. Another possible explanation for these coldspots in viruses could be that they are microenvironments that favour lysogeny over lysis, such as areas of low nutrient concentration

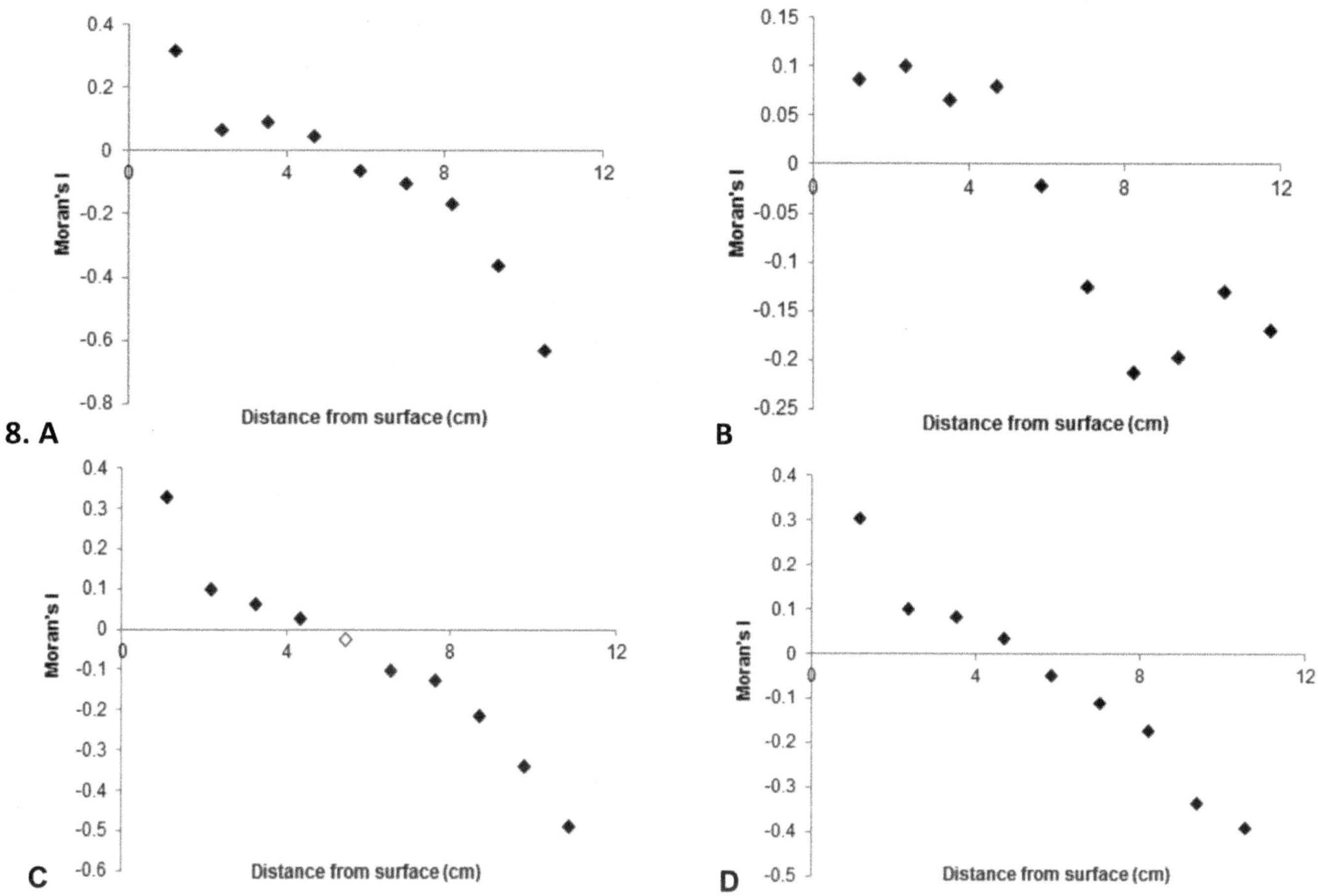

Figure 7. Significant Moran correlograms of non-randomly distributed bacterial subpopulations at St Kilda. **A** LDNA, microplate 2. **B** LDNA, microplate 3. **C** HDNA 2, microplate 2. **D** Total bacteria, microplate 2. Filled and unfilled data points indicate significant and non-significant Moran's I values ($p \leq 0.01$). Only sample points with ≥ 30 pairs of values were included.

[60]. The wide range of hotspot values (Figure. S4) suggests that all of these and possibly more mechanisms are generating hotspots. Of course, the actual processes occurring in these hotspots and coldspots cannot be determined with biomass values alone and call for microscale nutrient and species analysis.

At St Kilda, patchy viral and bacterial hotspots and coldspots were identified and, in the case of the hotspots, often coincided. These hotspots and coldspots resulted in changes in spatial heterogeneity of 10.5-fold cm^{-1} for viruses and 45-fold cm^{-1} for bacteria. These results are much higher than the previously found 1.4- and 1.5-fold cm^{-1} increases found in total virus and total bacteria by Seymour *et al.* [35]. As these hotspots and coldspots often occurred within single sampling points of 0.9 cm distance this highlights the importance of high sampling resolution and 2-dimensional arrays when studying the distributions of microbial communities.

The null hypothesis tested in this study was that patchiness has no effect on mean biomass estimates and therefore the mean field approach is an accurate estimate of biomass. This null hypothesis was rejected with the presence of patchy hotspots and coldspots in this study showing differences of up to four orders of magnitude in biomass. In the most extreme case, if the mean field approach had been employed without sufficient replication, estimates in microbial biomass would be off by a factor of 1,560.

Previous studies have also lacked a definition for hotspots and coldspots making comparative analysis impossible. In this study rank abundance graphs have been used as a method for determining what sample values are hotspots, coldspots or background. Background values fitted a linear trend as these values are random and are an indicator of equilibrated additive and reductive processes. Such additive processes in bacterial and viral communities could be reproduction or aggregation, whilst reductive processes could relate to grazing, decay or lysis events.

Hotspots were identified as the sample points that exceeded the linear fit and exhibited a steep power law trend, indicating their non-random nature whilst the coldspots were characterised by a linear trend with slope values much higher than those of the background values due to large differences seen between adjacent value points on the graph (Figure. 4A–D). The hotspots indicate a favouring towards additive processes and coldspots indicate a favouring towards reductive processes resulting in a non-equilibrated state when compared to the background values. Exponent values for the hotspots were similar at Noarlunga and St Kilda and ranged from -0.1 to -1.4 suggesting this is the level of structure seen within bacterial and viral communities. Exponent values at and below -1 indicated considerable self-organisation and strong patches [61].

The background values fit single or multiple linear trends with St Kilda being characterised primarily by single linear trends whilst Noarlunga was characterised primarily by two linear trends within its background values for bacterial and viral communities. This could be an indication of different randomisation processes present. For instance, when considering bacteria, the presence of two linear trends seen at Noarlunga could indicate equilibrium between grazing and aggregation in one linear trend whilst the second may indicate equilibrium between reproduction and lysis. By using these rank abundance graphs, not only are hotspots and coldspots identified but patterns within the background values are able to be observed. Experimental work will be required to distinguish between randomisation processes.

Bacterial and viral abundances

The LDNA and HDNA bacterial subpopulations observed have been described previously as different phylogenetic groups present within the environment [62] or as bacteria with differing activity levels [64–66]. In the former context, the total mean abundance of HDNA 2 at St Kilda and HDNA 1 at Noarlunga being higher than the total mean abundance of the other LDNA and HDNA subpopulations, may be explained by the species within these subpopulations being more efficient at utilising resources or avoiding predatory attack [62]. The total mean abundances also indicated HDNA 1 at St Kilda and LDNA at Noarlunga had the lowest abundances of the bacterial subpopulations which could be due to the species within HDNA 1 at St Kilda and LDNA at Noarlunga being more susceptible to infection or grazing [35,62,63]. In addition, higher concentrations of predators are found within the first few centimetres above the sediment-water interface, such as heterotrophic nanoflagellates [35]. However, in the context of differing bacterial activity, where the LDNA subpopulation represents dormant bacterial cells and the HDNA subpopulations represent active cells, the higher abundances of the HDNA subpopulations seen at both sites when compared to the LDNA subpopulation would be explained by the ability of the HDNA subpopulations to replicate and grow, hence leading to their higher abundances [35,64–66]. In the context of different species, the hotspots would represent accumulation of chemotactic bacteria and the coldspots areas they had left.

Another factor controlling the abundances of the bacterial subpopulations is the viral populations present. The total mean abundance of the VLP 1 subpopulation was larger than the total mean abundance of the VLP 2 subpopulation at both sites. This has been found in previous studies, with aquatic samples being dominated by the viral population that contains small viruses between 30 and 60 nm, which corresponds to the VLP 1 subpopulation within the cytograms (Figure.1A and C) [48,67].

Bacterial and viral coupling

Seymour *et al.* [35] did not find a correlation between total bacteria and viruses at this scale. In this study, a significant relation existed between total bacteria and total viruses at Noarlunga, whilst at St Kilda, a relation did not exist (Figure. 2A and B). This may be due to the presence of different microbial communities and the difference in environmental parameters between the two sampling sites. The dynamic nature of the environments may have led to short-lived events, such as nutrient patches, causing the aggregation and dispersion of bacteria over the course of minutes to hours. This initial increase in bacterial abundance would lead to increased viral production via lysis hence decoupling the relationship between viral and bacterial populations which may be why no relationship was seen between bacteria and viruses at St Kilda [15,38]. However, stages in between the bacterial aggregation and bacterial death via viral lysis may have been captured at Noarlunga and would explain the correlation between the bacteria and viruses in this environment.

While Seymour *et al.* [35] found no correlation between total bacteria and viruses; correlations were seen between the VLP 1, VLP 2 and LDNA subpopulations and the VLP 2 and HDNA 2 subpopulations. Therefore, showing subpopulations have independent and potentially more complex associations than what is seen in total bacteria and virus populations. These results imply that total populations reflect system biomass and the processes that may impact on this, such as grazing; whilst single subpopulations reflect the phylogenetic changes in the community from such processes as phage infection and lysis. As the individual subpopulations can be decoupled, with phylogenetic fluctuations being independent of the total system biomass, this explains why correlations were seen in the totals but not in the individual subpopulations. While the correlations are not mechanisms, they show that possible relationships exist that could be further investigated. These statements are based on biomass only. Taxonomic identification of the community will be part of a manuscript in preparation.

Spatial autocorrelation

Non-random vs random distribution. At St Kilda, non-random distributions were seen in the LDNA subpopulation of microplate 2 and 3 and the HDNA 2 subpopulation and total bacteria of microplate 2 due to these subpopulations having significant Moran's I values and z scores larger than the +1.96 cut-off value. Whilst all viral subpopulations were randomly distributed due to having non-significant Moran's I values. This may be due to motile bacterial subpopulations congregating around nutrient gradients and patches or resuspension of bacteria from the sediment leading to the observed surface gradients [68]. These non-random distributions may not have been seen in the viral subpopulations as viruses are not motile and are unable to respond to chemical signals.

The significant Moran's I values seen in the bacterial subpopulations at St Kilda were positive, indicating clustering and spatial dependence. This clustering indicates that high values are located close together and low values are located close together, indicating the presence of 'hotspots' and 'coldspots' or gradients of high or low values. Moran's I values ranged from 0.04 to 0.07 for clustered distributions. These values are low when considering a Moran's I value of +1 indicates perfect clustering. These values are also low compared to Waters *et al.* [17] which had significant Moran's I values between 0.082–0.180. These low values may indicate a low level of clustering present within these distributions.

At Noarlunga, all the bacterial and viral subpopulations were randomly distributed due to having non-significant Moran's I values. This indicates that no spatial autocorrelation was present amongst the bacterial and viral subpopulation distributions and that there is no spatial dependency within the distributions of the bacterial or viral subpopulations locally or globally. On a large scale this implies that environmental processes, such as turbulence, randomize the spatial distribution of bacterial and viral subpopulations at Noarlunga.

The non-random and random bacterial distributions found at St Kilda and Noarlunga coincide with the findings of Waters *et al.* [17], which found higher spatial complexity within low-energy, high-chlorophyll environments such as St Kilda mangroves than in high-energy, low-chlorophyll environments such as Noarlunga. As St Kilda mangroves is a low-energy system with high productivity, due to the presence of dense microbial mats and mangrove forests causing stagnation in the lagoons, this may allow the development and maintenance of nutrient surface gradients [12,13,35,46].

Correlogram autocorrelation patterns. At St Kilda, the positive spatial autocorrelation seen within 5.9 cm of the sediment-water interface in most bacterial subpopulations indicates the presence of potential nutrient gradients that lead to bacterial clustering and the formation of surface gradients. The negative spatial autocorrelation seen beyond 5.9 cm from the sediment-water interface shows the area at which such gradients dissipate, leading to dispersed bacterial distributions. This trend may be due to eddies, Nyquist frequency limitation or a nepheloid layer. The presence of eddies could cause resuspension of particles via eddy penetration leading to a short-time mixing effect and turbulence [69–71]. Whereas, Nyquist frequency, whereby sampling frequency should be at least twice the highest frequency present in the sample data, may have contributed to this trend as if the sampling frequency was insufficient this would lead to the presence of aliases and distortions in the dataset [72]. The presence of a 5.9 cm nepheloid layer would result in high levels of suspended particulate matter being present at sediment-water interface and the presence of such layer may largely depend on the time of sampling as nepheloid layers are subject to change depths depending on the season [73,74].

Taking into account the study by Sokal [57] which showed that high Moran's I values not accompanied by high Geary's C values are representative of deterministic patterning of extreme values and the reverse implies that extreme values are randomly distributed and are hence spatially independent. Then for the bacterial subpopulations which showed positive spatial autocorrelation 5.9 cm from the sediment-water interface and then negative spatial autocorrelation beyond 5.9 cm, with relatively low corresponding Geary's C values between 0.73 and 0.83 suggests there was deterministic patterning of extreme values. Whereas, the positive to negative spatial autocorrelation trend within the Moran correlograms of the bacterial subpopulations that were found to be randomly distributed and had low Moran's I values which were less than the estimated cut-off value for clustered distributions but also had relatively low Geary's C values between 0.85 and 0.89 may indicate that globally these populations are spatially independent and hence randomly distributed, which is why the Moran's *I* value is low, but locally they show non-random distributions. In addition, the LDNA bacterial subpopulation at St Kilda in microplate 3 showed non-random distributions in the Moran's I testing however it had a higher Geary's C value than the other non-randomly distributed populations (0.92 compared to 0.73–0.83) which was more similar to the Geary's C values of the randomly distributed subpopulations that showed local association of 0.85–0.89. According to Sokal [57] this suggests that this subpopulation may be randomly distributed as the non-random result from the Moran's I was skewed by extreme values within the data set. This again highlights the importance of using Moran's I in conjunction with Geary's C as globally bacterial subpopulations may be spatially independent however locally they may be spatially dependent. As Geary's C is able to account for extreme outliers that may skew the overall result it is important to use in microbial distribution studies which have shown increases of up to 45 and 2500 times cm^{-1} in bacteria and viruses.

'Mean field' approach limitations

The microscale spatial heterogeneity of bacterial and viral populations in this study would be underestimated, by more than 3 orders of magnitude in some cases, had the mean field approach been employed [75–77]. This is due to the presence of single point hotspots and coldspots containing more than 30% and less than 0.0008% of the total biomass within each microplate. As the distributions were random at Noarlunga but non-random at St Kilda the mean would not be representative of microbial biomass.

Conclusion

Spatial heterogeneity in microbial abundance has been found within the 11.3 cm directly above the sediment-water interface at Noarlunga estuary and St Kilda mangroves. This heterogeneity was present in the form of single and multiple point hotspots at Noarlunga and hotspots and surface gradients at St Kilda which resulted in high levels of viral and bacterial abundance being found near the sediment-water interface. These patterns in microbial distributions have been seen in previous studies but to a lesser extent. At St Kilda, gradients in bacterial and viral abundance were much higher than previously reported with a 45- and 10.5-fold change cm^{-1} found compared to the previously reported 1.5-fold change in abundance cm^{-1} in bacteria [35] and 1.4-fold change cm^{-1} in viruses [15, 35]. At Noarlunga, bacterial and viral abundance showed a 12.9– and 2584-fold change cm^{-1}. This large fold-change found cm^{-1} would have been missed if lower resolution sampling was used. This high level of microscale heterogeneity highlights the importance of analysing microbial spatial distribution on the level of what individual cells experience rather than using a 'mean field' approach. In addition, this high level of microscale heterogeneity also illustrates the necessity and applicability of the novel two-dimensional high resolution sampler used in this study which was capable of collecting eight vertical profiles. The use of this sampler allowed the discrimination of single and multiple point hotspots and/or surface gradients, which was not possible in past studies as one-dimensional samplers were used.

Bacterial subpopulations were found to be randomly distributed at Noarlunga and non-randomly distributed at St Kilda with reasons for this finding relating to the low energy-nature of St Kilda allowing the formation of sediment-water interface nutrient and microbial surface gradients. Viral subpopulations only showed random distributions at both sites. Moran's *I* and Geary's C reveal that spatial autocorrelation amongst bacterial subpopulations at St Kilda show a positive to negative trend 5.9 cm from the sediment-water interface, indicating a mechanism, whether this be sampling alias or particulate resuspension, is present leading to this clustering and dispersal of bacterial populations.

Despite hotspots being ubiquitous features of microbial distributions, past studies lacked a universal definition. They qualitatively classified patches as regions of high microbial abundance that 'exceed' or are 'elevated above' background values [12,13,43]. This paper provides a new quantitative method for determining hotspots, coldspots and background values within microscale microbial distributions. It is hoped that this method will be used in future microbial microscale distributions to allow comparative analyses among studies.

Supporting Information

Figure S1 Collection of Vertical Profiles. The use of an 8×12 96-well microplate allowed the two-dimensional array of 8 vertical profiles (row A, B, C, D, E, F, G and H) per microplate which consisted of 12 sampling wells per vertical profile.

Figure S2 Comparisons of bacterial and viral subpopulations and viral subpopulations at Noarlunga via vertical depth profiles. Bacterial and viral subpopulations of all three microplates (n = 270). **A** Total mean LDNA, HDNA 1 and HDNA 2; **B** Total mean VLP 1 and VLP 2. Error bars represent the 95% confidence intervals obtained from each subpopulation of all three replicates (n = 12).

Figure S3 Mean vertical profile of VLP 1 and LDNA, microplate one at Noarlunga (n = 90). Error bars represent the 95% confidence intervals obtained from one replicate (n = 12).

Figure S4 Adjacent orthogonal gradient distributions of bacteria and viruses as rank abundance. St Kilda: A VLP 1, **B** HDNA 2. **Noarlunga: C** VLP 2, **D** HDNA 2. Each distribution is ordered as a rank abundance. In each case the first rank was used for the 2 dimensional plots. A logarithmic trend-line was the best fit for each distribution, with the equations and r^2 for **A** $y = -4\times10^6 \ln(x)+2\times10^7$ ($R^2 = 0.97$), **B** $y = -7\times10^4 \ln(x)+4\times10^6$ ($R^2 = 0.99$), **C** $y = -2\times10^6 \ln(x)+8\times10^6$ ($R^2 = 0.98$), **D** $y = -8\times10^4 \ln(x)+4\times10^5$ ($R^2 = 0.92$). The p values are $<<0.05$ in all cases.

Table S1 Mean viral abundances per microplate at Noarlunga and St Kilda.

Table S2 Mean bacterial abundances per microplate at Noarlunga and St Kilda.

Table S3 Comparison of the Moran's I values and Geary's C values obtained for each bacterial subpopulation at Noarlunga and St Kilda.

Table S4 Comparison of the Moran's I values and Geary's C values obtained for each viral subpopulation at Noarlunga and St Kilda.

Acknowledgments

We would like to extend our gratitude to Dr P MacArdle and Dr S. Bailey from the Flow Cytometry Unit of the Flinders Medical Centre, South Australia. We also thank the wildlife rangers of the Salisbury Council, South Australia for allowing access to the sampling site on the St Kilda mangrove trail.

Author Contributions

Conceived and designed the experiments: LMD JGM PGS KN TJ JP. Performed the experiments: LMD KN JP. Analyzed the data: LMD JGM PGS KN JP. Contributed reagents/materials/analysis tools: LMD KN JP TJ. Wrote the paper: LMD JGM. Assistance with sample collection: TJ.

References

1. Azam F, Malfatti F (2007) Microbial structuring of marine ecosystems. Nature 5: 782–791.
2. Cotner JB, Biddanda BA (2002) Small players, large Role: microbial influence on biogeochemical processes in pelagic aquatic ecosystems. Ecosystems 5: 105–121.
3. Elser JJ, Stabler BL, Hassett PR (1995) Nutrient limitation of bacterial growth and rates of bacterivory in lakes and oceans: a comparative study. Aquatic Microbial Ecology 9: 105–110.
4. Gasol JM, Del Giorgio PA (2000) Using flow cytometry for counting natural planktonic bacteria and understanding the structure of planktonic bacterial communities. Scientia Marina 64: 197–224.
5. Pace N (1997) A molecular view of microbial diversity and the biosphere. Science 276: 734–740.
6. Pontes D, Lima-Bittencourt C, Chartone-Souza E, Nascimento A (2007) Molecular approaches: advantages and artifacts in assessing bacterial diversity. Journal of industrial Microbiology & Biotechnology 34: 463–473.
7. Rodriguez-Brito B, Li L, Wegley L, Furlan M, Angly F, et al. (2010), Viral and microbial community dynamics in four aquatic environments. International Society for Microbial Ecology (ISME), 4: 739–751.
8. Venter J, Remington K, Heidelberg J, Halpern A, Rusch D, et al. (2004) Environmental genome shotgun sequencing of the Sargasso Sea. Science 304: 66–74.
9. Azam F (1998) Microbial control of oceanic carbon flux: the plot thickens. Science 280: 694–696.
10. Long RA, Azam F (2001) Microscale patchiness of bacterioplankton assemblage richness in seawater. Aquatic Microbial Ecology 26: 103–113.
11. Mitchell JG, Okubo A, Fuhrman JA (1985) Microzones surrounding phytoplankton form the basis for a stratified marine microbial ecosystem. Nature 316: 58–59.
12. Seymour JR, Mitchell JG, Seuront L (2004) Microscale heterogeneity in the activity of coastal bacterioplankton communities. Aquatic Microbial Ecology 35: 1–16.
13. Seymour JR, Seuront L, Mitchell JG (2005a) Microscale and small-scale temporal dynamics of a coastal planktonic microbial community. Marine Ecology Progress Series 300: 21–37.
14. Seymour JR, Patten N, Bourne DG, Mitchell JG (2005b) Spatial dynamics of virus-like particles and heterotrophic bacteria within a shallow coral reef system. Marine Ecology Progress Series 288: 1–8.
15. Seymour JR, Seuront L, Doubell M, Waters RL, Mitchell JG (2006) Microscale patchiness of virioplankton. Journal of the Marine Biological Association of the United Kingdom 86: 551–561.
16. Suttle CA (2007) Marine viruses – major players in the global ecosystem. Nature Reviews 5: 801–812.
17. Waters RL, Mitchell JG, Seymour J (2003) Geostatistical characterisation of centimetre-scale spatial structure of *in vivo* fluorescence. Marine Ecology Progress Series 251: 49–58.
18. Angly FE, Felts B, Breitbart M, Salamon P, Edwards RA, et al. (2006) The marine viromes of four oceanic regions. PLoS Biology 4: 2121–2131.
19. Corinaldesi C, Crevatin E, Del Negro P, Marini M, Russo A, et al. (2003) Large-scale spatial distribution of virioplankton in the Adriatic Sea: testing the trophic state control hypothesis. Applied and Environmental Microbiology 69: 2664–2673.
20. Hillmer I, Reenen PV, Imberger J, Zohary T (2008) Phytoplankton patchiness and their role in the modelled productivity of a large, seasonally stratified lake. Ecological Modelling 218: 49–59.
21. Liu H, Imai K, Suzuki K, Nojiri Y, Tsurushima N, et al. (2002) Seasonal variability of picophytoplankton and bacteria in the western subarctic Pacific Ocean at station KNOT. Deep Sea Research Part II: Topical Studies in Oceanography 49: 24–25.
22. Pan LA, Zhang LH, Zhang J, Gasol JM, Chao M (2005) On-board flow cytometric observation of picoplankton community structure in the East China Sea during the fall of different years. FEMS Microbiology Ecology 52: 243–253.
23. Pan LA, Zhang J, Zhang LH (2007) Picophytoplankton, nanophytoplankton, heterotrophic bacteria and viruses in the Changjiang Estuary and adjacent coastal waters. Journal of Plankton Research 29: 187–197.
24. Strutton PJ, Mitchell JG, Parslow JS, Greene RM (1997) Phytoplankton patchiness: quantifying the biological contribution using Fast Repetition Rate Fluorometry. Journal of Plankton Research 19: 1265–1274.
25. Curtis TP, Head IM, Lunn M, Woodcock S, Schloss PD, et al. (2006) What is the extent of prokaryotic diversity? Philosophical Transactions of the Royal Society 361: 2023–2037.
26. Grunbaum D (2002) Predicting availability to consumers of spatially and temporally variable resources. Hydrobiologia 480: 175–191.
27. Karl DM, Dore JE (2001) Microbial ecology at sea: sampling, subsampling and incubation considerations. Methods in Microbiology 30: 13–39.
28. Waters RL, Mitchell JG (2002) Centimetre-scale spatial structure of estuarine in vivo fluorescence profiles. Marine Ecology Progress Series 237: 51–63.
29. Genin A, Greene C, Haury L, Wiebe P, Gal G, et al. (1994) Zooplankton patch dynamics: daily gap formation over abrupt topography. Deep Sea Research Part I: Oceanographic Research Papers 41 (5–6): 941–951.
30. Pickett S, Cadenasso M (1995) Landscape ecology: Spatial heterogeneity in ecological systems. Science 269 (5222): 331–334.
31. Levin S, Whitfield M (1994) Patchiness in marine and terrestrial systems: from individuals to populations. Phil. Trans R. Soc 343 (1303): 99–103.
32. Grünbaum D (2012) The logic of ecological patchiness. Interface Focus 2: 150–155.
33. Brentnall SJ, Richards KJ, Brindley J, Murphy E (2003) Plankton patchiness and its effect on larger-scale productivity. Journal of Plankton Research 25: 121–140.
34. Seymour JR, Mitchell JG, Pearson L, Waters R (2000) Heterogeneity in bacterioplankton abundance from 4.5 millimetre resolution sampling. Aquatic Microbial Ecology 22: 143–153.
35. Seymour JR, Seuront L, Mitchell JG (2007b) Microscale gradients of planktonic microbial communities above the sediment surface in a mangrove estuary. Estuarine, Coastal and Shelf Sciences 73: 651–666.
36. Mitchell JG (2002) The energetic and scaling of search strategies in bacteria. American Naturalist 160: 727–740.

37. Barbara G, Mitchell JG (2003) Marine bacterial organisation around point-like sources of amino acids FEMS Microbiology Ecology 43: 99–109.
38. Blackburn N, Fenchel T, Mitchell JG (1998) Microscale nutrient patches in planktonic habitats shown by chemotactic bacteria. Science 282: 2254–2256.
39. Seymour JR, Ahmed T, Durham WM, Stocker R (2010) Chemotactic response of marine bacteria to the extracellular products of Synechococcus and Prochlorococcus. Aquatic Microbial Ecology 59: 161–168.
40. Stocker R, Seymour JR, Samadani A, Hunt DE, Polz MF (2008) Rapid chemotactic response enables marine bacteria to exploit ephemeral microscale nutrient patches. Proceedings of the National Academy of Sciences of the United States of America (PNAS) 105: 4209–4214.
41. Stocker R (2012) Marine microbes see a sea of gradients. Science Review 338: 628–633.
42. Taylor JR, Stocker R (2012) Trade-offs of chemotactic foraging in turbulent water. Science 338: 675–679.
43. Seymour JR, Seuront L, Doubell M, Mitchell JG (2008) Mesoscale and microscale spatial variability of bacteria and viruses during a Phaeocystisglobosa bloom in the Eastern English Channel. Estuarine, Coastal and Shelf Science 80: 589–597.
44. Seymour JR, Humphreys WF, Mitchell JG (2007a) Stratification of the microbial community inhabiting an anchialine sinkhole. Aquatic Microbial Ecology 50: 11–24.
45. Seuront L, Lacheze C, Doubell MJ, Seymour JR, Van Dongen-Vogels V, et al. (2007) The influence of Phaeocystisglobosa on microscale spatial patterns of chlorophyll a and bulk-phase seawater viscosity, Biogeochemistry 83: 173–188.
46. Barbara G, Mitchell JG (1996) Microlaminations of motile bacteria in microbial mats. Applied and Environmental Microbiology 62: 3985–3990.
47. Brussaard C (2004) Optimisation of procedures for counting viruses by flow cytometry. Applied and Environmental Microbiology 70: 1506–1513.
48. Marie D, Brussaard C, Thyrhaug R, Bratbak G, Vauolt D (1999) Enumeration of marine viruses in culture and natural samples by flow cytometry. Applied and Environmental Microbiology 65: 45–52.
49. Brussaard C, Marie D, Bratbak G (2000) Flow cytometric detection of viruses. Journal of Virological Methods 85 (1–2): 175–182.
50. Schapira M, Buscot MJ, Leterme SC, Pollet T, Chapperon C, et al. (2009) Distribution of heterotrophic bacteria and virus-like particles along a salinity gradient in a hypersaline coastal lagoon. Aquatic Microbial Ecology 54: 171–183.
51. Marie D, Partensky F, Jacquet S, Vaulot D (1997) Enumeration and cell cycle analysis of natural populations of marine picoplankton by flow cytometry using the nucleic acid stain SYBR Green I. Applied and Environmental Microbiology 63: 186–193.
52. Wiebe PH (1970) Small-scale spatial distribution in oceanic zooplankton. American Society of Limnology and Oceanography 15: 205–217.
53. Holm S (1979) A simple sequentially rejective multiple test procedure. Scandinavian Journal of Statistics 6: 65–70.
54. Moran PAP (1950) Notes on continuous stochastic phenomena. Biometrika 37: 17–23.
55. Rossi RE, Mulla DJ, Journel AG, Franz EH (1992) Geostatistical tools for modelling and interpreting ecological spatial dependence. Ecological Monographs 62: 277–314.
56. Geary RC (1954) The contiguity ratio and statistical mapping. The Incorporated Statistician 5: 115–145.
57. Sokal RR (1978) Spatial autocorrelation in biology: 1. Methodology. Biological Journal of the Linnean Society 10: 199–228.
58. Griffith DA (1987) Spatial autocorrelation: a primer. Resource publications in Geography. Association of American Geographers, Washington, D.C.
59. Baric A, Kuspilic G, Matijevic S (2002) Nutrient (N, P, Si) fluxes between marine sediments and water column in coastal and open Adriatic. Hydrobiologia 475/476: 151–159.
60. Weinbauer MG, Suttle CA (1999) Lysogeny and prophage induction in coastal and offshore bacterial communities. Aquatic Microbial Ecology 18: 217–225.
61. Mitchell JG (2004) Rank-size analysis and vertical phytoplankton distribution patterns. In: Handbook of Scaling Methods in Aquatic Ecology: Measurement, Analysis, Simulation (L. Seuront, P.G. Strutton eds) CRC, Boca Raton, 257–278.
62. Wang Y, Hammes F, Boon N, Chami M, Egli T (2009) Isolation and characterisation of low nucleic acid (LNA)-content bacteria. The ISME Journal 3: 889–902.
63. Bouvy M, Troussellier M, Got P, Arfi R (2004) Bacterioplankton responses to bottom-up and top-down controls in a West African reservoir (Selingue, Mali). Aquatic Microbial Ecology 34: 301–307.
64. Gasol JM, Zweifel UL, Peters F, Fuhrman JA, Hagstrom A (1999) Significance of size and nucleic acid content heterogeneity as measured by flow cytometry in natural planktonic bacteria. Applied and Environmental Microbiology 65: 4475–4483.
65. Lebaron P, Servais P, Agogue H, Courties C, Joux F (2001) Does the high nucleic acid content of individual bacterial cells allow us to discriminate between active cells and inactive cells in aquatic systems. Applied and Environmental Microbiology 67: 1775–1782.
66. Lebaron P, Servais P, Baudoux A, Bourrain M, Courties C, et al. (2002) Variations of bacterial-specific activity with cell size and nucleic acid content assessed by flow cytometry. Aquatic Microbial Ecology 28: 131–140.
67. Bergh O, Borsheim KY, Bratbak G, Heldal M (1989) High abundance of viruses found in aquatic environments. Nature 340: 467–468.
68. Thar R, Kühl M (2003) Bacteria are not too small for spatial sensing of chemical gradients: An experimental evidence. Proceedings of the National Academy of Sciences of the United States of America (PNAS) 100: 5748–5753.
69. Boegman L, Ivey GN (2009) Flow separation and resuspension beneath shoaling nonlinear internal waves. Journal of Geophysical Research. 114: 1–15.
70. Braaten DA, Paw KT, Shaw RH (1990) Particle resuspension in a turbulent boundary layer – observed and modelled. Journal of Aerosol Science 21: 613–628.
71. Nicholson KW (1988) A review of particle resuspension. Atmospheric Environment 22: 2639–2651.
72. Kester W (2009) What the Nyquist criterion means to your sampled data system design. Analog Devices 1–12.
73. Rutgers van der Loeff MM, Meyer R, Rudels B, Rachor E (2002) Resuspension and particle transport in the benthic nepheloid layer in and near Fram Strait in relation to faunal abundances and Th depletion. Deep-Sea Research 49: 1941–1958.
74. Sandilands RG, Mudroch A (1983) Nepheloid layer in Lake Ontario. Journal of Great Lakes Research 9: 190–200.
75. Alonso MC, Jimenez-Gomez F, Rodriguez J, Borrego JJ (2001) Distribution of virus-like particles in an oligotrophic marine environment (Alboran Sea, Western Mediterranean). Microbial Ecology 42: 407–415.
76. Drake LA, Choi KH, Haskell AGE, Dobbs FC (1998) Vertical profiles of virus-like particles and bacteria in the water column and sediments of Chesapeake Bay, USA. Aquatic Microbial Ecology 16: 17–25.
77. Thomson PG, Davidson AT, Enden R, Pearce I, Seuront L, et al. (2010) Distribution and abundance of marine microbes in the Southern Ocean between 30 and 80°E. Deep-Sea Research II 57: 815–827.

Asexual Propagation of Sea Anemones That Host Anemonefishes: Implications for the Marine Ornamental Aquarium Trade and Restocking Programs

Anna Scott*, Jannah M. Hardefeldt, Karina C. Hall

National Marine Science Centre and Marine Ecology Research Centre, School of Environment, Science and Engineering, Southern Cross University, Coffs Harbour, New South Wales, Australia

Abstract

Anemonefishes and their host sea anemones form an iconic symbiotic association in reef environments, and are highly sought after in the marine aquarium trade. This study examines asexual propagation as a method for culturing a geographically widespread and commonly traded species of host sea anemone, *Entacmaea quadricolor*. Two experiments were done: the first to establish whether size or colour morph influenced survival after cutting into halves or quarters; and the second to see whether feeding was needed to maximise survival and growth after cutting. Survival rates were high in both experiments, with 89.3 and 93.8% of the anemones cut in half, and 62.5 and 80.4% cut in quarters surviving in experiments 1 and 2, respectively. Anemones that were cut in half were larger in size, and healed and grew quicker than those cut in quarters. However, even though survival was lower when the individuals were cut in quarters, this treatment produced the greatest number of anemones. Feeding increased oral disc diameter growth and reduced wet weight loss, but did not significantly influence pedal disc diameter. Given that the anemones took up to 56 d to form an off-centre mouth, it is highly likely that feeding may have produced greater effect if the experiment was run for longer. This low technology method of propagation could be used to produce individuals throughout the year and the anemones could then be used to supply the aquarium trade or restock depleted habitats, thus supporting biodiversity conservation in coral reef areas.

Editor: Wan-Xi Yang, Zhejiang University, China

Funding: Funding for this research was provided by the Associate Deans Research Support Scheme and the Marine Ecology Research Centre, Southern Cross University. The funders had no role in study design, data collection and analysis, decision to publish, or preparation of the manuscript.

Competing Interests: The authors have declared that no competing interests exist.

* Email: anna.scott@scu.edu.au

Introduction

The trade of marine ornamentals for aquariums is rapidly expanding, causing concerns about the sustainability and environmental impacts of the industry [1–3]. Collection often occurs on coral reefs in developing island nations, where it can be difficult to implement effective management [2,4,5]. The fishery is highly selective, and harvesting from limited areas has led to localised depletions or extinctions of target species in some cases [6–8].

Sea anemones that host symbiotic anemonefishes are highly sought after by aquarists [2,9]. An alteration in the population dynamics of either partner generally effects the other due to the obligate nature of the symbiosis [6,8,10]. These host anemones represent high-value species for collectors (e.g. in the Philippines the price paid to fisherman for anemones can be up to 13 times the amount for anemonefishes), which means they are often preferentially harvested [8]. The development of reliable and cost-effective methods for culturing anemones could facilitate the supply of animals for the aquarium trade, or the restocking of reefs that have already been impacted by natural or anthropogenic disturbances [11]. In doing so, it would create a new and ecologically sustainable industry, and thus support biodiversity conservation in coral reef areas [12,13].

Anemones could be cultured using sexual reproduction, which has been studied in two (*Entacmaea quadricolor* and *Heteractis crispa*) of the ten species of sea anemones that host anemonefishes [14–17]. Both of these anemones have separate sexes, and broadcast spawn gametes for external fertilization and development during predictable annual spawning periods [14,15,17,18]. The larvae settle onto a variety of surfaces, where metamorphosis and development occurs [16]. Alternatively, it may be possible to propagate anemones using asexual reproduction. *Entacmaea quadricolor* and *Heteractis magnifica* can reproduce asexually using longitudinal fission [19], where the anemone divides by stretching in opposite directions, thinning the tissue and tearing perpendicular to the axis of stretch [20,21]. Porat and Chadwick-Furman [22] cut six *E. quadricolor* in half to generate individuals for a split-pair laboratory experiment with anemonefishes, and found that the majority survived (67%). Likewise, anecdotal evidence from aquarium hobbyists suggests that a variety of anemone species might be able to be propagated by cutting them in half.

This study aimed to determine whether the most geographically widespread host anemone, *E. quadricolor*, could be cultured using fragmentation. We investigated: 1) whether anemones could survive being cut in halves or quarters, and which cut type would produce the maximum number of anemones; 2) if survival was influenced by initial size or colour morph; and 3) whether feeding was needed to maximise survival and growth. Given that this species naturally undergoes longitudinal fission [19], and there is some published [22] and anecdotal evidence to suggest these anemones can survive being fragmented, it was hypothesized that: anemones would survive being cut; that survival would be influenced by size (with larger individuals faring better than smaller individuals) but not colour morph; and that feeding would increase survival and accelerate growth.

Materials and Methods

Ethics statement

This work was done under the conditions specified by NSW Fisheries Permit P02/0025-4.0.

Experiment 1 – survival

Twenty-four *E. quadricolor* were collected on SCUBA from 15–18 m depth on 14 March 2012 at North Solitary Island, Solitary Islands Marine Park, New South Wales, Australia (29°55′S, 153°23′E). Anemones were gently removed from the substratum by hand, placed into a mesh bag, and transferred into a 70-L white polyvinyl chloride (PVC) tub on the boat. They were transported to the National Marine Science Centre, Coffs Harbour and maintained in a shaded rectangular 3 000-L outdoor tank with flow-through (10 L min^{-1}) coarse-sand filtered (~30 μm) seawater for 12 d before the start of the experiment. Seawater was sourced from Charlesworth Bay (30°15′S, 153°8′E) through a gravel filter-box system that was located 150 m offshore and at a depth of approximately 3 m. Half of the anemones had a purple column and brown tentacles with white tips (morph 1); while the other half had a red column and brown tentacles with a white ring below green tips (morph 2). It is likely that morph 1 was male, and morph 2 was female, as Scott and Harrison [17] found that the colour of the column and tentacles generally corresponded to a particular sex at the collection location.

The experiment was done in two 3 000 L tanks that were located outdoors to provide natural photoperiod and light. Shade cloth covers were placed over the tanks to ensure that light induced bleaching did not occur (light levels did not exceed 400 μmol photons m^{-2} s^{-1}). Twenty-four white PVC tubs (40 L, 29.5×39 cm wide, 30 cm deep) were evenly distributed between the tanks. Each tub had four circular outlets (2.5 cm diameter) located 3.5 cm below the top that were covered with 1 mm square mesh; a rock (approximately 15×10×6 cm) for habitat; and was supplied with flow-through seawater (sourced as described above but supplied at 0.5 L min^{-1}) to maintain ambient seawater temperature (20.5–24.5°C). Temperature was logged every 15 min in each of the tanks using a Thermochron iButton temperature logger (Maxim, USA).

The anemones were weighed on a top-pan balance after gentle squeezing and blotting with absorbent paper to remove as much water as possible (wet weight 61–274 g). Anemones were then randomly allocated to tubs and acclimatised for 3 d before the start of the experiment. Eight individuals (four of each colour morph) were randomly assigned to each of the following treatments: control (uncut), cut in half, or cut in quarters.

At the start of the experiment, the oral disc diameter (ODD) of each anemone was measured along the long and short axes to the nearest mm with vernier callipers and averaged (ODD 109–235 mm). Anemones were then removed from the tubs and fragmented with a 165 mm steel chefs knife. This was done on a plastic cutting board, whilst wearing latex gloves. Tentacles were moved toward the outer edge of the oral disc, and cuts were made through the anemones mouth (along a random plane to the directional mouth). The resulting fragments were immediately placed back into their respective tubs, with the pedal disc directed downwards. The knife and cutting board were cleaned between cutting each anemone. Control anemones were also removed from the tubs; however they were simply placed back into the tubs without cutting. If the water in the tubs became cloudy after cutting a ~80% water exchange was done to rapidly clear the water.

The number of alive fragments (i.e. survival) was recorded at 6 h, daily for the first week, and then on day 9, 11, 14, 21, 28, and 35. Anemones were fed small pieces of prawn on a weekly basis, starting 7 d after cutting. Tubs were cleaned every 3 d to remove boluses or mucus produced by the anemones and residual food, and to prevent algal build up on the tubs. This was done using a scouring pad and siphon hose.

Experiment 2 – survival, effect of feeding on size, and visual observations of recovery

The methodology for experiment 2 was similar to that described above, except for the following departures. Forty-two *E. quadricolor* of four colour morphs: i) red column, brown tentacles with white ring below green tips (n = 18); ii) purple column, brown tentacles with white tips (n = 13); iii) light brown column, brown tentacles white ring below brown tips (n = 7); and iv) pink column, brown tentacles with white tips (n = 4), were collected from 12–18 m depth on the 5 October 2012. The anemones were measured before cutting (ODD 98–215 mm, pedal disc diameter [PDD, which was measured using the same methods described for ODD] 45–108 mm, and wet weight 32–171 g). The linear measurements of ODD and PDD were correlated (correlation coefficient r = 0.77), but each only showed weak correlation with wet weight (r = 0.4 for both).

Anemones were randomly allocated directly into 42 separate tubs rather than held temporarily in a holding tank. The tubs were located in three 1 200 L tanks supplied with flow-through seawater (sourced the same way as described in experiment 1, but then filtered to 10 μm with Filtaflo sediment filters and supplied at 0.6 L min-1), which was heated in two 3 000 L header tanks using Aquahort heater/chiller units to maintain a similar temperature range (21–24°C) to that in experiment 1. Temperature was logged every 3 h in twelve tubs (four in each tank).

Fourteen individuals were randomly assigned to each of the treatments: control (no cut), cut in half, or cut in quarters. Half of the anemones from each of these treatments were fed weekly. This feeding frequency was selected as the growth rates of juvenile *Heteractis cripsa* fed once or three-times weekly have been shown to be equivalent [11]. Pieces of prawn flesh were placed onto the oral disc until each anemone was satiated, and consequently the amount of food received by each individual was proportional to its size. The remaining anemones were not provided with any additional heterotrophic food. Although individuals were randomly allocated to the various treatments, the treatment types were divided among the three experimental tanks.

The number of alive fragments (i.e. survival) and their recovery (cut healing status, and whether or not their oral disc was expanded and their pedal disc was attached to the tub) were recorded at 6 h, daily for the first week, and then on day 9, 11, 14, 21, 28, 35, 42, 56, 70, 84 and 105. Throughout the experiment ODD, PDD, and wet weight were used as indicators of anemone size. Measurements of ODD and PDD were first taken one week after cutting to allow the anemones time to re-expand, and were repeated on day 28, 42, 56, 70, 84 and 105. Fragments were weighed immediately after cutting, and then on d 28, 56, 84, and 105. Tubs were cleaned every 2–3 d.

Data analysis

Survival – experiments 1 and 2. Survival differences among the different treatments were investigated using the Fisher's exact test. Variations in survival over time among the different treatments were analysed for each experiment using nonparametric Kaplan-Meier (K-M) product-limit analyses. The census times at which fragments were last observed alive were used as the time of death, so estimates of survival times are slightly conservative. As non-parametric survival analyses cannot explicitly incorporate information from replication, survival curves were estimated for the total number of fragments in each treatment combination pooled across replicate tubs.

Data from experiment 1, were further analysed with a parametric survival analysis (Cox proportional hazards model), to determine whether pre-fragmented size or morph had a significant effect on survival over time. Pre-fragmented sizes were categorised as either small (<100 g) and large (>100 g) for analyses. Survival data were analysed using the 'survival' library in R (version 2.11.1, http://www.r-project.org).

Growth – experiment 2. After cutting, there were uneven numbers of anemones in each tub (one control, two halves, or four quarters). In tubs containing multiple fragments it was not possible to individually identify each fragment between measurements, so growth data (i.e. wet weight, ODD, and PDD) were averaged per tub for all cut anemones and tubs were used as replicate units for each treatment combination. Actual size measurements were used for all statistical analyses rather than percentage change to alleviate complications associated with negative growth and percentages for statistical computations. To account for the influence of initial size on subsequent growth, the first size measurement of each fragment (immediately after cutting for wet weight and on day 7 for ODD and PDD) was used as a covariate in all analyses. All growth variables were log-transformed to meet the assumptions of normality and homoskedacticity.

To assess the effects of cut types (no cut, half, and quarter), feeding (fed and unfed), and tanks (1, 2, and 3) on growth during the experiment, generalized linear models (GLM) were fitted with the 'lme4' package in R, with parameters estimated by restricted maximum likelihood (REML). Baseline models included the design/random variables of initial fragment size as a fixed covariate, and tank and days as random factors. Days were nested within tubs to account for the repeated measures experimental design. The treatment factors of cut types, feeding and days, and associated interaction terms were fitted in various combinations of fixed factors along with the design/random variables in separate GLM, and compared by maximum likelihood comparisons of the change in deviance from the baseline model using the Akaike Information Criterion (AIC).

Results

Survival

In experiment 1, all of the eight controls survived, along with 15 of the 16 halves (93.8%) (Figure 1). In contrast, significant mortality occurred in the quarters, with 20 of the 32 fragments (62.5%) surviving (Figures 1 and 2a, Fisher's, $\chi^2 = 10.6$, df = 2, p = 0.005). Nevertheless, anemones cut into quarters still produced the greatest number of anemones at the end of the experiment (20 quarters vs 15 halves). Among the quarters, fragments from morph 1 (75%) had greater survival than those from morph 2 (50%), but this result was not significant (Figure 2b, K-M, Wald = 1.7, df = 1,

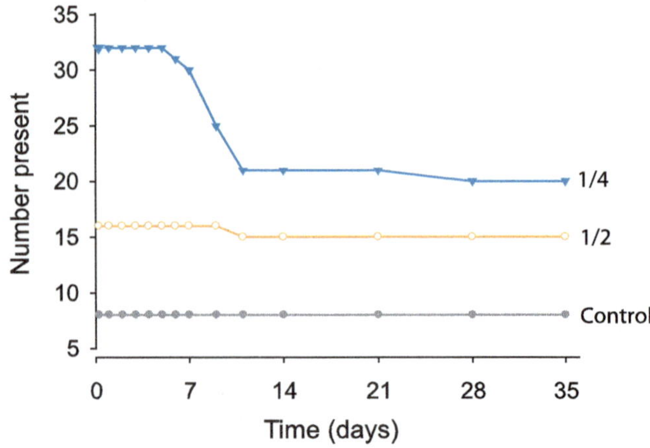

Figure 1. Number of *Entacmaea quadricolor* present in each treatment over time (in days) during experiment 1.

p = 0.197). Likewise, although more of the quarters from large anemones survived (75%), compared with those from small anemones (50%), the difference was not significant (Figure 2c, K-M, Wald = 2.3, df = 1, p = 0.131).

In experiment 2, no significant mortality was recorded among either halves or quarters relative to controls (Figure 3, Fisher's, $\chi^2 = 4.2$, df = 2, p = 0.124). All controls survived, and 89.3% and 80.4% of halves and quarters survived, respectively. Thus, the anemones cut in quarters produced almost double the number of anemones (n = 45 surviving fragments) than those cut in half (n = 25). Given the low mortality in this experiment, none of factors analysed (cut type, feeding or tanks) had a significant effect on the survival of fragments over time (Figure 4, K-M, p>0.05 in all cases).

Effects of feeding on size

Larger fragments resulted from cutting larger anemones for a given cut type. This initial fragment size had a significant influence on final fragment size as measured by all three variables, with anemones of a larger initial size remaining larger throughout the experiment (GLM; p<0.001 in all cases).

Although the ODD of the sea anemones showed no consistent positive or negative change over time across all treatments during the experiment (non-significant day effect, GLM, χ^2 deviance = 10.0, df = 5, p = 0.08), it did vary significantly among cut types and feeding treatments (Figure 5a–c). The ODD of quarters increased less during the experiment than that of halves or controls (significant cut type effect, GLM, χ^2 deviance = 9.9, df = 2, p = 0.007), even after the effect of initial fragment size was accounted for. The ODD of the fed anemones grew larger than unfed anemones (significant feeding effect, GLM, χ^2 deviance = 9.5, df = 1, p = 0.002), particularly among controls (Figure 5a) and to a lesser extent among halves later in the experiment (Figure 5b).

In comparison, feeding had no influence on the PDD of the anemones (non-significant feeding effect, GLM, χ^2 deviance = 2.6, df = 1, p = 0.108). Nevertheless, PDD generally increased throughout the experiment for all treatment groups (significant day effect, GLM, χ^2 deviance = 21.0, df = 5, p<0.001). Cut type also influenced PDD growth, with the controls and halves growing more than the quarters (significant cut type effect, GLM, χ^2 deviance = 12.5, df = 2, p = 0.002).

Although all of the anemones gradually decreased in wet weight, including the controls (significant day effect, GLM, χ^2

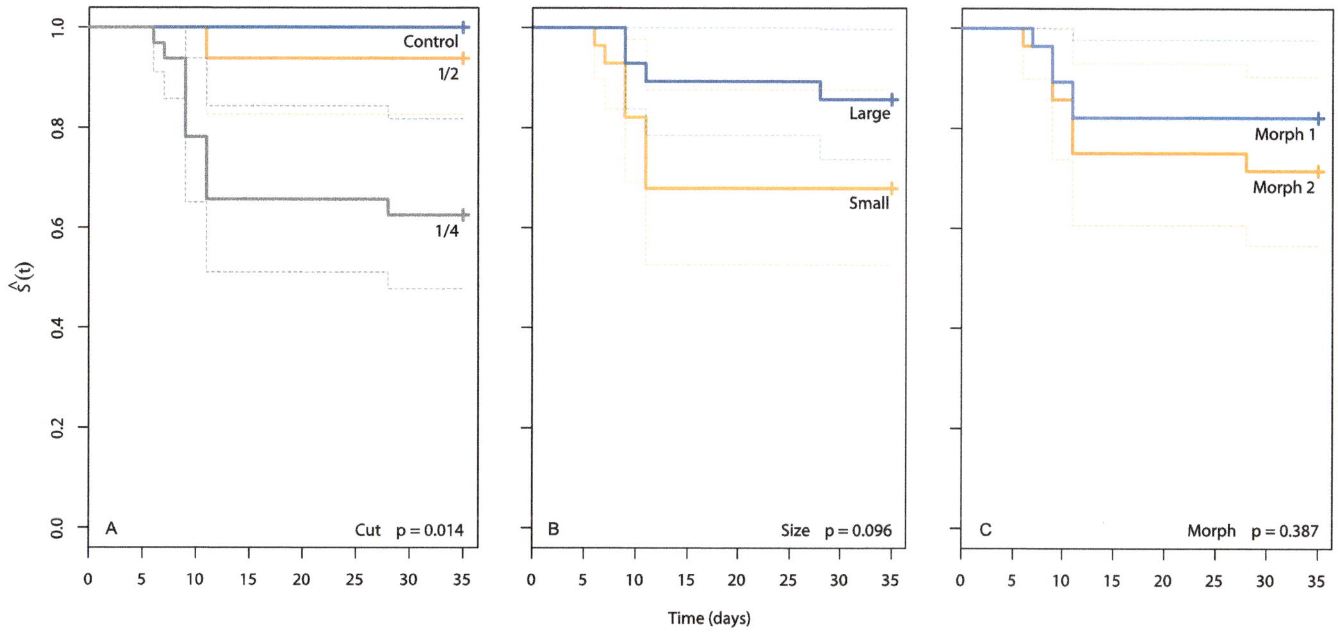

Figure 2. Kaplan-Meier survival [Ŝ(t)] plots for *Entacmaea quadricolor* among different: (a) cut types, (b) sizes, and (c) morphs during experiment 1. Dashed lines represent confidence intervals.

deviance $=68.6$, df$=3$, $p<0.001$), those that received food showed reduced wet weight loss in comparison to those that were unfed (significant feeding effect, GLM; χ^2 deviance $=5.3$, df$=1$, $p=0.022$). There was an interaction between these two factors, with feeding having a greater influence towards the end of the experiment than at the start (significant feeding × day effect, GLM, χ^2 deviance $=9.5$, df$=3$, $p=0.023$).

Visual observations

The majority of individuals in all of the treatments had attached to the bottom of the tubs 6 h after the start of experiment 2 (93% of the controls, 96% of the halves, and 89% of the quarters). Likewise, most of the anemones that weren't cut or were cut in half had re-expanded their oral disc and tentacles (93% and 73%, respectively); however only 32% of the quarters had re-expanded. Most anemones still had an open cut at this time (Figure 6a). While in others, the column had started to curl inwards towards the region of the cut (36% of the anemones cut in halves and 7% of those cut in quarters), or the column had completely covered the cut (39% of the anemones cut in half, Figure 6b). After 28 d, scar tissue covered the region where the cut was made in 96% of the cut individuals (Figure 6c).

All of the anemones accepted food 7 d after the start of the experiment (anemones that had been fragmented moved the food

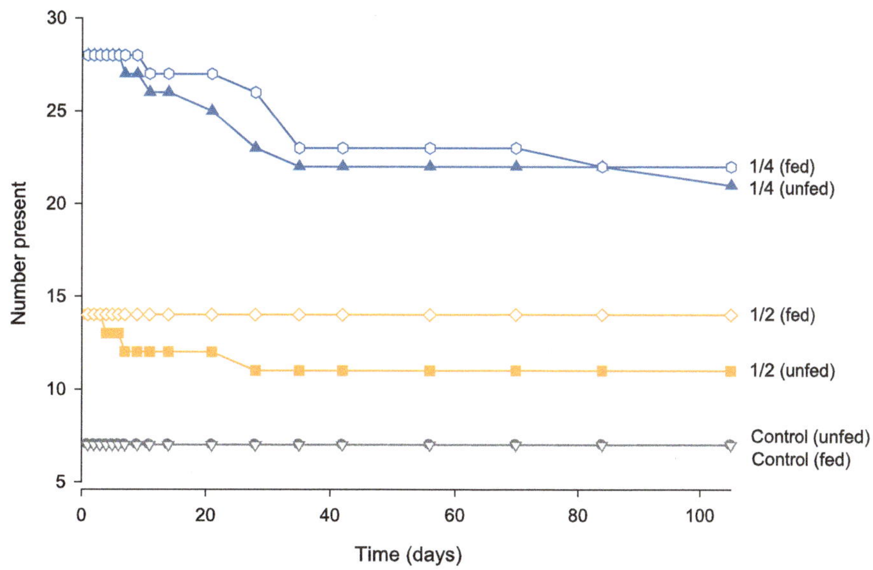

Figure 3. Number of *Entacmaea quadricolor* present in each treatment over time (in days) during experiment 2.

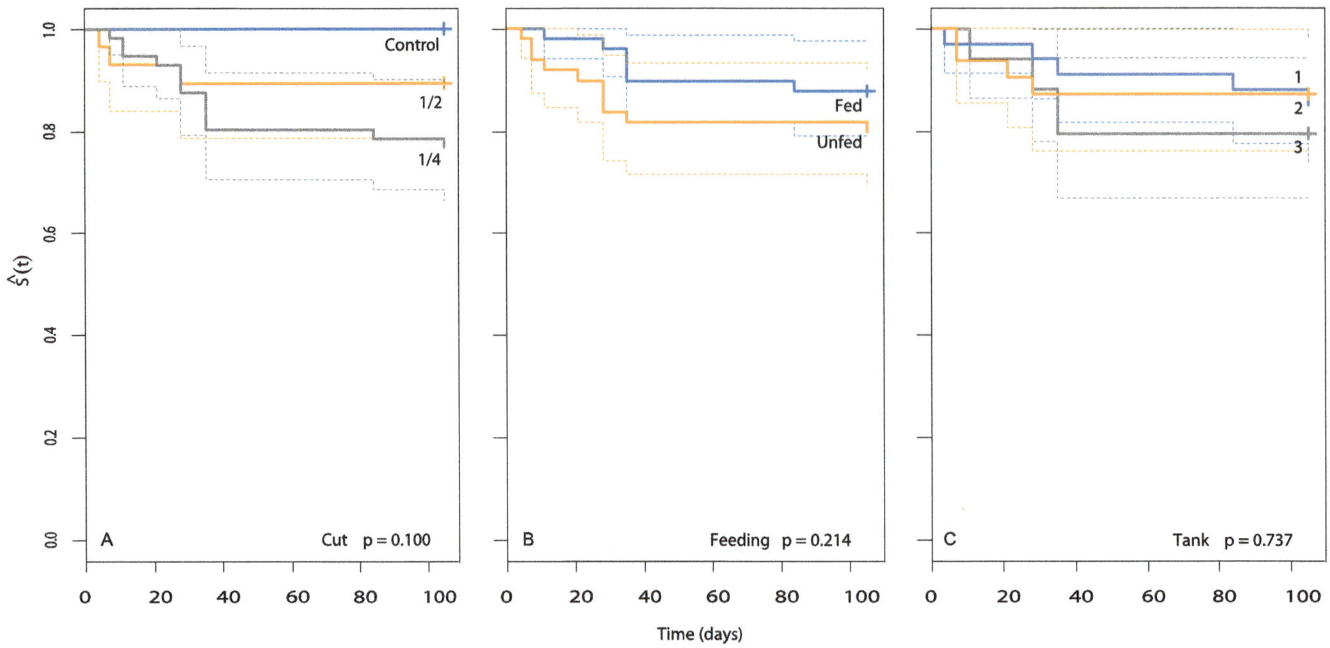

Figure 4. Kaplan-Meier survival [Ŝ(t)] plots for *Entacmaea quadricolor* among different: (a) cut types, (b) feeding treatments, and (c) tanks during experiment 2. Dashed lines represent confidence intervals.

towards the area of the cut). By day 56, off-centre mouths were clearly visible in all of the cut individuals (Figure 6d), except for one of the quarters, and by day 84, 68% of the anemones which were cut in half had mouths that were almost centrally located within the oral disc. This level of regeneration was not found for any of the anemones cut in quarters at any stage of the experiment.

Asexual reproduction did not occur naturally during the experiment; however, sexual reproduction did occur. Eggs were found in two tubs on the 16 January (19 nights after the full moon). One of these anemones had been cut in half, and the other had been cut in quarters 99 d earlier. It is not known if any males spawned, as sperm would have been rapidly removed from the tubs due to the flow-through seawater.

Discussion

E. quadricolor can be propagated asexually by cutting them in half or quarters. The survival rates of the fragments from the sea anemones that were cut in half were greater than has previously been reported (i.e. 89.3 and 93.8% for experiments 1 and 2 respectively vs. 67% in Porat and Chadwick-Furman [22]). Cutting the anemones in quarters produced more individuals; however mortality rates were comparatively higher, the individuals produced were smaller, and also tended to recover and grow more slowly. The latter finding contrasts other growth experiments with intact anemones, where small anemones often tend to grow faster than larger ones [23,24]. This difference likely arises from the smaller fragments needing more time to heal and grow a mouth. Irrespective of whether anemones are cut in halves or quarters, the methods detailed provide a low technology solution for culturing this species that would allow for the year round production of individuals for the aquarium trade. Furthermore, a major advantage of this technique for industry would be that once procured, desirable colour morphs that attract higher prices could be cloned thereby increasingly profitability.

Feeding increased ODD growth and reduced wet weight loss, whereas it did not significantly influence PDD growth. Ammonia supplements or the presence of anemonefish (which produce wastes that can be used by their host) have similarly been shown to reduce size loss in *E. quadricolor* [25,26]. It is possible that more frequent feeding may have been necessary to optimise growth during our study. For example, Chomsky et al. [24] found that *Actinia equina* needed to be fed twice a week for significant PDD growth. However, given that the size difference of fed and unfed anemones was greatest in individuals that were not cut, the similarity between feeding treatments in the cut anemones may have been due to the time it took for them to grow a fully functioning mouth (56 d). It is therefore likely that running the experiment for longer would have resulted in a greater feeding effect. Regardless of the treatment, all of the anemones would have received some nutritional benefits from their *Symbiodinium*, microscopic particulate matter, or dissolved organic matter that would have provided energy and potentially influenced size [26–28].

Precisely measuring size change in sea anemones using non-destructive methods is not simple given their ability to expand, contract, and store variable amounts of water in the body [29,30]. In this study, we found that wet weight showed greater variation and less correlation with the linear size measurements of ODD and PDD. Wet weight has been used in a number of studies on sea anemone growth (for example [11,23,24]), however these were generally on smaller anemones. Although we did try to standardise the amount of water contained within the anemones it is likely that some variation among individuals was amplified by their larger body size. Despite this, a significant positive influence of feeding on wet weight was still detected.

It is important to recognise that the findings of this study relating to survival might not apply to all sea anemone species. In anemones that use fission as a reproductive mode, wound healing and regeneration occur by rapid cell proliferation at the wound site [31]. Species that do not naturally asexually reproduce may

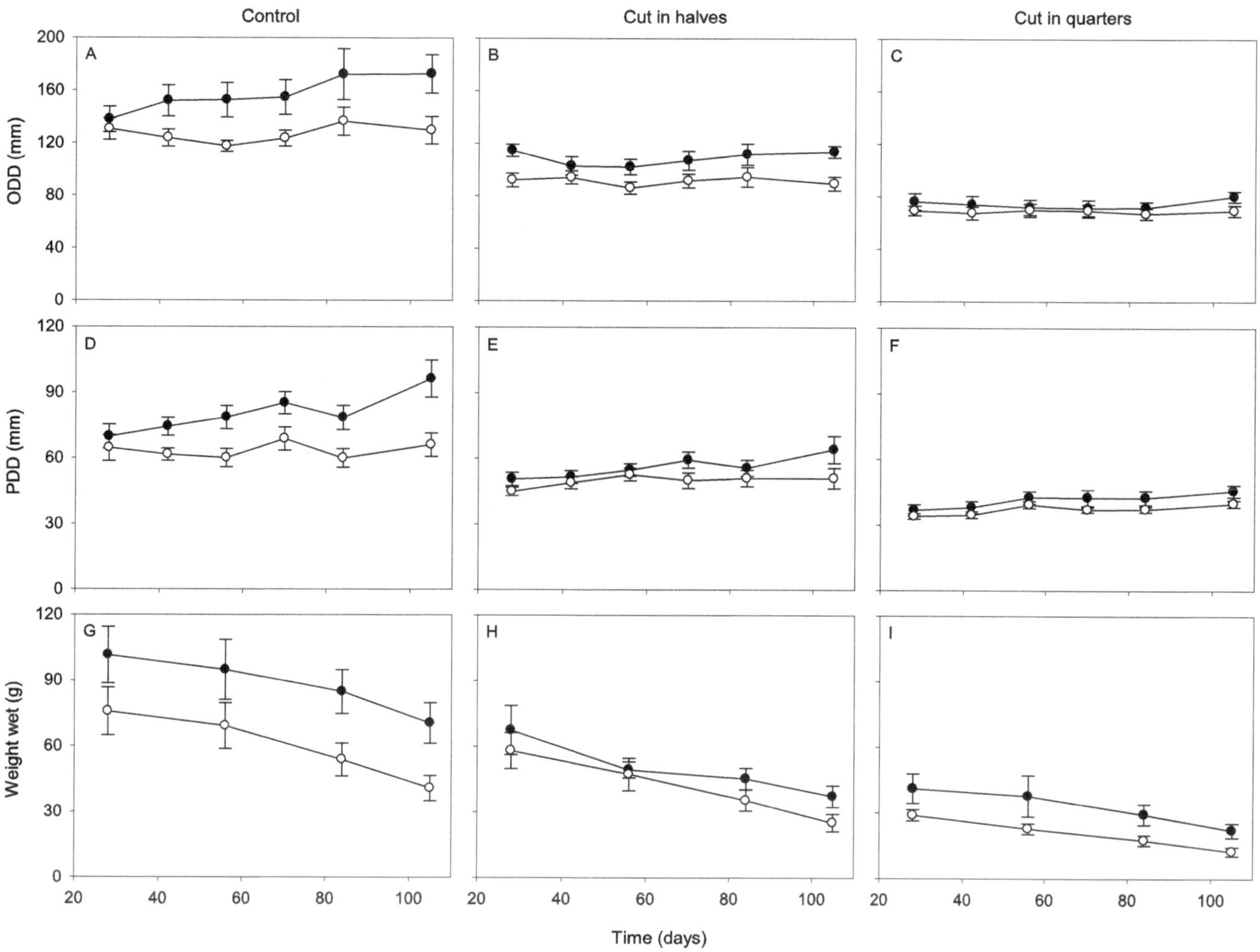

Figure 5. Change in size of *Entacmaea quadricolor* over time (in days) during experiment 2. Estimated means (±SE) of fitted (a–c) ODD, (d–f) PDD, and (g–i) wet weight from generalized linear models.

not have this capacity, and therefore may not be able to recover from cutting. Another important ethical consideration is whether cutting the anemones may cause them pain. Cnidarians possess a basic nervous system comprised of a diffuse nerve net with sensory neuron agglomerations at key structures [32]. Column stimulation produces nerve impulses that cause a closure reflex, which might appear to be a nociceptive response; however, nociceptors in Cnidaria only respond to mechanical stimuli and not heat stimuli and without a central processing unit they are unlikely to experience pain [32].

The practical application of the technique developed during this study would allow the creation of a new industry, which could have economic as well as environmental benefits. Collecting pressures on wild populations could therefore be reduced by supplying the aquarium trade with captive-bred anemones. Fragmentation is already used to supply a small percentage of hard and soft corals (which are closely related to sea anemones) for the trade [2,33,34]. The study species has been propagated by natural asexual reproduction for research purposes [Chadwick and Delbeek personal communication] and cutting has been used by aquarium hobbyists. Although marine ornamental aquaculture is still in its infancy, it is thought that the propagation of animals for the trade will become more important as further restrictions are placed on wild collection and consumers become more aware of the potential adverse impacts of these activities [12,13]. The potential for diversification and growth in this sector is promising due to high product value and the benefits to biodiversity conservation; however viability is highly dependent on the price at which individuals can be produced [2,13].

Because aquarium collecting provides employment in rural low-income coastal areas that have otherwise limited resources and economic options, it would be preferable if aquaculture occurred in the areas that are currently exporting anemones [2,5,12]. Given the simplicity of the technique developed during this study it could easily be used without personnel needing a large amount of training, thus allowing livelihoods to be maintained in a more sustainable manner. This may have flow-on benefits such as building the technical capacity of communities and fostering awareness for better stewardship of resources [35]. Furthermore, it may be possible to apply this technique in the field (either in cages or on the reef flat), which would negate the need for captive breeding facilities that are expensive to set up and maintain.

Asexual propagation could also be used for restoration programs and could potentially occur *in situ*. The restocking of areas would help restore the breeding populations needed to help replenish areas that have already been denuded by human or

Figure 6. Recovery of *Entacmaea quadricolor* fragments after cutting: (a) open cut, (b) column tissue covering the cut, (c) column coloured scar tissue, and (d) off-centre mouth.

natural disturbances [36]; and therefore have subsequent benefits for anemonefishes that cannot survive in the field without their hosts [10]. Given that this technique would produce individuals with limited genetic variability, the consequences of this would need to be considered [37], along with the causative factors that necessitated restocking in the first place. Management initiatives, such as better regulation and enforcement, may also be needed to address the issues that caused the reductions [36].

The findings of this study demonstrate a feasible solution for reducing aquarium collecting pressures that are impacting host sea anemone abundance in some areas of their distribution [8,38], and could also allow for the restoration of areas already impacted by human or natural disturbances. Culturing these species would offer economic and environmental benefits by providing an alternative source of these high-value species, which provide essential microhabitat for obligate symbiotic anemonefishes.

Acknowledgments

We would like to thank R. Edgar and S. Dalton for help in the field; T. Piddocke, K. Cowden, D. Eggeling and K. Kaposi for help in the laboratory; and N. Chadwick for comments on the manuscript.

Author Contributions

Conceived and designed the experiments: AS KH. Performed the experiments: JH AS KH. Analyzed the data: KH. Contributed reagents/materials/analysis tools: AS KH JH. Wrote the paper: AS KH.

References

1. Whitehead MJ, Gilmore E, Eager P, McMinnity P, Craik W, et al. (1986)Aquarium fishes and their collection in the Great Barrier Reef region. Townsville: GBRMPA. 1–39 p.
2. Wabnitz C, Taylor M, Green E, Razak T (2003) From ocean to aquarium: the global trade in marine ornamental species. Cambridge, UK: UNEP-WCMC.
3. Shuman CS, Hodgson G, Ambrose RF (2004) Managing the marine aquarium trade: is eco-certification the answer? Environ Conserv 31: 339–348.
4. Wood E (2001) Global advances in conservation and management of marine ornamental resources. Aquarium Sciences and Conservation 3: 65–77.
5. Wood EM (2001) Collection of coral reef fish for aquaria: global trade, conservation issues and management strategies. UK: Marine Conservation Society. 80 pp.
6. Edwards AJ, Shepherd AD (1992) Environmental implications of aquarium-fish collection in the Maldives, with proposals for regulation. Environ Conserv 19: 61–72.
7. Gasparini J, Floeter S, Ferreira C, Sazima I (2005) Marine ornamental trade in Brazil. Biodiversity and Conservation 14: 2883–2899.
8. Shuman CS, Hodgson G, Ambrose RF (2005) Population impacts of collecting sea anemones and anemonefish for the marine aquarium trade in the Philippines. Coral Reefs 24: 564–573.
9. Olivotto I, Planas M, Simões N, Holt GJ, Avella MA, et al. (2011) Advances in breeding and rearing marine ornamentals. Journal of the World Aquaculture Society 42: 135–166.
10. Fautin DG, Allen GR (1997) Anemonefishes and their host sea anemones: a guide for aquarists and divers. Perth: Western Australian Museum.
11. Scott A (2012) Effects of feeding on the growth rates of captive-bred *Heteractis crispa*: a popular marine ornamental for aquariums. Bull Mar Sci 88: 81–87.
12. Tlusty M (2002) The benefits and risks of aquacultural production for the aquarium trade. Aquaculture 205: 203–219.
13. Moorhead JA, Zeng C (2010) Development of captive breeding techniques for marine ornamental fish: a review. Reviews in Fisheries Science 18: 315–343.
14. Scott A, Harrison P (2007) Broadcast spawning of two species of sea anemone, *Entacmaea quadricolor* and *Heteractis crispa*, that host anemonefish. Invertebr Reprod Dev 50: 163–171.
15. Scott A, Harrison PL (2007) Embryonic and larval development of the host sea anemones *Entacmaea quadricolor* and *Heteractis crispa*. Biol Bull 213: 110–121.
16. Scott A, Harrison PL (2008) Larval settlement and juvenile development of sea anemones that provide habitat for anemonefish. Mar Biol 154: 833–839.
17. Scott A, Harrison PL (2009) Gametogenic and reproductive cycles of the sea anemone, *Entacmaea quadricolor*. Mar Biol 156: 1659–1671.
18. Scott A, Harrison PL (2005) Synchronous spawning of host sea anemones. Coral Reefs 24: 208.
19. Dunn DF (1981) The clownfish sea anemones: Stichodactylidae (Coelenterata: Actiniaria) and other sea anemones symbiotic with pomacentrid fishes. Trans Am Philos Soc 71: 1–115.
20. Stephenson TA (1929) On methods of reproduction as specific characters. J Mar Biol Assoc UK 16: 131–172.
21. Mire P (1998) Evidence for stretch-regulation of fission in a sea anemone. J Exp Zool 282: 344–359.

22. Porat D, Chadwick-Furman NE (2005) Effects of anemonefish on giant sea anemones: ammonium uptake, zooxanthella content and tissue regeneration. Mar Freshwat Behav Physiol 38: 43–51.
23. Tsuchida CB, Potts DC (1994a) The effects of illumination, food and symbionts on growth of the sea anemone *Anthopleura elegantissima* (Brandt, 1835). 1. Ramet Growth. J Exp Mar Biol Ecol 183: 227–242.
24. Chomsky O, Kamenir Y, Hyams M, Dubinsky Z, Chadwick-Furman NE (2004) Effects of feeding regime on growth rate in the Mediterranean sea anemone *Actinia equina* (Linnaeus). J Exp Mar Biol Ecol 299: 217–229.
25. Porat D, Chadwick-Furman NE (2004) Effects of anemonefish on giant sea anemones: expansion behavior, growth, and survival. Hydrobiologia 530/531: 513–520.
26. Roopin M, Chadwick NE (2009) Benefits to host sea anemones from ammonia contributions of resident anemonefish. J Exp Mar Biol Ecol 370: 27–34.
27. Fitt WK, Pardy RL (1981) Effects of starvation, and light and dark on the energy metabolism of symbiotic and aposymbiotic sea anemone, *Anthopleura elegantissima*. Mar Biol 61: 199–205.
28. Muller-Parker G, Davy SK (2001) Temperate and tropical algal sea anemone symbiosis. Invertebr Biol 120: 104–123.
29. Hirose Y (1985) Habitat, distribution and abundance of coral reef sea-anemones (Actiniidea and Stichodactylidae) in Sesoko Island, Okinawa, with notes on expansion and contraction behavior. Galaxea 4: 113–127.
30. Leal MC, Nunes C, Engrola S, Dinis MT, Calado R (2012) Optimization of monoclonal production of the glass anemone *Aiptasia pallida* (Agassiz in Verrill, 1864). Aquaculture 354: 91–96.
31. Passamaneck YJ, Martindale MQ (2012) Cell proliferation is necessary for the regeneration of oral structures in the anthozoan cnidarian *Nematostella vectensis*. BMC Dev Biol 12: 34.
32. Smith ESJ, Lewin GR (2009) Nociceptors: a phylogenetic view. J Comp Physiol A 195: 1089–1106.
33. Borneman EH, Lowrie J (2001) Advances in captive husbandry and propagation: An easily utilized reef replenishment means from the private sector? B Mar Sci 69: 897–913.
34. Olivotto I, Planas M, Simões N, Holt GJ, Avella MA, et al. (2011) Advances in breeding and rearing marine ornamentals. J World Aquacult Soc 42: 135–166.
35. Purcell SW (2012) Principles and science of stocking marine areas with sea cucumbers In: Hair CA, Pickering TD, Mills DJ, editors. Asia–Pacific Tropical Sea Cucumber Aquaculture: ACIAR Proceedings, 136. Australian Centre for International Agricultural Research, Canberra. pp.92–103.
36. Bell JD, Bartley DM, Lorenzen K, Loneragan NR (2006) Restocking and stock enhancement of coastal fisheries: potential, problems and progress. Fish Res 80: 1–8.
37. Baums IB (2008) A restoration genetics guide for coral reef conservation. Mol Ecol 17: 2796–2811.
38. Edwards AJ, Shepherd AD (1992) Environmental implications of aquarium-fish collection in the Maldives, with proposals for regulation. Environ Conserv 19: 61–72.

Permissions

All chapters in this book were first published in PLOS ONE, by The Public Library of Science; hereby published with permission under the Creative Commons Attribution License or equivalent. Every chapter published in this book has been scrutinized by our experts. Their significance has been extensively debated. The topics covered herein carry significant findings which will fuel the growth of the discipline. They may even be implemented as practical applications or may be referred to as a beginning point for another development.

The contributors of this book come from diverse backgrounds, making this book a truly international effort. This book will bring forth new frontiers with its revolutionizing research information and detailed analysis of the nascent developments around the world.

We would like to thank all the contributing authors for lending their expertise to make the book truly unique. They have played a crucial role in the development of this book. Without their invaluable contributions this book wouldn't have been possible. They have made vital efforts to compile up to date information on the varied aspects of this subject to make this book a valuable addition to the collection of many professionals and students.

This book was conceptualized with the vision of imparting up-to-date information and advanced data in this field. To ensure the same, a matchless editorial board was set up. Every individual on the board went through rigorous rounds of assessment to prove their worth. After which they invested a large part of their time researching and compiling the most relevant data for our readers.

The editorial board has been involved in producing this book since its inception. They have spent rigorous hours researching and exploring the diverse topics which have resulted in the successful publishing of this book. They have passed on their knowledge of decades through this book. To expedite this challenging task, the publisher supported the team at every step. A small team of assistant editors was also appointed to further simplify the editing procedure and attain best results for the readers.

Apart from the editorial board, the designing team has also invested a significant amount of their time in understanding the subject and creating the most relevant covers. They scrutinized every image to scout for the most suitable representation of the subject and create an appropriate cover for the book.

The publishing team has been an ardent support to the editorial, designing and production team. Their endless efforts to recruit the best for this project, has resulted in the accomplishment of this book. They are a veteran in the field of academics and their pool of knowledge is as vast as their experience in printing. Their expertise and guidance has proved useful at every step. Their uncompromising quality standards have made this book an exceptional effort. Their encouragement from time to time has been an inspiration for everyone.

The publisher and the editorial board hope that this book will prove to be a valuable piece of knowledge for researchers, students, practitioners and scholars across the globe.

List of Contributors

Troy C. Nelson
Fraser River Sturgeon Conservation Society, Vancouver, British Columbia, Canada

Phaedra Doukakis
Scripps Institution of Oceanography, University of California San Diego, San Diego, California, United States of America

Steven T. Lindley
National Marine Fisheries Service, Southwest Fisheries Science Center, Santa Cruz, California, United States of America

Andrea D. Schreier
Department of Animal Science, University of California Davis, Davis, California, United States of America

Joseph E. Hightower
U. S. Geological Survey, NC Cooperative Fish and Wildlife Research Unit, Department of Applied Ecology, NC State University, Raleigh, North Carolina, United States of America

Larry R. Hildebrand
Golder Associates Ltd., Castlegar, British Columbia, Canada

Rebecca E. Whitlock
Stanford University, Hopkins Marine Station, Pacific Grove, California, United States of America

Molly A. H. Webb
U. S. Fish and Wildlife Service, Bozeman Fish Technology Center, Bozeman, Montana, United States of America

Anping Lei, Wenlong Hao and Junda Yang
College of Life Sciences, Shenzhen University, Shenzhen, China

Chao Wang
College of Life Sciences, Shenzhen University, Shenzhen, China
Institute of Hydrobiology, Chinese Academy of Sciences, Wuhan, China

Zhengyu Hu
Institute of Hydrobiology, Chinese Academy of Sciences, Wuhan, China

Kai Zhou
Shenzhen Marine Environment and Resources Monitoring Center, Shenzhen, China

Annette Taugbøl, Tina Arntsen and Leif Asbjørn Vøllestad
Centre for Ecological and Evolutionary Synthesis (CEES), Department of Biosciences, University of Oslo, Blindern, Norway

Kjartan Østbye
Centre for Ecological and Evolutionary Synthesis (CEES), Department of Biosciences, University of Oslo, Blindern, Norway
Hedmark University College, Department of Forestry and Wildlife Management, Campus Evenstad, Elverum, Norway

Xuelu Gao
Key Laboratory of Coastal Environmental Processes and Ecological Remediation, Yantai Institute of Coastal Zone Research, Chinese Academy of Sciences, Yantai, Shandong, China

Wen Zhuang
Key Laboratory of Coastal Environmental Processes and Ecological Remediation, Yantai Institute of Coastal Zone Research, Chinese Academy of Sciences, Yantai, Shandong, China
College of City and Architecture Engineering, Zaozhuang University, Zaozhuang, Shandong, China

Jessica J. Meeuwig
School of Animal Biology and The UWA Oceans Institute, The University of Western Australia, Perth, Western Australia, Australia

Gabriel M. S. Vianna
School of Animal Biology and The UWA Oceans Institute, The University of Western Australia, Perth, Western Australia, Australia
Australian Institute of Marine Science, Perth, Western Australia, Australia

Mark G. Meekan
Australian Institute of Marine Science, Perth, Western Australia, Australia

Tova H. Bornovski
Micronesian Shark Foundation, Koror, PalauUniversity of Chinese Academy of Sciences, Beijing, China

Nadia S. Santini and Catherine E. Lovelock
The School of Biological Sciences, The University of Queensland, St Lucia, Brisbane, Queensland, Australia

Quan Hua
Institute for Environmental Research, Australian Nuclear
Science and Technology Organisation, Lucas Heights, New South Wales, Australia

Nele Schmitz
Plant Biology and Nature Management, Vrije Universiteit Brussel, Brussels, Belgium

Haiqiang Wu, Fang Zhang, Liao Zhang, Zeqiu Liang, Jinyu Wang and Yizhi Zheng
College of Life Sciences, Shenzhen University, Shenzhen, China

Neil Williamson and Alan Tunnacliffe
Department of Chemical Engineering and Biotechnology, University of Cambridge, Cambridge, United Kingdom

Jie Jian
Department of Chemical Engineering and Biotechnology, University of Cambridge, Cambridge, United Kingdom
College of Pharmacy, Guilin Medical University, Guilin, China

Linkun An
School of Pharmaceutical Science, Sun Yat-sen University, Guangzhou, China

Timothy A. Gowan and Joel G. Ortega-Ortiz
Florida Fish and Wildlife Conservation Commission, Fish and Wildlife Research Institute, St. Petersburg, Florida, United States of America

Michaela Schratzberger
Centre for Environment, Fisheries and Aquaculture Science, Lowestoft Laboratory, Lowestoft, Suffolk, United Kingdom

Piers Larcombe
RPS MetOcean, Subiaco, Western Australia, Australia

Yi-Ming Lin, Zhen-Hua Ding, Wei-Ming Chai and Shu-Dong Wei
Key Laboratory of the Ministry of Education for Coastal and Wetland Ecosystems, Xiamen University, Xiamen, China

Hai-Chao Zhou
Key Laboratory of the Ministry of Education for Coastal and Wetland Ecosystems, Xiamen University, Xiamen, China
Department of Biology and Chemistry, City University of Hong Kong, Hong Kong SAR, China
Futian-City Mangrove R&D Centre, City University of Hong Kong Shenzhen Research Institute, Shenzhen, China

Nora Fung-yee Tam
Department of Biology and Chemistry, City University of Hong Kong, Hong Kong SAR, China
Futian-CityU Mangrove R&D Centre, City University of Hong Kong Shenzhen Research Institute, Shenzhen, China

Luciano M. Chiaverano
Kewalo Marine Laboratory, Pacific Biosciences Research Center, University of Hawaii at Mānoa, Honolulu, Hawaii, United States of America

Brenden S. Holland
Center for Conservation Research & Training, Pacific Biosciences Research Center, University of Hawaii at Mānoa, Honolulu, Hawaii, United States of America

Gerald L. Crow
Waikiki Aquarium, University of Hawaii at Mānoa, Honolulu, Hawaii, United States of America

Landy Blair
Ocean Safety and Lifeguard Services, City and County of Honolulu, Honolulu, Hawaii, United States of America

Angel A. Yanagihara
Bekesy Laboratory, Pacific Biosciences Research Center, and Department of Tropical Medicine, John A. Burns School of Medicine, University of Hawaii at Mānoa, Honolulu, Hawaii, United States of America

Kevin Liebrand1 Harm G. van der Geest and Wim Admiraal
Department of Aquatic Ecology and Ecotoxicology, Institute for Biodiversity and Ecosystem Dynamics, University of Amsterdam, Amsterdam, The Netherlands

Brittany E. Alexander and Jasper M. de Goeij
Department of Aquatic Ecology and Ecotoxicology, Institute for Biodiversity and Ecosystem Dynamics, University of Amsterdam, Amsterdam, The Netherlands
Porifarma B.V. Poelbos 3, Ede, The Netherlands

Ronald Osinga
Porifarma B.V. Poelbos 3, Ede, The Netherlands

Jack P. M. Cleutjens
Department of Pathology, Cardiovascular Research Institute Maastricht, Maastricht University, Maastricht, The Netherlands

Bert Schutte
Department of Molecular Cell Biology, Research Institute Growth and Development, Maastricht University, Maastricht, The Netherlands

Fons Verheyen
Electron Microscopy Unit, CRISP, Maastricht, The Netherlands

Marta Ribes
Institut de Ciéncies del Mar-Consejo Superior de Investigaciones Científicas (ICM-CSIC), Barcelona, Spain

Emiel van Loon
Department of Computational Geo-Ecology, Institute for Biodiversity and Ecosystem Dynamics, University of Amsterdam, Amsterdam, The Netherlands

Chiara Maggi, Serena Lomiri, Bianca Di Lorenzo, Marco d'Antona and Maria Teresa Berducci
ISPRA, Institute for Environmental Protection and Research, Roma, Italy

Charitha Pattiaratchi
School of Environmental Systems Engineering, University of Western Australia, Perth, Western Australia, Australia
Oceans Institute, University of Western Australia, Perth, Western Australia, Australia

Julia Reisser
School of Environmental Systems Engineering, University of Western Australia, Perth, Western Australia, Australia
Oceans Institute, University of Western Australia, Perth, Western Australia, Australia
Wealth from Oceans Flagship, Commonwealth Scientific and Industrial Research Organisation, Floreat, Western Australia, Australia

Michele Thums
School of Environmental Systems Engineering, University of Western Australia, Perth, Western Australia, Australia
Oceans Institute, University of Western Australia, Perth, Western Australia, Australia
Australian Institute of Marine Science, Perth, Western Australia, Australia

Chris Wilcox and Britta Denise Hardesty
Wealth from Oceans Flagship, Commonwealth Scientific and Industrial Research Organisation, Floreat, Western Australia, Australia
Marine and Atmospheric Research, Commonwealth Scientific and Industrial Research Organisation, Hobart, Tasmania, Australia

Jeremy Shaw
Centre for Microscopy, Characterisation and Analysis, University of Western Australia, Perth, Western Australia, Australia

Maira Proietti
Instituto de Oceanografia, Universidade Federal do Rio Grande, Rio Grande, Rio Grande do Sul, Brazil

Christopher K. Pham, Eduardo Isidro, Telmo Morato and JoséNuno Gomes-Pereira
Center of the Institute of Marine Research (IMAR) and Department of Oceanography and Fisheries, University of the Azores, Horta, Portugal
Laboratory of Robotics and Systems in Engineering and Science (LARSyS), Lisbon, Portugal

Eva Joan B. Company
Institut de Ciéncies del Mar (ICM-CSIC), Barcelona, Spain

Ramirez-Llodra
Institut de Ciéncies del Mar (ICM-CSIC), Barcelona, Spain
Norwegian Institute for Water Research (NIVA), Marine Biology section, Oslo, Norway

Claudia H. S. Alt and Paul A. Tyler
Ocean and Earth Science, University of Southampton, National Oceanography Centre, Southampton, United Kingdom

Teresa Amaro
Norwegian Institute for Water Research, Bergen, Norway

Melanie Bergmann
Alfred-Wegener-Institut, Helmholtz-Zentrum für Polar- und Meeresforschung, Bremerhaven, Germany

Miquel Canals, Galderic Lastras and Xavier Tubau
GRC Geociéncies Marines, Departament d'Estratigrafia, Paleontologia i Geociéncies Marines, Facultat de Geologia, Universitat de Barcelona, Campus de Pedralbes, Barcelona, Spain

Jaime Davies and Kerry L. Howell
Marine Biology & Ecology Research Centre, Marine Institute, Plymouth University, Plymouth, United Kingdom

Gerard Duineveld
Netherlands Institute for Sea Research (NIOZ), Texel, The Netherlands

François Galgani
Institut Franc¸ais de Recherche pour l'Exploitation de la Mer (IFREMER), Bastia, France

Veerle A. I. Huvenne and Daniel O. B. Jones
National Oceanography Centre, University of Southampton Waterfront Campus, Southampton, United Kingdom

Autun Purser
OceanLab, Jacobs University Bremen, Bremen, Germany

Heather Stewart
British Geological Survey, Murchison House, Edinburgh, United Kingdom

Inês Tojeira
Portuguese Task Group for the Extension of the Continental Shelf (EMEPC), Paço de Arcos, Portugal

David Van Rooij
Renard Centre of Marine Geology (RCMG), Department of Geology and Soil Science, Ghent University, Gent, Belgium

Dominik Paprotny, Paweł Andrzejewski, Paweł Terefenko, Kazimierz Furmań czyk
Faculty of Geosciences, University of Szczecin, Szczecin, Poland

Karsten Laursen
Institute of Bioscience, Aarhus University, Rønde, Denmark

Anders Pape Møller
Laboratoire d'Ecologie, Systématique et Evolution, CNRS UMR 8079, UniversitéParis-Sud, Orsay, France

Nathalie Pülmanns and Inga Nordhaus
Department of Ecology, Leibniz Center for Tropical Marine Ecology (ZMT), Bremen, Germany

Karen Diele
School of Life, Sport and Social Sciences, Edinburgh Napier University,
Edinburgh, United Kingdom

Ulf Mehlig
Instituto de Estudos Costeiros, Universidade Federal do Pará- Campus de Bragança, Bragança, PA, Brazil

Richard M. Allen and Dustin Marshall
School of Biological Sciences, University of Queensland, St Lucia, Queensland, Australia

Lisa M. Dann, James G. Mitchell, Peter G. Speck, Kelly Newton and James Paterson
School of Biological Sciences, Flinders University, Adelaide, South Australia, Australia

Thomas Jeffries
School of the Environment, University of Technology Sydney, Sydney, New South Wales, Australia

Anna Scott, Jannah M. Hardefeldt and Karina C. Hall
National Marine Science Centre and Marine Ecology Research Centre, School of Environment, Science and Engineering, Southern Cross University, Coffs Harbour, New South Wales, Australia

Index

A

Acipenser Medirostris, 3, 19-22

Acipenser Oxyrinchus Desotoi, 1-3, 19, 22

Acipenseridae, 1-3, 18, 21

Acoustic Telemetry, 6, 19, 22, 59-61, 63, 66, 69, 133

Ammonium, 23-27, 29, 31-32, 237

Anadromous, 1-2, 5, 7, 16-19, 40

Arid, 71, 77

Avicennia Marina, 71, 74-75, 77-78, 203

B

Biomass, 12, 15-16, 22-23, 25-26, 31, 60, 77-78, 112, 124, 126, 128-132, 135, 137, 139, 142, 188, 193-194, 216-217, 225-227

Bloom Formation, 23-24

C

Carcharhinus Amblyrhynchos, 59-60, 70

Cdna Synthesis, 35

Citizen Science, 59-60, 66, 68-69

Climatic Gradients, 71

Communities, 71, 93-94, 96, 100-101, 103, 105, 107-108, 110-113, 135, 144, 178-179, 192, 216-217, 224-226, 228-229, 235

Coral Reefs, 59-60, 68-69, 113, 134, 143, 230, 236

Cystic Fibrosis Transmembrane Regulator, 33-35

D

Data Analysis, 9, 11, 25, 63, 72, 80, 165, 171, 205, 210, 215, 218, 232

Data Collection, 1, 23, 33, 42, 59-60, 68-69, 71, 93, 114, 124, 127, 134, 145-146, 156, 167, 178, 180, 188, 194, 207, 216, 230

Degradation, 1, 7, 115, 121, 144, 158, 169, 175, 177, 194, 217

Densitometry, 71-72

Density, 15-16, 31, 60, 71-78, 83, 86, 90-91, 167-169, 171-173, 175, 177, 196-197, 201-203, 214-215, 218, 224

Dive Guides, 59-60, 62-66, 68-69

Dive Sites, 59-61, 63-64, 68

E

Ecological Risk Factor, 42, 45

Electronic Tagging, 1, 8, 12, 17, 19

Enrichment Factor, 42, 44

Environments, 1-2, 12, 14-17, 19, 27, 31, 33-35, 39-40, 51, 56, 60, 69-70, 93-94, 99, 101, 103, 105, 107, 110-113, 124-125, 157, 164, 167, 178, 205, 207, 214, 216-217, 226, 228-230

Estuarine, 1, 3-5, 9, 13, 15-17, 19-23, 26-27, 29-32, 41, 56-58, 71, 77, 112, 166, 228-229

Estuarine Environment, 23, 29-31

Eubalaena Glacialis, 79, 91

Euryhaline Fish, 33

F

Feeders, 93, 101, 103, 105-108, 110, 141

Fish Osmoregulate, 33

Fishes, 1-2, 13-14, 18, 21, 37, 40-41, 69, 153, 165, 178, 214, 236

Freshwater, 1, 3, 5, 8, 11, 14-15, 17, 22, 26, 31, 33-41, 70-72, 76-77, 112-113, 133, 135, 144

G

Gene Expression, 33-37, 39-41

Genetics, 1, 12-13, 17, 20-21, 41, 215, 237

Gill Tissue, 33-34, 40

Gracilaria Tenuistipitata, 23, 30-32

Grey Reef Sharks, 59, 61, 63-66, 69-70

Growth Rates, 7, 12, 25, 71-74, 76-77, 105, 131, 135, 141, 231, 236

Gulf, 1-4, 17-22, 71-74, 76-78, 82, 84, 86-87, 89, 91, 112, 132, 155-156, 162, 168-176, 179, 193

H

Habitat, 1-2, 4-12, 14-22, 39, 59, 69-70, 79-81, 88-91, 94, 107, 110, 112-113, 125-126, 139, 145, 150, 154, 179, 231, 236-237

Heat Shock Protein Gene, 33-34, 41

M

Macroalgae, 23-24, 27-28, 31-32

Mangrove, 71-74, 77-78, 114-115, 118, 122, 134-135, 137-139, 141, 143-144, 194-197, 200-206, 217, 226, 228

Marine, 1-5, 7-10, 13-19, 21-24, 27, 29-43, 45, 47, 51, 56-57, 59-60, 66, 69-70, 77, 79-80, 91-94, 103, 111-113, 122, 124, 131-135, 137, 143-146, 148, 151, 153-160, 162, 164-167, 169-170, 173, 175, 177-180, 186, 188, 190, 192-194, 205, 207-210, 214-217, 219, 228-231, 235-237

Marine Megafauna, 59, 157

Maturation, 1, 12, 133

Meiofaunal Nematode, 93-94, 105, 111

Metabolism, 28, 32-33, 237

Metal, 8, 42-48, 50-51, 55-58, 112, 153-155, 171-172, 175-177

Metal Fractionation, 48, 57

Microchemistry, 1, 7, 13-14, 17, 21

Migrations, 1-3, 5, 7, 9, 11, 14, 17-19, 21-22

Monitoring, 1, 7, 12, 15, 18-20, 22-23, 25, 59-60, 68-69, 80, 90, 92, 105, 112, 145-146, 153-155, 165, 173, 179-182, 189-191, 193

N
North Atlantic Right Whale, 79-80, 91

O
Observatory, 1, 167, 171-172, 178

Oceanic Environment, 79

Osmoregulatory Genes, 33-34, 37, 41

Overfishing, 1, 132, 196, 205

P
Pollution, 1, 32, 42, 44, 52, 56-57, 112, 132, 146, 155-158, 160, 162, 164-166, 169, 177-179

Populations, 1-2, 4-7, 9, 11-13, 15-18, 21, 33-34, 37, 39-41, 59-60, 69, 91, 94, 99, 124-125, 128, 132-133, 140, 142-144, 191, 193, 208, 217-222, 224, 226-229, 235

Predators, 66, 83, 93, 101, 106, 108, 130, 175, 196, 226

Professional Dive Guides, 59-60, 66

Q
Quality Control, 44

Quality Indices, 42

R
Radiocarbon Dating, 71-73

Radiocarbon Techniques, 71-72

Recreational Divers, 59-60, 68-70

Research Tools, 1, 17

Risk Assessment Code, 42, 45

Rna Isolation, 35

S
Saltwater, 1-2, 8, 14, 33-35, 37-40

Sampling, 8, 11-13, 15, 17-18, 20-21, 25, 34, 43-44, 46-47, 49, 51-52, 59, 68-69, 72, 77, 81-84, 90-91, 94-97, 103, 112, 124, 126, 134, 146, 148, 151, 154, 157-158, 165, 167-173, 175, 177, 189, 194, 196, 198, 200, 203-204, 216-219, 222, 224-225, 227-228

Sandbank, 93-94, 96-97, 99-100, 112

Sea, 3, 5-8, 11-15, 19-21, 34, 37, 40-41, 43, 46-47, 51, 56-57, 70, 73, 77-84, 86, 88-91, 93-96, 98, 103, 112-113, 124-126, 128, 131-133, 135, 143-146, 148-158, 160-162, 164-167, 169-170, 173-175, 177-183, 185-193, 205, 214-215, 228-230, 232, 234-237

Sea Surface Temperature, 8, 20, 79, 82, 84, 86, 89, 124-126, 128, 193

Seawater, 14, 19, 21, 25-26, 33, 35, 37, 41, 43, 135-137, 156-157, 162, 197, 200-201, 208-210, 228-229, 231, 234

Sediments, 42-49, 51-58, 93-94, 96, 99, 104-106, 111-113, 153-154, 156, 166, 173, 177, 179, 194-195, 201, 204-206, 229

Sharks, 59-70

Spawning, 1-7, 9-20, 22, 43, 77, 91, 208-210, 215, 230, 236

Species, 1-8, 10-19, 21-25, 27-33, 37, 40-41, 43, 45, 59-60, 62-74, 76-80, 83, 86, 88, 90-91, 93-94, 96, 99-101, 103, 105-114, 116, 122-125, 128, 130-148, 150-151, 153-154, 160, 169, 179, 188, 192-193, 196, 203, 205-209, 214, 217, 224-226, 230-231, 234-236

Statistical Analysis, 36, 46, 116, 126, 137

Sturgeons, 1-3, 5-7, 11-14, 16-22

Subtropical, 29, 71-72, 122, 125-126, 132-133, 156, 162, 164-165

T
Telemetry Data, 59-60, 62, 65-66, 69

Time Scales, 59, 128, 186

Tourist Divers, 59-60, 63, 66-68

Tree Age, 71-72, 76

Trees, 71-74, 76-78, 189, 202-203

Tropical, 70-72, 77-78, 112-113, 122, 124-125, 128-130, 132-135, 137-139, 141-143, 194, 205, 237

Tropical Latitudes, 71

V
Voltage Gated Potassium, 33-35

W
Water Depth, 79, 94, 100, 104, 172

Wintering, 5, 79-80, 83, 88-91, 188-189, 191-193

Wood, 71-78, 91, 132, 171-172, 214, 236

Wood Formation, 71

X
X-ray, 71-72